Climate Change Negotiations

As the Kyoto Protocol limps along without the participation of the US and Australia, ongoing climate negotiations are plagued by competing national and business interests that are creating stumbling blocks to success. *Climate Change Negotiations* asks how these persistent obstacles can be down-scaled, approaching them from five professional perspectives: a top policy maker, a senior negotiator, a leading scientist, an international lawyer and a sociologist who is observing the process.

The authors identify the major problems, including great power strategies (the EU, the US and Russia), leadership, the role of NGOs, capacity and knowledge-building, airline industry emissions, insurance and risk transfer instruments, problems of cost benefit analysis, the IPCC in the post-Kyoto situation, and verification and institutional design. A new key concept is introduced: strategic facilitation. Strategic facilitation has a long time frame, a forward-looking orientation, and aims to support the overall negotiation process rather than individual actors.

This book is aimed at academics, university students and practitioners who are directly or indirectly engaged in the international climate negotiation as policy makers, diplomats or experts.

Gunnar Sjöstedt is Director of Studies at the Swedish Institute of International Affairs, Associate Professor of Political Science at the University of Stockholm, and a member of the steering committee of the Processes of International Negotiations Program at IIASA. He has published extensively on international negotiation on environmental and economic affairs.

Ariel Macaspac Penetrante is a research fellow at the Institute for Infrastructure and Resources Management of the University of Leipzig in Germany.

Climate Change Negotiations

A guide to resolving disputes and facilitating multilateral cooperation

**Edited by
Gunnar Sjöstedt and
Ariel Macaspac Penetrante**

Routledge
Taylor & Francis Group

LONDON AND NEW YORK

First published 2013
by Routledge
2 Park Square, Milton Park, Abingdon, Oxfordshire OX14 4RN

Simultaneously published in the USA and Canada
by Routledge
711 Third Avenue, New York, NY 10017

First issued in paperback 2015

Routledge is an imprint of the Taylor & Francis Group, an informa business

British Library Cataloguing in Publication Data
A catalogue record for this book is available from the British Library

Library of Congress Cataloging-in-Publication Data
    Climate change negotiations : a guide to resolving disputes and
    facilitating multilateral cooperation / edited by
    Gunnar Sjostedt, Ariel Penetrante.
    pages cm
    Includes bibliographical references and index.
    1. Climate change mitigation–International cooperation.
    2. Climatic changes–Government policy–International cooperation.
    I. Sjöstedt, Gunnar, editor of compilation.
    II. Penetrante, Ariel, editor of compilation.
    QC903.C5627 2013
    363.738'74561–dc23
    2012041164

ISBN13: 978-1-138-92671-4 (pbk)
ISBN13: 978-1-84407-464-8 (hbk)

Typeset in Baskerville by
Swales & Willis Ltd, Exeter, Devon

To Bert Bolin

Contents

Figures, Tables and Boxes

Figures

Tables

Boxes

About the Processes of International Negotiation (PIN) Network at the Netherlands Institute for International Affairs, Clingendael

Since 1988, the PIN Program, formerly at the International Institute for Applied Systems Analysis (IIASA) in Laxenburg, Austria, and now located at Clingendael, The Hague, Netherlands, has been sponsoring research, publications, and international conferences on the subject of negotiation, analysed as a process. PIN is conducted by an International Steering Committee of scholars and practitioners, meeting three times a year to develop and propagate new knowledge about the analysis and improvement of the processes of negotiation. The Steering Committee conducts one to two workshops every year devoted to a given topic, involving scholars from a wide spectrum of countries, in order to tap a broad range of international expertise. It also offers mini-conferences on international negotiations in order to disseminate and encourage research on the subject. Such "Road Shows" have been held at the Argentine Council for International Relations, Buenos Aires; Beida University, Beijing; the Center for Conflict Resolution, Haifa; the Center for the Study of Contemporary Japanese Culture, Kyoto; the School of International Relations, Tehran; University of World Economy and Diplomacy, Tashkent; The Swedish Institute of International Affairs, Stockholm; the University of Cairo; Pepperdine University, California; The Johns Hopkins University, Washington, DC; Nelson Mandela Metropolitan University, Port Elizabeth; University Hassan II, Casablanca; the University of Helsinki; Bahçesehir University, Istanbul; and the UN University for Peace, Costa Rica. The PIN Network publishes a semi-annual newsletter, PINPoints, and sponsors a network of more than 4,000 researchers and practitioners in negotiation. Past projects and the Program have been supported by the William and Flora Hewlett Foundation, Smith Richardson Foundation, the US Institute of Peace, UNESCO, International Research Development Institute, Carnegie Corporation and Carnegie Commission for the Prevention of Deadly Conflict.

Members of the PIN Steering Committee

Cecilia Albin, Uppsala University
Mark Anstey, Nelson Mandela Metropolitan University, Port Elizabeth
Guy Olivier Faure, University of Paris V–Sorbonne
Paul Meerts, Netherlands Institute of International Relations, Clingendael

Mordechai Melamud, Comprehensive Nuclear-Test-Ban Treaty Organization, Vienna
Valerie Rosoux, Catholic University of Louvain
Rudolf Schüssler, Bayreuth University
Mikhail Troitskiy, Moscow State Institute of International Relations
I. William Zartman, The Johns Hopkins University, Baltimore

Members Emeriti

Rudolf Avenhaus, The German Armed Forces University, Munich
Gunnar Sjöstedt, The Swedish Institute of International Affairs, Stockholm

Coordinator

Wilbur Perlot, Netherlands Institute of International Relations, Clingendael
www.pin-negotiation.org

PIN Publications

Melamud, M., Meerts P. and Zartman I. W. (eds.) (2013) *Banning the Bang or the Bomb? Negotiating the Nuclear Test Ban Regime.*Cambridge, UK: Cambridge University Press.

Faure, G. O. (ed.) (2012) *Unfinished Business: Why International Negotiations Fail.* Atlanta, GA: University of Georgia Press.

Faure, G. O. and Cede, F. (eds.) (2012) *Unfinished Business: Saving International Negotiations from Failure.* Atlanta, GA: University of Georgia Press.

Zartman, I. W., Anstey, M. A. and Meerts, P. (eds.) (2012) *The Slippery Slope to Genocide: Reducing Identity Conflicts and Preventing Mass Killings.* New York, NY: Oxford University Press.

Zartman, I. W. and Faure, G. O. (eds.) (2011) *Engaging Extremists.* Washington DC: USIP.

Faure, G. O. and Zartman, I. W. (eds.) (2010) *Negotiating with Terrorists.* New York/London: Routledge.

Aleksy-Szucsich, A. (ed.) (2009) *The Art of International Negotiations.* Zurawia Papers.14. Institute of International Relations, University of Warsaw, Poland.

Avenhaus, R. and Sjöstedt, G. (eds.) (2009) *Negotiated Risks: International Talks on Hazardous Issues.*Heidelberg, Germany: Springer-Verlag.

Bercovitch, J., Kremenyuk, V. A. and Zartman I. W. (eds.) (2008) *The SAGE Handbook of Conflict Resolution.* London, UK: Sage Publications.

Avenhaus, R. and Zartman, I. W. (eds.) (2007) *Diplomacy Games. Formal Models and International Negotiations.* Heidelberg, Germany: Springer-Verlag.

Zartman, I. W. and Faure, G. O. (eds.) (2005) *Escalation and Negotiation in International Conflicts.* Cambridge, UK: Cambridge University Press.

Zartman, I. W.and Kremenyuk, V. A. (eds.) (2005) *Peace versus Justice: Negotiating Forward- and Backward-Looking Outcomes.* Lanham, MD: Rowman and Littlefield Publishers Inc.

Spector, B. I. and Zartman, I. W. (eds.) (2005) *Getting It Done: Post-Agreement Negotiations and International Regimes.* Washington, DC: United States Institute of Peace Press.

Meerts, P. and Cede, F. (eds.) (2004) *Negotiating European Union.* Basingstoke, UK: Palgrave Macmillan.

Sjöstedt, G. and Lang, W. (eds.) (2003) *Professional Cultures in International Negotiation. Bridge or Rift?* Lanham, MD: Lexington Books.

Faure, G. O. (2003) *How People Negotiate: Resolving Disputes in Different Cultures.* Dordrecht, Netherlands: Kluwer Academic Publishers.

Avenhaus, R., Kremenyuk, V.A. and Sjöstedt, G. (eds.) (2002) *Containing the Atom: International Negotiations on Nuclear Security and Safety.* Lanham, MD: Lexington Books.

Kremenyuk, V.A. (ed.) (2002) *International Negotiation. Analysis, Approaches, Issues.* 2nd edn. San Francisco, CA: Jossey-Bass.

Zartman, I.W. (ed.) (2001) *Preventive Negotiation: Avoiding Conflict Escalation.* Lanham, MD: Rowman & Littlefield.

Sjøsted, G. and Kremenyuk, V. (eds.) (2000) *International Economic Negotiation: Models versus Reality.* Cheltenham, UK: Edward Elgar.

Zartman, I.W. and Rubin, J. Z. (eds.) (2000) *Power and Negotiation.* Ann Arbor, MI: University of Michigan Press.

Berton, P., Kimura, H. and Zartman, I.W. (eds.) (1999) *International Negotiation: Actors, Structure/Process, Values.* New York, NY: St. Martin's.

Zartman, I. W. (ed.) (1994) *International Multilateral Negotiation: Approaches to the Management of Complexity.* San Francisco, CA: Jossey-Bass.

Spector, B. I., Sjöstedt, G. and Zartman, I.W. (eds.) (1994) *Negotiating International Regimes: Lessons Learned from the United Nations Conference on Environment and Development* (UNCED). Dordrecht, Netherlands: Graham & Trotman/Martinus Nijhoff.

Spector, B. I. (ed.) (1993) *Decision Support systems in Negotiation.* Special issue of *Theory and Decision* XXXIV 3.

Faure, G.O. and Rubin, J.Z. (eds.) (1993) *Culture and Negotiation. The Resolution of Water Disputes.* London, UK: Sage Publications.

Sjöstedt, G. (ed.) (1993) *International Environmental Negotiation.* London, UK: Sage Publications.

Kremenyuk, V.A. (ed.) (1991) *International Negotiation: Analysis, Approaches, Issues.* San Francisco, CA: Jossey-Bass.

Mautner-Markhof, F. (ed.) (1989) *Processes of International Negotiations.* Boulder, CO: Westview Press.

Acknowledgements

This book is a product of the fruitful interdisciplinary and international collaboration of researchers that has taken place in the Program on International Negotiations (PIN) for more than twenty years. At present, PIN is housed in the Clingendael Institute in the Netherlands. However, when this project was initiated, PIN was part of the International Institute for Applied Systems Analysis (IIASA) situated in Laxenburg outside Vienna, Austria. All the research work for the project was carried out in an IIASA context. IIASA also gave the book project organizational, editorial and other forms of assistance, as a supplement to PIN's own budget. We want to give special thanks to a number of people at IIASA, who were especially important for this project on the UN negotiations on climate warming.

We are grateful for the continuous support of Leen Hordijk who was Director of IIASA when the project started.

We are in great debt to the IIASA editorial team. Iain Steward, Head of IIASA Communications, gave us indispensible support and help with practical matters related to communication with publishing houses and the whole editing process. Special thanks go to Kathryn Platzer (IIASA Communications) who copy edited the original manuscript of this book. Her contribution to the project is simply invaluable. Kathryn did a good job eliminating small errors in chapters written by English speakers from the United Kingdom, the United States and Kenya. Her editing of chapters written by authors from other countries, including the editors of this book, was more far-reaching. Kathryn polished the text and, by means of small interventions, she managed to give the text more flow and elegance.

Support to all our project activities and careful and pleasing attention to every detail of our work was provided by Tanja Huber, our project administrator who served as PIN coordinator at IIASA.

We are grateful for the support from the Austrian Ministry of Environmental Affairs that made it possible for a group of authors of chapters in this book to organize a side event at COP11 in Montreal. We are equally grateful to IIASA which supported a similar side event at COP10 in Buenos Aires.

As always, the PIN group of researchers has represented an extremely stimulating reference group that has given us sensible advice, suggested creative ideas and confronted us with beneficial critique.

The Editors

Foreword

I. William Zartman

The road to the publication of the Climate Project of the Processes of International Negotiation (PIN) Program has been almost as long as the process of negotiating a climate change regime itself. The difference is that this publication has arrived at its conclusion. The road has its gateway in the general interest of the International Institute of Applied Systems Analysis (IIASA), which formerly housed PIN in Laxenburg, Austria, in climate issues and the role of PIN as consultant for the Secretariat of the Rio UNCED Conference in 1992. Later, in 2004 and 2005, a PIN team under the direction of International Steering Committee member Gunnar Sjöstedt, of the Swedish Institute of International Affairs presented side events at the UN Framework Convention on Climate Change (UNFCCC) Conference of Parties (COP10 and COP11/MOP1 in Buenos Aires and Montreal, respectively), supported by the Austrian Ministry of Environmental Affairs. The "mini-road-show" was enthusiastically received by the practitioners involved in the climate talk review. Soon after, Katherine Calvin was a PIN fellow at IIASA's Young Scientists' Summer Program, with a study of her own for Stanford University on the climate regime.

PIN projects generally follow themes that build collective knowledge on related topics. One of PIN's first projects was a work on *International Environmental Negotiations* (Sage 1993), edited by Gunnar Sjöstedt, followed in the next year by *Negotiating International Regimes: Lessons from UNCED* (Nijhoff), edited by Bertram Spector, Gunnar Sjösted, and I. William Zartman, and *International Multilateral Negotiations: Approaches to the Management of Complexity* (Jossey-Bass), edited by I. William Zartman. Later in the series came *Getting It Done: Post-Agreement Negotiations in International Regimes* (USIP 2003), edited by Bertram Spector and I. William Zartman. Thus, a study of the removal of obstacles and stumbling blocks in the climate change regime, at a time when the 1997 UNFCCC Kyoto Protocol was facing its second commitment period in its Meeting of Parties (MOP), was topically important.

The workshop for the Climate Change Project was held in IIASA in January 2005 under the PIN grant from the Hewlett Foundation, and work continued on it over the following years. However, the MOPs at Copenhagen and Cancun in 2009 and 2010 introduced changes in the regime – although perhaps not as much as desired – that required changes in the book, delaying publication until the present time.

The climate talks are not only enormously important but also extremely complex. Efforts need to be made to facilitate the climate negotiations and remove obstacles to its progress. Academia has a strong responsibility in this regard. There is a large literature on facilitation that, however, is usually thought of as quick fixes in a short time frame. Such facilitation can useful in particular situations but there is also great need to develop approaches to facilitate a long term strategic perspective, manage regime complexity and consummate post-agreement negotiations. This is the mission of this project.

Notes on Contributors

Steinar Andresen (Norway) has been research professor of political science and international relations at the Fridtjof Nansen Institute in Oslo since 1979. He was a professor of the Department of Political Science at the University of Oslo (2002–2005) and a visiting research fellow in Princeton University (1997–1998), International Institute of Applied Systems Analysis (IIASA) (1992–1996), and University of Washington, Seattle (1987–1888). His main research interests include international agreements, regimes and international organizations, particularly in the area of environment.

Joanne Linnerooth-Bayer (United States) leads IIASA's Program on Risk, Policy and Vulnerability. Her current research focuses on financial solutions for low-income households and businesses to help them cope with their catastrophe risk exposures, a topic she pursues with many collaborators, including insurers, NGOs and partners in the developing world. She has over 100 publications in the area of risk participation, communication and management, and she is on the editorial board of three international journals.

Lucas Bobes (Spain) is Group Environmental Officer, Industrial Affairs of the Amadeus IT Group SA, which is a leading transaction processor for the global travel and tourism industry, providing transaction processing power and technology solutions to both travel providers and travel agencies. The company acts as a worldwide network connecting travel providers and travel agencies through a processing platform for the distribution of travel products and services, and as a provider of a comprehensive portfolio of IT solutions which automate certain mission-critical business processes.

Bert Rickard Johannes Bolin (Sweden) served as the first chairman of the Intergovernmental Panel on Climate Change (IPCC) (1988 to 1997), which he helped to establish. A meteorologist by training, he was professor of meteorology at Stockholm University from 1961 until his retirement in 1990. He served as scientific director of the European Space Agency.

Urs Steinar Brandt (Denmark) is currently associate professor at the Department of Environment and Business Economics at the University of Southern Denmark. He specializes in risk perception and climate change, environmental economics, analysis of tradable permit systems, game theoretic analysis of international environmental problems and decision making under uncertainty. He has published more than 20 articles in refereed journals.

Franz Cede (Austria) has been professor for diplomacy at the Andrássy University in Budapest, Hungary since 2008 and a former member of the PIN Steering Committee. He served as Austrian ambassador to Belgium and NATO (2003–2007), to the Russian Federation (1999–2003) and Zaire (1985–1988). He was also the Austrian general consul in Los Angeles (1988–1991). He has published numerous works in international and European law as well as security and negotiations.

Guy Olivier Faure (France) is professor of sociology at the Sorbonne University, Paris V. He is a member of the editorial board of three major international journals dealing with negotiation theory and practice: *International Negotiation* (Washington), *Negotiation Journal* (Harvard, Cambridge), *Group Decision and Negotiation* (New York). His major research interests are business and diplomatic negotiations, especially with China, focusing on strategies and cultural issues. He has authored, co-authored and edited fifteen books and more than eighty articles.

Dirk Hanschel (Germany) is a postdoctoral scholar (Privatdozent) at the University of Mannheim, which granted him the *venia legendi* for public, international, European and comparative law. In addition, he holds various other lecturing appointments at German and foreign universities. His main research interests include environmental and energy law, human rights law, the law of international institutions, and the law and negotiation processes as well as comparative constitutional law.

Helmut Hojesky (Austria) is a meteorologist and currently serves as head of the Department of Ambient Air Pollution and Climate Protection of the Austrian Federal Ministry of Agriculture, Forestry, Environment and Water Management, dealing with technical and legal matters associated with air pollution control and climate protection. He previously held several positions at the Central Institute for Meteorology and Geodynamics (ZAMG). Moreover, he is chairing various national coordinating bodies concerned with air pollution control and climate protection, and he bears chief responsibility, *inter alia*, for coordinating Austria's air pollution control and climate protection policy as well as for introducing the EU emission trading system in Austria.

Angela Churie Kallhauge (Sweden) is senior policy advisor at the International Climate Policy Unit of the Swedish Energy Agency. She represents Sweden at high-level climate change negotiations and has her roots in a pastoralist community in Kenya.

Norichika Kanie (Japan) is a research fellow in the Sustainable Development Governance (SDG) initiative at the United Nations University Institute of Advanced Studies (UNU-IAS). He is also an Associate Professor at the Graduate

School of Decision Science and Technology, Tokyo Institute of Technology. His research focuses on international environmental governance. Currently he is a vice chair of the Working Party on Climate, Investment and Development (WPCID) at OECD, and a member of the UNEP International Environmental Governance Advisory Group. Since 1996 he has been a representative of Japan to air pollution regime negotiations in East Asia (EANET). From August 2009 to July 2010 he was a Marie Curie Incoming International Fellow of the European Commission and based in Sciences Po, Paris.

Bo Kjellén (Sweden) is a senior research fellow at the Stockholm Environment Institute. He is an author on the topic of climate negotiations, policy and diplomacy including the book *A new diplomacy for sustainable development: The challenge of global change*. He served as the chief climate negotiator for Sweden. He led part of the international negotiations for the United Nations Framework Convention on Climate Change (UNFCCC) and had played a key role in the negotiations for the Kyoto Protocol. He was chairman of the international negotiations on the UN Convention to Combat Desertification (UNCCD), which was adopted in 1994.

Gerhard Loibl (Austria) is currently professor of international and European law at the Diplomatic Academy of Vienna and at the University of Vienna. He is a consultant at the Austrian Federal Ministry of Agriculture, Forestry, Environment and Water Management. His research interests include international and European environmental law, international economic law, the law of international organizations and multilateral diplomacy.

M. J. Mace (United Kingdom) currently works as senior teaching fellow at the School of Law, University of London. She has provided legal advice and assistance to the Alliance of Small Island Developing States (AOSIS) in the international climate change negotiations for more than ten years and has been a member of the Kyoto Protocol Compliance Committee since 2006. During 2003–2008 she headed the Climate Change and Energy Programme at the Foundation for International Environmental Law and Development (FIELD) in London. During 1995–2001 she worked for the National Government of the Federated States of Micronesia, first for the FSM Supreme Court and then as an Assistant Attorney General.

Larry MacFaul (United Kingdom) is a senior researcher at the Verification Research, Training and Information Centre (VERTIC), an organization that supports the development and implementation of international agreements and related initiatives in the areas of arms control, disarmament and the environment. He manages VERTIC's environment programme, carries out research and analysis into environmental agreements and related areas, and also engages in capacity building initiatives and project development. He has written, published and spoken widely at national and international forums in collaboration with other organizations. He acts as joint editor for VERTIC's publication series and is on the editorial board of the journal *Climate Law*.

Reinhard Mechler (Germany) is an economist and is leading the Group on Disasters and Development. Specific interests of his include catastrophe risk modelling, the longer-term impacts of disasters and climate change on development, the use of novel risk financing mechanisms such as microinsurance, sovereign risk transfer and catastrophe bonds for sharing disaster risks, and the assessment of the efficiency and equity of risk management and adaptation measures. He has published widely in journals such as *Science, Climate Policy, Atmospheric Environment* and *Natural Hazards*. He acted as a reviewer of the Fourth Assessment Report of the Intergovernmental Panel on Climate Change. He has been leading and contributing to research projects and consultations for organizations including the European Commission, DFID, UN, ProVention Consortium, World Bank, Inter-American Development Bank, Caribbean Development Bank and GTZ.

Charles Pearson (United States) is Professor of Economics and Environment, Diplomatic Academy of Vienna, Austria, and Professor Emeritus, Johns Hopkins University School of Advanced International Studies (SAIS), Washington, DC, where he is a former director of the International Economics Program. He has taught international and environmental economics in several universities, including Thammasat (Thailand), Nanjing University (China), the International University of Japan and the Diplomatic Academy of Vienna. In addition, he has served as a senior staff economist for the Commission on International Trade and Investment Policy as well as a consultant to the World Bank, the US Department of Commerce and other organizations.

Ariel Macaspac Penetrante (Philippines/Germany) is a research fellow at the Institute for Infrastructure and Resources Management of the University of Leipzig in Germany. He currently serves as a chapter scientist (Chapter 7, Working Group III) of the upcoming Fifth Assessment Report (AR5) of the Intergovernmental Panel on Climate Change which is to be published in 2014. Furthermore, he conducts lectures in international negotiation and sustainable development in various universities. He was programme coordinator of the Program on International Negotiations at the International Institute for Applied Systems Analysis (IIASA) (2008–2010). His research interest focuses on environmental negotiations, North-South relations, conflict management, stakeholders' dialogue and energy systems.

Josef Proell (Austria) served as Vice Chancellor and Minister of Finance from 2008 until 2011, while also chairman of the Austrian People's Party (Österreichische Volkspartei – ÖVP). Prior to this, he was Federal Minister of Agriculture, Forestry, Environment, and Water Management (2003–2008).

Tora Skodvin (Norway) is senior research fellow at the Center for International Climate and Environmental Research, Oslo (CICERO). Her main research interests include international environmental negotiations, regime formation, implementation and effectiveness, and non-state actors in international environmental management. She is currently editorial member of *Global Environmental Politics* (since 2008).

Gunnar Sjöestedt (Sweden) is former research director at the Swedish Institute of International Affairs where he is presently an associated researcher. His research has covered different areas such as European integration, trade policy, the exercise of non-military power and international regime building. During his 20 years participation in the PIN programme, Gunnar Sjöstedt has edited or co-edited a number of books addressing different topics pertaining to multilateral negotiations. Recent work focuses on non-state actors – NGOs and business companies – particularly their role in peace processes.

Vasily Sokolov (Russia) is director of the Canadian Department of the Institute of US and Canada Studies of the Russian Academy of Sciences. He is also professor at the Moscow-based Higher School of Economics.

Lisa van Well (Sweden) is senior research fellow at the Nordic Centre for Spatial Development. She has specialized in European regional development, institutions and territorial cohesion, multi-level governance, territorial cooperation and climate change adaptation and mitigation policy. She held positions as researcher at the KTH, Division of Urban Planning (2001–2006) and the Swedish Institute of International Affairs (1993–1998).

Werner Wutscher (Austria), a lawyer by training, is an expert on sustainability issues and currently serves as steering committee member of respACT, the Austrian Business Council for Sustainable Development. He also works as a consultant for sustainability and change management. He was chairman and member of the Executive Board of the REWE International AG (2007–2011) and general secretary of the Austrian Federal Ministry of Agriculture, Forestry, Environment, and Water Management (2000–2007).

Tanja K. Huber (Austria) is a former coordinator of the PIN Program. She currently coordinates the Young Scientist's Summer Program (YSSP) and Postdoc Program of IIASA.

Part I

Introduction

Strategic Facilitation of Climate Change Negotiations

An Introduction

Gunnar Sjöstedt and Ariel Macaspac Penetrante

The UN negotiations on climate change remain complex and difficult, as they have been for thirty-odd years. A major inquiry addressed in this book is *if, to what degree*, and *how* obstacles confronting negotiators in the climate talks can be reduced, or perhaps even entirely eliminated, with the help of external facilitators. The focus is not only set on how to cope with diverging party interests. Primarily, the project addresses technical negotiation issues such as the framing of issues, capacity building in weak nations, institutional reform or process redesign. The key concept of the book is *strategic facilitation*, as seen in a long time perspective.

Objectives of the study and its design

There are many impediments to effective climate change negotiation; some are simply incidental, in the sense that they are entirely tied to a particular situation and therefore difficult to predict and prevent. One example of this would be the unexpected outcome of a national election in a key country, giving power to a new administration opposing internationally coordinated measures to reduce green-house gas (GHG) emissions. Incidental impediments are difficult to foresee for practitioners, and also demanding to cope with in negotiation analysis.

Other negotiation obstacles than such incidental problems are of a quasi-structural nature, although they can be modified or change gradually over time. Moreover, such *stumbling blocks* that will have an important impact on future negotiations may already be discernible in the present. An example would be mounting shortcomings on the part of one of the organizations supporting the climate talks, say, the Intergovernmental Panel on Climate Change (IPCC). In this case, institutional reform could represent a useful facilitation approach, which will have an impact only in the longer term.

The main objective of this book is to suggest and assess useful methods to facilitate the UN negotiations on climate change to consider a strategic and forward-looking perspective. This approach presupposes the discovery of principal obstacles in the climate talks – stumbling blocks – that are meaningful targets for strategic facilitation efforts for a longer period of time.

The center of attention of this study will be long-term facilitation approaches related to a continuous regime-building process, unfolding at the "macro level" above particular negotiation rounds or sessions that are integrated into the macro process by various "continuities" linking consecutive rounds. One example of such couplings is the preparatory work being carried out for the negotiations in the capitals of states that are parties to the climate negotiating process. Such ground-work for future climate talks may often develop from an evaluation of an earlier negotiation round, joining forward linkages to backward couplings.

Strategic facilitation is very different from troubleshooting in a current situation. This new facilitation concept implies that it is useful to try to plan and structure the future negotiation process in advance in such a way that at least some enduring stumbling blocks can be managed, circumvented, or even eliminated. Technically, strategic facilitation is a form of external intervention in a multilateral negotiation. It is designed by actors who are not direct parties to the climate talks, such as independent consultants, research institutes, universities, and nongovernmental organizations (NGOs), but are still believed to have a potential capacity to make the climate negotiations more effective.[1] Obviously, negotiation parties such as a national government or an international secretariat can do various things to aid a negotiation, but such measures fall outside the definition of strategic facilitation used in this project. Strategic facilitation is "operationalized" as coping with *stumbling blocks* which represent persistent obstacles in the climate talks.

One of the motives for this project is the observation that the concept of strategic facilitation of regime-building through multilateral negotiation is not common in the literature, although it is clearly required in numerous issue areas, for example, the environment and international trade. Another reason is that the exceptionally weighty contribution by the world scientific community to the climate talks is heavily skewed in favor of natural scientists such as physicists and meteorologists. This project assumes that social scientists also are in a position to considerably increase the contribution of the international scientific community to the development of the climate negotiations towards a fruitful outcome.

Social scientists have certainly been involved in the worldwide mobilization of scientific knowledge that has taken place through the IPCC. For example, senior economists have been engaged in developing and refining economic policy measures to cope with climate warming. However, social scientists with a special focus on the processes of international negotiation have to a great extent been overlooked by the IPCC and the organizers of the recursive climate talks. Such process specialists should be given a much larger role in the planning and facilitation of the complex international negotiation on climate change. This is an argument developed and responded to in this book.

While the project is concerned with how the UN negotiations on climate warming will develop in the years to come, its general outlook is both forward- and backward-looking from point zero; here and now. The effects of the facilitation measures that will be discussed pertain to the future. The knowledge basis for the determination of facilitation approaches is founded in the past – in the progress of a regime-building process that began almost a quarter of a century ago.[2] In one respect, this project can be regarded as a historical study essentially covering the period from the Copenhagen Climate Change Conference in December 2009

(COP15) for the purpose of looking forward at likely negotiation obstacles and possible facilitation strategies in future negotiations. The logic we plead is that such a systematic historical case is an instrumental point of departure for educated guesses about negotiation problems and their solutions in the future climate talks. It is better to have this point of reference in a clear near past than in a current more obscure situation.

The Copenhagen Conference has a central function in the design of this project. It is an example of how the climate negotiation unfolds at a particular time and in a particular setting. However the historical case study of the Copenhagen Conference does not only include the events that took place in the Danish capital in late November and early December 2009. The general background to and the preparations for the Copenhagen Conference are also a part of the case as well continuities from Copenhagen to consecutive large negotiation rounds in Cancun (2010) and Durban (2011).

Note that the case of the Copenhagen Conference is here regarded as but one element of a much wider understanding of the overall climate negotiation in the UN, which includes a multitude of other meetings and activities. Other types of cases pertaining to the broad understanding of the climate talks will be addressed in the project. These other cases are of a different character, as they represent important themes pertaining to the climate talk as a whole and to a specific event in this process.

The Copenhagen Climate Conference

The 2009 climate meeting had a number of special traits, some of which were simply due to its location in Denmark. All the 17 major UN conferences addressing the issue of climate change have had a special history and have brought about somewhat different end results. At the same time, the Copenhagen meeting, or any other climate conference, exhibited important similarities with the16 other climate conferences that have taken place. Therefore, it is possible to draw lessons from Copenhagen that can be expected to be relevant for other grand climate conferences in the future.

The road to Copenhagen

From its inception, the climate change negotiation process has included hundreds of meetings of various kinds, for example, ministerial meetings, professional diplomatic encounters, and workshops attended by large numbers of scientists and other experts. However, for the purposes of this project the Climate Conference in Copenhagen that took place during 7–19 December 2009 needs to be especially highlighted, for several reasons:

1. It was the first attempt to achieve a binding agreement on climate change to come into effect after the first commitment period of the Kyoto Protocol to the UNFCCC ends in December 2012.

2.   Copenhagen was seen as a point of departure for the future negotiation and regime-building process, which is addressed in this project.
3.   Its perceived political importance was high, as indicated by the presence in Copenhagen of more than 100 heads of state or government.
4.   It created a broad awareness that the climate negotiations as a whole will need to be of long duration if they are to have a satisfactory braking effect on climate warming.

The history of international negotiation on climate change dates back to the mid-1980s. These meetings were first organized by the international scientific community and attended by policymakers and international civil servants. Informal agenda setting was eventually drawn into the United Nations system and generated the 1992 United Nations Framework Convention on Climate Change (UNFCCC).

The Convention itself contained no legally binding commitments by governments to reduce GHGs. It is a framework convention, intended to serve as a platform for continued negotiation to establish effective international regulations on GHG reductions. After five years of negotiation, regulatory instruments were indeed established under the 1997 Kyoto Protocol to the UNFCCC. By signing the Kyoto Protocol, developed countries committed themselves to decrease their atmospheric GHG emissions according to an agreed schedule of relatively modest emission reductions. The heaviest mitigation burden among industrialized countries was that of the European Union (EU) which pledged an 8 per cent average reduction – well below the 60–80 per cent reductions requested by the international scientific community.[3] While not solving the problem of climate warming, the Kyoto Protocol did indicate a strategy to address it, namely, via the framework/protocol approach driven by recurrent interlinked negotiations.

The Kyoto Protocol was intended to remain in force until 2012 and to be followed and further developed by a new and more ambitious global climate agreement. Thus, shortly after its entry into force in February 2005, post-Kyoto climate talks were signalled. These negotiations began formally at the eleventh Conference of Parties (COP11) to the UNFCCC at the Climate Change Conference in Montreal in 2005, which adopted more than 40 decisions to strengthen global efforts against climate change.[4] The Canadian Environment Minister described the situation as follows: "The Kyoto Protocol has been switched on, a dialogue about the future action has begun, parties have moved forward to work on adaptation and have advanced the implementation of the regular work programme of the Convention and of the Protocol."[5]

The 2006 Climate Conference in Nairobi represented the second major session of the post-Kyoto talks but did not produce a breakthrough in the regime building process. Nevertheless, Nairobi focused on long term matters and action, continued the "multi-track" approach to these issues that had been established at COP11/MOP1 in Montreal, and reflected on the development of a framework for action once the Kyoto Protocol's "first commitment period" would be finished in 2012 (IISD 2006).

More concrete results were attained at COP13 the following year. On 3–14 December 2007, more than 10,000 delegates from 180 nations, governmental and nongovernmental organizations, and global media took part in the United Nations Climate Change Conference in Bali, Indonesia, at which the thirteenth Conference of Parties (COP13) adopted the "Bali Road Map" designed to guide a two-year process toward finalization of a binding agreement in Copenhagen 2009. The Bali Road Map included an Action Plan and set up two new negotiation institutions, the Ad Hoc Working Group on Further Commitments for Annex I Parties under the Kyoto Protocol (AWG-KP) negotiations and the Ad Hoc Working Group on Long-Term Cooperative Action under the Framework Convention (AWG-LCA).[6]

The first comprehensive round of negotiations framed by the Bali Road Map took place in Bangkok in March 2008. This meeting further specified the work program for post-Kyoto talks, focusing on the five main components of the agenda: adaptation to climate warming, mitigation of emissions of greenhouse gases, technology, finance, and the vision for long-term international cooperative action in the climate area.

In 2008 the fourteenth Conference of Parties (COP14) met at the UN Climate Conference in Poznan, Poland (1–12 December). The COP welcomed the progress made with respect to the Bali Action Plan.[7] Similarly noted was the determination of negotiating parties "to shift into full negotiating mode in 2009" and an invitation was made to all Parties to put forward further proposals regarding the content and form of the desired outcome as early as possible "in order to have them processed and considered in good time before the Copenhagen conference in December 2009."[8]

Some progress was made in various climate meetings following the 2007 Bali Conference (IISD 2007). Forward movement took place on a number of specific issues, including the establishment of an Adaptation Fund, and with regard to technology transfer, the Clean Development Mechanism, capacity building in developing countries, and financial support (IISD 2008).

The fifteenth Conference of Parties in Copenhagen

The fifteenth Conference of Parties (COP15) to the 1992 UNFCCC met on 7–18 December 2009 at the United Nations Climate Change Conference in Copenhagen. One principal aim was to reach an agreement on a new framework for tackling rising GHG emissions that would enter into effect at the end of 2012 after the expiry of the first commitment period of the Kyoto Protocol to the Climate Convention. Binding commitments regarding cuts of GHG emissions were meant to be linked to the would-be Copenhagen framework.

Before Copenhagen, there were various differing but often high expectations as to what should come out of the COP15 meeting. To cite one example among many, the Caribbean Community (CARICOM 2009) expected to come away with an accord that would help the Caribbean's capacity to reduce its vulnerability to the effects of climate change through a framework of support for adaptation prioritizing the needs of the most vulnerable countries. The Executive Director of the Caribbean Community Climate Change Centre (CCCCC), Dr. Kenneth Leslie, expressed the need to reach an agreement that would provide financial flows of US$ 80 billion per year to developing countries and provide access to new "green" technologies, both for climate change adaptation and mitigation purposes (CARICOM 2009). Other countries and groups of nations had other, but equally specific, demands related to the same general prospects regarding, for example, mitigation, adaptation or financial arrangements.

However, as COP15 approached, expectations were lowered in many quarters. There had been conspicuous lack of progress in pre-Copenhagen talks, notably in Bangkok in September/October and in Barcelona at the beginning of November. Delays by US legislators in passing a climate bill was likewise an ill-boding signal.

The 15 November 2009 Leaders' Statement, issued after the Asia-Pacific Economic Cooperation (APEC), left the impression that only a "political framework" was possible in Copenhagen. Asia Pacific leaders backed away from their original target of halving greenhouse gas emissions by 2050, pledging instead to "substantially" slash them by that date (Ministry of Foreign Affairs of Japan 2009). Many delegations anticipated correctly an arduous future negotiation process and sensed that there was a need for a "two-step" process to reach a final climate treaty "at the earliest" in 2010 (Schuenemann 2009). The Swedish Prime Minister, Fredrik Reinfeldt, who was President of the European Council in the autumn of 2009, hinted that the EU should see COP15 as a starting point and warned that European countries would have to make do with a less ambitious global deal than they were hoping for (Groen and Niemann 2011). With both state and non-state stakeholders now divided about the extent or even the possibility of a deal, in November 2009 around 60 Nobel Prize laureates united to appeal to the governments of the world to reach a sustainable agreement in Copenhagen to confront climate-change-related problems (*Der Spiegel* 2009a).

A main reason for the "failure" of the talks in Copenhagen to reach the goals adopted in the Bali Road Map was the inability (or unwillingness) of the major emitters such as Brazil, China, South Africa and the United States to reach a compromise regarding their opposition to commitments to reduce GHG emissions. However, the "failure" of the talks occurred also in various other respects, for instance, effectiveness of the climate negotiation process.

In the view of both parties and observers, the organization of the Conference by its Danish hosts was "chaotic" and "under overwhelming pressure" (*Die Zeit* 2009), and thus not likely to be conducive to a harmonious accord. The hosts were furthermore reported to have been involved in leaking a secret "Danish text" to the *Guardian*, purportedly drawn up by some of the developed countries, which would see effective control of climate change finance passing to the World Bank. This would effectively make grants of money to help poor countries adapt to climate change dependent on their taking a range of specified mitigation actions (*Guardian* 2009; Whitemann 2009). Such a change remains unsupported by developing countries who try hard to preserve their right to exception from rules regarding reductions of greenhouse gases.

Conflict between developed and developing countries escalated further when the chief negotiator of the Group of 77, Lumumba Di-Aping of Sudan, compared the apparent reluctance of the developed countries to provide assistance to developing countries with the "holocaust" (Wetzel and Lachmann 2009).Governments in developed countries were annoyed by this ideological rhetoric.

The Copenhagen Accord, 2009

The Copenhagen Conference is far and wide considered to be a failure because no binding commitments for the post-2012 future were agreed. However, at the final plenary session on 18 November 2009, the meeting accepted "to take note of" the so-called Copenhagen Accord. This text had been drafted by the heads of state of the United States and the BASIC bloc countries (Brazil, South Africa, India and China). While many delegations were disappointed by the outcome of COP15, there was also hope that the Copenhagen Accord could become a stepping stone in the pursuance of a fair, large-scale, and binding agreement to solve the climate crisis.The main issues covered by the Copenhagen Accord can be summarized as follows:

- Reaffirmation of the ultimate objective of the UNFCCC that greenhouse gas concentrations in the atmosphere should be stabilized at a level that would prevent dangerous anthropogenic interference with the climate system.
- Recognition of the scientific view that the increase in global temperature should be maintained below 2° Celsius on the basis of equity and in the context of sustainable development.
- Call for an assessment of the implementation of the Copenhagen Accord to be completed by 2015, including strengthening the long-term goal in relation to limiting temperature rises to 1.5° Celsius.
- Commitment sought from Annex I Parties to mitigate emissions, by implementing individually or jointly "economy-wide emissions targets for 2020" by 31 January 2010.
- Delivery of reductions and finance by developed countries to be measured, reported, and verified (MRV) in accordance with COP guidelines. However, this strategy is constrained by the lack of binding quantitative commitments with respect to emission reductions in the post-Kyoto period.
- Provision of scaled up, new and additional, predictable and adequate funding to be provided to developing countries to enable and support enhanced action on mitigation, including substantial finance to reduce emissions from deforestation and forest degradation (REDD-plus), adaptation, technology development and transfer and capacity-building, for enhanced implementation of the Convention.
- Short- and long-term financing: US$ 30 billion for the period 2010–2012, and long-term finance of a further US$ 100 billion a year by 2020 to be mobilized from a variety of sources. Four new bodies established to mobilize financial resources: a mechanism on REDD-plus, a High-Level Panel under the COP

to study the implementation of financing provisions, the Copenhagen Green Climate Fund, and a Technology Mechanism.[9]
•   Provision for international consultations and analysis under clearly defined guidelines that will ensure that national sovereignty is respected.

The Durban conference in 2011 represents still another step forward in the direction of the "soft" deadline of 31 January 2010 which was set under the Accord for countries to submit emissions reduction targets. At this point, 114 countries representing 87 per cent of the global GHG emissions had made pledges regarding the reduction of these releases. In contrast, 8 countries representing some 2 per cent of GHG emissions had declared that they would not engage in the Copenhagen Accord. The pledges related to the Accord are not legally binding and do not commit countries to agree to a binding successor to the Kyoto Protocol (Wynn 2009). However, the Copenhagen Accord and the pledges related to it should be regarded as a considerable development of the Bali Road Map, confirming the long term direction of the climate talks and putting together tentative commitments pertaining to the cutback of greenhouse gases. It is in this light that the two yearly major climate conferences following the Copenhagen meeting in Mexico (Cancun) and South Africa (Durban) should be seen.

In Cancun (2010), 193 countries came together and demonstrated a renewed commitment to the struggle against global warming. The Cancun Agreements are a detailed set of visionary, yet pragmatic, principles that make important strides to begin implementing the accord reached in Copenhagen the year before. The countries gathered in Cancun made progress on emissions reductions, greater transparency, forest preservation, and the creation of the green fund to help mobilize much needed investments throughout the world.[10]

Durban produced a document whose character is similar to that of both the Bali Road Map and the Copenhagen Accord but under a new headline, the Durban platform.[11] This non-binding agreement calls for revitalized negotiations for the new agreement on emission reductions which should not be concluded later than 2015, resulting in a new binding agreement that will take effect from 2020. This framework agreement was top-down, linked to an agreement to a second commitment period of the Kyoto Protocol from 2013. One objective was to preserve what in the climate negotiation jargon has been called the Kyoto architecture, formal rules for managing emissions. The Parties to the Durban Conference declared that the negotiations for a new agreement replacing the Kyoto Protocol should be concluded not later than 2015, and that the commitments in the new agreement should take effect from 2020. This agreement preserved the legal framework of the Kyoto Protocol, while at the same time opening the path to a new more comprehensive and more ambitious global agreement. The Conference formally recognized that existing emissions reduction pledges up to 2020 had to be considerably upgraded if the global goal of limiting average temperature increases to below 2° above pre-industrial levels were to be realized.

Durban also produced institutional/organizational developments which may become important in the longer term; the creation of the Adaptation Committee, which will provide advice and ensure coherent action on adaptation, and the establishment of a Technology Executive Committee, to facilitate the development of low-carbon technologies.

The meeting in South Africa also tackled the conflict between developed and developing countries in a constructive way. It decided to establish the Green Climate Fund (US$ 100 billion per year or more by 2020) whose principal function would be to support climate policies and activities in developing countries.

In Durban emerged significant shifts in the political landscape of the climate negotiation. China displayed a more positive attitude towards binding regulations of GHG emissions. A large 'coalition of high ambition', including more than 120 countries, emerged for the purpose of supporting a decisive progress towards a global and legally binding agreement on emission cuts. This grouping of states represented a new development in the relationship between developed and developing countries that had been so problematic during the 2009 Climate Conference. The coalition of high ambition was joined by many African and Latin American states, the group of least developed countries, as well as by the Alliance of Small Island States and the EU.

Negotiation problems in the Copenhagen talks: Agenda setting for strategic facilitation

This project wants to propose approaches and methods to facilitate the UN negotiations on climate change. For this reason it is important to establish clear guidelines for an evaluation of progress and failure in the climate regime building process. Ultimately, useful facilitation measures should promote success and help to lessen the risk for failure in the climate talks.

The Copenhagen Conference: event or process stage?

A critical problem illustrated by the Copenhagen Conference, and the negotiations that preceded and followed it, concerns the time perspective in which progress and failure should be seen. Short-term and long-term assessments, respectively, are likely to yield somewhat different if not completely contradictory results.

A common view in the debate about the United Nations negotiation on climate change is that COP15 in Copenhagen clearly represented a fiasco, simply because the Copenhagen Accord following from it is a much weaker document than the agreement with binding commitments regarding cutbacks of GHG emissions that many delegations wanted and had expected. After Copenhagen, the prospect of replacing the Kyoto Protocol with a new treaty after 2012, which would strengthen the UN climate regime, appeared remote.

However, if the Copenhagen Conference is not regarded only as a separate and autonomous event but also as a phase of a long term regime building process, the assessment of it becomes more complicated and also more favorable. The Copenhagen Accord did not produce binding commitments to reduce GHG emissions but neither did it stop a future development of the climate regime in that direction.

This was clearly indicated by developments and achievements associated with the Climate Conferences in Cancun (2010) and Durban (2011).

For example, the period of validity of the Kyoto Protocol has been extended beyond 2012, many countries have upgraded their pledges to reduce GHG emissions, and support for a new binding agreement to replace the Kyoto Protocol has seemingly increased in the last few years.

The point to make here is that strategic facilitation of the climate talks should not only target elements of separate negotiations such as the grand yearly Climate Conferences. The designers of strategic facilitation measures and strategies also need to consider events and circumstances related to continuities between climate conferences. Strategic facilitators should strive to develop and employ concepts pertaining to the regime building process at a macro level which envelops individual meetings and other events occurring in the context of the UN climate talks.

Linked to the negotiations that preceded and followed it, the Copenhagen Conference displays a number of negotiation problems that, first, may have an impact on the long term climate regime building negotiations unfolding at the macro level of the regime building process and, second, are likely to represent suitable targets for strategic facilitation measures.

Extreme magnitude of the climate talks

Like other multilateral negotiations in the UN context, the climate talks have an huge magnitude with regard to both participants and agenda. The number of official participants in Copenhagen was 33,526 persons, including 126 heads of state.[12] The conference was serviced by around 6,000 staff and included 2,500 meetings of different kinds.[13] Including formal, organizational matters, the agenda specified almost eighty items that needed to be addressed during the Conference.

Some of the issues on the agenda were highly complicated from a technical negotiation point of view and required careful studies in participating countries. This issue complexity clearly impeded the search for a constructive and effective negotiation outcome, both at the Climate Conference in Copenhagen but also with regard to the long term regime building process of which COP15 was a part.

Patterns of conflict and cooperation

Conflict and cooperation involving participants in the climate talks (national delegations, inter-governmental organizations, and nongovernmental organizations) are often generated and developed in particular situations and events, for example, as a reaction to a move by one of the players.

There are, however, also more long term patterns of conflict and cooperation recurring in one negotiation round after the other, including at the 2009 Climate Conference in Copenhagen. For example, patterns of conflict among leading actors such as the United States, the EU, China, Japan, Russia, India and Brazil, which were discernable in the pre-Kyoto negotiation, have more or less been transferred to the current post-Copenhagen stage of the climate regime-building process (Pan 2006). Thus, the EU has continued to be a strong proponent of far-reaching formal international regulations to cut emissions, although its position does seem to have changed somewhat in connection with the Copenhagen meeting, whereas the United States and some other countries persist in opposing this approach.

For a long time, developing countries have refused to accept binding mitigation commitments. Although some change in their position can be noted, they still demand exceptional treatment, economic adaptation assistance, and technology transfer in their favor (Najam *et al.* 2002: 3). At the same time, the negotiation strength of many developing countries is increasing, which makes their conflict with developed nations more complex and unpredictable in the longer term. There is no simple political formula in sight that has a clear potential to easily bridge the differences of interest between leading OECD nations and coalitions of participating developing countries in the climate talks.

Tactical facilitation, for example in the form of mediation, can be attempted in order to cope with a current conflict between two or more parties in a particular situation, such as an ongoing meeting in a negotiation group. Patterns of conflict like those that are discernable in the climate talks can also represent obstacles in the negotiation that are possible to demote by means of facilitation with a more strategic direction. The conflict between developed and developing countries at the Copenhagen Conference is one example. Many developing countries were provoked by the way in which the text to what became the Copenhagen Accord was put together. In order to make negotiations more effective, developing countries were kept outside this process and their response was their walkout from the conference room. In Cancun (2010) and Durban (2011), the organizers of the Climate Conferences were careful to use different procedures than in Copenhagen in order to give developing countries more access to important meetings. As a result, the relationship between developed and developing countries became more harmonious and cooperative as compared with COP15 in Copenhagen.

Knowledge diplomacy: effective use of science

Policymakers and diplomats in many countries find it difficult to fully understand the causes and consequences of climate warming because such complete comprehension needs to be expressed in the language of natural scientists. *Knowledge diplomacy* and the use of scientific information are important elements of the climate negotiation (Sjöstedt 2009). An inflow of scientific knowledge and information into the climate talks has represented a prerequisite for effective negotiation, not least in attempts to reach a costly and binding agreement (notably regarding cuts of GHG emissions).

Science must assist in the various preparatory efforts that are needed to pave the way for success in the UN negotiations on climate warming (Lanchberry and Victor 1995). It is, however, not always obvious how this can be achieved. As issues are processed through the negotiation machinery from one negotiation round to another, the focus on needed knowledge often shifts. For example, as compared with the pre-Copenhagen talks, there is an increased need for scientific knowledge about adaptation issues in the post-Copenhagen negotiation. Processes and institutions feeding scientific knowledge into the climate talks must be sensitive and flexible enough to be able to respond effectively to such changes of demand. They must furthermore be aware of the often considerable differences among negotiating parties with regard to their capacity to generate scientific knowledge and to pursue effective knowledge diplomacy. One reason is that most governments in the developing world can send only very small delegations, with few or no experts, to climate conferences.

The challenge of new or reframed issues

When issues such as *adaptation* or *forestry and land use and land use change ("sinks")* were upgraded on the agenda of the climate talks, along the lines established in the Bali Road Map, the climate negotiations became more difficult to handle for many countries. In certain ways, the issues of *"sinks" and adaptation* are examples of new negotiation trials, one of which is the need for new scientific knowledge/information. However, the challenge of new or reframed issues in the climate talks has broader implications than that.

In the pre-Kyoto talks, adaptation and sinks had been addressed in a number of climate meetings. But, after Copenhagen, these topics were still "new" in the sense that they had not been prepared for negotiation purposes to the same high degree as mitigation questions (emission reductions). An important negotiation problem is that adaptation needs a quite different approach than mitigation in the climate talks because it cannot equally easily be expressed and measured quantitatively. Like tariffs in trade talks, quantified emission cuts in the climate talks represent an excellently framed stake from a technical negotiation point of view. Notably, exchange of concessions in bargaining for a binding agreement is enormously facilitated.

Adaptation issues cannot be handled in the same simple way at the negotiation table. A different approach needs to be developed. Another challenge is that scientific networks built up to support the pre-Kyoto negotiation on mitigation cannot in the same way elucidate all relevant aspects of adaptation that concern social scientific issues, such as development assistance, poverty, or urbanization. This is not only a matter of analytical quality. Will national governments and intergovernmental organizations accept the advice given by social scientific researchers to the same degree as they listened to natural scientists in the pre-Kyoto talks?

The gap between need and feasibility

Table 0.1 exhibits the pledges twelve countries made in association with the 2009 Copenhagen Accord.[14]

The promises exhibited in Table 0.1 are not easy to compare. For example, the 12 nations use four different reference points; business as usual and actual

Table 0.1 National pledges to reduce GHG emissions associated with 2009 Copenhagen Accord

Compared with 1990	Compared with 2000	Compared with 2005	Compared with business as usual
EU: 20%–30%	Australia: 5%–25%	Canada: 17%	Brazil: 36.1%–38.9%
Japan: 25%		US: 17%	Indonesia: 26%
Russia: 15%–25%			Mexico: 30%
Ukraine: 20%			South Africa: 34%
			South Korea: 30%

emissions in 1990, 2000 and 2005, respectively. The national pledges are not part of a binding treaty so there is no guarantee that they will be realized. Even if they will be, there is still a huge "gap between need and feasibility" to consider as the world scientific community recommends cuts of greenhouse gas emissions in the range of 60–80 per cent (Meehl *et al.* 2007; Grubb *et al.* 1992). It is unlikely that the gap can be closed in a single successful negotiation round. Recall that the emission rates included in the binding Kyoto Protocol were extremely difficult to attain although they were quite modest; 8 per cent for the EU, 7 per cent for the United States, and 6 per cent for Japan (Enzler 2008). A long term perspective and stepwise regime building process through a sequence of negotiation rounds and agreements seems to be required in the post-Copenhagen period.

Institutional support to the climate negotiation: design problems

A general observation from the climate talks, as well as from other environmental negotiations, concerns the great importance of institutional support. The value of such external – contextual – assistance is recalled in observations of the Copenhagen Climate Conference. One example is the importance of the knowledge input provided by the IPCC into the climate negotiation, which has been manifested at every major climate conference.

However, the Copenhagen meeting also demonstrated risks due to malfunctioning institutional support. The employment of procedural rules that led to the half day walkout of developing countries in Copenhagen is one example. Another case in point is the slow process of admitting nongovernmental organizations into the conference building, which caused considerable frustration in the global civil society.

The above observations do not represent a comprehensive analysis of the Copenhagen Climate Conference and its associated pre- and post-negotiations. Neither does Copenhagen give complete picture of the climate talks. Still, the Copenhagen case has an important function in this project. It serves as an empirical platform for the design of the study in three important respects.

First, the Copenhagen case indicates areas in which stumbling blocks and strategic facilitation are important to discuss and assess.

Second, it gives direction to the development of a conceptual framework,

Third, the Copenhagen case indicates guidelines for the selection of concrete elements of the climate negotiation that give a reasonable picture of the great variation of negotiation problems confronting parties to the climate talks.

A conceptual framework

Policy advice based on social science for facilitation of the negotiations on climate change requires prior systematic analysis. The special framework for analysis and consequential prescription that has been constructed for this project consists of three principal parts. The first offers *a long-term macro perspective* on the climate negotiation, the second explains the meaning of *stumbling block* and the third part describes the essence of *strategic facilitation* and how it might generate positive effects in the climate negotiation.

Recurrent climate talks: A macro perspective

A national government and its delegation which wants to influence a session of the climate talks must prepare itself carefully. Policy makers and diplomats have to gain access to the latest scientific knowledge and information pertaining to the various dimensions of the complex climate issues relating to causes, manifestations, and consequences of climate warming. They must be well informed about effective measures to mitigate global warming, the conditions that would foster these, and ways of adapting to the disasters and problems resulting from the warming of the atmosphere. They also need reliable intelligence about the interests, capabilities, and positions of other significant actors in the process. The critical balance between problem solving and conflict resolution varies at different stages of a long-term regime-building process.

In order to maneuver effectively in the regime-building process, governments need to develop detailed climate policies in advance, and diplomats must repeatedly reconsider their positions in the climate talks and determine tactics in the continuous interaction with other players. However, tactical considerations, that typically have a short time frame, are not sufficient. As seen in a historical perspective, the UN climate talks can be looked at as a sequence of major rounds of negotiation largely corresponding to the rounds of multilateral trade negotiation in GATT/WTO.[15] This situation signals a need for a long term, strategic perspective on the climate talks.

A basic concept in this study is *negotiation session*: A general definition is *a major event in a progression of a regime-building process*. In the climate negotiation, a number of particularly important negotiation sessions are discernable; examples of such intensive negotiation sessions are UNCED producing the UNFCCC in Rio de Janeiro 1992, COP3 establishing the Kyoto Protocol in 1997, COP13 drawing up the Bali Road Map, and the Climate Conference in Copenhagen 2009 resulting in the Copenhagen Accord.

In the climate talks, separate negotiation sessions have been interconnected by various kinds of backward and forward linkages. The connection between the

1992 UN Framework Convention on Climate Change and the 1997 Kyoto Protocol can be described as an important forward linkage. New issues on the negotiation table, or issues that are further developed between rounds, represent other types of *forward linkages*.

A *backward link* occurs when negotiators in a negotiation session refer back to the outcome of an earlier session, for example, in order to strengthen their position in the current negotiation.[16]

To date, the UN climate negotiation has included a multitude of negotiation sessions in various settings that, however, in the final analysis were all related to three principal rounds pertaining to the macro-level of the climate talks. Each round can be identified with the help of an important intermediary outcome of the macro-process.

The first of these rounds started with informal consultations in the middle years of the 1980s and ended with the establishment of UNFCCC in 1992. The second round was concluded with the Kyoto Protocol, which was adopted in December 1997 and came into force on 16 February 2005. The current post-Kyoto talks, which include the 2009 Copenhagen Meeting, represent a third major round.[17] A logical future transition to a new round can be expected to begin if the Kyoto Protocol can be replaced by a new binding agreement on emission cuts.

This complex character, stepwise development, and long time frame are not exceptional for the process of the climate talks. For example, the dynamics of the international trade regime in the WTO has similar features.[18] In fact, a continuing recursive process with backward and forward loops can be seen as a general model for complex multilateral negotiations, particularly in the environmental area (Crump and Zartman 2003). In both the climate talks in the UN and the trade negotiations in WTO, each round can be regarded as a separate episode with a beginning and an end. Seen in this perspective, each episode is an autonomous negotiation process starting with *pre-negotiation* then evolving into *agenda setting, negotiation for a formula, bargaining about detail*, and ending with *agreement* and *post-negotiation*.[19] This is a usual analytical perspective on the climate talks, as well as on other multilateral negotiations.

However, to be fully understood and accurately assessed, each of the separate rounds of the climate talks also have to be seen as stages of continuous developments at "the macro level"; an incessant regime-building process, which incorporates all negotiation episodes, be they rounds or particular sessions.

Unless this macro approach is adopted, mistakes will be made in both analysis and outcome evaluation when the climate talks are assessed. For example, important results from one round of negotiation may not become visible until they materialize in the outcome of a following round. Although UNFCCC (1992) did not contain binding commitments to reduce emissions of greenhouse gases, it was a prerequisite for the 1997 Kyoto Protocol, which did include compulsory schedules for emission cuts for developed countries. Therefore, it would be misleading to say that, in contrast to the Kyoto Protocol, which was a success, UNFCCC was a relative failure because it did not include binding regulations about emission reductions. In reality, binding regulations about emission reductions did represent one important result of the negotiation creating UNFCCC, although this outcome did not materialize fully until five years later when the Parties to the Framework Convention met in Kyoto in 1997. An assessment of what the 2009 Copenhagen Climate Conference achieved should have the same forward-looking perspective. Some of the Copenhagen achievements, such as regulations concerning emission cuts, may not come into view until a following COP Meeting sometime after the 2011 Durban Conference.

Hence, an important lesson for practice is that long-term strategic planning of the climate negotiation needs not only to be forward-looking in the context of an ongoing negotiation round, such as a particular Conference of the Parties to UNFCCC. Strategists engaged in the climate talks, be they analysts, policymakers or facilitators, should also consider "the macro level"; the negotiated long-term, continuous, regime-building process. Similarly, efforts to facilitate the climate negotiation should not be limited to easing a particular meeting or negotiation round but should also strive to find ways of easing the continuous regime-building process unfolding at the macro level.

Stumbling blocks

In this project, strategic facilitation is understood as measures suggested by external actors trying to help negotiating parties to cope with stumbling blocks in the climate negotiation. A *stumbling block* is here defined as an impediment that, to a certain degree, has a structural character and is hence not tied exclusively to one single event or situation. A particular person not living up to the requirements for good chairmanship in a negotiation committee at a particular meeting does not represent a stumbling block; however, unsatisfactory chairing in a more general sense does, if it tends to recur from session to session. An important consequence of the structural character of stumbling blocks is the possibility of drawing lessons from the present for future negotiations. For example, by noting and understanding a problem such as failing chairing in negotiation groups, it may be possible to cope with it through measures that are planned or taken in the present. In turn, such lessons may become the basis for long-term approaches to easing the climate talks, namely, *strategic facilitation*.

The critical criterion for the detection of stumbling blocks is their negative impact on the climate talks. Essentially, stumbling blocks can generate two main types of obstructing effect. Ultimately, a stumbling block 1) prevents or delays the attainment of satisfactory agreements in the climate talks (*process effects*) or 2) hampers the quality of such accords (*outcome effects*).

Negotiated agreement may mean different things. A multiparty negotiation like the climate talks can, if it is successful, be regarded as the sequential establishment of different kinds of negotiated agreements concerning, essentially, the initiation of the process, the setting of the agenda, the creation of consensual knowledge, the choice of negotiation approach (formula), the settlement of separate disputed issues, and eventually the acceptance of the whole package of issues addressed in the negotiation (the final negotiated text) (Kremenyuk 2002).

Thus, it is important to bear in mind that outcome effects of stumbling blocks do not only relate to the closing agreement reached in a negotiation round, such as the Kyoto Protocol. The problem is that other agreements are usually more difficult to discern and describe, because they are not made explicit and formalized to the same degree as the final accord of a negotiation. One may also argue that any significant difficulties in reaching intermediary agreements will eventually affect the final accord. However, the distinction between different types of agreement along the process development of a multilateral negotiation is important in the search for a facilitation approach and concrete facilitation methods. Combinations of causative factors can be expected to vary, depending on what kind of outcome is obstructed. For example, the conditions for the signature of a formal final agreement are likely to be different from the prerequisites for agreement on consensual knowledge. Consequently, an external party needs to use other methods to facilitate a formal endgame of than if the purpose is to support the establishment of consensual knowledge.

As seen in a somewhat different perspective, stumbling blocks may also be linked to elements of the negotiation per se, or functions it performs. Generally speaking, a stumbling block may contribute to making the interaction between negotiating parties unnecessarily ineffective, time-consuming and costly in terms of human, technical, financial, or other resources invested into the negotiation (*negotiation effect*).

Stumbling blocks generating a negotiation effect may pertain to various elements of a negotiation. They may, hence, relate directly to the complex problem area of climate change and how it has been framed for negotiation purposes (*issue*). They may have to do with how individual negotiating parties perform (*actors/ strategies*). The difficulty many developing countries have in terms of participating actively and effectively in the climate talks because of lack of expertise and other resources is a well-known example. Some ways in which the negotiation has been organized and functions (*process*) as well as certain features of the context (e.g. organization) in which the climate talks take place (*structure*) may similarly represent stumbling blocks.

A major part of this project has been to identify critical stumbling blocks in the climate talks related to each of these elements of the climate talks; the issue, actors/ strategies, process and structure (Sjöstedt 1993; Susskind *et al.* 1993; OECD 1999; Victor 2001).

A taxonomy of stumbling blocks

A systematic means of describing the complexity in the climate change negotiations is to use a taxonomy covering its complexity categories: *issue complexity, actor complexity, structure complexity, outcome complexity*, and *process complexity*. The difficulties involved in clearly separating actors, issues, structure, outcome, and process from one another implies that there are interlinkages between them and that the elements interact in a dynamic way, which in itself is a source of complexity. A taxonomic approach makes it possible to clearly identify the issues involved in a conflict, and also to determine *zones of possible agreements* (ZOPAs).

Stumbling blocks related to issues

Issues typically represent stumbling blocks because they epitomize diverging interests in different countries. For example, island states in the Pacific Ocean, whose existence is threatened by a rising sea level, want to limit emissions of CO_2 as much as possible. In contrast, some oil producing countries resist binding international regulations prescribing cuts of CO_2 emissions. Looking for issue-related stumbling blocks, it is important to take note of other dimensions of negotiated matters than the interests to which they relate.

As Bercovitch *et al.* (2009) argue, parties in conflict differ so widely in terms of their values, beliefs, and goals, that they can also be expected to differ with respect to their perception of the issues underlying the conflict.

Perceptions, in turn, are embedded in various social and societal circumstances. In the climate change negotiations, the issues center on the means of distributing "responsibilities" in terms of greenhouse gas emissions. The delays and even breakdowns in this process could be due to the lack of a normative framework on justice and fairness. The contesting notions of justice and fairness between the developed and developing countries make it almost impossible to find a "just and fair" procedure of the distribution of climate change "burdens." Whereas the "North" follows a "forward-looking" notion of justice and fairness, the South adheres to the "backward-looking" notion of justice and fairness (see Chapter 8). The gap between these perspectives is, to a significant extent, linked to a mindset or cognitive structure that influences decisions and behavior relating to negotiation procedure.

Another way of analyzing issues is to calculate the rewards, or cost, that can accrue to each party from the issues defining the conflict. If the only possible outcome to each party is either victory or defeat, then the conflict is a "zero-sum game" (one party gains what the other party loses) (Bercovitch *et al.* 2009: 6). The understanding of conflict as a source of reward or punishment is necessary for its management, and this is not possible under the current structures of the UNFCCC/Kyoto Protocol.

The issues addressed in the climate change negotiation can be classified in terms of their contents; they include 1) resource issues (e.g. funds or technologies), 2) sovereignty issues (e.g. verification measures), and 3) security issues (e.g. natural disasters). At COP15, states wished to bring different priority issues to the negotiation table. For example, China prioritized resource and sovereignty issues, whereas Bangladesh was primarily concerned with security issues, further adding to the complexity of the climate talks. This variation in perception of issues represents a stumbling block whose importance should not be underestimated.

The diversity of issues discussed above is not just a source of complexity but also causes changes in the influences that bear on issues' nature and scope. The intense public attention focused on the COP15 meeting led to an increase not only in the participation of stakeholders but also of spoilers, some of whom were creative enough to push forward the agenda or increase public awareness of the climate issues, and some (fewer) of whom rioted or destroyed property. Between 40,000 and 100,000 people attended a march in Copenhagen on 12 December calling for a global agreement on climate change (BBC 2009a). Some 968 protesters were detained at the event, including 19 who were arrested for carrying pocket knives and wearing masks during the demonstration. The increased focus on the climate change negotiations led to greater security measures being taken to protect them. At the Copenhagen summit, Per Larsen, the chief of police coordinating the security measures, stated that this was "surely the biggest police action ... in Danish history" (Zeller 2009). The transformation of the climate issues from "low politics" to "high politics" adds to the complexity confronting negotiating parties. With more stakeholders wanting to influence the decision-making process, there needs to be greater coordination efforts between national and international agencies.

Inclusion of air transport into the international climate regime highlights the problems of dealing with *new issues* in the regime-building process (Chapter 6). The difficulty of making cost/benefit assessments regarding negotiated climate policy options is likewise linked to the climate issue, particularly the extreme uncertainty problems it evokes (Chapter 12).

Stumbling blocks related to actors

One of the key factors in the analysis of any conflict is the identity of the parties. As a global deal can be reached only by states, these will be the focus of this chapter. As Bercovitch *et al.* (2009) state, "parties to a conflict" may refer to an entire scale of entities ranging from the individual to national and international organizations, with different parties to the conflict and different levels of analysis occurring at different aggregation levels. Indeed, the diversity of actors is in itself a source of complexity.

Each nation has its own means, strategies, approaches, and procedures for dealing with other nations involved in the process. Sovereignty can be an obstacle to a global deal, as states are accountable to their constituents. For example, the coal dependence of Australia is a powerful determinant of its position in the climate talks. Australian leaders have declared that they cannot make commitments in a global agreement to reduce emissions of greenhouse gases unless advanced economies take on comparable commitments and major developing economies begin to substantially curb their emissions (UNFCCC 2009; Rudd *et al.* 2009).

Because of states' differing means and situations, their preferred procedures and approaches for coping with climate warming also differ. The domestic situation of a country, which is important for external action, also varies across nations, making it difficult for other nations to understand and anticipate its performance. Although the administration of US President Barack Obama could have regulated the emissions of greenhouse gases without the approval of the US Senate and the House of Representatives through an endangerment finding on the part of the US Environmental Protection Agency with respect to CO_2 emissions, the President preferred to integrate this decision into the democratic process, which was one reason for the protracted negotiation process at COP15 (*Der Spiegel* 2009b). At a press conference during COP15, Todd Stern, the US Special Envoy for Climate

Change, highlighted the importance of the US climate bill for the decisions that would be taken at the international level.

In many multilateral talks, the positions of many nations remain surprisingly unaltered from one negotiation session to the next. This state of affairs can be seen as a facilitating factor. If governments are familiar with what their opponents and allies prefer and think, this is helpful in a negotiation. In contrast, when nations or groups of states begin to change policies and positions, this development may become a stumbling block.

The requirement of, for instance, Australia, Russia, the European Union and other developed nations, that major developing countries with emerging economies commit themselves to carbon emissions cuts is one indicator of a new development in the international system. Developing countries moved from the periphery to the center of the climate negotiations during the Copenhagen meeting, gaining issue-specific power (for example, "no solution without China"). The walkout of the delegates from African nations over the leaked "Danish text" led to a temporary suspension of talks on the eighth day of COP15, indicating a growing "veto or blocking power," on the part of the developing countries, an ability to delay the process, and ultimately prevent an agreement (see Chapter 8).

Clearly, the diversity of actors and the notion of sovereignty are among the constraints to reaching a global deal. Parties in conflict determine their positions within a social context, in which specific decisions and behavior on the part of others condition one's own strategies, procedures, and approaches (see Chapter 2).

Selected themes in the book directly related to the *actors* of the climate negotiation are great power policies (Chapter 2), leadership (Chapter 3), capacity-building in weak developing countries (Chapter 5), the role of NGOs (Chapter 4) and the private sector (Chapter 6).

Stumbling blocks related to process

An important category of stumbling blocks concerns the progress of the negotiation process. One process dimension pertains to the planned time frame of a particular meeting, something which became problematic during the Copenhagen Conference. The inherent inflexibility in the way the conference organizers managed the negotiation proved to be a major stumbling block. Their fixed plan that a final agreement would be in place by the last Saturday of the meeting forced the formal negotiation leaders (notably the chair of the plenary) to manage the process badly. For example, to save time, there were insufficient consultations with states outside the inner circle of the leading nations. This poor process management appears to be one of the reasons for the developing countries walkout on 14 December (Johnson 2009).

The "time frame paradox" with which negotiators were confronted in Copenhagen represents another temporal problem. Negotiators tend to address the climate change issue within a short-term time frame because of domestic constraints such as national elections. Negotiations also focus on the costs of mitigation that start to accrue in the short term. However, climate negotiators need to have a long-term perspective, because mitigation measures taken today will produce effects only in the long term. The essence of the "time frame paradox" is that the long-term consequences of climate warming require short-term policy action. According to the IPCC Fourth Assessment Report: Climate Change 2007, many long term impacts of climate warming be reduced, delayed or avoided by mitigation if only appropriate measures are undertaken in the shorter term. Delayed emission reductions significantly constrain the opportunities to achieve lower stabilization levels and increase the risk of more severe climate change impacts (IPCC 2007).

Another dimension of process complexity refers to the problem of "sectoral arrangements," that is, the subdivision of complicated climate issues in order to simplify the negotiation. However, as Copenhagen recalled, issue interlinkages are generally so dense in the climate negotiations that separate sectoral agreements would be almost impossible to attain and implement. This is not necessarily a bad thing, as sectoral agreements can sometimes impede overall "package deals" being made among the parties.

The linkages between negotiation sessions and rounds represent another aspect of process complexity. The COP15 summit should not be evaluated as a single event, but rather as a step in the process that began in earnest with COP11 in Montreal. The failure to reach a binding agreement on mitigation measures at COP15 shows there is a need to develop a framework for bargaining on emission reductions, and particularly the procedures regarding the participation of developing countries in the climate talks. Work leading in this direction was carried out at the Climate Conferences in Cancun and Durban. Although COP15 failed to produce a binding agreement, it represented a necessary step forward in the regime-building process.

The "exceptional procedure" through which the Copenhagen Accord was developed represented an inadequate effectiveness model. This approach made it possible for a small group of leading countries to work out an agreement which will probably have positive effects on the post-Copenhagen talks. However, this so-called achievement came at a substantial cost, because it contributed to undermining the multilateral character of the UN negotiation on climate change. Copenhagen highlighted the urgent need to better integrate developing countries in the multilateral negotiation process, as developing countries are in general more vulnerable than developed countries to climate warming. Moreover, developed countries have an interest in increasing the active participation and influence of developing countries at all stages of the climate talks, from agenda setting to bargaining on detail in a final agreement, in order to encourage developing countries to begin reducing their emissions of CO_2 and other greenhouse gases. A negotiated agreement is the only way to a binding commitment.

The building and management of scientific knowledge in the negotiation (Chapter 7) and land use/forestry (sinks) (Chapter 11) represent critical *process-related* themes.

Stumbling blocks related to structure

The structure of a negotiation is a configuration of enduring circumstances influencing actors and process. This pattern of structural components may differ between negotiations, depending on, for example, issues and participants. In the case of the climate talks, conceivable elements of the structure of the negotiation are, for example, the distribution in the world of oil and coal extraction, of emissions of greenhouse gases, and of industrial production. The distribution of power in the world conditioning the climate talks is an important part of the negotiation structure.

Support institutions represent elements of the negotiation structure, which are particularly close to the negotiation process. The main institutions which formally service the climate negotiation are related to the UNFCCC. This is a complex system of committees and secretariat units (notably the UNFCCC Secretariat in Bonn), as well as special bodies like the IPCC (see Chapter 7).

The effectiveness of the negotiation on climate change depends on the well-functioning of its support organizations. Therefore, if problems with the support bodies become too large, they may come to represent stumbling blocks that are impeding the climate talks. Deficiencies in the support organizations can be expected to have different characteristics. They may, for example, concern the competence and relevance of particular organizational bodies, or they may have to with a capacity to adapt to new situations and demands. Coordination capabilities within the whole network of support organizations are likewise a critical factor (see Chapters 5, 8, 9).

Stumbling blocks related to outcomes

It is clear that COP15 should be regarded as but a single step in a long regime-building process but it is far from obvious what this movement achieved. A string of questions come to mind. How binding is "soft law," that is, non-compulsory political guidelines as represented in the Copenhagen Accord? How significant was the Copenhagen conference if it is regarded only as a point of departure for a continued regime-building process rather than as an end point? To what extent did the Copenhagen Accord reinforce or develop the aims and plans represented in the Bali Road Map?

Negotiation is an instrument intended to move parties to a specific state of affairs that is estimated to be better than the status quo. However, in the climate change context, negotiation itself has in certain ways become the "end" to the process. One example is the gaps in the procedure regarding notions of justice and fairness as applied to the climate talks. Particularly diffuse in the climate change context is the approach being used to reach an outcome. While, at the COP15 summit, the European Union pursued a multilateral approach, the United States uses bilateralism as an instrument of diplomacy, negotiating bilaterally with China, India, Brazil, and South Africa to find a compromise and, according to them, a "meaningful" agreement (BBC 2009b). Several stakeholders blamed the United States for the failure of the conference to achieve a binding deal because, by negotiating with only a select group of nations, most states were excluded from the critical decision-making process. For instance, the Bolivian delegate called the way the Copenhagen Accord was reached "anti-democratic, anti-transparent and unacceptable" (BBC 2009c). Bolivia and other developing countries believed it to be an "unfair procedure" which also included an element of blackmail, because access to the funds for climate measures that were to be provided by developed countries was contingent on signing the agreement that had been worked out by the leading nations.

The bottom line is that bilateral negotiations at Copenhagen, which took place within a UN climate negotiation system designed to have a genuinely multilateral character, represented a noteworthy departure from the norm. In fact, it was argued that "the future of the UN's role in international climate change negotiations is in doubt" (BBC 2009c; Hamilton 2009). It is questionable if the United Nations can still deliver substantial measures to confront climate change. Bilateralism and multilateralism are two contesting approaches, and the introduction of bilateralism risks destabilizing the climate negotiation system. In the eyes of many governments, particularly in the developing world, the system lost some of its legitimacy at Copenhagen. It remains to be seen what the enduring effects of the repair work carried out in Cancun and Durban will be.

In a multilateral negotiation, an obvious problem related to the outcome is how to attain it. Other outcome-related difficulties are associated with the final agreement once it has been achieved. To a great extent these have to do with implementation, including verification of agreements, which is one of the themes addressed in this project (Chapter 10).

The outline here does not offer a comprehensive inventory of stumbling blocks confronting parties to the climate talks. One aim of the review has been to demonstrate the great variability of the basic character of stumbling blocks. Another objective is to categorize stumbling blocks in terms of a conceptual framework, which link them to basic elements of a multilateral negotiation. These connections give guidance to the development of facilitation approaches and methods.

Strategic facilitation

Facilitation is external – third-party – intervention in the climate negotiation for the purpose of making it easier for the states and organizations involved as parties in the process to reach a satisfactory agreement, or to move the talks in a satisfactory direction. A *facilitator* is a kind of external consultant at the service of the entire negotiation process. Her or his task is not to help an individual government realize its separate interests, unless this is in line with the common interests of all negotiation parties. The facilitator serves the common interests of negotiating parties as defined in the UNFCCC and the final texts of the COP meetings that have occurred since 1992, and particularly the 1997 Kyoto Protocol. The task of strategic facilitators is to use social scientific knowledge to propose or design measures that are likely to help negotiating parties reach these collective goals. In some regards, as the available literature is insufficient, the fulfilment of this task may require the building or reinterpreting of scientific knowledge about what complex multilateral negotiations do and achieve.

Strategic facilitation aims to create conditions that are "conducive to reaching agreement" in the future climate talks (Hopmann 1996: 231). This may mean the undertaking of measures during the ongoing post-Kyoto negotiation for the purpose of attaining facilitation effects in expected future stages of the climate talks. Facilitation measures are quite costly. For example, institutional reform within the UN system to empower weak developing countries will take time and the allocation of necessary material resources may be substantial. Therefore, it is logical to look at binding decisions to undertake strategic facilitation measures as something akin to investments that are made now to gain interest and income only in coming years or even decades. Consequently, strategic facilitation involves a substantial element of risk taking. A facilitation measure such as institutional reform may, or may not, yield the expected positive effects.

A facilitator may be a practitioner, such as an experienced diplomat, or an academic researcher familiar with the climate negotiation. Indeed, the social scientific community has a significant potential capacity to make an important contribution to a constructive discussion about strategic facilitation.

Voids in the literature on facilitation of multilateral negotiation

The literature on how to negotiate better is large. Often such counsel is given although the term facilitation is not used. In some works the ambition of facilitation is more explicit than in others. There is, for example, a special branch of literature striving to develop the *art of negotiation* (Raiffa *et al.* 2002; Rose *et al.* 2002). Some authors have provided handbook texts for negotiators that they claim are relevant to any talks (Zartman and Beerman 1982; Fisher and Ury 1996; Kolb and Williams 2003; Watkins 2002). Other negotiation analysts focus on a particular context such as a certain problem area, for example, international business or intercultural interaction (Brett 2001; Cleary 2001; Susskind 2001; Watkins 2002; Goldman and Rojot 2003; Salacuse 2003; Sjöstedt 2003). Many works dealing with negotiation in specific issue areas (e.g. environmental negotiation) offer concrete advice on this particular context (Sjöstedt 1993; Susskind 2001). Some of these studies pertain specifically to negotiations on climate change (Gupta 2000; Johansson and Lindholm 2002; Penetrante 2012).

There are hence numerous academic books and articles that in different ways have something to offer with regard to how a negotiation may be facilitated; how the process can become smoother and more effective, and how a better outcome can be achieved. However, much of this large and varied literature has two important shortcomings as far as this book project is concerned, which relates to facilitation with long-term forward-looking ambition.

A major deficiency is the strong focus on a bilateral relationship that characterizes not only the explicit discussion about negotiation support and facilitation but the entire theory-oriented literature on negotiation. There are numerous descriptive accounts of particular multiparty negotiations, for example the Uruguay Round of trade negotiations in GATT or the negotiations on "acid rain" within the UN system. However, relatively few theory-oriented works have striven to depict general features of multilateral negotiation, which is the type of setting in which the climate talks are carried out.[20] The climate talks certainly include numerous episodes of two-party confrontations or dialogs that are important for the negotiation as a whole. However, multilateral talks represent a qualitatively different form of state interaction than bilateral negotiation. These special features need to be highlighted when facilitation approaches are designed and used in the climate talks with their extreme multilateral structure.

A second important inadequacy for this project is the typical short-term time perspective prevailing in most of the facilitation literature. Short-term facilitation typically targets acute immediate problems "on the table" requiring "quick fixes." Examples of such creative interventions to deal with current problems are 1) to help negotiation parties come out of a "hurting stalemate" (Zartman 2000); 2) to reformulate interests so that they are not threatening to the "core values" of other parties; 3) to help give credence to positive gestures by "the other side"; 4) to help assure fairness and balance in the negotiation process; 5) to help avoid the emergence of pseudo conflicts due to communication failures; 6) to reframe an issue in order to clarify joint party interests (a so-called win–win solution); and finally, 7) to provide immediate substantive support to individual parties (e.g. technical assistance or coaching) to give them a better actor capability in the negotiation (Zartman and Beerman 1982; Mnooking and Lee 1995; Fisher *et al*. 1995; Fisher *et al*. 1997).

Similar inventories of measures of long-term facilitation are not available. Nor does a general analytical/theoretical framework exist pertaining specifically to strategic facilitation of complex multilateral talks. It is for this reason that a theoretical approach has been determined in this project for analysis and prescription.

A theoretical approach to strategic facilitation

A conceptual framework regarding strategic facilitation needs to give clarity as well as direction to this project in at least two regards. First, it needs to specify how the criteria for successful facilitation should be understood; how effectiveness should be assessed. Second, it must shed more light on exactly how strategic facilitation measures have an impact on the climate talks: what facilitation attempts do when they influence a negotiation positively.

Effectiveness of strategic facilitation

A logical criterion for the assessment of the effectiveness of specific facilitation measures in the UN negotiation on climate change is the degree of collective goal achievement. Strategic facilitation measures are successful to the extent that they will contribute to helping the totality of the negotiating parties (or a critical mass of them) attain goals that they have jointly set up for the climate negotiation.

Unfortunately, this commonsense approach is hard to use in this project for practical reasons.

There are two main impediments to the application of the "goal achievement approach," given that strategic facilitation represents measures undertaken in the context of the climate negotiation.

First of all, it is not perfectly clear how all negotiation parties in actual fact understand the main long-term objective guiding the climate talks. Moreover, after the 2009 Climate Conference in Copenhagen, this issue is still debated and remains an important topic in the negotiation.

Since 1992 there has been a general agreement among the parties to the climate talks about Article 2 of the UNFCCC. This states that the ultimate aim of the Framework Convention is to:

> achieve stabilization of greenhouse gas concentrations in the atmosphere at a level that would prevent dangerous anthropogenic interference with the climate system. Such a level should be achieved within a time frame sufficient to allow ecosystems to adapt naturally to climate change, to ensure that food production is not threatened and to enable economic development to proceed in a sustainable manner.

A problem with Article 2 of UNFCCC is that it neither specifies how "stabilization of greenhouse gas concentrations in the atmosphere" shall be achieved nor exactly when "anthropogenic interference with the climate system" becomes "dangerous." The permitted maximum increase of 2° Celsius stipulated in the Copenhagen Accord represents an important clarification of Article 2 of UNFCCC, but does not eliminate all uncertainties. For example, it remains somewhat uncertain how the temperature objective relates to other goals for the climate talks that have been debated: the stipulated reductions of greenhouse gas emission and the maximum tolerable magnitude of greenhouse gas concentrations in the atmosphere.[21]

Although these goal formulations are evidently closely associated with one another, they still are important to distinguish in an analysis/assessment of possible facilitation approaches. Different goal formulations imply somewhat dissimilar guidelines in the search for effective facilitation approaches applicable to the climate talks.

The forward-looking and long-term aim of strategic facilitation is a particularly formidable obstacle to the use of the goal-achievement approach for the assessment of facilitation measures. Coping with uncertainty is the main problem. It is impossible to foresee how large future emission reductions will result from the post-Copenhagen negotiation. It is even more uncertain what future effects a given emission reduction will have on the average temperature in the atmosphere, and even more difficult to predict how reduced or stabilized climate warming will affect the frequency and seriousness of storms and inundations in, say, 15 or 30 years. Because of these and other difficulties, it is extremely difficult to relate strategic facilitation measures directly to the desired final results of the climate regime-building process driven by multilateral negotiation.

A less satisfactory but feasible approach has to be used: to look for direct outputs from the climate talks that are likely to eventually affect the problem of climate warming indirectly. A common approach in the general literature on the climate negotiation is to look at negotiated consensual intentions whose purpose is to abate the problem of climate warming. Such authoritative plans have been formulated in legal or quasi-legal instruments formally agreed upon in the negotiation, for example, the 1992 Framework Convention (UNFCCC), the 1997 Kyoto Protocol, and the 2009 Copenhagen Accord. One idea is hence that facilitation measures targeting the climate negotiation are effective to the extent that they will help the parties to the climate talks establish formal plans that are likely to lessen global warming.

However, this project has taken the position that this essentially legal outlook is too narrow for the proper identification and assessment of long-term approaches to facilitating the climate negotiation. Formal goals in international agreements may have two different functions as criteria for strategic facilitation. One is commitment by signatory states to undertake measures that will have an impact on climate change. The other is to pave the way or prepare for "future commitments to undertake measures that will have a favorable impact on climate change" (Bodansky 2003), for example, to reduce emissions of certain greenhouse gases. Thus, the UNFCCC conditioned the Kyoto Protocol, which in turn will be a foundation for further agreements on emission cuts in the post-Kyoto talks and thereafter if they can be realized.

Now, factors other than formal international accords like the 1997 Kyoto Protocol are in a similar way "conducive to reaching agreement" in future climate talks (Hopmann 2001: 231). The conception of an international regime can be used to capture such influencing factors that do not have the form of legally binding commitments. For the purposes of this study, a fairly simple and straightforward conceptualization of an international regime is sufficient. Stephen Krasner (1986) and others have proposed such an approach: a regime is a configuration of "norms, principles, rules and procedures, around which actor expectations converge" (Krasner 1986).

A typical feature of such an international regime is dynamics. Regime elements are potentially in continuous development. A regime may become reinforced over time, but it may also decline. For example, the consensual knowledge embedded in regime principles may become outdated. The respect for regime norms may weaken and the compliance with regime rules may decrease over time. Different elements of an international regime may move in different directions, making it hard to determine if the regime as a whole is becoming stronger or weaker.

Rules are precise policy prescriptions having the status of international law. Rules have a structural character in the sense that noticeable revisions are likely to occur comparatively seldom and by means of formal collective decisions. In the context of the climate regime, the Kyoto Protocol has become the backbone of the regime rules concerning mitigation of climate warming with its commitments of some 30 states to reduce emissions of CO_2 and other greenhouse gases into the atmosphere (UNFCCC 1998).

With regard to rules, facilitation measures are effective to the extent that they ease the establishment or implementation of regime rules which have a credible potential to positively affect climate warming.

There is not total clarity in the distinction between rules and *norms* of a regime. Norms may also contain policy prescriptions, but of a less formal and binding character than those of the rules of international law. Usually, norms give fairly general and "soft" direction to policymaking in a particular issue area, whereas rules prescribe in detail what policy action is prohibited or desirable.[22] Regime norms reflect widespread or universal beliefs held by the members of the regime that give direction to their performance. The stronger the norm, the more direction it gives to performance. The belief component of the norm may have an ideological or doctrinal character and contain policy instructions, for instance, *conflicts should be solved by peaceful means* or *human rights should be respected in all countries*. Norms may, however, also pertain to specific issue areas, such as that of biological diversity: *as many different species as possible should be permitted to survive; in the long run a broad biological diversity will be of great importance to mankind*.

In the climate regime, issue-specific norms pertain particularly to problem solving regarding the warming problem. They give direction to the general approaches that should be used to stabilize and ultimately reduce the concentrations of greenhouse gases in the atmosphere. For example, the clear prioritization of mitigation before adaptation in the pre-Kyoto talks was norm-driven. A tendency to change in this respect has occurred since COP10, in the sense that adaptation has been relatively upgraded compared with mitigation.

Principles represent a body of consensual knowledge about the issues covered by the regime. In the climate regime, principles are largely based on scientific knowledge (Haas 1990). However, scientific knowledge and consensual knowledge in a negotiation are not identical. Consensus about knowledge does not emerge automatically. In a multilateral negotiation, it is the result of negotiation between many parties, some of which have differing and important interests to defend. Knowledge becomes consensual when it represents an accord among "a critical number" of negotiating parties who are influential enough to drive the process. When an agreement is established, the input of scientific knowledge and information becomes authoritatively interpreted and structured by a majority of state actors in line with their joint interests.[23]

The contents and form of scientific knowledge represent an extremely powerful constraint in the negotiation about consensual knowledge, but do not determine the outcome entirely. A discrepancy between scientific and consensual knowledge may have various explanations. For example, for political reasons negotiating parties may choose to refer to a minority view in the scientific community rather than to mainstream thinking. Or negotiating parties may decide to reduce complexity in scientific knowledge in order to make these ideas manageable at the negotiation table. Such politically motivated simplification may lead to some distortion of scientific knowledge.

In common with norms, principles are not likely to be comprehensively described in the text of a treaty. The introductory section of international conventions (the preamble) often contains direct or indirect references to relevant quasi-institutionalized bodies of consensual knowledge. For example, the Preamble to the convention of the WTO contains references to neoclassic free trade theory. The UNFCCC and the Kyoto Protocol contain references to climate consensual knowledge that has been systematically pulled together within the IPCC.

Regime principles may have various functions in a regime-building process, as well as actually helping parties to understand the issues in a general sense.

Regime principles in the climate talks represent a knowledge-based frame of reference simplifying inter-party communication and making it more effective, for example, by preventing the occurrence of pseudo-conflicts (Sjöstedt 1994).

Consensual knowledge may serve as a face-saving authoritative mediator in interpretation disputes between parties with diverging interests (such as the EU and US).

Political guidance in knowledge building has led to the diagnosis and assessment of the climate problem being structured in such a way that an approach to problem solving is clearly indicated. Thus, as consensual knowledge explains that emissions of CO_2 and other greenhouse gases cause climate warming, then, clearly, an important part of the solution must be the reduction of emissions into the atmosphere. Note that well established consensual knowledge tends to acquire normative qualities.

Procedures pertain to the institutions that are set up to manage a regime, and their formal work methods. How institutions are designed and how they actually function is important not only for the effectiveness of a regime but also for how regime building functions. The climate regime shares some features with many other international institutions that have been developed within the UN system. With its "framework-to-detail" approach, the climate institutions have to date been very similar to those of other earlier environmental regimes, from which useful lessons have been drawn. Institution architects in the climate area have particularly looked to the regime to cope with ozone depletion in the stratosphere.

However, the institutions of the climate regime also have some fairly unique properties, notably those of the IPCC. As scientists and scientific knowledge have played a crucial role in the regime-building process, special procedures have been developed.

In the short term, the four types of regime element (rules, norms, principles, and procedures) represent a stable structure; at any particular point in time they are set conditions for the climate negotiation (Krasner 1986). Each element of the regime generates its own particular stream of influence at the same time, as there are various combined effects. For example, rules may give support to procedures. At the same time, norms and principles may increase the probability for full implementation of regime rules, such as commitments to reduce emissions of CO_2 or other greenhouse gases according to a negotiated schedule.

How strategic facilitation can have an impact on the climate talks

Ultimately, effective measures of strategic facilitation have a positive impact on the climate regime or one of its constituent elements, rules, norms, principles, or procedures. However, this impression is indirect, as strategic facilitation targets not the climate regime but the negotiation on the climate regime.

Strategic facilitation may target different dimensions, or elements, of a multilateral negotiation. For the purposes of this project, five such categories have been distinguished:

1. **The climate issues:** Various approaches to facilitation related to issues come easily to mind. One contingency is simply the input of additional scientific knowledge. Another possibility might be that a facilitator is engaged in producing suggestions as to how a particular issue can be reframed. The general aim of using strategic facilitation to target issues is to help make negotiated more manageable in the climate talks.

2. **Actors and their strategies:** Actor-directed facilitation measures can be designed to help parties to perform better in the climate talks so that they contribute more effectively to achieving common goals and to strengthening the climate regime. Institutions for capacity building in weak developing countries are part of the UN where the negotiation on climate change takes place.[24] Another form of strategic facilitation might be to support coalitions and coalition building (Gupta 2000).

3. **The negotiation process:** In facilitation of the climate talks, a distinction needs to be made between different basic process stages: pre-negotiation, agenda setting/issue clarification, formula negotiation, negotiation on detail, decision on a final agreement, and post-negotiation (Zartman 1978). The process stages represent a kind of negotiation function and can be thought of as basic types of collective action performed by negotiation parties that may emerge at different times, as the negotiation moves forward over time. Different types of such collective activities may emerge at the same event, such as at a conference. For example, new issues may be drawn into the process in the midst of the negotiation, unleashing agenda-setting activities in spite of the talks being dominated by, say, formula negotiation.

    The distinction between process stages is important because they tend to generate dissimilar patterns of party interaction associated with somewhat divergent aims. For example, the role of scientists, and science, differ across process stages. Hence, in the climate talks, scientists and nongovernmental organizations have been more involved and also more influential in agenda setting/issue clarification than in formula negotiation. Consequently, the effectiveness and weight of a given facilitation approach is likely to fluctuate across process stages. A facilitation method that is important in agenda setting/issue clarification may be more or less irrelevant in a negotiation on formula or detail.

    Process can also be looked at in a functional perspective. Functions are understood here as important types of collective activity in the context of the climate talks that may become the target for facilitation measures. For the purposes of this project, a distinction between three particular such functions is useful: *issue clarification*, *problem solving*, and *exchange of concessions*.

The general aim of facilitation targeting the process of the climate talks is to make this overall pattern of party interaction more effective, less time-consuming, and less costly in terms of financial, human, and other resources.

4. **The structure of the climate negotiation:** A potential major target for facilitation measures in the structure of the global negotiation on climate change is the institutional setting in which it unfolds (Watkins 2006; Fisher 2007). Accordingly, facilitation linked to the negotiation structure can have somewhat different directions, one of which is changes in specific working methods and procedural rules. Strategic facilitation can also have a grander design and represent proposals for the creation of a new organisational body or reform of existing institutions. The general aim of facilitation targeting the structure of the climate negotiation is to increase its capacity to give positive support to both individual actors and the whole negotiation process.

5. **The outcomes of the climate talks:** Facilitation which is meant to help parties to reach an agreement concerning the outcome of a session or round of climate talks pertains to actors, process, or structure. Facilitation targeting the outcome directly is here thought of measures to ease or support the implementation of agreement. One example that will be discussed in the book concerns improved verification techniques.

Looking for means of strategic facilitation: Summary of the conceptual framework

As summarized by Table 0.2, the research strategy of this project is to use a case study approach to identify measures of strategic facilitation to cope with stumbling blocks in the climate talks (see lowest box of Table 0.2). Effective facilitation will eventually lead to reduced climate warming. All facilitation measures represent some kind of intervention in the climate negotiation conceived of as a long-term regime-building development, targeting one, two or several of its principal elements – climate issues, actors/strategies, process, structure, or outcome.

Table 0.2 Summary of conceptual framework

Direct goal achievement: climate cooling
Indirect goal achievement: Development of the climate regime (Rules, norms, consensual knowledge, procedures)
Positive effects on stumbling blocks
Facilitation targets in the climate talks (Climate issues, actors/strategies, process, structure, outcome)
Means of strategic facilitation (to be identified in case studies)

Successful facilitation contributes to lessening stumbling blocks and manifests itself in positive effects on some basic element(s) of the evolving climate regime: rules, norms, consensual knowledge (regime principles), or procedures. A positive impact on the regime-building process will eventually contribute to reducing the climate problem, as indicated by at least one of the three climate indicators: 1) the average atmospheric temperature, 2) the magnitude of greenhouse gas concentrations, 3) the amount of greenhouse gas emissions.

The project is concerned with long-term, strategic facilitation. Table 0.3 illustrates the meaning of this direction by means of a comparison between strategic and short-term, tactical facilitation targeting actors, strategies, process, or structure.

Research design: A case study approach

This project has taken on two major research tasks.

The first task is to identify, classify and assess stumbling blocks in the various case studies addressed in the project.

The second research task is to propose approaches to cope with the kind of stumbling blocks that have been discovered in the empirical studies.

Ideally, the empirical studies searching for stumbling blocks and strategic facilitation measures should represent a comprehensive analysis of the entire negotiation on climate change. However, such an all-inclusive analysis of the climate talks has been beyond the means available to this project. Therefore a case study approach has been employed. Cases pertain to different kinds of developments in the climate negotiation in which it has been deemed interesting and useful to identify stumbling blocks and to assess possible approaches and methods of strategic facilitation

Cases of negotiation topics have been selected in such a way that they can give insight into how to cope with stumbling blocks related to all the principal dimensions of the climate negotiation: the climate issue, actors and their strategies, the negotiation process, the structure of the negotiation and its outcome. The climate issue looked at as a dimension of the climate talks is primarily addressed in the Conclusion but is also looked at in other various case studies.

The list of chapters identifies the sections of the book relating to particular dimensions of the climate talks.

Table 0.3 Difference between tactical and strategic facilitation: Illustrative examples

	Direction of Facilitation	
	Tactical outlook	*Strategic outlook*
Strategies	understanding institutions	changes in actor relations
Actors	improved communication	changes in actor relations
Strategies	smart fixes	upgrading joint interests
Process	good chairmanship	process design
Structure	understanding institutions	structural reform

Actors and their strategies

Urs Steinar Brandt: *Defining a Politically Feasible Path for Future Climate Negotiations: the EU–US divide over the Kyoto Protocol* (Chapter 6)

Vasily Sokolov: *Between Two Giants: Lessons from the Russian Policy on the Kyoto Protocol* (Chapter 7)

Steinar Andresen: *Leadership and Climate Talks: Historical Lessons in Agenda Setting* (Chapter 8)

Norichika Kanie: *NGO Participation in Global Climate Change Decision-making Process: A Key for Facilitating Climate Talks* (Chapter 9)

The structure of the climate talks

Angela Churie Kallhauge and Lisa van Well: *Institutional Capacity for Facilitating Climate Change Negotiations: A Need for New Thinking* (Chapter 10)

Lucas Bobes: *Stumbling Blocks in a Sectoral Approach: Addressing Global Warming through the Airline Industry* (Chapter 11)

Tora Skodvin: *Overcoming Stumbling Blocks: Can the Intergovernmental Panel on Climate Change deliver on Adaptation?* (Chapter 12)

Ariel Macaspac Penetrante: *Common but Differentiated Responsibilities: The North South Divide in Climate Change Negotiations* (Chapter 13)

Dirk Hanschel: *Developing a Legal Toolkit: Institutional Options to Remove Stumbling Blocks in the Climate Negotiations* (Chapter 14)

Outcome-related stumbling blocks

Larry MacFaul: *Verification as a Precondition for Binding Commitments: Facilitation through Trust* (Chapter 15)

Process-related stumbling blocks

Charles Pearson: *Difficulties of Benefit-Cost Analysis in Climate Policy: Stumbling Blocks for Reaching an Agreement* (Chapter 16)

Joanne Linnerooth-Bayer, M.J. Mace and Reinhard Mechler: *Proposal for Insurance for Facilitation of Adaptation* (Chapter 17)

After the Introduction, the first section of the book gives an overview of how representatives of five different professional cultures look at climate negotiation: A politician, a diplomat, a scientist, an international lawyer represent participants of the climate talks whereas a sociologist looks at it from the outside.

The Conclusion compares the cases and puts forward generalized observations about stumbling blocks in the climate talks and assesses the prospects for making a difference with the help of strategic facilitation.

Notes

1 For a discussion of effectiveness in multilateral negotiation, see Albin (2012).
2 See Annex for a brief description of the yearly principal climate conferences that have taken place so far.
3 In certain ways the figure of 8 per cent conceived of as a "burden measure" is misleading

because it represents an average for the whole European Union. A row of north European countries accepted emission cuts of around 30 per cent.

4 The COP Meeting in Montreal was the first Conference that included talks in both the Conference of the Parties to the UN Framework Convention on Climate Change (COP) and the Meeting of the Parties to the Kyoto Protocol (MOP).

5 http://unfccc.int/meetings/cop_11/items/3394.php.

6 Recall that, in relation to UNFCCC and the Kyoto Protocol, the participating nations are separated into three groups: *Annex I countries*: industrialized countries and economies in transition; *Annex II countries*: developed countries which pay for costs of developing countries; *Non Annex I countries*: developing countries.

7 Decision, 1/CP.14.

8 Ibid.

9 Reducing Emissions from Deforestation and Forest Degradation.

10 Cancun Climate Conference, November 2010. United Nations Framework Convention on Climate Change.

11 "Working Together. Saving Tomorrow Today". COP17/CMP7.United Nations Climate Conference in Durban South Africa. November/December 2011.

12 It should, however, be noted that the number of participating heads of state was unusually high in Copenhagen due to the expectations of a new treaty with binding commitments designed to replace the Kyoto Protocol after 2012.

13 COP15, Logistik/www.cop15.dk.

14 Copenhagen Accord, Annex I and non-Annex I pledges.

15 Recall that the General Agreement on Tariffs and Trade (GATT) was transformed into the World Trade Organization (WTO) in 1994/1995, a development that may be described as institutional change of the international trade regime established after World War II. If the current Doha round is included, 9 rounds have taken place in the last 50 years.

16 For a more extensive introduction to this *continuous process perspective*, see Sjöstedt (2005).

17 The time schedule for the movement towards Copenhagen was included in the Bali Road Map. COP/SB Archives COP13 CMP3 SB 27&AWG 4.

18 The current Doha round is carried out in the World Trade Organization (WTO). The eight earlier rounds took place under WTO's predecessor, the General Agreement on Tariffs and Trade (GATT) (website: Understanding the WTO: Basics. *The GATT years. From Havana to Marrakesh. WTO*).

19 Recall that *pre-negotiation* decides on the start of negotiation. *Agenda setting* determines what issues will be negotiated, indicates goals and sets a time table for the negotiation and usually also establishes a setting for it. A *formula* represents a concrete and specified approach, or method, for the negotiation. *Bargaining on detail* represents the application of the formula and agreement has the form of an international treaty. *Post-negotiation* deals with left-overs from earlier stages of the negotiation. See, for example, Kremenyuk (2002); Zartman (1994).

20 Two exceptions are Zartman (1994) and Lall (1985).

21 These different goal formulations point in the same general direction: they express a corresponding aspiration to cope with the global problem of climate warming. However, the three approaches (emissions, greenhouse gas concentrations, and atmospheric temperature) have somewhat different implications for the conduct and closure of the climate talks.

22 A *norm* may prescribe: Thou Shall Not Pollute the Atmosphere! A *rule* specifies what quantities of hazardous substances (pollutants) may be let out or not let out into the atmosphere.

23 NGOs and other representatives of the civil society may certainly influence this choice, which, however, at the end of the day can only be made by signatory governments.

24 See, for example, United Nations climate Change Conference in Poznan 1–12 December 2008: Decision/CP.14; Grindle (1997).

References

Albin, C. (2012) (ed.) Improving the Effectiveness of Multilateral Trade Negotiations. *Special issue of International Negotiation: A Journal of Theory and Practice*, 17 (1).

Alvin L., Goldman, A. and Rojot, J. (2003) *Negotiation: Theory and Practice*. The Hague: Kluwer Law International.

British Broadcasting Corporation (BBC) (2009a) Climate Activists Copenhagen Police Tactics. *BBC News*. 13 December, available through http://news.bbc.co.uk/1/hi/world/europe/8410414.stm, last accessed 20 January 2010.

British Broadcasting Corporation (BBC) (2009b). Key Powers Reach Compromise at Climate Summit. *BBC News*, 19 December, available through http://news.bbc.co.uk/2/hi/europe/8421935.stm, last accessed 24 August 2012.

British Broadcasting Corporation (BBC) (2009c). Harrabin's Notes: After Copenhagen, *BBC News*, 21 December, available through http://news.bbc.co.uk/2/hi/science/nature/8423822.stm, last accessed 24 August 2012.

Bercovitch, J., Kremenyuk, V. and Zartman, I. W. (eds.) (2009) *The SAGE Handbook of Conflict Resolution*. Los Angeles, London, New Delhi: SAGE Publications.

Bodanksy, D. (2003) Introduction, In Aldy, J., Ashton, J., Baron, R., Bodansky, D., Charnovitz, S., Diringer, E., Heller, T., Pershing, J., Shukla, P.R., Laurence, T., Tudela, F. and Wang X. (eds.) *Beyond Kyoto. Advancing the International Effort against Climate Change*. Pew Center on Global Climate Change. 37–59, available through http://www.pewclimate.org/docUploads/Climate%20Commitments.pdf, last accessed 24 August 2012.

Brett, J. (2001) *Negotiating Globally: How to Negotiate Deals, Resolve Disputes, and Make Decisions across Cultural Boundaries*. San Francisco: Jossey-Bass.

Carribean Community (CARICOM) Secretariat (2009) *Adaptation to Climate Change in the Carribean (ACCC) Project*, available through http://www.caricom.org/jsp/projects/macc%20project/accc.jsp, last accessed 26 January 2010.

Cleary, P. (2001) *The Negotiation Handbook*. Armonk, NY: M.E. Sharpe.

Crump, L. and Zartman, I. W. (2003) Multilateral Negotiation and the Management of Complexity. *International Negotiation*, 8 (1), 1–5.

Der Spiegel (2009a) *Memorandum: Nobelpreisträger fordern Klimaabkommen*, 10 November 2009, available through http://www.spiegel.de/wissenschaft/natur/0,1518,660332,00.html, last accessed 26 January 2010.

Der Spiegel (2009b) *Treibhausgasse: Obama hat freie Hand beim Klimaschutz*, 8 December 2009, available through http://www.spiegel.de/politik/ausland/0,1518,665749,00.html, last accessed 26 January 2010.

Die Zeit (2009) Kopenhagen: Die Klimakonferenz einigt sich auf unkonkrete Minimalziele, 19 December, available through http://www.zeit.de/politik/ausland/2009-12/klima-gipfel-kopenhagen, last accessed 24 August.

Enzler, S. M. (2008) *Overview of the Emission Reductions Required by the Kyoto Protocol*, available through http://www.lenntech.com/greenhouse-effect/kyoto-emission-reductions-overview.htm, last accessed 24 August 2012.

Fisher, R. (2007) The Structure of Negotiation. An Alternative Model, *Negotiation Journal*, 2 (3), July 1986, 233–235. doi: 10.1111/j.1571-9979.1986.tb00360.x.

Fisher, R. and Ury, W. with Bruce Patton (1996) (eds.) *Getting to Yes: Negotiating an Agreement Without Giving In*. London: Arrow Business Books.

Fischer, R., Ury, W. and Patton, B. (1997) *Getting to Yes. Negotiating Agreement without Giving In*, 2nd edn, London: Arrows Business Books.

Grindle, M. (ed.) (1997) *Getting Good Government. Capacity Building in the Public Sectors of Developing Countries*. Cambridge, MA: Harvard University Press.

Groen, L. and Niemann, A. (2011) Challenges in EU External Climate Change Policy-Making in the Early Post-Lisbon Era: The UNFCCC Copenhagen Negotiations, In Cardwell, J. (ed.) *EU External Relations Law and Policy in the Post-Lisbon Era*. The Hague: T. M. C. Asser Press. 315–333.

Grubb, M., Sebenius, J., Magalhaes, A, and Subak, S. (1992). Sharing the Burden, In Mintzer, A. (ed.) *Confronting Climate Change. Risks, Implications and Responses*. Cambridge, UK: Cambridge University Press. 302–322.

Gupta, J. (2000) *On Behalf of my Delegation – A Survival Guide for Developing Country Climate Negotiators*. New York: International Institute for Sustainable Development.

Haas, E. (1990) *When Knowledge is Power. Three Models of Change in Organizations*. Berkeley, CA: University of California Press.

Hamilton, C. (2009) History's Long Shadow, *Australian Broadcasting Corporation*, 21 December 2009, available through http://www.abc.net.au/unleashed/stories/s2777595.htm, last accessed 26 January 2010.

Hopmann, T, (1996) *The Negotiation Process and the Resolution of International Conflict*. Columbia, SC: University of South Carolina Press.

Hopmann, T. (2001) An Evaluation of the OSCE's Role in Conflict Management, In Gärtner, H., Hyde-Price, A. and Reiter, E. (eds.) *Europe's New Security Challenge*. Boulder, CO: Lynne Reinner. 219–254.

IISD (2006) Summary of the Twelfth Conference of the Parties to the UN Framework Convention on Climate Change and Second Meeting of the Parties to the Kyoto Protocol, 6–17 November 2006, *Earth Negotiations Bulletin*, 12 (318), 20 November, available through http://www.iisd.ca/vol12/enb12318e.html, last accessed 25 August 2012.

IISD (2007) Summary of the Thirteenth Conference of the Parties to the UN Framework Convention on Climate Change and Third Meeting of the Parties to the Kyoto Protocol, 3–15 December 2007, *Earth Negotiations Bulletin*, 12 (354), 18 December, available through http://www.iisd.ca/download/pdf/enb12354e.pdf, last accessed 25 August 2012.

IISD (2008) Summary of the Fourteenth Conference of the Parties to the UN Framework Convention on Climate Change and Fourth Meeting of the Parties to the Kyoto Protocol, 1–12 December 2008, *Earth Negotiations Bulletin*, 12 (395), 15 December, available through http://www.iisd.ca/vol12/enb12395e.html, last accessed 25 August 2012.

Johansson, S. and Lindholm, H. (2002) *Role-play on Climate Change: An Exciting Game of Hot Discussions and Cool Negotiations*. Stockholm: Naturvårdsverket.

Johnson, K. (2009) Copenhagen Walk-out. Poor Countries Bail on Climate Talks. *Wall Street Journal*, 14 December.

Kolb, D. and Williams, J. (2003) *Everyday Negotiation: Navigating the Hidden Agendas in Bargaining*. San Francisco: Jossey-Bass.

Krasner, S. (1986) *International Regimes*. Ithaca, NY: Cornell University Press.

Kremenyuk, V. (ed.) (2002) *International Negotiation: Analysis, Approaches, Issues*. San Francisco: Jossey-Bass.

Lall, A. (1985) *Multilateral Negotiation and Mediation Instruments and Methods*. London: Pergamon.

Lanchberry, J. and Victor, D. (1995) The Role of Science in the Global Climate Negotiations, In Bergeseon, H. (ed.) *Green Globe Yearbook*, Oxford: Oxford Universuty Press. 29–39.

Meehl, G. A., Stocker, T. F.,Collins,W. D., Friedlingstein, P., Gaye, A.T. Gregory, J.M. and Kitoh, A. (2007). Global Climate Projection, In Susan, S., Qin, D., Manning, M., Chen, Z.,

Marquis, M. (eds.) *Climate Change. Physical Science Basis. Contribution of Working Group 1 to the*

Fourth Assessment Report of the Intergovernmental Panel on Climate Change. Cambridge: Cambridge University Press. 747–845.

Ministry of Foreign Affairs of Japan (2009) The 17th APEC Economic Leader's Meeting, Ministerial Statement, Singapore, 14–15 November, available through http://www. mofa.go.jp/policy/economy/apec/2009/state1.pdf, last accessed 25 August 2012.

Mnookin, R. and Lee, B. (1995) *Barriers to Conflict Resolution.* New York: W. W. Norton & Company.

Najam, A., Huq, S. and Sokona, S. (2002) Climate Negotiations Beyond Tokyo: Developing countries concerns and interests, *Climate Policy* 3, 221–231.

OECD (1999) *Action Against Climate Change: The Kyoto Protocol and Beyond.* Paris: Organization for Economic Co-operation and Development.

Pan, J. (2006*) Road Map to Post-Kyoto Climate Agreements.* Beijing: Research Centre for Sustainable Development, Chinese Academy of Social Sciences.

Penetrante, A.M. (2012) Simulating Climate Change Negotiations: Lessons from Modeled Experience, In *Negotiation Journal*, 27 (3), 279–314. doi: 10.1111/j.1571-9979.2012.00341.x.

Raiffa, H., Richardson, J. and Metcalfe, D. (2002) *Negotiation Analysis: The Science and Art of Collaborative Decision Making.* Cambridge, MA: Belknap.

Rose, J., McLean, J. and Conrad, A. (2002) *The Art of Negotiation: A Simulation for Resolving Conflict in Federal States.* Orchard Park, NY: Broadview Press.

Rudd, K., Wong, P. and Swan, W. (2009) *A New Target for Reducing Australia's Carbon Pollution.* Department of Climate Change, available through http:///www.climatechange.gov. au/~/media/Files/minister/wong/2009/media-releases/May/mr20090504c.ashx, last accessed 25 January 2010.

Salacuse, J. (2003) *The Global Negotiator: Making, Managing and Mending Deals around the World in the 21st Century.* New York: Palgrave Macmillan.

Schueneman, T. (2009) Managing Expectations for COP15. Global Warming is Real. 16 November, available through http://globalwarmingisreal.com/2009/11/16/managing-expectations-for-cop15/, last accessed 24 August 2012.

Sebenius, J. and Lax, D. (1986) *The Manager as Negotiator. Bargaining for Cooperation and Competitive Gain.* New York: Free Press.

Sjöstedt, G. (ed.) (1993) *International Environmental Negotiation.* London: Sage.

Sjöstedt, G. (1994) Issue Clarification and the Role of Consensual Knowledge in the UNCED Process, In Spector, B., Sjöstedt, G. and Zartman, W. (eds.) *Negotiating International Regimes: Lessons learned from the United Nations Conference on Environment and Development (UNCED).* London: Graham Trotman/Martinus Nijhoff.

Sjöstedt, G., (ed.) (2003) *Professional Cultures in International Negotiation – Bridge or Rift?* Lanham, MD: Lexington.

Sjöstedt, G. (2005) What did the World Summit on Sustainable Development (WSSD) accomplish? In Churie Kallhauge, A., Sjöstedt, G. and Corell, E. (eds.) *Global Challenges. Fostering the Multilateral Process for Sustainable Development.* Sheffield: Greenleaf Publishing. 31–56.

Sjöstedt, G. (2009) Knowledge Diplomacy: The Things We Need to Know to Understand It Better, *PinPoints* 33/2009, Laxenburg: IIASA/The Processes of International Negotiation Network. 4–7.

Susskind, L. (ed.) (2001) *Reforming the International Environmental Treaty-making System.* Cambridge, MA: PON Books.

Susskind, L., Moomaw, M. and Gallagher, K. (eds.) (2002) *Transboundary Environmental Negotiation New Approaches to Global Cooperation.* San Francisco, CA: Jossey-Bass.

The Guardian (2009) *Copenhagen Climate Summit in Disarray After 'Danish Text' Leak,* 8 December 2009.

The United Nations Framework Convention on Climate Change (UNFCCC) (2009) *Joint Submission: Information relating to possible quantified emissions limitation and reduction objectives as submitted by Parties, Australia, Belarus, Canada, EU, Iceland, Japan, New Zealand, Norway, Russia, Switzerland, Ukraine*, available through http://unfccc.int/files/kyoto_protocol/application/pdf/jointsubmission290409.pdf, last accessed 20 January 2010.

The United Nations Framework Convention on Climate Change (UNFCCC) (1992) *The Convention*, FCCC/INFORMAL/84, GE.05-62220 (E) 200705, available through http://unfccc.int/resource/docs/convkp/conveng.pdf, last accessed 25 August 2012.

The United Nations Framework Convention on Climate Change (UNFCCC) (1998) The Kyoto Protocol, available through http://unfccc.int/resource/docs/convkp/kpeng.pdf, last accessed 25 August 2012.

Victor, D. (2001) *The Collapse of the Kyoto Protocol and the Struggle to Slow Global Warming*. Princeton, NJ: Princeton University Press.

Watkins, M. (2002) *Breakthrough Business Negotiation: A Toolbox for Managers*. San Francisco: Jossey-Bass.

Watkins, M. (2006) *Shaping the Game. The New Leader's Guide to Effective Negotiating*. Cambridge: Harvard Business Press.

Wetzel, D. and Lachmann, G. (2009) Kopenhagen gescheitert, US-Präsident Obama stürzt vom Klima-Gipfel, 19 December, available through http://www.welt.de/politik/ausland/article5581658/US-Praesident-Obama-stuerzt-vom-Klima-Gipfel.html, last accessed 26 January 2010.

Whiteman, H. (2009) Poor Nations' Fury over Leaked Climate Text, *CNN*, available through http://www.cnn.com/2009/WORLD/europe/12/09/danish.draft.climate.text.0850/, last accessed 20 January 2010.

Wynn, G. (2009) What was agreed and left unfinished in U.N. climate deal. *Reuters*, 20 December, available through http://in.reuters.com/article/2009/12/20/idINIndia-44872920091220?sp=true, last accessed 14 May 2011.

Zartman, I. W. (1978) *The Negotiation Process. Theories and Applications*, Beverly Hills: SAGE Publications.

Zartman, I.W. (ed.) (1994) *International Multilateral Negotiation: Approaches to the Management of Complexity*. A publication of the Processes of International Negotiation (PIN) Project of the International Institute for Applied Systems Analysis. San Francisco: Jossey-Bass.

Zartman, I. W. (2000) Ripeness: The Hurting Stalemate and Beyond, In Stern, P. and Druckman, D. (eds.) *International Conflict Resolution After the Cold War*. Washington, DC: The National Academies Press. 225–250.

Zartman, I. W. and Berman, B. (1982) *The Practical Negotiator*. New Haven: Yale University Press.

Zeller, T. (2009) Copenhagen Talks Tough on Climate Protest Plans, *New York Times*, 6 December 2009, available through http://www.nytimes.com/2009/12/07/science/earth/07security.html, last accessed 20 January 2010.

Part II

Professional perspectives

1 The Perspective of a Politician

How Decisions are Made

Josef Proell, Helmut Hojesky and Werner Wutscher

Climate talks in a historical policy context

Climate talks have a comparatively long history in international environmental negotiations, dating back to the late 1970s. It was the 1972 conference in Stockholm that first addressed climate change as a serious environmental threat. In the course of the 1980s, important meetings took place in Villach (Austria) in 1985 and 1987, and in Bellagio (Italy) in 1987; these led to the foundation of the Intergovernmental Panel on Climate Change (IPCC) jointly by the United Nations Environment Programme (UNEP) and the World Meteorological Organization (WMO). Although these conferences and the discussions clearly had a scientific background, political interests marked the talks from the very beginning.

In the Second World Climate Conference in Geneva, in 1989, there was even greater emphasis on the political context. It became clear that the international community needed to react to the dangers of a changing global climate, and a Climate Convention was seen as the appropriate tool for this. Tensions among different interest groups arose at this conference, immediately giving the task a serious political aspect. The United Nations established a broadly based Intergovernmental Negotiating Committee (INC) to negotiate a climate convention in late 1990. The process leading to the United Nations Framework Convention on Climate Change (UNFCCC), the Kyoto Protocol and the Marrakesh Accords is now continuing, with discussions about further steps in a post-2012 regime under the Kyoto Protocol and in long-term cooperative action under the Convention. However, these negotiations are overshadowed by political struggles between those who wish to preserve their own (mostly economic) interests and those for whom the basis of life itself is under threat, such as people living in low-lying island states who fear catastrophic rises in sea levels.

Given these diverging views, it is only natural for the climate talks to be much more complex than other international environmental negotiation processes. The complexity results from the multiplicity of issues touched upon in the course of the negotiations:

1. The necessity for sound scientific input, which is delivered by thousands of international scientists for the IPCC Assessment Reports.

2.   The global impacts of climate change, which affect every part of the world in some way.
3.   The fact that the industrialized world has contributed most to global climate change in the past 150 years through greenhouse gas emissions, while the developing world will suffer most in the future from global warming.
4.   The common but differentiated responsibilities and respective capabilities (as stated in Article 3 of the UNFCCC) which underpin discussions between North and South regarding the development of common strategies.
5.   How to achieve a trade-off between environmental and economic concerns in efforts to combat climate change.

All these questions and their possible solutions are a major challenge for politicians. On the other hand, it is extremely attractive for a politician to deal with climate talks, because they touch the most important aspects of mankind within the next hundred years.

Because of the climate talks' complexity, and especially because of the tensions they cause, the process needs to be facilitated. The entry into force of the Kyoto Protocol has not really achieved this; instead, the tensions between some parties seem to be even greater than in the past. Before the Montreal Climate Talks (COP12), there was danger of deadlock, but some "ice-breaking" moves made it possible to restart talks in a meaningful way.

Major political stumbling blocks in the climate talks

The climate talks are much more complex than other multilateral environmental agreements (MEAs). Treaties like the *Montreal Protocol on Substances That Deplete the Ozone Layer* and the *Basel Convention on the Control of Transboundary Movements of Hazardous Wastes and their Disposal* deal with rather narrow issues in the environment field. There are only few producers of ozone-depleting substances, while greenhouse gases are "produced" by almost every individual. The limited scope of the MEAs means that negotiation groups do not take such divergent positions as they do in the climate talks.

There is one general stumbling block (not only in the climate talks, but also in biodiversity and other talks). All conclusions and decisions in the climate process have to be taken by a consensus of all parties, as no rules of procedure –,especially voting rules – were adopted at the start of the process. This takes time and may not always lead to optimal results.

Negotiating groups or blocs have been formed to strengthen group advocacy. The main blocs in the climate change deliberations are 1) the countries of the Organisation for Economic Co-operation and Development (OECD: the developed world) and 2) the Group of 77 and China (G77 + China: the developing world). The North–South conflict is reflected in the frequently conflicting views between these two blocs. However, the major blocs themselves are not in themselves homogenous. In both, forward-looking positions can be found, represented *inter alia* by the Alliance of Small Island States (AOSIS) in the G77 + China group, or the EU and Switzerland in the OECD group. Other groups of countries, exemplified mainly by the *Organization of the Petroleum Exporting Countries* (OPEC) countries on the one hand and the US on the other hand, tend to slow down the process. There is also the "special treatment" (in fact, exceptions) foreseen within the UNFCCC and the Kyoto Protocol for Least Developed Countries (LDCs) and for countries with economies in transition (Eastern Europe).

Another important stumbling block is the double braking mechanism laid down in Article 25 of the Kyoto Protocol, whereby the Protocol enters into force:

> on the ninetieth day after the date on which not less than 55 Parties to the Convention, incorporating Parties included in Annex I which accounted in total for at least 55% of the total carbon dioxide emissions for 1990 of the Annex I countries, have deposited their instruments of ratification, acceptance, approval or accession.

This gave two major players, namely the Russian Federation and the US, a special role, as acting in concert would have enabled them to delay the Protocol's entry into force. With the US announcing its retreat from the Protocol in early 2001, the entry into force was then in the hands of the Russian Federation. With the Russian ratification in late 2004, the Kyoto Protocol entered into force on 16 February 2005.

Other examples of "stumbling blocks" are 1) widespread and different economic interests within the groups, 2) the historic responsibility of industrialized countries for emitting greenhouse gases, and 3) the right to development for developing countries. As all these positions have to be brought together to find consensus, "package solutions" are often the only way of moving ahead.

Finally, a fundamental political stumbling block has to be borne in mind, namely, that climate change is a long-term environmental threat, which cannot be reversed, nor dealt with by short-term solutions. It can only be mitigated, which is difficult for a politician to "sell" to the public. A direct "dose-response" effect of mitigation measures cannot be shown. The effects of mitigation measures on emission reductions cannot immediately be seen in the GHG emission inventories, and may take some years to show their full potential. Projections do, however, help to estimate the effects of measures.

The role of politicians in the climate talks

The most important role for politicians in the negotiations is that of "trouble-shooters" when negotiations get stuck . While negotiation officials often have strict mandates to follow, political weight often makes it easier to find a compromise, as politicians tend to have more flexibility to negotiate and a greater appreciation of the negotiating stances of other groups. It is easier for politicians to modify negotiating positions in order to get a "deal". Important decisions in the process, such as the adoption of the Kyoto Protocol at the Third Conference of the Parties (COP3), the Bonn Agreement at COP6 or the Marrakesh Accords at COP7, would not have been possible without the political level becoming involved.

Because of the complexity of the negotiation process, political support is required. The decisions taken may often have economic impacts or touch upon financial questions, thus affecting the national budget. Such responsibilities cannot be taken by public servants. It is the politician who has to argue the results within the government back home.

Another invaluable element of politicians being involved is the informal talks that arise between ministers during COPs. There is enough time for bilateral meetings where questions of mutual interest can be discussed informally and personal relations between ministers of different parts of the world can be developed.

Of course, some parts of the climate negotiations are very technical. It is the role of the officials to ensure that the open questions in a package are limited to just a few and that they really are policy-relevant "big issues". Otherwise it would be very difficult for politicians to have meaningful discussions and to make a deal.

Expectations regarding future process

How the regime after the first Kyoto commitment period, which ends in 2012, will look is a major question. The negotiation process for the second (and following) commitment period was begun at COP11/CMP1 in Montreal at the end of 2005. At COP12 in Bali, it was specified that post-Kyoto talks should be concluded in 2009 at COP15 in Copenhagen. Two tracks, the dialogue on long-term cooperative action under the UN Framework Convention on Climate Change and the negotiations based on Article 3.9 of the Kyoto Protocol on post-2012 commitments, were decided upon. The outcome of the negotiations under Article 3.9 were to be adopted within a reasonable time frame before the end of the first commitment period to give certainty to the business sector. The dialogue has been held in workshops (two per year) and guided by two co-facilitators, one from Annex I Parties (the industrialized countries) and one from non-Annex I Parties (the developing world). The negotiations on Article 3.9 have been handled by an ad hoc Working Group, with a chair (predominantly from an Annex I country) supported by a vice chair. Both the co-facilitators and the chair must report results regularly to COP/CMP.

The discussions in the ad hoc Working Group are held among the Parties to the Kyoto Protocol (most of the industrialized countries and many developing countries), thus excluding the US. However, the dialogue takes place among all Parties to the UN Framework Convention, which includes the US.

Both processes are guided by the ultimate objective of the UN Framework Convention on Climate Change laid down in its Article 2: stabilization of greenhouse gas concentrations in the atmosphere at a level that would prevent dangerous anthropogenic interference with the climate system, within a time frame sufficient to allow ecosystems to adapt naturally to climate change, to ensure that food production is not threatened and to enable economic development to proceed in a sustainable manner. To meet this objective, overall global mean surface temperature increase should not exceed 2°C above pre-industrial levels; significant global emission reductions will also be necessary in accordance with the principle of common but differentiated responsibilities and respective capabilities. This will require global greenhouse gas emissions to peak within two decades, followed by substantial reductions of at least 15 per cent and perhaps as much as 50 per cent by 2050 compared with 1990 levels. Failure to achieve such reductions would lead to an increased risk of abrupt climate change.

The dialogue especially provides an appropriate means of discussing with all parties the ways and means of achieving this objective. Without prejudice to new approaches for differentiation between Parties in a future fair and flexible framework, parties explore strategies for achieving the necessary emission reductions. To be most effective in tackling climate change, strengthened further actions by Annex I Parties must form part of a global effort by all parties in accordance with the principle of common but differentiated responsibilities and respective capabilities to enhance the implementation of the UN Framework Convention.

In this context, what has to be highlighted is the crucial role of a global carbon market and the continuity and effective functioning of the flexible mechanisms of the Kyoto Protocol, to deliver the necessary deep emission cuts in a cost-effective manner and to stimulate the development, deployment and transfer of climate-friendly technologies, practices and processes.

The fact must be faced that the stumbling blocks and tensions described above will also be present in the coming deliberations. There is some danger that the different views with regard to priority for short-term and/or long-term measures might lead to a deadlock. But there is some hope as well – the seriousness of the problem of climate change and the first signs of climate change impacts should lead to an open exchange of views in both processes and to the desired results in the short-term and the long-term time frame.

The distrust that still exists between negotiating groups has to be turned into a positive relationship, otherwise the coming climate talks might slow down again. There is still some distrust between the G77 + China on the one hand and the OECD countries on the other. This is partly due to discussions on the possibility of involving certain developing countries in a future mitigation regime, keeping in mind the fact that a global problem needs a global solution, but simultaneously following the principle of common but differentiated responsibility.

Focus must now be given to the question how to combine future commitments with other policies in the economic and financial area and to define how these areas can contribute to overall climate policy. The interaction between the various political issues must be strengthened; otherwise an agreement on the next steps could be rather difficult. It is likely that the political representatives of all negotiating groups have learned their lessons from the past and are willing to contribute constructively in the coming talks under the two tracks. They would definitely benefit from facilitation.

2 The New Diplomacy from the Perspective of a Diplomat

Facilitation of the Post-Kyoto
Climate Talks

Bo Kjellén

Background

The perspective of the author of this chapter is that of the practitioner, although I
am now retired. As chief negotiator for Sweden from the beginning of the climate
negotiations at Chantilly, outside Washington, in February 1991, I was engaged
in all the major negotiations on climate change until late 2001, when I retired
from active negotiation. However, as an adviser in the Ministry of the Environment, I kept in close contact with the negotiating scene and also ventured into
academia, participating in projects undertaken by various institutions, in particular, the Tyndall Centre for Climate Change Research at the University of East
Anglia, which uses integrated assessment to support negotiations for the post-2012
climate regime.

Over the years I have become increasingly convinced that the negotiations on
sustainable development require a deep analysis, both of process and of content.
I have also coined the expression "A New Diplomacy for Sustainable Development," a concept which was first expressed in a speech I gave in New York in
1999. Others might argue that our negotiations on normative issues in the Rio
– Johannesburg process, or in the climate or desertification negotiations, are not
fundamentally different from the traditional bargaining in multilateral fora (Kjellén 2007). There are long night sessions, wrangling over words and expressions,
papering over of differences in obscure language, much as one finds in other UN
meetings or in trade negotiations.

These arguments have to be taken seriously, but I would nevertheless insist that
we are in the presence of a new branch on the very old tree of diplomacy. As we
try to analyze the stumbling blocks ahead for the climate negotiations, I therefore
believe that it is helpful to elaborate on the nature of this new diplomacy, and on
its relationship to more traditional international relations.

There are hundreds of environmental treaties in the world today, bilateral and
multilateral; but I wish to concentrate here on the Rio process and the major multilateral conventions that are directly linked to the Rio Conference on Environment
and Development in 1992. These are the Montreal Protocol on Substances That
Deplete the Ozone Layer (1989), the Framework Convention on Climate Change
(1992), the Convention on Biodiversity (1992), the Convention to Combat Desertification (1994), and the Convention on Persistent Organic Pollutants (2001). To

this list should also be added the important Kyoto Protocol to the Climate Change Convention, concluded in 1997, which entered into force in 2005 after a long-delayed Russian ratification. These instruments all concern major global issues, and they go beyond the environment in their ambition to meet the definitions of sustainable development, as contained in the report of the World Commission on Environment and Development (Brundtland Commission 1987). The report's central language on sustainable development is very carefully drafted:

> Sustainable development seeks to meet the needs and aspirations of the present without compromising the ability to meet those of the future. Far from requiring the cessation of economic growth, it recognizes that the problems of poverty and underdevelopment cannot be solved, unless we have a new era of growth in which developing countries play a large role and reap large benefits.

This is the background to the now generally accepted idea that sustainable development contains three main components: ecology, economic factors, and social sustainability. Several comments could be made regarding this statement. First, it is clear that governments and the world community have, for a very long time, sought to achieve economic and social sustainability as essential elements in institutional and political security. What is new is really the environmental component. Second, this concern for the environment was highlighted by the increased interest in global problems in the 1980s with climate, ozone layer, and water stress becoming important new threats. Third, the triangular relationship between the environment, the economy, and social issues, which is given prominence in the political discourse, is of course much more complicated, involving a large number of parameters linked in very complex patterns. Fourth, scarcity of non-renewable resources, as highlighted by the Club of Rome in the late 1960s, represents a limitation, which has a bearing on all components of sustainability.

It is within this emerging consensus on a new type of world problem, going well beyond the traditional security concerns, that the new diplomacy evolves. Of course, this does not mean that the classical topics of international relations have lost their dominating role, be they political, economic, or military. The end of the cold war was not the end of history. Today, the problem of terrorism occupies the minds of leaders all over the world; and traditional power relations, or the effects of economic globalization, continue to appear on the headlines of newspapers. Nobody, however, can deny that concerns related to global environmental change – and the consequences for what is being more and more seen as an integrated Earth System under pressure – are being given increasing attention. Paul Crutzen and Eugene Stoermer have coined the expression the *Anthropocene Era* to suggest that we have entered a new geological epoch, characterized by the central impact of humans on the whole system. As this insight gains ground, it is quite clear that global negotiations have to reflect a new awareness. As we have sometimes said about the climate negotiations: we can negotiate *about* climate change, but we cannot negotiate *with* climate change.

This new diplomacy has a number of distinct characteristics, which I will present in the following section, while at the same time focusing on the climate negotiations and highlighting some of the stumbling blocks of a general nature that have appeared during the negotiation process.

I will then make some reflections on the nature of the negotiating process itself, in particular, its relationship with national interests and policies.

The final section of this chapter will then give some concrete examples of specific negotiating situations, based on my own experiences of stumbling blocks, and ways in which negotiators have dealt with them.

Characteristics of the New Diplomacy and general stumbling blocks in climate negotiations

We have already noted that traditional international relations have an important bearing on negotiations on sustainable development. Climate is certainly a case in point, with the policy stance of the United States (US) until now as a good illustration.

From the very beginning of the climate negotiations, there were tensions between the European Union (EU) and the US, even if negotiators were also able to develop a good working relationship within the member countries of the Organisation for Economic Co-operation and Development (OECD). The underlying factors certainly had to do with high politics at superpower level, but probably still more with the different economic and social structures and traditions on the two sides of the Atlantic. As climate policy has important domestic components, it is not surprising that issues related to taxation, gasoline price, or lifestyles (illustrated by the fact that US per capita emissions are twice as high as those of Europe), have led to different negotiating positions on different issues.

With the advent of the George W. Bush administration in 2001, however, the nature of the tensions changed. We can now clearly see that the blunt refusal of the US to ratify the Kyoto Protocol in March 2001 was the first example of the then new administration's negative policy toward multilateral cooperation, and in particular toward the United Nations. The increasingly military discourse of the US administration after 9/11, underlining the need for unconditional support in the war on terrorism, created a new atmosphere in world politics, which did not help in the climate negotiations.

Another example of stumbling blocks of a general character relates to the North–South issues. The divide between rich and poor countries has been a constant source of friction in all UN negotiations on economic and social matters over the last 50 years, and it is one of the central elements in the new diplomacy, underlined by the very expression "environment and development," used both by the Brundtland Commission and in the official title of the Rio Conference in 1992: UN Conference on Environment and Development. The main issue here is obviously that the group of 77 developing countries (G77) rightly points out that the present increase in the CO_2 concentration in the atmosphere (from 280 ppm in pre-industrial times to more than 370 ppm at present) has been due mostly to emissions from the present developed countries; it would therefore be unreasonable to force developing countries to take on quantified reduction targets. This line of reasoning has found its strongest expression in the labeling of G77 emissions as "survival emissions," while the emissions of the industrialized countries would be "luxury emissions." It is not surprising that North–South issues are often seen as the main stumbling blocks in the climate negotiations, further amplified by the many diverging interests in the G77 itself, and by the constant opposition of Saudi Arabia and other countries of the Organization of the Petroleum Exporting Countries (OPEC) to measures to reduce emissions of CO_2, with the associated claim for compensation for the potential loss of oil markets.

With these various tensions in different directions, it is natural that the lofty vision of inter-generational equity combined with intra-generational equity and international solidarity, seen as central components of the Rio process and associated negotiations, often clash with hard-fought national positions reflecting serious differences of interest. Action to combat climate change goes straight into the heart of the modern industrial civilization: energy and transport. Enormous economic interests are at stake, and it is difficult for negotiators to live up to the ideal of a global and long-term vision which would integrate all dimensions of sustainability.

One characteristic of the new diplomacy is its reliance on science and research. In the case of climate, the role of the Intergovernmental Panel on Climate Change has been decisive. The contribution of this international network of more than 2000 scientists of different disciplines was instrumental in launching the negotiations for the Framework Convention in 1991; the First Assessment Report of the IPCC had been launched in September 1990, giving clear evidence of the risks of business as usual. The following assessment reports have clarified the threats and underlined that there is a now a clear anthropogenic impact on the climate system.

The strength of IPCC arguments has been underlined by the fact that the Panel is not just a group of scientists, but that it operates in close contact with representatives of governments; furthermore, the negotiators have been regularly presented with summaries for policymakers, for several years introduced to the Conference of Parties (COP) by the forceful first Chairman of the IPCC, Bert Bolin. The assessment reports have also been subject to careful scrutiny by the COPs to the UN Framework Convention on Climate Change (UNFCCC), in particular through the Subsidiary Body on Scientific and Technological Advice (SBSTA).

However, it is in the nature of research to focus on uncertainties, and it is not surprising that one of the main stumbling blocks in the climate negotiations has been the perceived weakness of the IPCC models. A number of scientists question the very threat of global warming, or feel that mainstream scientists exaggerate the risks of emissions of greenhouse gases. It is clear that public opinion is influenced by such reports, and that governments feel that strong and costly action, with potentially negative impacts on economic growth and employment, is more difficult to motivate because of uncertainties in the nature and dimensions of the threat. At one stage the Bush administration made efforts to discredit the scientific basis for climate change concerns, but that line now seems to have been abolished. And certainly the issue of the credibility of climate science loomed large in the protracted Russian hesitations over the ratification of the Kyoto Protocol.

It is fair to say that the negotiations themselves have not been seriously influenced by scientific uncertainties. Rather than a stumbling block, it has been an irritant in the background. And one can probably maintain that, among the positive achievements of the process, the signal given by the adoption of the Kyoto Protocol – so dependent on scientific evidence – has made it possible to remove other stumbling blocks, both nationally and internationally.

The special character of these negotiations forces the negotiators to acquire new skills. The negotiators have to understand the scientific evidence and they have to grasp complex issues, which are not normally part of what diplomats learn. A growing number of experts are involved, and they sometimes have difficulties in understanding all the complexities of multilateral procedures and practice. This warrants particular efforts in terms of capacity building, particularly with regard to the least developed countries. It should also be underlined here, however, that the particular interests of some small developing countries, such as those belonging to the Association of Small Island States (AOSIS) or some West African countries have led them to involve particularly talented representatives in the negotiations, frequently from their UN delegations in New York. These delegates have often been involved in the negotiations over a long period, and they have thus been able to contribute very actively to the progress of the negotiations, for example, through chairmanships of committees, or as coordinators for crucial negotiations.

In negotiating circles, particularly among industrialized countries, there has sometimes been criticism of the so-called "New York gang" for their capacity to take advantage of their knowledge of procedures and other technicalities to actually create stumbling blocks in the negotiations. Personally I think that this criticism has been rather exaggerated; on the whole I feel that the people from UN delegations have generally helped to move the negotiation forward. The G77 has great need of this expert knowledge, as they are at a serious disadvantage in following all negotiations with their small delegations. Another matter is the general difficulty of the G77 – which now numbers more than 130 countries – to agree on common positions, or to show flexibility once a G77 position has been fixed.

For all negotiators today, new technologies have radically changed the conduct of negotiations. The immediate communication with capitals through e-mail and mobile phones in certain ways has reduced the independence of negotiators, while at the same time multiplying their opportunities to argue directly with the home base on possible changes in instructions and thereby facilitate agreement. It is worth noting that the diplomacy for sustainable development has emerged during the period in which these new technical facilities became available.

Another new element is the changed role of the European Union after the latest enlargement. It is obvious that the organization of EU pre-negotiations will be still more important in a Union with 25 members. The EU clearly has ambition to play a leading role in the climate negotiations, but we have already seen that there is a tendency for EU procedures to be too heavy, reducing the capacity of EU negotiators to adapt to changing circumstances or even to have sufficient time for consultation with other actors. Streamlined procedures will have to be developed by negotiators who understand the intricacies of EU decision making. Furthermore, an analytical capacity is needed to discern possible solutions as well as the political clout to carry them out.

Among the salient features of the new diplomacy is also the active involvement of NGOs and other stakeholders. This could have been a problem for the progress of negotiations if government representatives had felt uncomfortable with the presence of different lobbying groups within the meeting rooms. The UN Conference on Environment and Development established a procedure. At the very first meeting of the Preparatory Committee for the Rio Conference, clear rules were laid down which stated the conditions for NGO participation. In particular it was decided that NGOs were not negotiating parties, but that they had the right to attend meetings with the consent of the chair of a given meeting. The practice has evolved in the direction of great openness, which means that today NGOs are present at most meetings up to the final bargaining sessions, which take place in closed rooms. Since the number of stakeholders participating began to steadily increase, quite a lot of lobbying has been taking place; I have never felt this to be an obstacle to the negotiation. Rather, it adds a sense of urgency to proceedings, as most NGOs are pressing for more stringent commitments. Increasingly NGOs and academic institutions are organizing side events, which often offer considerable interest. A possible drawback in that connection could be that delegates do not have enough time to follow these events. Among the stumbling blocks, I would rank the extremely hard time pressure throughout the climate negotiations, which reduces the possibilities for delegates to interact outside the conference premises and to discuss possible compromises in a calmer atmosphere than the conference centre itself. The value of traditional diplomatic receptions or dinners should not be underestimated.

Negotiating positions and the definition of national interests

During the Cold War, with its characteristic bipolar relationship, analysis of international negotiations was greatly influenced by game theory. There is no doubt that this continues to be an important tool for understanding the mechanics of negotiation; however, in the framework of diplomacy for sustainable development, it is important to reflect on the extreme complexity of the issues and actors involved. Some of these elements have been highlighted in the previous section. As a practitioner of multilateral diplomacy, I find it more natural to put greater emphasis on the position of the negotiating team, both in relation to negotiating partners and to my government which provides the instructions for the negotiations.

It is also true that there are many different stakeholders involved in present-day international negotiations. NGOs and representatives of business and industry are present in the meeting rooms; different forms of lobbying seek to influence the negotiators. However, from the point of view of the negotiating team, national governments maintain control of events. They issue the instructions and they decide on the latitude given to negotiators in the bargaining process. Of course, within the EU, national positions are constantly refined and pre-negotiated before they are taken into the international arena as EU positions. Nevertheless, it is in the national capitals that different national interests are transformed into negotiating positions.

These national interests are, however, defined in a complicated web of political, social, and economic influences. Every senior civil servant and every student of government is therefore well aware of the role of different stakeholders at the national level. These interests are to a certain extent represented in the government itself. Ministries of industry or finance often take positions that are different from those of the ministry of the environment, requiring the intervention of the prime minister's office to settle issues. And beyond the government offices, the parliament as a reflection of the overall domestic political situation can carry considerable weight. In the United States, with the division of power between the executive and the legislative branch, there is the constant threat of non-ratification by Congress as an additional instrument of pressure wielded by US negotiators.

All these complexities take on special dimensions as we discuss issues of global sustainability. These issues are often long-term or diffuse, with conflicting scientific evidence, and it is difficult to grasp their significance in a broader perspective. A case in point is the question of gasoline taxes. The immediate problem is the rising price of gasoline and its consequences for citizens; it is not easy to explain the links to climate policy. What is sometimes perceived as "public opinion" is a compound of regional or local interests, NGOs of different types, religious institutions or movements, trade unions, the corporate sector, and not least the media. Processes involving these actors in government decision making constitute the very essence of the functioning of a modern democratic society. Negotiators have to understand that their work is now carried out in full openness, and that they are all part of a kind of public diplomacy.

We also have to realize that in the world of today there are often no barriers between the local and the global in terms of attitudes and support for the policies of governments. Many people are certainly more skillful than their grandparents were, as analytical skills and emotional capacities have expanded in response to new educational opportunities, new information and communication technologies, and evolving values. But this does not mean that they have better judgment or a better capacity to deal with all the stresses and dangers of living in a partially globalized world. The very complexity of the global challenges means that even well-informed citizens or civil society have difficulties in grasping the full significance of global change. Information is not sufficient as a driver of a paradigm shift which would enable nations or the international community to move to a sustainable world. Something deeper is involved, a new perception of the extraordinary responsibility of this generation, a capacity to understand and to have an overview of the changes that must be accepted as part of the quest for global sustainability.

These are essential issues for international negotiations on climate change, as we move to the determination of the post-2012 climate regime. Success in this endeavor will depend on the attitudes of governments, which in turn depend on the state of public opinion and the positions of central national stakeholders. These elements create the "enabling conditions" at the national level, which are needed for successful international negotiations. A transition to global sustainability will only be possible if the prevailing paradigm of thought will permit it. The time at which paradigm shifts will appear cannot be foreseen with certainty, but their appearance seems to be a necessary precondition for efficient global action. Given the present state of knowledge and the perceived need to move to a low-carbon world, the next 15 to 20 years may well be decisive for the prospects of long-term success.

Dealing with stumbling blocks in concrete negotiation situations: Personal experiences

Against the background of these general considerations on negotiations for global sustainability and the perceived problems of such negotiations, I now wish to turn to some more specific stumbling blocks and the ways of dealing with them, in particular with reference to climate change, based on my own experiences. It is obviously not possible to go into the details of all these negotiations. I limit myself to some central aspects, aimed at illustrating the most important points. However, even if there are already quite a large number of books written on these subjects, I feel that further research is most probably warranted.

The negotiation of the Framework Convention on Climate Change itself would have seemed almost impossible to conclude in the time available; and it is true that many of the negotiators, as they met outside Washington in February 1991 at the invitation of the George H.W. Bush administration, were rather ignorant of the issues involved. There had to be a steep learning curve as successive frequent meetings of the Intergovernmental Negotiating Committee (INC) were held in the period leading up to final agreement in New York in May 1992, just a few weeks before the Rio Conference.

The need for a quick conclusion of the negotiation and the lack of knowledge of the issues involved were certainly stumbling blocks, which could have derailed the negotiation at an early stage. Other problems were related to the very organization of the process: this was a new kind of treaty to be concluded.

Not surprisingly, the first negotiating session was also both difficult and confusing, but the pre-negotiations on who should chair the INC had resulted in unanimous agreement to elect the French high official, Jean Ripert. This soon proved to be a very successful decision: Ripert, who for me embodies the French ideal of "un grand commis de l'État", had all the credentials to lead this difficult negotiation. He had been responsible for the extensive French planning system for part of the post-World War II period, and he had for a long time been the highest official of the UN dealing with economic and social matters. He had also, since his return to France, been involved in IPCC work. He was quite an impressive person and nobody questioned his authority.

Furthermore, the United Nations decided to appoint the Maltese Michael Zammit Cutajar to the post of Secretary-General of the INC. Like Ripert, he knew the UN system well, having been an official of the UN Conference on Trade and Development (UNCTAD) for a long time.

Thus, the management of the negotiation was in very good hands. In my opinion it is too easy to underestimate the danger of confusion, bad management, or bad organization as major stumbling blocks at any negotiation, in particular when delegates themselves are uncertain about the issues at stake or the goals to be achieved. At the same time, it is of course necessary to underline that even the best chair in the world cannot achieve success if the underlying conflicts of interest are too serious.

At the second meeting of the negotiating committee, a few months after the Washington meeting, delegates could already feel that the organizational conditions for success had been created. Confidence in the management was established, and the sense of urgency was ensured because of the need to conclude a deal enabling the Convention to be signed at the Rio Conference in early June 1992.

At the same time, all delegates were of course well aware of the substantive difficulties involved in the negotiations. Several of these problems are well documented, *inter alia*, in *Negotiating Climate Change*, compiled by the Stockholm Environment Institute and published by Cambridge University Press in 1993. Here, I just wish to focus on some of the stumbling blocks which were present as the final agreement in the INC was being worked out in May 1992, and which involved the type of conflicts – between the EU and the US, and between South and North – which have been present in the climate negotiations ever since.

It was clear at an early stage that the G77 would not accept the same kind of commitments that the industrialized countries would undertake. The notion of "common but differentiated responsibility" had surfaced in the early stages of the Rio process and in the climate negotiation itself. But the OECD countries insisted that there had to be some commitments which were common to all countries; and in the Convention these are formulated in Article 4.1. However, this deal was clearly not enough for the G77. There had to be additional language, ensuring that the development imperative be safeguarded and, furthermore, that there would be provisions underlining the need for development assistance and transfer of technology, as well as special rules for least developed countries and other particularly vulnerable countries. In addition, we needed agreement on tricky institutional questions, in particular the role and location of the Financial Mechanism, designed to transfer resources to developing countries. On this latter issue, the OECD countries insisted on the World Bank–based Global Environment Facility, which was not acceptable to the G77.

At the same time, on the OECD side, things were not going well. Canada had taken an early initiative to hold regular meetings in what became known as the "common interests group," but it soon became clear that interests were not that common. The EU was applying pressure to get hard and fast targets and timetables of emissions reductions, whereas the United States was talking in terms of recommendations for policies and measures, while Japan tried to find compromise language, with voluntary commitments submitted to some form of review.

At the next but last INC session in New York in March 1992, the EU took the initiative to ask Sweden (not yet a member of the European Union) to chair an informal OECD group to try to work out an agreement. I took on the task with some trepidation; the stumbling blocks seemed too formidable. And we did not make much progress; for most of the session the OECD countries were locked in interminable and frustrating discussions, while the G77 were just waiting for us to agree and move forward in the negotiation.

We were three months away from Rio. The negotiating texts were full of square brackets, indicating disagreement. The organizational stumbling blocks had been removed, but the substantive ones had hardly moved at all. Chairman Ripert took the decision to organize an informal consultation with some of the leading delegations in Paris in the first half of April.

I seized the opportunity to call an informal OECD meeting in the days preceding the Ripert consultation. This OECD meeting turned out to be very difficult, and progress was marginal. However, at an informal dinner, the chief US delegate, the witty and constructive Robert Reinstein, launched ideas for a solution, on a purely personal basis. Perhaps there was an opening.

Ripert's consultation did not lead to any substantial breakthrough either. But there was now a real sense of urgency, and a general understanding that the prospects for the Rio Conference itself would be endangered if a substantive Framework Convention on Climate Change could not be signed. Discussions were at times rather confused, but finally the participants agreed that there was simply not enough time to finalize negotiations aimed at removing the square brackets in the existing text. Jean Ripert was mandated to prepare a wholly new Chairman's text, and to organize discussions at the final INC session in a way that could lead to agreement.

This was a way to solve the Gordian knot, which carried great dangers. There were risks that non-participants in the Paris meeting would revolt; there were risks that the US would not support Reinstein's proposals; there were risks that Ripert could not strike the right balance. The stumbling blocks were still there, but Ripert now had the tools to deal with them. He tackled the task with great energy, mobilizing the highest level of the French Republic for discussions with US and other OECD leaders. Reinstein managed to whip up support at high level in the US administration for a deal. As negotiators gathered in New York in early May to work toward a deal, elements for a compromise were in place.

However, it remained for the Chairman to convince all delegates that a small negotiating group was necessary. With the help of the chair of G77, Pakistan, and a particularly talented Indian diplomat, Chandrasekhar Dasgupta, who became the leading South negotiator, he managed to get a small group of 25 countries operative; and it was in this setting that the Convention was finally crafted and later agreed by most countries participating in the INC. One of the exceptions was Saudi Arabia, although they have ultimately ratified; and Saudi Arabia has ever since carried out a skilful and relatively successful rearguard action against the Convention and the Kyoto Protocol.

The Convention was signed by 153 countries in Rio, and entered into force in March 1994, 90 days after the deposit of the fiftieth instrument of ratification. (It is worth noting that the US was among the first countries to ratify). It can be argued that there are many weaknesses in the text, and that in particular Article 4.2, which defines the stabilization commitment of industrialized Annex I countries for the period 1990–2000, is far too convoluted to be really operative, but nevertheless the Convention remains the best example to date of a legally binding document, laying the basis for action on a serious global threat.

The hero of the FCCC negotiation was in my view its chairman, Jean Ripert. Because of the complexities of these negotiations, the chair has a central role as manager, facilitator, and inspirer. The chair has to be the driving force and the manager of the negotiating process, and has the central responsibility to draft compromise texts at critical points: in that context, timing is of primordial importance. A chair's final draft must not come too early, because then it might not rally the necessary support among negotiators who still feel that they have better ideas; but it must not come so late that there is not sufficient time for the small adjustments that will be necessary to carry consensus. (One characteristic of the new diplomacy so far is that decisions are normally taken by consensus; no votes are taken.) It is not necessary that this final draft should cover the whole treaty text, and the chair often has to rely on the work of coordinators or small group chairs, who report to him: but the final responsibility rests with the chair. It is also essential that the chair is impartial, so that he inspires complete confidence. This obviously means that he has to make it very clear to the national authorities that the chair has no possibilities of promoting national interests. Normally this is also well understood and appreciated in capitals, as the appointment of a chair is also an opportunity to promote other more general national interests than those related to the specific negotiation.

In the decisive moment of the FCCC negotiation, Ripert was able to present the complete draft that led to agreement, after long negotiations, in a small group; and this was later confirmed by the plenary. But other negotiations have shown different styles and different approaches to leadership in the management of that crucial instrument: the chair's final draft. In the final negotiation of the Kyoto Protocol in December 1997, the ebullient chair of the Main Committee (and previously Chairman of the Ad Hoc Group on the Berlin Mandate, which prepared Kyoto), Ambassador Raúl Estrada of Argentina, operated in another style than Ripert, skipping the small group approach and conducting the final negotiation in an all-night plenary. This was a high-risk operation, but it worked, and the negotiation could be concluded, albeit with a rather incomplete protocol, not yet ready for ratification. However, the main component of Kyoto, the quantified commitments of industrialized countries, was agreed after a separate bargaining session at high ministerial level.

We had hoped that the completion of the Kyoto Protocol would take place one year later in Buenos Aires, but all that could be agreed then was a schedule for further negotiations, called the Buenos Aires Plan of Action, which was supposed to lead to final agreement by COP6, to be held in late 2000. The process now changed character, with a more visible involvement of the political level; strong senior officials were no longer in the driver's seat. There were various reasons for this; the main underlying factor was most probably that the global climate issue had moved higher on the political agenda in many countries. This was already evident in Kyoto: Vice-President Gore paved the way for American agreement – and ultimately signature – of the Protocol, and European Ministers, with the United Kingdom Deputy Prime Minister John Prescott in the lead, were instrumental in securing consensus on the quantified commitments.

But personalities also played a role. Once it was decided that the crucial COP6 meeting would take place in The Hague, the high-profile Dutch Minister of Environment, Jan Pronk, prepared his chairmanship in a very active way, with continuous consultations with ministerial colleagues as a major component. In The Hague, Pronk had a firm grip of the proceedings, and activated the political level to an extent not before seen in the climate negotiations: ministers were called upon to chair contact groups at short notice on an almost 24-hour schedule. At the end there was considerable fatigue, and as Pronk presented a compromise text with just few hours of negotiation remaining, the final negotiations became rather confused, and ultimately failed.

The failure of The Hague COP6 was serious, but not fatal, and now Pronk's energy and contacts were decisive factors in putting the negotiation back on track. Several high-level multilateral consultations were organized during the spring of 2001, which was also characterized by the US withdrawal from the Protocol. The climate issue had moved to the very highest levels of government.

The US position could have proved to be the ultimate stumbling block for the Kyoto Protocol, but a firm EU position in favor of ratification of the Protocol and a major effort on the part of EU to rally support from Japan, Russia, other industrialized countries, and the Group of 77, enabled the process to move forward. Here was another example of successful political intervention led by the Swedish Presidency of the EU and, in particular, the then Minister of Environment of Sweden, Kjell Larsson, who spared no efforts in meeting ministers of key countries. It was of special importance that Larsson established a good working relationship with the Minister of Environment of Iran, the 2001 Chair of G77.

In fact, it might be argued that EU/G77 relations had seldom been as good as during this period. These two central groupings in the climate negotiations have often had difficulties in coming to terms, not least because of the difficulty of establishing common positions within the groups; and certainly within the EU there is a constant concern that too much time is spent in internal coordination, leaving too little space for contacts with other countries and groups. However, during 2001, relations were good and stable, which assisted Minister Pronk in his efforts to resume negotiations after the failure at The Hague.

So when the time came for the resumed COP6 in Bonn in July 2001, Pronk had managed to create a rather favorable outlook for the negotiations, even if there were still many uncertainties. In the delicate balance to be struck with the G77, a number of industrialized countries, with the EU and Canada in the lead, had worked out a financial package which was presented at an early stage in the negotiation; and within the Annex B group of industrialized countries, difficult negotiations led to satisfactory deals on such tricky issues as the calculation of sinks, the modalities for trading, and the implementation of compliance rules. This time Pronk chaired the negotiations with efficiency and skill; EU coordination worked well under the leadership of Olivier Delheuze, Minister of Environment of Belgium, and the final deal was hailed as a considerable success and greeted with strong applause.

As is often the case, however, the devil is in the details, and it soon turned out that some points needed further clarification; difficult consultations had to be carried out over the coming months to confirm the Bonn deal at COP7 in Marrakesh in November 2001.

We now know that the Marrakesh Accords did not lead to quick ratification by all countries and that in particular Russian hesitations postponed the entry into force of the Kyoto Protocol. But the EU has decided to launch its system of emission trading, and the process of climate negotiations was not floundering.

Scientific assessments of the dangers of climate change continue to become more precise; most governments, including that of the US, now seem to agree that climate change is a major political issue. In that sense, the political commitment, mainly launched by Jan Pronk, has been a success. But the complexity of the problems already discussed in this paper, and in particular the societal consequences of the necessary changes in the energy and transport systems, continue to complicate the international process; as does the necessary quest for fairness and justice, which are related to development and to combating poverty. As was pointed out in the first part of this chapter, these broad problem areas continue to be major stumbling blocks as the world community now faces tackling the post-2012 climate regime. But as I have tried to show here, there are also a number of concrete issues related to the management of process which are instrumental in removing the minor stumbling blocks that appear in any negotiation.

There are important lessons to be drawn from the saga of the climate negotiations when it comes to removing stumbling blocks.

First, deficient organization and management of a negotiation can be a serious stumbling block. The early climate negotiations could have failed from the beginning without a highly respected and experienced chair and an efficient Secretary-General, who both knew UN systems well. They managed to set up a negotiating structure, which could accommodate the many pressures sustained in the process.

Second, the standing of the chair should enable him or her to approach personalities at very high level to argue for a solution to problems of substance. The chair should also inspire so much confidence among the negotiators that they are ready to argue in capitals for his proposed solutions.

Third, the active engagement of senior delegates with a good standing and good contacts in their capitals helps greatly to remove stumbling blocks; it is sometimes necessary to argue forcefully for a compromise.

Fourth, obstacles can be removed only if delegates show not only competence but also constructive thinking. As chair, it is striking how often one has to rely on good ideas coming from the floor; similarly, constructive proposals tend to move the negotiation forward, wherever they come from. Stumbling blocks are there, but if all delegates feel really involved in a meaningful process, there is something of a collective will to succeed.

Fifth, in a complicated negotiation, it is sometimes necessary to hold restricted meetings in order to explore the removal of stumbling blocks; it is simply too difficult to negotiate on some things in a plenary with hundreds of delegates present. For the chair to organize such groups, and for the delegates left out to accept them, requires diplomatic skill, an atmosphere of confidence, and maximum transparency.

Sixth, a strict timetable helps to reduce the importance of stumbling blocks, as countries have to weigh their positions against the risk of a total breakdown of the negotiation. It is also important for internal preparatory processes to be well aware of the timetable for negotiation. As said earlier, there is no doubt that the FCCC was assisted by the obligation to have a Convention ready for signature at the Rio Conference. I myself had a typical experience of the way in which an absolute time limit can force an agreement when, as chairman of the INC for the Convention to Combat Desertification, I had to tell delegates at 1a.m. on the last night that time had run out and that the negotiated package had to be agreed, otherwise the whole negotiation would have failed. Agreement was reached, but it was a close call.

Seventh, it is obvious that the FCCC negotiation would not have succeeded, had not all parties shared the view that the political importance of an agreement was such that failure to agree would be negative for everybody concerned. There is obviously in every negotiation a tipping point when non-agreement becomes a preferable option for some participants, particularly if the proposed text is considered too ambitious. This is an issue that needs to be taken seriously, and an important lesson for the future. Because of the risks for climate change, we need to push for a regime that is ambitious, but negotiable. This will require a very significant effort on the part of all parties concerned, not least scientific researchers, as there has to be guidance to the policymakers and to the negotiators, when the components of a post-2012 regime are defined.

Concluding remarks

I have no doubt that a global climate regime will be a major issue for the diplomacy on sustainable development in the years and decades to come. I am also convinced that the societal enabling conditions will ultimately appear, as scientific evidence of a changing world climate strengthens. This means that some major stumbling blocks to successful negotiations will be removed. The question is how much time this will take, and how much time we have. Action to combat climate change will have an impact on all of us: a major question will be how *fairness* and *justice* can be ensured. How will the situation of especially vulnerable countries or groups be handled? How will other world events and conflicts influence the negotiations? How will the globalized world economy with its high degree of interdependence react to measures which would radically change energy production and distribution?

All these are legitimate questions, but one thing remains clear: we need to reflect very carefully on issues related to the organization and conduct of negotiations. Over the next few years a major effort has to be undertaken to work out the post-2012 climate regime. This regime will have to be diversified and flexible, yet sufficiently firm to establish confidence and pave the way for restructuring of societies in a low-carbon direction. I have tried here to use past experience to assist the analysis of how best to organize future climate negotiations. But we have to realize that new problems and challenges will always appear; there is scope for fresh thinking and innovation as we consider the future anatomy of multilateral negotiations on climate change.

References

Bodansky, Daniel (2004) International Climate Efforts beyond 2012: A Survey of Approaches. Arlington, VA: Pew Center on Global Climate Change.

Buchner, B. and Lehmann, J. (2005) Equity Principles to Enhance the Effectiveness of Climate Policy: An Economic and Legal Perspective. In: Bothe, M., Rehbinder, E. (eds.) Climate Change Policy. Utrecht: Eleven International Publishing. 45–72.

Kjellén, B. (2007) *New Diplomacy for Sustainable Development. The Challenge of Climate Change.* London: Taylor & Francis Ltd.

World Commission on Environment and Development (The Brundtland Commission) (1987) *Our Common Future.* Report of the World Commission on Environment and Development. Published as Annex to General Assembly document A/42/427, 2 August 1987.

3 Costs and Uncertainties in Climate Change Negotiations

A Scientist's Perspective

Bert Bolin[1]

The early days of climate negotiations

Even though the issue of possible future human-induced global climate change was discussed widely among scientists in the 1970s and 1980s, general political awareness about this serious environmental issue did not emerge until late in the 1980s. *Our Common Future*, the report by the United Nations Commission on Environment and Development (UNCED), chaired by Dr. Gro Harlem Brundtland of Norway and published in 1987, dealt with this issue, and it was further discussed at the UN General Assembly in 1987. The UN Commission expressed the view that the climate issue might very well require special attention in the future. A UN resolution was adopted a year later, requesting the Intergovernmental Panel on Climate Change (IPCC), recently formed by the UN Environment Programme (UNEP) and the World Meteorological Organization (WMO), to present a broad scientific assessment of current knowledge in time for the 1990 General Assembly.

The IPCC response was quick; an elaborate report was completed in a matter of just 18 months and presented to the UN in late 1990. It became obvious during the UN discussions that the political dimensions of this matter required further attention. An Intergovernmental Negotiation Committee (INC) was created, tasked with proposing a text for a Climate Convention to the UN Conference on the Environment and Development scheduled for June 1992 in Rio de Janeiro, Brazil.

This series of quick decisions was very important. The scientific community sensed an urgency and also a willingness among politicians to engage with the issue. A number of leading scientists made a significant time commitment to meet the challenge, and the ensuing analyses of the fundamental scientific issues were well supported by the international scientific community. On the other hand, the impacts of climate change on the environment, and its possible threats to the wellbeing of people and their countries, were not well thought through, nor were analyses of possible measures to protect against climate change. Nevertheless, an intergovernmental institution had been established that had the confidence of both the scientific community and United Nations agencies.

On the basis of the First IPCC Assessment Report, the INC was able to produce an agreed text for a Framework Convention on Climate Change (FCCC) during the following year and a half, and 156 countries signed the proposed convention in Rio. The key idea was to aim for a framework convention, to which protocols on specific commitments by countries could be added when needed. This approach had already been successful in UN engagement regarding protection of the ozone layer. The detection of the hole in the ozone layer had triggered the development of a common view that far-reaching protective measures were required to address it. Agreement on a protocol to the 1985 Vienna Convention for the Protection of the Ozone Layer agreement was reached in 1987 in Montreal, Canada.

Ratification of the Climate Convention proceeded quickly during the next two years, and the Convention entered into force in 1994. The first Conference of the Parties to the Convention was arranged in Berlin the following year, where it was agreed that a protocol with binding commitments for the period 2008–2012 should be a first step toward slowing down and ultimately stopping the climate changes which were undoubtedly under way.

Kyoto was the venue for the Third Conference of the Parties, and an agreement on a protocol was reached there. US Vice-President Al Gore took part in the final negotiations and, without doubt, contributed greatly to reaching an agreement. His message was that the US was willing to reduce its emissions by 7 per cent by 2008–2012, a bold proposal in view of the growing resistance to US emission reductions in the US Congress. Indeed, the protection of national interests, rather than the threats from climate change impacts, was at the forefront of the Kyoto negotiations. Countries were positioning themselves, envisioning even tougher negotiations in the future. The uncertainties inherent in the scientific conclusions were used by some countries as an argument that more far-reaching measures were as yet premature.

The fate of the Kyoto Protocol

The Kyoto Protocol specifies that, at the time of the 2008–2012 commitment period, industrialized (Annex I) countries should have decreased their emissions by on average 5 per cent in comparison with their 1990 emissions. Specific quotas were assigned to each country or region of the industrialized world. The European Union (EU) was willing to decrease its emissions by 8 per cent and Japan by 6 per cent, while generous emission quotas were allotted to the Russian Federation and Ukraine. In spite of the decline in their emissions by 25–30 per cent by 1996, these two countries were permitted to return to their 1990 emission levels. As the Protocol allows for the introduction of tradable emission permits, this implied that Russia and Ukraine would be able to sell some of these "hot air" emissions to other industrialized countries that were having difficulties in fulfilling their Protocol requirements. This was clearly in the interest of most developed countries, not least the EU, Japan, and the United States.

Developing (non-Annex-I) countries, on the other hand, simply refused to take on any specific commitments because of their obvious need for more energy for development. After all, cheap energy was one of the primary sources of wealth and prosperity in industrial countries during the twentieth century. At the time of the

Kyoto Conference, the average per capita emissions of industrialized countries were about six times greater than those of developing countries (cf. Bolin and Kheshgi 2001; see Figure 3.1). Almost two-thirds of global emissions in 1990 still came from the industrialized world. Developing countries argued successfully that industrialized countries were responsible for taking the lead in getting to grips with the climate change issue by reducing their emissions, as already agreed under the Convention.

The carbon dioxide emission allowances for different countries during the 2008–2012 commitment period had become the central political issue during the Kyoto Protocol negotiations. They remained that way during the years that followed and, indeed, became a stumbling block to further progress. Attempts to reach an agreement on implementation procedures failed at the Sixth Conference of the Parties in Amsterdam in 2000. There was unease on the part of the US that the Kyoto goals were not reachable; moreover, the EU was not willing to allow the US to be potentially credited for forestry sector sinks. Soon after the Sixth Conference, the newly elected US President, George W. Bush, declared that the US would not ratify the Kyoto Protocol. As a treaty of this kind needs formal approval by a qualified majority of the US Senate, there was in fact little chance that ratification could be achieved. Lobbyists (in particular the Climate Council and the Global Climate Coalition in the US) convinced the Senate that US negotiators should not be permitted to return from the negotiating table with commitments

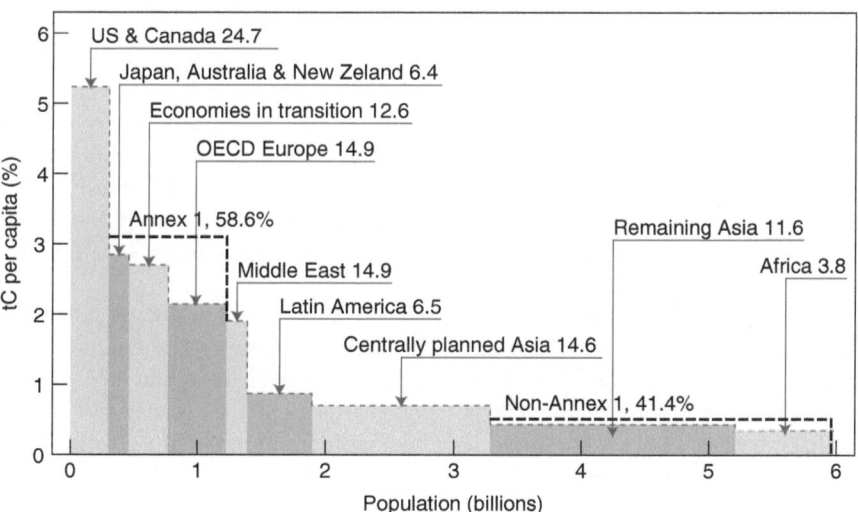

Figure 3.1 Per capita fossil fuel emissions in 1999 averaged for nine geographical regions and grouped into Annex-I and non-Annex-I countries

Note: The height of the bars gives the average per capita emissions for each region. The width of the bars gives the population. The area of the bars is proportional to the 1999 carbon dioxide emissions from fossil fuels and cement production. The percentages indicate the 1999 total emissions attributed to each region (Bolin and Kheshgi 2001).

for the US unless developing countries had also formally accepted some responsibilities for combating climate change.

Progress in the climate negotiations was surprisingly rapid until the mid-1990s, with the climate change issue obviously being seen as a serious global environmental threat. However, at the turn of the new century, it had become increasingly clear to politicians and stakeholders in the energy sector that major changes in the industrial structure might be required to slow down and ultimately stop climate change. While considerable costs would be incurred right away, it would take decades for any potential benefits from avoided climate-change-related damage to accrue. Domestic controversies between environmentalists, industrial leaders, and the public sharpened, and hesitation grew. The aim for industrial countries to reduce their emissions by about 5 per cent on average by 2008–2012 was increasingly judged to be unrealistic. Emissions had actually increased significantly at the turn of the century. Yet, the target was only a very modest first step toward the scientifically agreed effort required to secure a stabilization of greenhouse gases and, through this, of the climate. A gap grew between visions about progressive environmental goals, on the one hand, and short-term economic interests, on the other.

However, while ratification of the Kyoto Protocol progressed, the reluctance of Russia and the US to join meant that few steps toward implementation were taken. The Protocol did actually not enter into force until early in 2005 after Russia signed it in late 2004. Still, nobody knows if, when, and how the US might change its mind and join the other 140 or so countries that are willing, several even anxious, to work toward combating human-induced climate change.

Some primary concerns

The early eagerness on the part of politicians to take action was largely genuine and in line with the increasing overall focus on environmental issues that lasted until about the middle of the 1990s. However, there was early reluctance to proceed quickly on the part of industry and other stakeholders who feared that actions to protect the current climate (i.e. reduced use of fossil fuels) might be a threat to their economic growth.

Other major global issues were also brought into focus that greatly influenced world politics in the early years of the twenty-first century. The conflict in the Middle East was to a considerable degree due to the realization that conventional oil reserves and resources might dwindle during the next few decades and that the global energy supply system might have to change. International terrorism made the international situation even more complex, and the climate change issue was simply given less attention.

Even so, in the struggle between those giving priority to securing control of Earth resources and those anxious to protect the global environment, there was a need for trustworthy information. Polarizations of this type had also developed around other environmental issues. For example, for several decades overfishing in the oceans had led to rapidly decreasing fish stocks, in spite of scientific efforts to point out the emerging critical situation. The threat to the ozone layer was dealt with successfully because the measures envisaged were much less costly and the issue was seen as a more imminent threat. The words "cancer due to enhanced UV radiation from the sun because of a hole in the ozone layer" were a powerful argument.

The climate change issue, on the other hand, is complex, and synthesizing available knowledge on it much more difficult. Recognition of the differences that exist between developed and developing countries, as well as the different approaches to finding solutions, raised key issues that are not so easy to resolve. One can ask: How well are the key facts about climate change in fact understood by politicians and the public in general? What are the particularly important things to know and be aware of when developing a long-term climate policy? How can a better awareness about the most urgent issues be created? Might a more penetrating analysis of the available scientific knowledge be of value, and how can it be better presented to serve political negotiations?

The following aspects of the climate change issues particularly need our attention:

- It is vitally important to show the importance of robust scientific findings as a basis for the crucial political issues being discussed; scientific findings must also be presented in a way that recognizes the key political issues at stake.
- We will never be able to make adequate detailed predictions of the regional and especially local characteristics of human-induced climate change. We should also not expect much to gain more accurate projections of the future climate than we have now; as the global climate system is chaotic, we can only predict the gross features of climate changes for a quite limited time into the future. This must not, however, lead to paralysis and stop sensible preventive actions being taken. There are a number of robust findings that need to be spelled out more convincingly as a basis for such efforts.
- Although there may be some winners as a result of climate change, no country will remain unaffected by its main negative impacts. Global society, during its development, has become used to the type of climate that has prevailed for centuries or even millennia. Making an adjustment to a markedly different environmental setting is not simple, will cause considerable social unease, and will also be costly, particularly if it had to be accomplished quite quickly, which might be necessary in view of the slow start made.
- There is an obvious need to develop new energy supply systems for the world, not only to slow down and ultimately stop the increase in concentrations of atmospheric carbon dioxide, the most important greenhouse gas emitted as a result of human activities, but also because of changes in the availability of fossil fuels expected in the course of the twenty-first century. Ultimately we will have to resort to renewable energies to secure a long-term sustainable future for global society. The threat of global climate change makes this transition more urgent.
- To be successful in such efforts, cooperation must be fostered among the countries of the world and similarly among scientists, politicians, stakeholders

in industry, agriculture and forestry, and the public at large. Competition is still an important way of increasing efficiency and reducing costs, but it must not defeat the overall purpose of taking adequate action to slow ongoing climate change.

- Political solutions cannot be found without progress being made toward a more equitable world based on democratic values and institutions. This process is under way, but it is taking its time.
- The scientific and technical findings need to be analyzed in terms of their importance for all parties; this will enable everyone to understand the factual background and to judge the seriousness of the issues. On the other hand, the natural tendency for any country to protect its own national interests by peaceful means must obviously also be recognized.
- Getting to grips with the issue of climate change is a prerequisite for long-term sustainable development.

A more effective interplay between scientists and country delegates engaged in the climate change issue has been achieved during the last 15 years under the aegis of the IPCC. Its summaries have been of great importance in providing an authoritative presentation of the available knowledge, while the IPCC summaries for policymakers have played a crucial role in the establishment of a scientific basis for the Climate Convention. The Third IPCC Assessment Report (TAR) consisted of three volumes of altogether more than 2600 pages (IPCC 2001a, 2001b, 2001c) and a Synthesis Report (IPCC 2001d). Still, I sense that a key stumbling block is the fact that the scientific community does not yet fully understand the way in which politicians make use of, and the general public interprets, the information that scientists provide. An easier-to-understand story needs to be told that sets out the key climate change issues in a simpler way than has hitherto been the case and that people trust.

The first part of the analysis follows aims to achieve this kind of synthesis, but it is an effort that certainly needs further elaboration. The aim is not to contradict the reports of the IPCC to date, but to tell the story in a manner that comes closer to people's view of the interplay of climate change and life on Earth.

Key scientific findings seen in a political perspective

The following key conclusions regarding the climate system, as given in the IPCC's Third Assessment Report (TAR), seem to be the most important for the negotiation process. They are all quite robust and not very controversial, but they have not been brought together into a short set of key conclusions that do justice to the story to be told.

On the time scale of a human individual, the climate system is responding slowly to human interference. This implies that a significant part of the ultimate warming resulting from the increase in greenhouse gas concentrations in the atmosphere is unknown at any given time, including the present (see IPCC 2001a).

- The global mean surface temperature has risen by 0.7±0.2°C during the last century. Over the continents, where humans live and work, the average increase has been about 0.9°C; over the oceans, only about 0.6°C. This temperature increase cannot be explained unless the human-induced enhancement of greenhouse gas concentrations in the atmosphere and the associated increase in the amount of water vapor in a warmer world are factored in (IPCC 2001a; Moberg *et al.* 2005). There are still a few scientists in relevant fields of research who openly doubt that the global warming observed during the twentieth century is out-of-the-ordinary, and who also reject the idea that warming has been caused by increased greenhouse gas concentrations in the atmosphere. Their arguments are scientifically weak; however, the groups maintaining this view are vocal, and some are undoubtedly supported by stakeholders who feel that their own activities are threatened by measures aimed at preventing "imaginary" climate change.

- There are a number of other observations supporting the conclusion that the ongoing change is caused primarily by human-induced activities. Stratospheric temperatures have decreased by 1–2°C, which is to be expected from greenhouse gas warming of the lower atmosphere (i.e. the troposphere). The daily temperature range has also decreased over some parts of the continents. The global hydrological cycle has intensified. Heavy precipitation has become more common in some continental regions. The extent of sea ice in the Arctic has decreased, and the ice has become significantly thinner.

- Warming has occurred despite a significant increase in atmospheric air pollution, mainly aerosols that cool the atmosphere by increasing the reflection of solar radiation back to space. It is therefore likely that global warming due to increasing concentrations of carbon dioxide and other greenhouse gases in the atmosphere is presently 20–40 per cent higher than we believe, so that when measures are taken to reduce air pollution, an enhanced greenhouse effect will be seen.

- The response of the climate system to increased greenhouse gas concentrations is being delayed because energy is required to warm ocean waters. The combined effect of the cooling due to higher levels of atmospheric aerosols and the inertia of the climate system implies that perhaps only about half of the ultimate greenhouse gas warming might yet have occurred. There is an imbalance between incoming and outgoing radiation that heats the oceans (see also Hansen *et al.* 2005).

- The transfer of atmospheric carbon dioxide from the atmosphere to the oceans is slow. About 40 per cent of annual emissions are accumulating in the atmosphere, in spite of the sea being a huge carbon dioxide reservoir that actually contains about 50 times more carbon (as carbon dioxide and carbonate ions) than the atmosphere. The fairly rapid buildup of carbon dioxide in the atmosphere is primarily due to the slow transfer of carbon dioxide to deeper layers of the oceans. A warmer climate will increase the stable stratification of the oceans and thereby possibly further decrease the rate of transfer of the excess carbon dioxide into the deep oceans.

- The IPCC has developed a wide range of emission scenarios (further discussed below) to analyze the future rate of increase of carbon dioxide concentrations in the atmosphere. It is interesting to note that the differences between the most expansive and the most constrained projections of future atmospheric carbon dioxide concentrations will hardly be noticeable until a decade or two have gone by (Figure 3.2). Accordingly, we will not be able to detect the results of even a very constrained use of fossil fuels, and thus also a change in the rate of climate change, for about twenty years. A very substantial and sustained decrease in emissions is required to achieve stabilization of atmospheric greenhouse gas concentrations and thus slow down global climate changes.

Stabilizing the climate and ultimately reducing global climate change does not mean, however, that the global climate system would return to preindustrial conditions.

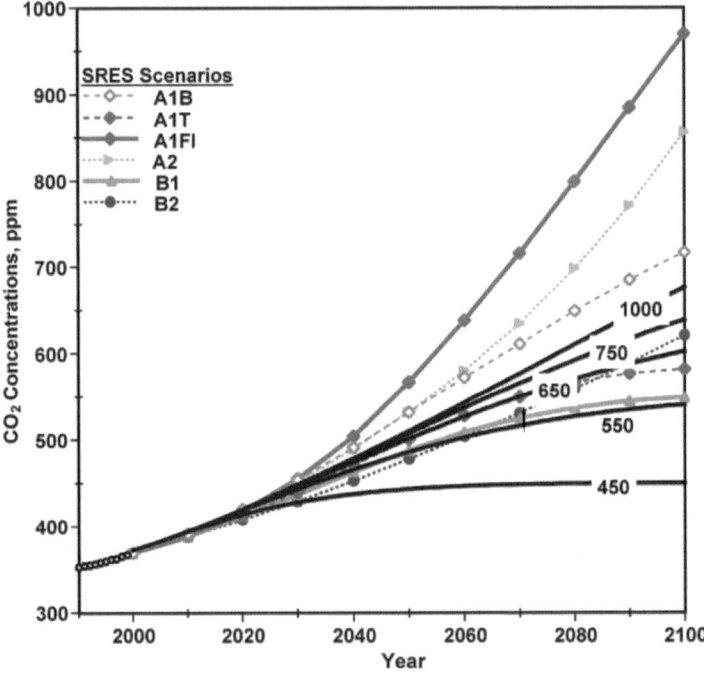

Figure 3.2 Average annual concentrations of atmospheric concentrations of carbon dioxide 1990 projected to 2100

Note: The average annual concentrations of atmospheric concentrations of carbon dioxide measured at the Mauna Loa (Hawaii) and South Pole observatories (1990–1999) are shown by circles. Specified concentration pathways leading to stable concentrations ranging from 450 to 1000 ppm are shown by labeled solid lines. Curves with symbols show concentrations from 2000 to 2100 resulting from six specified emission scenarios from the IPCC's Special Report on Emissions Scenarios (Bolin and Kheshgi 2001).

- Because of the inertia of the climate system and of human society, it will probably be impossible to stabilize the concentrations of greenhouse gases in the atmosphere until sometime during the latter part of the present century (see later in chapter for more details). The twenty-first century (and probably also a century or more thereafter) will thus be characterized by a global climate that is quite different from that experienced during the twentieth century. The later that stringent restrictions of emissions are imposed, the more pronounced the difference will be.

- The climate system is in principle chaotic (i.e. to a considerable degree unpredictable). Nevertheless, it does respond regularly, but not identically, every year to the seasonal variations in the distribution of solar radiation over the Earth. The sustained change in the heat balance between Earth as a whole and space, which we are now undoubtedly causing, will similarly lead to changes in climate; however, only changes in its gross features can be foreseen. These will in turn be associated with secondary changes, such as changes in global atmospheric circulation patterns, the distribution of ecosystems, ice conditions in the Arctic Sea, a slow melting of the Greenland ice sheet, and so on. Most of these are irreversible, seen from a human perspective. Even if the greenhouse gas concentrations were ultimately brought back to preindustrial concentrations, which would take a century or more, the climate would not return to the state that prevailed before human global impacts became significant.

- This makes it more important to keep the changes that have begun within "reasonable" bounds, even though we are not sure what is "reasonable" in this regard, nor what efforts are required to do this. We note, however, that the atmospheric carbon dioxide concentration is now about 376 ppm. If one adds in the greenhouse effect due to the increased concentrations of other greenhouse gases (particularly methane and nitrous oxide), the present total human-induced greenhouse effect corresponds to a carbon dioxide concentration in the atmosphere of more than 445 ppm, which is a reasonably accurate assessment. We are well beyond half way toward doubling the equivalent atmospheric carbon dioxide concentrations.

Human beings will experience a changing climate primarily in terms of extreme and destructive events potentially becoming more common. It is important to clarify what can and cannot yet be concluded with some confidence about the characteristics of climate change in this regard. It is then useful to distinguish between three kinds of extreme events.

- A warmer atmosphere will probably contain more water vapor, which is the major energy resource for the development of storms and hurricanes. Extreme weather events of this kind may therefore become more frequent than today and possibly also more destructive. But extreme weather has always occurred, and it has not yet been possible to establish statistically with confidence the extent to which an increase in the frequency and intensity of storms and floods has occurred; nor is it possible to be more precise about the effects of future global climate. However, some increase in storms and flooding seems likely, especially in regions with increasing precipitation.

It is often claimed in the media that storms and floods are already more common and that they have become more intense than was previously the case, implying that they might become much worse in the future. However, a firm conclusion on this is still difficult to reach. Statistical analyses of where and to what extent such changes might be expected, during the next decades in particular, would be very valuable, but precise answers are unlikely because of the random nature of extreme events phenomena and the limited spatial resolution of the climate models being used today.

- It has long since been recognized that the global circulation of the atmosphere may have several preferred states that occur with varying probabilities. Projections of likely future changes in the general circulation of the atmosphere in a warmer world using global climate models confirm that although the frequency of their occurrence may change, precisely when and how still cannot be determined.

In this context, however, it is interesting to note the extended period of hot and dry weather in parts of the Mediterranean region, particularly in France, in the summer of 2003, which shows clear similarities with regional model projections of what might happen more frequently in a warmer world (see Solomon *et al.* 2007: Chapter 10). In comparison with past years in the twentieth century, 12,000–15,000 premature deaths, particularly among elderly people, were recorded in France alone during that summer. Similar situations have been observed elsewhere, particularly in the Mediterranean area and in Africa, but are less well documented. It is still difficult, however, to judge the extent to which this extreme situation in 2003 may have been a result of ongoing climate change, but it may well have contributed. Statistical analyses of the variability of the weather show that such weather conditions might be more common, and even more extreme, in a warmer world.

If such extended periods of hot and dry weather were to become more common polewards of the dry subtropics, agricultural regions with large populations (e.g. the Mediterranean, the Middle East, India, and possibly parts of China and the southern US) might be expected to become more vulnerable. Admittedly, the likelihood of this occurring has not been established. However, more extended periods of hot and dry weather could be the most threatening feature of a warmer world at these latitudes, as the availability of adequate water supplies is already an increasing concern in many of these regions.

- A third feature of significant changes in the climate system might be brought about by interactions between the atmosphere and the other components of the climate system (i.e. the oceans, ice sheets, sea ice, and terrestrial systems). For example:

  a) Changes in the so-called thermohaline circulation of the oceans, especially in the North Atlantic region, could occur based on a gradual decline in the intensity of the Gulf Stream. This has happened in the past, when the last ice age was gradually replaced by the interglacial period that still prevails. A very remarkable past incident was the rapid return to a cold climate in the region between approximately 12,800 and 11,500 years ago (the Younger Dryas) that lasted for about a thousand years and during which major ice sheets dominated the climate in the northern parts of the Atlantic Ocean and surrounding continental areas until general warming began and the present interglacial period resumed (see National Academies 2004). We are now in an interglacial epoch. Whether similar changes might now be on the way cannot be determined. Changes observed during the last century in the region are probably rather associated with the so-called North Atlantic Oscillation, which influences the variations of climate significantly on a decadal time scale.

b) The El Nino phenomenon is characterized by close interaction between the ocean and the atmosphere in the equatorial regions of the Pacific Ocean. Its features have changed significantly during the last few decades, which has influenced the quasi-regular variations between the two opposite phenomena, El Nino and La Nina. This in turn has been associated with marked changes in the weather along the west coasts of both North and South America, which has had quite significant socio-economic consequences in some of the areas hit. Are they the result of global warming? We do not really know, but similar changes were not observed before global warming began to appear during the later decades of the twentieth century, and modeling experiments indicate that there might be some connection.

c) Changes in the extent of sea ice in the North Polar Sea, and particularly its possible disappearance in the future in outlying regions (or even total loss) during part of the summer season, might have far-reaching consequences for adjacent regions. The likelihood of such changes occurring, and when, cannot be foreseen with much certainty, but they could occur during this century.

d) Similarly, the Greenland ice sheet is currently melting more quickly than it was a few decades ago. If this continues, there will be a point at which the melting of the ice sheet can no longer be stopped by climate stabilization, which is not likely to take place for a number of centuries (ACIA 2004).

The message of this section is a mixed one because of the difficulty of precisely pinpointing what is going to happen in the future, and the list above is certainly not exhaustive. The IPCC has also constantly emphasized that surprises cannot be excluded. So what might humans expect in a warmer world? How serious might a global change in climate be? And how urgently is action needed?

First and foremost, it is important to realize that, even with substantially greater research efforts, it will not be possible to predict in much more detail how the climate and weather will change in a long-term perspective. We will not know how trustworthy such projections might be until validation is possible using the data that are currently accumulating. But changes have already occurred and further significant changes cannot be avoided. We may identify areas of greater risks (vulnerability) where there could be serious consequences, but the key message must be that significant change in Earth's climate is already under way. Even if more far-reaching mitigation efforts are initiated soon, it will take perhaps more than half a century to stop these developments. Early and sustained preventive actions are needed to keep expected changes within "reasonable" bounds. Measures to adapt to changes will therefore also be required. The delay in the entry into force of the Kyoto Protocol has, simply put, meant a considerable delay in the time available for stabilizing the climate. Stabilization at a relatively low level of carbon dioxide concentration has become significantly more difficult to attain as a result.

What is required to stabilize the climate?

The goal of the Climate Convention, namely, to stabilize the global climate, must of necessity be a major global and cooperative undertaking. It is essential that major countries take part, but it is also important to recognize the differences between countries in terms of their ability to contribute. This is well expressed in Articles 3 and 4 of the Climate Convention:

> The Parties should protect the climate system for the benefit of present and future generations of humankind, on the basis of equity and in accordance with their common but differentiated responsibilities and respective capabilities. Accordingly, developed countries should take the lead in combating climate change and the adverse effects thereof.
>
> (A3)

> The developed country Parties ... commit themselves specifically as provided in the following:
>
> (a) Each of these Parties shall adopt national policies and take corresponding measures on the mitigation of climate change, by limiting its anthropogenic emissions of greenhouse gases and protecting and enhancing its greenhouse gas sinks and reservoirs. These policies and measures will demonstrate that developed countries are taking the lead in modifying longer-term trends in anthropogenic emissions ...
>
> (A4)

These principles are reflected in the Kyoto Protocol. However, the efforts to stabilize the carbon dioxide concentration in the atmosphere and, through this, the climate will only just be beginning at the end of the first commitment period, that is, almost 20 years after the Climate Convention was opened for signatures in Rio in 1992.

The global carbon cycle is reasonably well understood (Prentice *et al.* 2001; Field and Raupach 2004). As already pointed out, the inertia of the climate and of global socio-economic systems makes quick reductions in the rate of increase of atmospheric carbon dioxide concentration difficult. From the middle of the nineteenth century to 2004, about 320 Gt^2 C (as carbon dioxide) was emitted into the atmosphere from fossil fuel burning and cement production, and about 130 Gt C from deforestation and changing land use (i.e. in total about 450 Gt C). The increase in atmospheric carbon dioxide content amounts to about 200 Gt C (i.e. about 45 per cent of total emissions), while the remainder, about 250 Gt C, was taken up by the oceans and terrestrial ecosystems, the latter partly because photosynthesis is more intense in an atmosphere enriched with carbon dioxide.

We cannot predict the future of humankind, the future emissions of greenhouse gases, or the way in which the global climate might change as a result of further human interference. It is for these reasons that the IPCC developed a set of scenarios to illustrate different possible future pathways. The different assumptions made allow very different scenarios of energy supply systems and future world population development to emerge. Some of the scenarios display an extensive use of fossil fuels in the future; others depict a gradual changeover to other sources for primary energy and the development of more sustainable energy supply systems. Approximate emissions of carbon dioxide, as well as air pollution in general, can be assessed for these different scenarios, as can the changes in atmospheric greenhouse gas concentrations. However, these are scenarios, not predictions.

Figure 3.2 shows the changes in atmospheric carbon dioxide derived in this way. We note that all scenarios show a carbon dioxide concentration in excess of 550 ppm by 2100, in one extreme case even exceeding 900 ppm. Special efforts are obviously needed to keep future concentrations below 550 ppm, (i.e. doubling concentrations).

Stabilization at 450 ppm would require very major, quick, and therefore costly efforts. Global emissions would have to start declining within about a decade. Though technically possible, it would be very difficult to achieve in practice. Extensive early retirement of previous capital investments would be required, as well as the establishment of an energy supply system primarily based on non-carbon-dioxide-emitting sources of primary energy. If the effects of other greenhouse gases (some very long-lived) are also included, then the equivalent carbon dioxide concentration would be very likely to exceed 500 ppm, corresponding to an ultimate global warming of at least about 1.5°C and perhaps as much as 4°C. The EU has agreed that 2°C would be the maximum acceptable warming of the global climate system. One may indeed ask if the EU goal is still attainable. Moreover, surprises can never be excluded – some of a beneficial nature resulting from human ingenuity, others destructive because of the chaotic characteristics of nature and human society.

It is, however, essential at this stage of the analysis also to consider the implications of these conclusions for developing countries. Figure 3.1, referred to earlier, shows the annual per capita emissions for developed and developing countries in 1999 in nine world regions. The differences are huge. Global annual average per capita emissions in developed countries were about 3 tons C, with the US in the top emitting position with about 5.2 tons C per capita. Developing countries, on the other hand, emitted on average only about 0.6 tons C per capita. Annual global emissions are increasing by about 1 per cent in developed countries (in the US, by about 2 per cent); EU emissions are slightly decreasing, while emissions in the Russian Federation are increasing again (after a 25–30 per cent drop during the 1990s). Emissions in developing countries, on the other hand, are increasing quickly, some by about 5 per cent annually, and by significantly more in China.

Figure 3.3 depicts the differences between developed and developing country future annual emissions (again in emissions per capita terms) as a result of different assumptions about the future implied by the scenarios.[3] Figure 3.3 also shows the global emissions per capita required to stabilize atmospheric carbon dioxide concentrations at 450, 550, 650, 750, and 1,000 ppm (assuming an increase in the

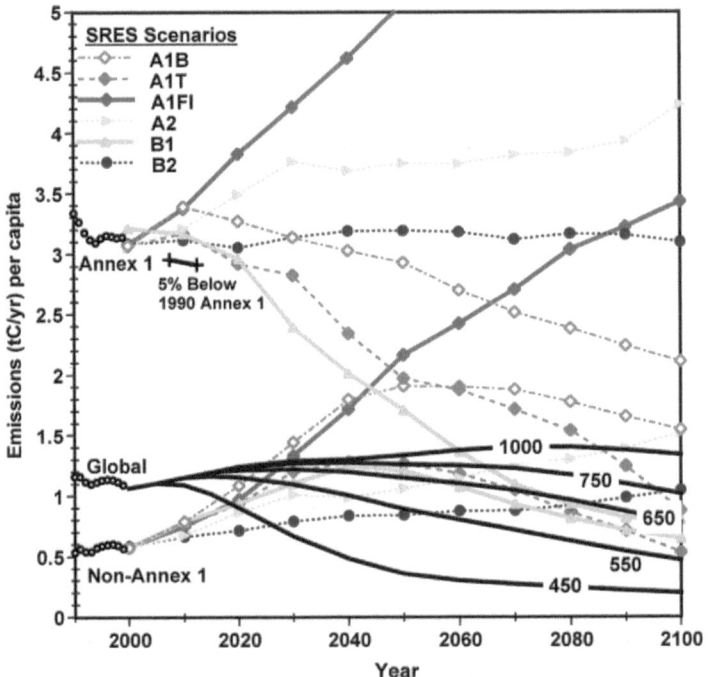

Figure 3.3 Per capita fossil fuel emissions of carbon dioxide

Note: The small black circles show emissions per capita from 1990 to 1999. A line segment shows emissions to be 5 per cent below the Annex-I rate during the 2008–2012 period as aimed for in the Kyoto Protocol. Deduced fossil fuel emissions leading to the specific pathways toward stabilization of atmospheric concentrations from 450 to 1,000 ppm are shown as solid black curves. Per capita fossil emissions for Annex-I and non-Annex-I regions according to six key IPCC scenarios are shown by the curves and symbols as given (Bolin and Kheshgi 2001; IPCC 2000). To reach stabilization, the emissions pathways for Annex-I countries (at present well above any stabilization curve) and those of non-Annex-I countries (at present consistently below) must asymptotically and simultaneously approach the particular stabilization curve being aimed for.

world population to about 11 billion by 2100). The details are uncertain because of the assumptions made, but the gross features of the analysis are still informative.

Stabilization of atmospheric carbon dioxide concentration at 550 ppm or below implies that global average per capita emissions must never exceed about 1.3 tons C per year (it is currently [2005] about 1.1 tons C per year), unless the global population increases only slowly in the future and stabilizes well below 11 billion during the latter part of the century. Only one of the IPCC scenarios, B1, is close to stabilization at the end of the century. This is because per capita carbon dioxide emissions from developed countries in that case have decreased by about 60 per cent in 2100, but are still somewhat above the corresponding emissions of presently developing countries, even though the latter have about doubled their per capita emissions. Developed countries would have to decrease their emissions very substantially even if they accept a stabilization of carbon dioxide concentrations at 600–700 ppm. Figure 3.3 further shows that in order to achieve stabilization, developing countries would never be able to emit on average more than 1.3–1.5 ton C per capita, far less than developed countries emitted on average during their rapid industrialization in the latter part of the last century.

The Kyoto Protocol does not require developing countries to take on any quantitative commitments to reduce greenhouse gas emissions for the 2008–2012 commitment period, as greater energy use is deemed to be of basic importance for their development and fossil fuels would probably be their primary energy source for decades to come. It is, however, obvious that climate stabilization will require the engagement of all countries to keep the emissions per capita below 1.3–1.5 tons C. How this might be accomplished will be crucial in upcoming negotiations. Both developing and industrial countries will have difficulty in accepting these conclusions even though they are strictly scientific.

There is an obvious need to find a way to avoid this stumbling block because of the very definitive limit it places on how far nature can be exploited. Admittedly, this conclusion is somewhat uncertain quantitatively, but most probably correct qualitatively. What does this mean for the future?

Transformation of the world energy supply system

Fossil fuel reserves and resources

We need to look more closely at the global energy system and the future likely use of fossil fuels as our basic resource in the light, not only of the threat of global climate change, but also given the declining reserves and resources of conventional oil and, later this century, of natural gas resources. It is important to recognize the likelihood of ongoing changes in the fossil fuel mix and that unconventional oil sources might become increasingly important. This would imply that carbon dioxide emissions per unit of energy produced would become significantly larger. Some additional information will be given here, but detailed analyses are needed, particularly with regard to geopolitical issues, which are already becoming acute.

The decisions taken in Kyoto in 1997 that industrial countries should aim to reduce their greenhouse gas emissions by about 5 per cent by the end of the first decade of the twenty-first century and to distribute this commitment among themselves, were not based on detailed analysis but should rather be seen as *ad hoc* decisions and as simply the first steps toward mitigating global climate change. Discussions have now begun regarding the next steps, namely, what can be achieved during a second commitment period, which will presumably extend until at least about 2020. Let us for the moment focus only on carbon dioxide emissions, given that fossil fuels will have a part to play in any possible future changes in the world energy supply system.

The global reserves, resources, and additional occurrences of fossil fuels are summarized in Table 3.1 in terms of the potential carbon dioxide emissions if fossil fuels are used to provide energy. The estimated reserves are reasonably trustworthy, while the resources[4] listed are quite uncertain (Data from IPCC 2000; cf. UNDP 2004). Nevertheless, the table provides an adequate overview for present purposes. The available reserves and estimated resources of conventional oil and gas and known coal reserves suffice to raise carbon dioxide concentrations to well above 600 ppm and, with further technical development and exploitation of unconventional resources, far above that level. On the other hand, conventional oil resources are not abundant, and their use might peak within a decade or two. Similarly, natural gas resources, although considerably larger, may decline in use by the middle of the century because of the increasing exploitation of natural gas as a substitute for conventional oil. Coal resources, on the other hand, would last well into the twenty-second century, even if future energy use were to expand.

It is clear that major changes to the global energy supply system will be required in the course of the twenty-first century and that these need to be achieved alongside steps to mitigate climate change. This is the challenge that humanity is now facing. Country negotiators, politicians, and stakeholders in particular, will have to decide how to proceed to resolve the issue of exploitation of future primary

Table 3.1 Global reserves, resources and additional occurrences of fossil fuels: potential carbon dioxide emissions compared with past emissions

	Emissions 1860–2004 2004		*Reserves identified*	*Resources Estimated*	*Additional occurrences*
Oil					
Conventional	100	2.9	100	70	
Unconventional			130	130	>250
Natural gas					
Conventional	42	1.4	80	220	>150
Unconventional			100	300	330
Clathrates				>12,000	
Coal	182	2.9	1050	2500	>4,000
TOTAL	324	7.2	1460	3220	>16,700

Sources: As of 2004 (in Gt C). Data from IPCC 2000, cf. UNDP 2004.

energy sources. The measures used must also be acceptable to the general public (i.e. to provide energy for development in poor countries and at least to maintain the standard of living in industrial countries). This will become a central issue for consideration and will have a serious North–South dimension.

Even though the gradual changes in the global energy supply system are a long-term concern, short-term issues will probably, to a considerable degree, determine politics. A few examples can illustrate this, although a more careful analysis of such aspects of the issue is desirable. The starting point will be the issue of the decreasing availability of conventional oil. Table 3.1 shows that a total of roughly 170 Gt C of reserves and resources in the form of conventional oil may be available, compared with the total amount of carbon that has been burned to date, namely, about 100 Gt C. Within a decade or two, more than half the reserves and estimated resources will have been used up. Oil production will decrease and the price will presumably rise further. There will be increasing use of gas and coal, and new technology may become available to exploit unconventional resources of oil at competitive prices. The patterns of supply and demand will change, particularly because of the increasing demands for primary energy in China and India, home to about 40 per cent of the world population.

This changing scene is already under way. The oil reserves and resources are very unevenly distributed, as can be seen from Table 3.2. The dominant role of the Middle East is remarkable. Many of the Russian reserves are also located in regions around the Caspian Sea and in the southern parts of Siberia. The assets in Europe are quite modest. We note that the annual use of oil in the three large industrialized regions (North America, Asia Pacific, and Europe) currently correspond to as much as 7–10 per cent of their domestic reserves and resources. Their present use of fossil fuels would thus be covered for only 10 to 15 years, assuming the same use of oil as today. All are becoming rapidly and increasingly dependent on the assets in the Middle East, and the supply side will become even

Table 3.2 Reserves of conventional oil as distributed in six major regions as a percentage of total and in absolute amounts

Region	Reserves as % of total	Reserves in Gt C	Consumption as % of total	Consumption in Gt C per yr
North America	5.5	9	30.0	0.81
South & Central America	8.9	15	6.0	0.16
Europe, Russian Federation, including Asia	9.2	16	25.9	0.69
Africa	8.9	15	3.3	0.09
Middle East	63.3	108	6.0	0.16
Asia Pacific	4.2	7	28.8	0.79
TOTAL	—	170	—	2.70

Source: Underlying data from BP Statistical Review of World Energy, 2004.

Note: Reserves of conventional oil as distributed in six major regions as a percentage of the total (about 170 Gt) and in absolute amounts (Gt); the annual consumption (2004) in these regions as a percentage of total emissions (about 2.7 Gt C per year) and in absolute terms (Gt per year).

more monopolistic than today. OPEC is strengthening its dominant position. The politics in this region, as played out at present in Iraq and neighboring countries, should be seen in this perspective. The future supply of conventional oil is, and will remain, a major geopolitical issue. In this context it is also interesting to note that unconventional oil resources (e.g. tar sands in Canada) are becoming increasingly exploited, but at a higher cost and with larger emissions of carbon dioxide per unit of energy when used as the primary source for energy.

The exploitation of natural gas began comparatively late. The total reserves and estimated resources are at present considerably larger than those of conventional oil and are being exploited less rapidly. It is not likely that their exploitation will peak much before the middle of the twenty-first century. Reserves and resources of natural gas are, however, also very unevenly distributed over the Earth. About 50 per cent are found in Russia. The Middle East also possesses substantial reserves and estimated resources, while again quite limited resources are located in rich industrial countries.

Coal reserves and resources, on the other hand, are at least five times larger than those of conventional oil and natural gas, and they are also distributed more evenly around the globe. Large assets are found in Australia, China, India, and the US, as well as in some European countries that all have limited reserves of oil and natural gas. Coal is the key primary energy source for production of electricity in most countries (its share is on average about 45 per cent and its use for this purpose is increasing). It should be recalled that for the production of one unit of electricity in a fossil-fuel-based power station, the emissions of carbon dioxide using coal are 40–50 per cent and 60–80 per cent greater than if oil and natural gas, respectively, are used. Changing the fuel used from oil and gas to coal would increase carbon dioxide emissions substantially, and this process seems to be under way in several industrial countries.

Huge reservoirs of methane hydrates (clathrates) are buried mainly in permafrost regions in Canada and Russia and in bottom sediments of the North Polar Sea (cf. Table 3.1). To date, they have received little attention as a future energy source. It is not clear how they might be exploited safely so that methane, also a greenhouse gas, would not accidentally leak into the atmosphere. Issues of this kind can usually be handled if the technical and financial resources are made available. This might be the case when energy prices have risen sufficiently above their present still quite low levels. It will, however, take time to provide the logistics for methane hydrate exploitation.

Renewable energy

In the long term, renewable energy, above all solar energy, will need to be the basic primary energy source. Lack of sufficiently developed technology and comparatively high costs are still the main reasons why limited efforts have been made to date to exploit renewable energy resources. The role of the competitive energy market is to maintain efficiency and provide energy to consumers as cheaply as possible. Renewable energy technology suffers because of the tough competition with cheap fossil energy. Almost 80 per cent of the primary energy used in the world still comes from fossil fuels (cf. Table 3.3), a figure that has not decreased during the last few decades.

We note, however, that:

- Hydropower has been exploited successfully as a source of electricity, but in 2001 produced only about 2.3 per cent of the total energy and just over 5 per cent of the total world electricity use. Total natural resources are limited and their contributions can probably not be much more than doubled. In the meantime the global demand for electricity is increasing by 2–3 per cent per year.
- Globally, photosynthesis locks energy into biological structures (i.e. biomass) in an amount that is annually 8–10 times the total energy used today. However, only about a quarter ends up in wood and about 1 per cent of the total is presently used as a source of energy in modern society. This can certainly be increased. Traditional biomass remains an important source of energy, primarily in developing countries, and contributes about 9 per cent to the global total energy use.

We note, however:

- Biomass might become a more significant energy source in northerly latitudes, namely, in Canada, the Scandinavian countries, and Russia, as well as in moist tropical countries where photosynthesis is rapid because of the warm and humid climate. Plantations of rapidly growing species (e.g. *Salix*) supply energy in increasing amounts in, for example, Sweden. Burning of waste, particularly wood and paper products, also provides increasing amounts of primary energy. Its contribution to the total amount of primary energy is 1–2 per cent, and this is included under the heading of "new renewables." It is hardly likely that these sources will play a major role in the early future, particularly not in the densely populated regions that are presently undergoing industrialization (e.g. China, India), but the potential is still considerable.

Table 3.3 World primary energy use, 2001

Source	Primary Energy, EJ	Primary Energy 10^9 tons oil equiv.	Percentage of total
Oil	147	3.51	35.2
Natural gas	91	2.16	21.7
Coal	94	2.26	22.5
Large Hydro	9	0.23	2.2
Traditional biomass	39	0.93	9.3
"New renewables"	9	0.21	2.2
Nuclear	29	0.69	6.9
Total	418	9.99	100.0

Source: excerpt, UNDP 2004.

- The harvesting of solar energy for heating and electricity generation and the use of wind energy and other renewable energy sources still play a small role in the global energy supply system (<1 per cent and included under the heading of "new renewables"), and their contributions in absolute terms are increasing quite slowly. Wind energy is, however, becoming competitive with the more expensive fossil fuels and its potential resources should not be ignored, even though its present contribution to meeting total energy needs is merely a few per mille of the total.

- Because of public resistance, it is unlikely for the time being that a major expansion in nuclear power will occur. It should be noted that the amount listed in Table 3.1 includes waste heat; the amount of electricity produced is commonly assumed to be about 33 per cent of the total – approximately the same as that provided by hydropower. The development of new technology would also be required to secure a long-lasting use of nuclear energy. Modest resources are at present set aside for such research and development.

Thus the main conclusion is simply that fossil fuel resources in the world are large and that an expanding world economy could most certainly be based on energy supply from fossil fuels well beyond the present century. The increasing emissions of carbon dioxide might, however, in such a case lead to a doubling, or even tripling, of atmospheric carbon dioxide concentrations that would most certainly engender unacceptable changes in the global climate. Measures need therefore be taken to limit and then reduce emissions, if climate change is to be kept within "reasonable" bounds. Technological prospects are good. It is thus crucial to ascertain if there can be a sufficiently quick transformation to an almost fossil-fuel-free society when there are still large reserves and resources of fossil fuels in the world that can be exploited at quite modest cost.

It is clearly necessary to change the world energy supply system, and not only to limit global climate change. Climate changes to date have been quite slow, and the use of fossil fuels continues to increase. The process of rapid industrialization underway in many developing countries means that, in the decades to come, there will not be enough energy for development unless fossil fuels remain a primary source of energy. A very important political conclusion is then obviously that there should be increased investment in research and technical development of renewable energy resources. The initiatives to date by the private sector have been modest, and they decreased substantially during the last decades of the twentieth century.

It therefore seems likely that coal will be increasingly used during a good part of the next half-century unless there are special efforts to avoid this. It might even be viewed by some as an attractive temporary solution in the sense that coal would make it easier for many countries to achieve energy supply security. It would, however, be among the least desirable solutions with regard to the climate change issue. How can this obstacle be overcome and the present trend changed?

Increase in energy end-use efficiency

There are other means of slowing down the rapid increase in atmospheric carbon dioxide. Energy is not used very efficiently at present. Increasing efforts are being made to avoid energy wastage, and there is progress on this. The likely, even desirable, price increase will further stimulate a more efficient use of energy. Nevertheless, for example, the transportation sector is rapidly increasing its share of the total energy use because of the increasing world population. More automobiles will be on the roads, more airplanes in the air; improving living standards will encourage people in both industrialized and developing countries to travel. There is no sign of a decrease in energy use in the transport sector. Energy efficiency in road transportation of goods has increased substantially during recent decades but international trade and travel in general have expanded more quickly, and this against the background of a substantial increase in fossil fuel use in this expanding sector of society.

Electricity generation has become more efficient in industrialized countries, but much more can be done, particularly in developing countries, where it is the basis of ongoing industrialization. Co-generation of power and heat is expanding, which also implies an increase in end-use efficiency; it is seldom used in tropical countries where heating is not a prime cause of energy demand.

In the present context, however, it is important to recognize that the lack of energy end-use efficiency has, until quite recently, been due in large part to the low price of fossil energy. There are many signs that this situation now is gradually disappearing. Economic incentives are powerful mechanisms, and should be used, but because of the lack of equity in the world, such schemes will not find broad international approval. Rich countries have a major advantage, and it is difficult to reach agreements on more precise goals to be striving for because of the very different development aims in industrial and developing countries.

The carbon dioxide concentration is at present still increasing at just about the same steady rate as during the last two decades of the twentieth century (i.e. by about 1.5–2.0 ppm per year or about 3.5 Gt per year) and, if anything, this rate of increase may actually be slowly growing.

Sequestration of carbon dioxide

The Climate Convention repeatedly states that reducing emissions of greenhouse gases and enhancing carbon sinks are the main means of stabilizing the global climate. It is only recently that carbon sinks have come under the spotlight. While it is obvious that enhancing the uptake of carbon dioxide by the oceans, terrestrial systems, deep geological structures, or aquifers could reduce the rate of increase of atmospheric carbon dioxide concentrations, long-term storage must also be secure.

Enhanced ocean storage is technically difficult and costly, and may also be elusive. Its efficiency depends on ocean circulation that may change as a result of climate change. Further, warmer surface waters will increase the vertical stability of the oceans and thus possibly decrease the rate of carbon dioxide exchange with deeper strata, which might slow down the oceanic uptake.

Terrestrial ecosystems, including soils, have probably served as a significant natural sink for atmospheric carbon dioxide, particularly during the latter part of last century, and have maintained the airborne fraction of the annual emissions at quite a low level (presently about 45 per cent). The Kyoto Protocol recognizes the possibility that measures to increase terrestrial storage are a way for developed countries to fulfill their commitment to reduce emissions. It will, however, be necessary to adopt an accounting system that ensures that efforts to enhance sinks do not bring about return flows of carbon dioxide to the atmosphere. Nor must it allow loopholes in the reporting system that is being established by the Climate Convention. It is actually questionable if this uptake will remain significant for a long time into the future. In any case, the long-term stability of an increased terrestrial carbon reservoir cannot be guaranteed when the climate is changing.

Geological storage in deserted oil and gas fields as well as aquifers below 800–1000 m in depth offer interesting prospects, but have not yet been considered carefully with regard to stability and other possible environmental implications. The magnitudes of such storage opportunities are probably large and, although long-term stability may be adequate, it should be further assessed. The geographical locations and capacities of potential storage sites have to be carefully assessed to permit a more accurate evaluation of the amounts of carbon dioxide that might be stored and the logistics that would be required to transfer it to the storage sites. An analysis is at present being undertaken by the IPCC and will become available in the course of 2005. This possibility might also be technically attractive for the energy industry, as the exploitation of deep storage fits well with its traditional structure and technology. Rough estimates of the costs for this kind of mitigation indicate costs increases by 50–100 per cent for the provision of energy if coal is used as the source for primary energy (i.e. about the same size as the increase in price of crude oil during the last decade, to which an adjustment is underway).

Is it costly to change the present global energy supply system?

The transition to a sustainable global energy supply system would certainly imply additional annual costs of many hundred billions of dollars, accumulating over years and decades. This is seemingly a frightening sum of money, even for large industrial countries. The costs were presented in the analyses by Illiaronov, economic advisor to President Putin, in 2003, and also in a report from Russian Academy of Sciences, when Russia was still hesitant to ratify the Kyoto Protocol. These views were obviously not based on the IPCC reports but dictated by political considerations.

Costs may well appear as a major stumbling block in the negotiations, but the prospects of a change of such an essential component of the modern industrialized society must be seen in a long-term and broad societal context and, for example, be compared with the annual increase in national and global gross domestic product (GDP) (see Azar and Schneider 2002). The IPCC has assessed the costs for transition to and maintenance of a new sustainable energy system and has arrived at an annual expenditure of the order of 1 per cent, at most 2 per cent, of the GDP for industrialized countries, and to begin with, perhaps about 5 per cent or temporarily even more for developing countries. Although this expenditure will not arise suddenly, there is obviously a need to allocate the required resources gradually over several decades.

In this context it should be recalled that most countries have the goal of increasing their GDP by about 2 per cent annually, or preferably even more, and commonly succeed in doing so. This means that less than 10 per cent of the annual increase in their productivity may for some decades have to be used for the purpose of transforming the energy system, in order ultimately to reach an annual allocation of about 1–2 per cent for the establishment and maintenance of a new energy supply system, in addition to the costs for the present system which is primarily based on the use of fossil fuels. This means that only a small part of the increasing economic resources becoming available annually would be needed to undertake this major task. As pointed out by Azar and Schneider (2002) this might mean a delay in productivity increase by merely about a year, perhaps somewhat more for developing countries.

This is by no means a simple task. The crucial matter is that the economic rewards of the increased productivity of an industrialized society usually do not end up in the hands of those are responsible for securing the future energy supply system. Rather, in a free market economy, the earnings of increased productivity are primarily shared between profitable firms and their employees. The key political issue thus becomes one of facilitating the channeling of a (small) part of the increased national productivity to support efforts to change the energy supply system and thereby reduce carbon dioxide emissions to the atmosphere.

Some countries might consider raising taxes in order to secure the funds required and make the government the prime mover in terms of ascertaining what development is desirable. On the other hand, the introduction of tradable emission permits now underway in the EU might have a dual benefit. The price for energy would increase and also be a stimulus to improve, for example, energy end-use efficiency. Although the emission permits now being introduced in the EU are being handed out free of charge to the users (the grandfathering principle), in the next stage, the overall emissions will have to be restricted and the total number of emission permits therefore reduced, resulting in higher price for the permits. The price of energy would increase accordingly and the funds acquired by the energy sector used for basic investments required in the new energy system. To date, however, the EU trading system only includes some energy-related activities and must be further expanded to do its job properly.

The importance of trustworthy scientific knowledge in climate negotiations

It is clear from the above that an inadequate basic knowledge and understanding of continuing human-induced climate change, both in terms of its environmental and societal consequences in industrial and developing countries, may prove a stumbling block to negotiations. This may prevent a more rapid engagement by countries to reach international agreements on how and when to act. Building awareness will be a major responsibility for the scientific community.

It is difficult to analyze the issue more closely because it is interwoven with a number of other political issues that often differ from one country to the next. We need to recognize the following:

- Scientific knowledge should be relevant and trustworthy, and should be presented in a way that clarifies key political issues without compromising the credibility of scientific efforts to provide the information. Negotiators need a common factual basis with regard to the main environmental issues. This has largely been achieved by the IPCC, although its focus on information of political relevance could still be sharpened to emphasize issues where reasonably robust information can be provided.

- Some of the continuing changes in the climate are masked and largely irreversible: the changes observed to date could be just a modest fraction of what greenhouse gas emissions might ultimately bring about. Certainly, slow or delayed actions will, within a few years, severely constrain ambitions to limit the enhanced greenhouse effect due to human-induced emissions to the equivalent of a doubling of preindustrial carbon dioxide concentrations.

- Given that the climate system is basically chaotic, only changes in its gross features resulting from human intervention are really predictable. Further research should focus on determining short- and intermediate-term risks at the local and regional levels and on the gross features of long-term changes in climate.

- The expected impacts of climate change on people and societies need to be viewed against a prevailing background of lack of equity and social justice, in particular vis-à-vis the differences between poor and rich and between developing and industrial countries.

- Socio-economic issues will increasingly be a focus of climate change negotiations. The protection of productive land is likely to be a central theme in a world with developing country populations that are still rapidly increasing, albeit more slowly than during the twentieth century. Developing countries will be aiming for higher standards of living, even though this implies increasing requirements for energy that will continue to be provided by increased use of fossil fuels for quite some time. Industrial countries are likely to carry on striving for increased productivity, particularly in the transport sector, and adjust only slowly to the requirements of decreasing demands for fossil fuels. The inertia of society remains a major stumbling block to negotiations.

- Increasing trade and rapid globalization will change the structure of global enterprises; the global economic powers will play an increasingly important role. They need to address the following kinds of questions in a cooperative spirit. How will priorities be set? How will the increasing demands for energy be met? How will environmental protection issues, sustainability, and less destruction of the global environment be achieved? Will it be possible to formulate the tasks ahead in terms of attractive future business opportunities, particularly for industry, thereby fostering a cooperative attitude to climate change mitigation?

- It has often been stressed that developing countries are more vulnerable to climate change, while industrial countries have much greater resources to protect themselves. It follows that rich industrial countries might prefer a strategy that is more focused on adaptation to climate change, seemingly a rather common view in the US. On the other hand developing countries are more anxious to see industrial countries, with their much higher greenhouse gas per capita emissions, focus on mitigation. This fundamental difference in attitudes is likely to be a key issue and a stumbling block in negotiations.

- The analysis shows very clearly that the necessary long-term transition to a global energy supply system must be built around solar energy as the primary energy source. However, solar energy will probably not be competitive with the cheap fossil fuels for quite some time. The exploitation of the rather large assets of unconventional oil might also slowly become an attractive alternative to the market (cf Table 3.1). Fossil fuels will remain the prime source of energy for decades to come. To overcome these challenges, international agreements and government interventions will be required.

- Sequestration of carbon dioxide emissions will be required for an extended transition period to limit the increase in atmospheric carbon dioxide concentrations during the first half of the twenty-first century and keep the global changes of climate within bounds. Energy price increases are here to stay, which in turn will stimulate the more rational use of energy, a goal that must be achieved in cooperation with industry. However, sequestration is only a temporary measure, as a sustainable supply of renewable energy is the only long-term solution. The energy industry would be able to switch over to this transient technology quite easily. Countries with deposits of coal and unconventional oil might be able to exploit their domestic assets for some time and be less dependent on imports, while renewable energy is further developed.

It is hoped that this analysis shows that in-depth analyses of the factual situation is an absolute necessity so that we can get to grips with the main societal and political issues that will confront us as we work to resolve the global climate change issue. Further, analyses must be trustworthy and carried out as independent scientific efforts, free from government representatives involved in the climate negotiations. As a scientist who has been involved in the interplay of politics and scientific analysis in the case of climate change, I have tried to present an impartial analysis, but this can only be ascertained by more comprehensive efforts of the kind practiced by the IPCC. Short oversights of the key issues, of the type attempted here, could be valuable additional contributions from the scientific community to the present IPCC procedure.

Notes

1   Professor Bert Bolin wrote this chapter in 2004–2005, two years before his death. He stipulated that it could be used in a future publication only if it were not factually updated in any way. It thus remains as he wrote it, and reflects the status quo of the time.
2   Gt = 10^9 ton.

3   The results are obviously dependent on the assumptions made about the future changes of world population. It has been assumed that the world population in 2100 would be at least 7 and at most 14 billion.
4   Estimated reserves that can be exploited, and estimated resources to be discovered and developed as reserves.

References

ACIA (2004) *Impacts of a Warming Arctic*. Cambridge, UK: Cambridge University Press.

Azar C. and Schneider S. (2002) Are the Economic Costs of Stabilizing the Atmosphere Prohibitive? *Ecological Economics* 43. 73–80.

Bolin B. and Kheshgi H. (2001) On Strategies for Reducing Greenhouse Gas Emissions. *Proceedings of the National Academy of Science*. 4850–4854.

BP (2004) *Statistical Review of World Energy*. London: BP.

Field C. B. and Raupach M.R. (2004) *The Global Carbon Cycle*. Washington, DC: Island Press.

Hansen J., Nazarenko L., Ruedy R., Sato M., Willis, J. Del Genio A., Koch D., Lacis A., Schmidt G. and Tausnev N. (2005) The Earth's Energy Imbalance. Confirmation and Implications. *Science*. 308 (5727). 1431–1435.

IPCC (2000) *Special Report on Emissions Scenarios*. Cambridge, UK: Cambridge University Press.

IPCC (2001a) *Climate Change 2001. The Science of Climate Change*. Cambridge, UK: Cambridge University Press.

IPCC (2001b) *Climate Change 2001. Impacts, Adaptation and Vulnerability*. Cambridge, UK: Cambridge University Press.

IPCC (2001c) *Climate Change, 2001. Mitigation*. Cambridge, UK: Cambridge University Press.

IPCC (2001d) *Climate Change 2001. Synthesis Report*. Cambridge, UK: Cambridge University Press.

Moberg A., Sonechkin D.M., Holmgren K., Datsenko N.M. and Karlén W. (2005) Highly variable Northern Hemisphere Temperatures Reconstructed from low- and high-Resolution Proxy Data. *Nature* 3265, 433 (7026). 613–617.

Prentice, I.C., Farquhar G.D., Fasham M.J.R., Goulden M.L., Heimann M., Jaramillo V.J., Kheshgi H.S., LeQuéré C., Scholes R.J. and Wallace D.W.R. (eds.) (2001) The Carbon Cycle and Atmospheric Carbon Dioxide, In Houghton, J.T., Ding, Y., Griggs, D.J., Noguer, M., van der Linden, P.J., Dai, X., Maskell, K. and Johnson, C.A. (eds.) *Climate Change 2001: The Scientific Basis*. Contribution of Working Group I to the Third Assessment Report of the Intergovernmental Panel on Climate Change. Cambridge, UK: Cambridge University Press. 183–237.

Solomon, S, Qin, D., Manning, M., Chen, Z., Marquis, M., Averyt, M., Tignor, M. and Miller, H.L. (eds.) (2007) Climate Change 2007. Contribution of Working Group I to the Fourth Assessment Report of the Intergovernmental Panel on Climate Change, Cambridge, UK/New York: Cambridge University Press.

The National Academies (2004) *Abrupt Climate Change: Inevitable Surprises*, Washington, DC: National Academies Press.

UNDP (2004) *World Energy Assessment*: Overview 2004 Update, Goldemberg, J. and Johansson, T. (eds.).

4 The Observing International Lawyer

Franz Cede and Gerhard Loibl

Introduction

In recent decades, the focus on environmental issues has grown rapidly, with ozone depletion, loss of biodiversity, air pollution, soil pollution, and water pollution being among the topics of concern. These topics have been discussed and analyzed in international governmental and nongovernmental fora, and this has led to a considerable growth in the number of international agreements and institutions dealing with environmental issues. The evolution of international treaties over the last decades has had a positive impact on the international community's awareness of environmental issues and has spearheaded environmental improvements, such as reductions in air pollution and protection of the ozone layer. But the rapid growth of international agreements and institutions has also raised questions as to how effective or coherent the international response to environmental problems really is. The evolution of the climate change regime, in particular, demonstrates the institutional and legal constraints to elaborating international environmental agreements, and the challenges that this involves.

Specific international legal challenges with regard to climate change

International environmental law is part of international law. However, international environmental agreements contain inherent "stumbling blocks" to their own further elaboration. This is particularly so of the climate change regime.

As a very complex issue, climate change is not limited to clearly discernable and easily definable issues. It poses a broad range of difficult scientific questions – some of them disputed and still unresolved – regarding not only the economic and social impacts of climate change but also the policies and measures needing to be adopted to reduce its impacts. It must be borne in mind that people see climate change not as an issue that affects them today, but as one whose negative impacts will be felt only by future generations.

Because of the enormous complexity of the climate change issue, public awareness of the economic and social effects of many measures and policies required to effectively combat climate change is lacking, and international law experts are no exception. While the evolution of the climate regime has demonstrated the specific character of international environmental agreements, it has also shown the major differences between the subject matter of climate change and that of other fields of international environmental regulation. Therefore, international regulations addressing climate change are more complex than other international environmental agreements which deal with very specific issues, such as the depletion of the ozone layer or the protection of water resources.

Many international environmental agreements have followed a "framework" approach; as a first step, a "general treaty" was adopted which only involved very general principles and institutions, which would discuss the issues further and elaborate more precise rules in the future (such as by adopting protocols). The Convention of the United Nations Economic Commission for Europe (UNECE) on Long-Range Transboundary Air Pollution (which addresses air pollution), the Biodiversity Convention and its Cartagena Protocol on Biosafety, and the United Nations Framework Convention on Climate Change (UNFCCC) and its Kyoto Protocol are examples of this approach to addressing environmental issues. We thus have a clear picture of the pros and cons of the framework concept as applied to environmental agreements.

Some of the stumbling blocks result from an overly narrow approach to the establishment of internationally agreed rules on combating climate change. In general, the framework approach has been used with great success in environmental agreements. In contrast, the international climate regime has, in the course of its evolution, run up against a number of institutional and legal problems that have resulted from a failure to address the complexity of issues raised by the scientific, economic, and political questions related to climate change. These shortcomings on the political level to recognize the complexity of the questions raised by climate change have made negotiations very difficult.

Institutional shortcomings

The institutional framework of climate negotiations is intricate. The network of global, regional, and domestic institutions, all dealing in one way or another with environmental protection, is so complicated that even experts have difficulty determining where exactly the decisive developments are taking place.

Before the UN Conference on Environment and Development (UNCED), held in Rio in 1992 (the Earth Summit), the UN Environmental Programme (UNEP) was the lead institution for environmental issues within the UN system. It was UNEP that provided the framework for many negotiations that led to the elaboration and adoption of soft law instruments, that is, non-binding guidelines on certain aspects of the protection of the environment (e.g. the World Charter for Nature 1982) or instruments that resulted in legally binding regulations (e.g. the Convention on Biological Diversity). However, UNEP appears no longer to be the main institutional framework provider for global environmental negotiations. The shift away from UNEP has gone hand in hand with the adoption of a more assertive role by new institutions created for example by UNFCCC in elaborating rules addressing climate change. This also holds true for other international environmental agreements which have created their own institutions, such as the Conference or Meeting of the Parties and secretariats.

A cursory look at the main bodies within the UN system provides the following picture: as well as UNEP, the main bodies responsible for promotion of environmental protection are the Commission on Sustainable Development, established after the Rio Conference, the regional commissions of the UN (for example, the Economic Commission for Europe), and the institutions established under international environmental agreements. Because of growing concern regarding environment deterioration, a number of UN bodies – *United Nations Educational, Scientific and Cultural Organization* (UNESCO, the World Health Organization (WHO), the *Food and Agriculture Organization* (FAO), the International Maritime Organization (IMO) – are also increasingly active in the environmental field, even though their primary mandates are not environmental matters. In principle, the fact that environmental matters are so high on the agendas of so many international organizations is to be welcomed wholeheartedly. Nevertheless, the proliferation of interest has led to a high degree of fragmentation within the institutional framework. When considering environmental issues, international organizations frequently fail to take account of concurrent developments in other international forums; they appear to proceed rather haphazardly, ignoring achievements already made on the very same issues to which they are devoting time and energy. Unfortunately, the lack of communication, and absence of coordination and transparency, have sometimes clouded their well intentioned commitment to the environmental cause.

An effective interface allowing relevant international bodies to exchange information at the executive level appears to be missing, and this translates into a lack of coherence in global climate talks.

The shortcomings of inter-agency communication are compounded by similar problems at the national level. National administrators involved in shaping a government's position ahead of and during an international environmental conference know only too well the difficulty of bringing the various ministries into line on any given negotiating point. More often than not, the elaboration of a position is hampered by the divergent interests of the ministries or bureaucratic departments involved or because of their sheer disregard for, or ignorance of, internal policy considerations.

It is suggested that the international institutional framework be strengthened by improving the inter-agency flow of information and communication regarding the respective organization's past and current environmental activities. An exchange of views on these matters should become a regular part of the agenda of the annual meetings of the heads of the specialized agencies and other bodies within the UN family. In the same spirit, efforts should also be increased at the national level to provide for improved information and communication flow between the ministries concerned. This will allow positions to be elaborated that are in tune with both the latest international developments and domestic climate policy.

Linkage of the climate change arrangements to other international organizations

It should be further noted that specific climate-change-related issues are generated by the large number of committees and subcommittees set up under the UNFCCC and the Kyoto Protocol to deal with, for instance, reporting requirements and methodologies or the operation of Kyoto Protocol mechanisms. Although this complex institutional structure has frequently helped negotiators find solutions, it has also burdened the overall negotiation process. If many negotiations are held in parallel, a strain is placed on the resources of delegations. Small delegations, in particular, are not in a position to closely follow the negotiating process, but can deal only with the specific issues that concern them most. This reduces the transparency of the decision-making process and may slow down the progress of the negotiations, as delegations that have not followed the entire negotiation process may wish to make their views known or request changes to compromises reached in the small negotiation groups.

Regional groupings and interest groups of parties, such as the "umbrella group" the Small-Island Developing Countries (SIDS), and G77 + China, are an important instrument in the negotiations. They help establish transparency and enable smaller delegations to make their views known; they also contribute to shaping the group negotiation position so that group representatives can negotiate on their behalf. But their mandate in the negotiations, and thus their room for maneuver, is often limited, and consultations within the various groupings of parties might have to take place before a new step is taken in the negotiations and a compromise is achieved.

The large number of bodies participating in the climate change negotiations is due to the complexity of the issues. A more structured agenda for each negotiating group should be developed, each focusing on key issues. Such an institutional improvement in the negotiating process can be achieved only if the main political consent to the urgency of the problem is given and participants are willing to accept that not all issues can be discussed at the same time.

Specific issues under international law

The evolution of traditional international law unfolds within the framework of "codification and progressive development of international law" as envisaged in Article 13 of the UN Charter. However, because of the distinctive nature of environmental issues, the rules that apply to international regulation do not always obtain when environmental issues are being dealt with. International environmental agreements create rules of their own to deal with new challenges, leading to a different approach not only to the elaboration of law but also to questions of implementation and compliance.

The complexity of the issues at stake and the lack of scientific certainty regarding the environmental problems needing to be addressed have led to negotiations following a "framework approach." A central issue in international negotiations is the procedures and mechanisms to be adopted to ensure that parties to international environmental agreements implement the agreements and comply with their obligations.

Framework approach

The framework approach was also chosen by the international community to address the issue of climate change (cf. UN General Assembly 45/851 dated 21 December 1990). The UNFCCC, which was adopted during the United Nations Conference on Environment and Development in Rio in 1992, contains principles and institutional arrangements that set out a general framework for dealing with the issue of climate change. Article 4 paragraphs 2(a) and (b) provides that Annex I Parties (i.e. the developed countries) should adopt national policies and measures to limit their anthropogenic emissions "with the aim of returning individually or jointly to their 1990 levels." Furthermore, Article 17 of the UNFCCC states that the Conference of the Parties may adopt protocols to provide for further measures to address the issue of climate change.

The first Conference of the Parties (COP), held in Berlin in 1995, adopted the so-called Berlin mandate, providing for a process to enable the COP:

> to take appropriate action for the period beyond 2000, including the strengthening of the commitments of the Parties included in Annex I of the Convention (Annex I Parties) in Article 4 paragraph 2(a) and (b), through the adoption of a protocol or another legal instrument
>
> (Decision 1/CP.1)

The subsequent negotiations led to the adoption of the Kyoto Protocol at the third session of the 1997 COP. The Kyoto Protocol contains specific obligations for Annex I Parties to reduce or limit their anthropogenic emissions (Article 3 paragraph 1 and Annex B). However, it leaves a number of issues to be further specified by decisions of the Conference of the Parties serving as the Meeting of the Parties (COP/MOP). Such "mandates" (i.e. authorization provided to the COP/MOP to adopt decisions containing detailed provisions) are contained in 1) Articles 5, 7, and 8 of the Kyoto Protocol relating to reporting and review, 2) the mechanisms (i.e. Joint Implementation, Clean Development Mechanism, and Emissions Trading, under Articles 6, 12, and 17, respectively), and 3) procedures and mechanisms on non-compliance (Article 18).

An example of such a mandate is the provisions establishing the mechanisms. The Kyoto Protocol states that Annex I Parties may – in addition to using national policies and measures to achieve their commitments – make use of the mechanisms. Although these mechanisms are established by the Kyoto Protocol, the relevant articles state that COP/MOP may adopt further guidelines, modalities and procedures, principles, and rules on their operation. Thus, it is only when decisions are adopted on the operation of the mechanisms that Annex I Parties will know the exact circumstances under which they are to fulfill their Kyoto Protocol commitments.

These mandates, contained in several provisions of the Kyoto Protocol, raise several questions. A number of international agreements have established mandates for international institutions to determine specific questions. For instance, Article 24 of the Convention on Biodiversity decides the seat of the secretariat set up by the treaty. The Kyoto Protocol, however, provides for much broader mandates to COP/MOP concerning issues central to implementation and operation. This uncertainty, in particular about the operation of the mechanisms and the non-compliance procedures, has been an obstacle to the ratification of the Kyoto Protocol by Annex I Parties; the latter have been reluctant to ratify the Kyoto Protocol until the specific regulations concerning the mechanisms and the non-compliance procedure are known. Thus, only after agreement was reached on these issues, *inter alia*, in the Marrakesh Accords at COP7 in 2001 (the decisions were adopted formally by COP/MOP1), did Annex I Parties ratify the Kyoto Protocol. The mandates provided by the Kyoto Protocol to COP/MOP were thus an obstacle to its speedy entry into force.

The mandates provided by the Kyoto Protocol also raise questions regarding the status of these decisions at the international and national level. At the international law level, the legal status of the decisions remains unclear, as the Kyoto Protocol mandates do not clearly state that the decisions adopted by COP/MOP are legally binding. The language used in the relevant provisions range from "guidelines" to "rules." While "rules" implies that they are legally binding, "guidelines" would imply only a recommendation to the Parties. This legal uncertainty about the status of "guidelines" adopted by COP/MOP under international law has led to decisions which provide for Parties having to fulfill specific requirements to allow them to make use of the mechanisms. Thus, only if a Party meets the requirements provided in the guidelines may they use the mechanisms to reach their commitments under the Kyoto Protocol.

On the national law level, the broad mandates may also be seen as an obstacle to the ratification of the Kyoto Protocol, which itself determines only to a limited extent the scope of the obligations. It could be argued that, at the time of ratification, not all relevant provisions were known to the legislature, and also that the necessary legal provisions can be established only when the relevant decisions have been adopted by COP/MOP.

Thus, on the one hand, the framework approach has been an effective means for the elaboration of international climate change regulations, as it led – within a relatively short period after the start of the negotiations – to the adoption of the UNFCCC in 1992 and to the adoption of the Kyoto Protocol in 1997. On the other hand, the unclear language used in mandates provided in the Kyoto Protocol raises a number of concerns. More precise terminology would have helped to avoid legal uncertainties and might have led to a speedier implementation process.

Compliance with and enforcement of international obligations

The question of compliance and enforcement has been a central issue in the negotiations to establish international climate change regulations. This issue was a focal point in the negotiations leading to the adoption of the UNFCCC, whose Article 13 gives a mandate to COP to:

> consider the establishment of a multilateral consultative process, available to Parties on their request, for the resolution of questions regarding the implementation of the Convention.

Such a process was not established until now, as no agreement could be reached concerning the size and membership of a Committee entitled to deal with questions concerning UNFCCC implementation. However, the negotiations did underline the importance for compliance procedures and mechanisms under the Kyoto Protocol.

The establishment of compliance procedures under international environmental agreements has been a general practice in recent years. Although nearly all international environmental agreements adopted recently contain dispute settlement provisions, to date they have not been used by Parties in practice. Therefore, an alternative was sought to ensure that Parties implement and comply with their obligations under international environmental agreements. The first compliance procedure was established under the Montreal Protocol of 1991. Since then such procedures have been established, for example, under the UNECE Convention on Long-Range Transboundary Air Pollution and its Protocols, as well as the Cartagena Protocol on Biosafety. Under the Kyoto Protocol, the provision of an effective mechanism to ensure that Parties fulfill their obligations was even more central than in the negotiation of other international environmental agreements. In particular, it was argued that the mechanisms made a compliance system an essential part of the Kyoto Protocol, and that an emission trading system would be functional only if fulfillment of obligations by Parties could be assured.

The Kyoto Protocol contains several provisions addressing the question of implementation and compliance. Article 8 provides for the review of the annual reports of the "expert review teams" of Annex I Parties. Article 18 gives a mandate to COP/MOP1 to:

> approve appropriate and effective procedures and mechanisms to determine and to address cases of non-compliance with the provisions of this Protocol, including through the development of an indicative list of consequences, taking into account the cause, type, degree and frequency of compliance. Any procedures and mechanisms under this Article entailing binding consequences shall be adopted by means of an amendment to this Protocol.

At COP7 an agreement was reached to establish a compliance system consisting of a facilitative branch and an enforcement branch. The provisions concerning the facilitative branch follow the example of other compliance systems, providing advice and assistance to Parties that face difficulties in fulfilling their obligations under the Kyoto Protocol. The provisions concerning the enforcement branch go beyond the traditional compliance systems. The enforcement branch is responsible for determining whether Annex I Parties are in compliance with their emission limitation commitments, their reporting obligations, and are fulfilling the requirements so that they can make use of the mechanisms. One of the most controversial issues in the negotiations was the consequences to be applied by the enforcement branch to Annex I Parties found to be in non-compliance. The second sentence of Article 18 stating that if "binding consequences" were to be applied, then an amendment to the Kyoto Protocol would be necessary, proved a stumbling block to the negotiations. The consequences agreed at COP7 to be applied by the enforcement branch consist of a declaration of non-compliance, the deduction from the Party's assigned amount for the second commitment period of 1.3 times the amount of tonnes of excess emissions, the development of a compliance action plan, and suspension of the eligibility to make use of the mechanisms.

No agreement could be reached at COP7 as to whether these were to be regarded as "binding consequences." It was argued that the consequences did not add new obligations for Annex I Parties as they were within the general law of state responsibility for breaches of international law. Parties are obliged under general international law to cease the internationally wrongful act and to make full reparation for the injury it has caused. This legal argument was supported by the political argument that it was essential for the compliance system for it to apply to all Parties to the Kyoto Protocol and not only to those ratifying the amendment. COP/MOP1 decided to adopt the compliance system by way of a decision, but also decided to consider the need for amendment in the future.

Concluding remarks

The framework approach used in the elaboration of the climate change regime has proved to be successful. The UN General Assembly set up the International Negotiating Committee for a Framework Convention in 1990. While the risk of global climate change at that time raised public concerns, views differed greatly about the likely effects of climate change and what measures should be taken to deal with the issue. The complexity of climate change issues only became clear to a wider public during the negotiations. The growing knowledge about the reasons for it and its effects, in particular through the work of the International Panel on Climate Change, underlined the need to address global climate change and to provide an international legal instrument to deal with it In contrast to other areas of environmental regulation, such as the protection of the ozone layer, measures against climate change affect all sectors of society. Thus, the negotiations are of interest to a large number of players on the national and international level, ranging from environmental interest groups to business representatives and labor unions. Establishing rules on the international level needs more time so that a compromise can be found between the different players.

The framework approach has helped to establish a "permanent negotiating process" on global climate change which has moved from general principles to specific commitments. Although the framework approach raises a number of concerns, it has established an international regime which provides for measures to address climate change that have occurred within the last 15 years. Moreover, it has helped to concentrate negotiations on climate change within the institutions set up by the United Nations Convention on Climate Change and the Kyoto Protocol. One should not, however, overlook the fact that, to deal effectively with global climate change, measures have to be taken by all sectors of society, which makes it important for all international organizations and institutions to take into account the effect that their activities are having on the world climate.

5   The Observing Sociologist

Guy Olivier Faure

Climate talks, in common with all environmental issues, are not just a scientific matter, nor just an economic issue, nor merely a series of political discussions. Dealing as they do with societal problems – people's needs, fears, political choices, and beliefs – climate talks also have a sociological dimension.

Moreover, the issues that come under scrutiny in the course of climate talks are perceived through the numerous and varied cultural lenses of the parties to the negotiations, who are from widely different professional backgrounds and/or countries. Although, during lengthy and recurring negotiations, parties come closer, both in terms of how to set problems and also operationally, each profession/culture nevertheless has its own way of framing and dealing with issues, coping with obstacles, and managing the negotiation process. Thus, scientists, diplomats, politicians, industrialists, and militant environmentalists will each have highly contrasting views as to what issues should be dealt with and how. This holds equally true for national cultures, especially clusters of nations such as Europe, North America, the Arab world, Asia-Pacific, Latin America, and Africa.

Among the many lessons that can be drawn from past experiences of climate negotiations is that science alone cannot provide the definitive solution. The ways in which scientific concepts are viewed and interpreted may cause misunderstandings and deadlocks, and also allow decisions to be made that are irrelevant. Scientific data may be instrumentalized by some parties to bolster their own position, for instance, through specific information selection, by resorting to value-laden concepts as a basis for analyzing problems, or by using biased criteria to evaluate options.

A number of stumbling blocks to successful negotiations are rooted in this professional and cultural diversity, such as differences in the perception of a problem, its nature, and urgency. There may be difficulties in actually expressing perceptions; negotiators may fall into the classic trap of using the same words to express quite different concepts. Difficulties may also arise from differences in negotiating and procedural behavior during the negotiation process. Thus, while our quest is ultimately for decision-making consensus, we may come across the sort of conflicting values that divide traditional and modern societies. "Fair" and "just" outcomes are relative to social values and moral judgments, and this also raises thorny issues.

Considering the specific character of climate negotiations, sociological and cultural obstacles to reaching an agreement fall into the following categories: the various actors of the process, their objectives, the structural components of the situation, the conditions that prevail at the very start of the negotiations, the process itself – how it encourages progress and the techniques used for building a consensus, the concrete outcome that is reached, and the consistency and feasibility of the outcome.

The actors

A number of actors are directly or indirectly involved in this type of negotiation, and they introduce to the debate a great variety of views and a multiplicity of interests. Among them are governments, international bodies, think tanks, nongovernmental organizations (NGOs), media organizations, industries, acknowledged scientific authorities, and the public. Within each of these groups, there are divisions, making the overall structure even more complex. For instance, governments are split into two main categories, developed countries and developing countries. Within the first group we can occasionally observe diverging positions between the European Union (EU) and Canada on one side and the United States (US), Australia, and Russia on the other. Some situations may be quite fluid, as Russia does not systematically side with the US. Within the G77, there are also important divisions: small island states, oil-producing countries, least developed countries, major powers such as India and Brazil. China, though not a formal member of G77, is closely associated with it.

As far as industry is concerned, there is a similar conflict of interests between carbon-based or carbon-dependent industries and non-carbon-based activities such as insurance, banking, and so on. Even NGOs are not immune from this major problem. On one side stand groups such as Greenpeace, the World Wide Fund for Nature (WWF), and humanitarian groups, and on the other side, for example, *The Advancement of Sound Science Center* (TASSC), a group funded by Exxon, Amoco, Chevron, GM, among others.

Voters should also be seen as being among the stakeholders, at least in countries that have a democratic system of government. Consumers may have also a word or two to say, for example, on specific issues such as CFC aerosol sprays. Even religious leaders may play a role (Benedick 1993).

A number of government ministries are involved in the process to set up a national position. No fewer than 18 departments and agencies participated in the preparation of the US position for the Montreal Protocol Negotiations. One might also add interested parties such as the pharmaceutical industry, an extremely powerful lobby group. In Washington in 2002, this pressure group employed 675 lobbyists, that is, more than one for each member of the Congress.

Once a position, resulting from a complex balance of interests, has been decided, it becomes extremely difficult to move away from. Thus, negotiators may not even have the minimum flexibility required to make an agreement. This is the case for negotiations within the EU, where the views of 25 member states have to be accommodated.

Another factor contributing to the current imbalance is the influence of environmental organizations. "Greens" are rather weak in most of the developing countries and cannot play the same role as they do in industrialized countries. Often initiated or supported by Western groups, they are even accused of being "unpatriotic," which may get some of them into serious trouble.

This impressive array of stakeholders translates into a great variety of attitudes toward climate change knowledge. Knowledge is a social construct, rooted in society, whose utility is to provide answers to problems considered important by a society at any given stage of its evolution (Faure 1995). Once acquired, knowledge can be used politically, instrumentalized, and manipulated in order to bring influence to bear on situations that have social consequences. Knowledge, such as technical expertise, is far from being universally shared, even though there are remarkable technologists in any country. Moreover, the enormous amount of data that has been collected on climate change is difficult not only to absorb but also to repurpose into truly cogent arguments. Scientific knowledge is in many ways the language of the dispute in climate change talks, but as such it does not play as essential a part as it could, namely, as an objective arbitrator simply telling the facts.

Climate negotiations are open to a broad audience of onlookers, and negotiators are tempted to add another goal to their task, which is to play to public opinion. Climate issues are likely to have consequences for the lives of many citizens; thus, public opinion is a major factor in the game and adds to the complexity of the negotiation process management. Research by Faure and Rubin (1993) indicates that the presence of onlookers increases the tendency of negotiators to take extreme positions and to stick to them, in order to avoid looking weak or willing to betray their own cause. To counter such a risk, and as negotiations often become tense in the final stages, the chairperson may take the decision to call closed meetings. On the other hand, the presence of NGOs at this very moment can be very valuable by reducing the need for street demonstrations.

Another question is the degree to which the negotiation system established on climate change reflects the macro-system. On the one hand, there are strong players such as the US or the EU or the "Group of 77 and China." On the other hand, there are extremely small social groups, which are genuine civilizations from an anthropological standpoint, such as Amazonian tribes, Papuan groups, or Bushmen clans. Such people are probably the most vulnerable to climate change, but although some of them have managed to create a reasonable platform in the current debate, they are not sufficiently represented. If, when facing the consequences of climate change, all humans are equal, then some are indeed more equal than others.

Ultimately, we face a paradox. To be as fair a forum as possible and to have the greatest possible acceptance as an arena for problem solving, the climate change negotiations have to include a wide selection of stakeholders; however, the more stakeholders there are, the greater the number of vested interests represented and the lower the chances are of reaching a good rather than just suboptimal outcome. What is more, every aspect of the debate has its own specific logic, and this in itself adds directly to the complexity. How then can common terms of reference be found between the political, social, and economic dimensions of the debate, when behind the ambiguous or sometimes empty wording of proposals, there is so often cacophony—the Babel effect.

The objectives

In climate negotiations, the social interests at stake are dramatic: nothing less than the socio-economic development of developing countries through industrialization. Here we face the North–South conflict of interest, epitomized in the unequal level of economic development between developed and developing countries. One of the desired outcomes of climate change negotiations is to promote responsibility now and accountability to the generation to come. To achieve such a goal, a common vision and a common framing of the climate problem must first be established. Here too, because of the number and variety of parties involved, with their different cultures and thus their diverging understanding of what may or may not be desirable for themselves and for others, that "common vision" becomes refracted through a prism of diverging values and conflicting interests.

The ethical dimension of the issues being dealt with is becoming increasingly important in public opinion terms. When evaluating a decision, populations are more sensitive to morality than to legality. Even the most cynical governments or companies need to take this into consideration before adopting any public position. At the end of the day, any kind of double-speak will only serve to complicate the negotiation process, increase its fuzziness, and, if discovered, prove counterproductive.

The measures agreed upon also need to be sustainable if an effective agreement is to be reached. The cultural/social question then arises as to what is a sustainable solution. What time span does it cover? A human being's time frame is not the same as a country's or a company's. If we consider only human beings, then we encounter highly contrasting cultural views regarding time spans. To take a simple example drawn from the housing market: when a French person buys a house, this represents full ownership for ever, while a British property buyer is sometimes the owner of the property only for the duration of the lease, which would make a French person feel extremely uncomfortable.

The relationship of urban dwellers to the climate is also different from that of rural populations, whose fate depends strongly upon the weather conditions. This can be reflected in the position some of the parties take during the discussions. What is more, although there may be tremendous negative consequences for large natural ecosystems that everyone has to bear, global warming is not viewed everywhere as a catastrophe. Whenever a situation changes, there are losers and winners. Cold countries in line for milder winters may not see global warming as very harmful.

Thus, to establish a negotiation process that produces useful and sustainable results, the first stage of any negotiations must be to identify the objectives preoccupying multiple parties and where these overlap or coincide.

Structure

Multiple parties, multiple roles, and multiple issues are among the distinctive attributes of climate negotiations. The Kyoto agenda, with its many different kinds of issues, makes the negotiation process very difficult to deal with. Any simple environmental decision has implications at the economic, social, political, legal levels, which in turn has a multiplier effect in terms of the number of issues to be dealt with. It is like a medical prescription for more than two or three medicines, where no one really knows what their combined effect will be or even if the death of the patient will be the end result. No one can really say what the final outcome of making a set of decisions, sometimes loosely related and without precedent in history, will be.

In a conference of this type, there is always the major stumbling block of opposing collective and individual interests. In greenhouse effect terms, emissions from a single country may harm the whole world. The harm for the country responsible for those emissions may be evaluated quite differently by the various actors. A consensus on remedies must always be preceded by a consensus on causes and on how far-reaching the consequences may be.

The issue of power in multilateral negotiations is a highly complex one in that it requires all parties to seek allies. Usually, countries operating independently do not make a significant impact on the final text (Széll 1993). Thus, the first stage of the negotiation process is to build or to join a coalition, and hammer out a compromise. As the final text will also be a compromise – a compromise based on more compromises – negotiation dynamics will lead the parties toward an outcome that addresses their initial concerns on an ever-diminishing basis. Thus, the incentives to sign a common text grow fewer as an agreement comes nearer.

The size of the conference is an important factor in how effective the process turns out to be. Huge meetings attended by scores of delegations, industry representatives, environmental groups, lobbyists, and observers are just an arena for dogmatic rhetoric and apocalyptic speeches. Consequently, plenary sessions are more of a recipe for going nowhere than an enabling structure for consensus building. These oversized conferences have a feeling of United Nations culture about them, but without the remedies offered by the Security Council. They can only be used as a departure point for the real negotiation. The actual value is in the small-scale working groups that allow room for informal discussions with negotiators away from exposure to public scrutiny. This is the ideal formula for disaggregating complex problems and dealing with them properly. However, some countries try to prohibit meetings of this kind in case they miss out on any important information or to avoid facing unexpected coalitions being put together behind the scenes.

Conditions

Of all the conditions affecting attitudes to climate change negotiations, scientific uncertainty plays a special role. Where there may be a risk, but this is not clearly assessed, scientific knowledge loses its pivotal function. There were no questions about uncertainty during negotiations over pollution in the River Rhine or following the Chernobyl disaster, as the measures taken were in response to damage that had already been done. In contrast, climate change negotiations are based on a fuzzy hypothesis of what may happen after one or two generations. Moreover, uncertainty tends to somewhat freeze the dynamics of the negotiation process within the relevant forums. Even though uncertainty in climate change has been substantially reduced following more than two decades of scientific research, it is nonetheless embedded in the social and strategic context, which makes it highly vulnerable to external considerations. Knowledge becomes the basis of scientific politics, and scientific complexity then becomes social and political complexity.

In climate talks, the risks under discussion are not the consequences of a natural event but the result of human actions. How to define risk is a crucial issue in the debate because it leads to definitions of liabilities and responsibilities, and can consequently point to culprits and victims, creating a situation that does not make for ease of cooperation.

Demonization of another party is a common way of producing protracted deadlocks (Zartman and Faure 2005). This is a process that is easily triggered when a strong disagreement occurs, for instance, between industry representatives and environmentalists. It may result, for example, in environmental organizations raising a problem and industries being supposed to find a solution. Moreover, all parties tend to act strategically in exaggerating the possible risks, underestimating the seriousness of a problem, or greatly overstating the potential costs of remedial actions.

The greenhouse effect is not a new discovery, having been explained by the French scientist Fourier in 1827. However, the heated debate about it began only in the second half of the twentieth century. The first world conference on climate was held in 1979 to look at how climate change might affect the wellbeing of humanity. However, the conference's outcomes did not reach the ears of policymakers. Public awareness has grown steadily, in parallel with the growing sense of unpredictability regarding possible consequences. Though public anxiety is on the rise, it is greatly lessened by the weight of uncertainty about what the outcomes might be. When people are not certain whether a phenomenon is irreversible or not, nor of the magnitude and effect of the possible changes it entails, they do not put as much pressure on their political representatives to "perform" in negotiation forums. Environmental negotiations have been defined (Faure and Rubin 1993) as "a process aimed at distributing misfortune among people." The measures that need to be taken will generate costs, trouble, and unwanted changes, with no visible result in the short term. Public opinion is far more sensitive to issues such as peace, security, employment, and helping the poor, where there is a sense of urgency, than to long term issues for which no immediate visible results can be expected.

Another stumbling block lies in the relatively new concept of deterritorialization, which has still not been easily understood or taken on board by the public. Global problems call for global solutions, but how should issues involving national contributions that are designed to be non-territorial be dealt with? Globalization mainly tends to dilute responsibilities. In the common view, "global" means the problems and burden of the measures needing to be taken to solve them rather than the benefits that are there to be shared, especially if the time span is a number of generations. Political globalization is not taking place alongside economic and ecological globalization. The national interest still prevails, and a system of negotiation such as that which led to the Kyoto Protocol is not the best means of enhancing a broader view of the problem at hand.

Moreover, as far as national views are concerned, such agreements represent a challenge to state sovereignty. Nation states, especially the relatively new ones, are often anxious not to give away what they have had to struggle for so long to obtain. The South is anxious to keep as much control as possible of its recently acquired rights over its own resources and development opportunities.

Even where there is public awareness of climate change, parties to climate talks do not always have the same sense of urgency and thus do not all act at the same pace. Some Western countries tend to consider the climate issue as absolutely crucial, but this view is not shared by all. Moreover, developing countries for the time being have other priorities.

Accountability, if not a sense of obligation, may also play a role, but the question is to whom? Probably the human race itself. However, where there is a sense of accountability, this is directed toward voters, clients, sponsors, and institutions. The next generation is still very much a "virtual constituency."

Process

A negotiation process is not just a way of reaching some kind of agreement, but first and foremost a means of enabling parties to develop consensual knowledge. Even just defining what is meant by climate change is a major issue. Furthermore, an understanding of climate change is not just about pulling a set of conclusions from scientific research, but is, to some extent, a negotiated product. Problem framing is a second step that is done partly within the negotiation process. Without a common understanding of a problem, no common answer can be provided. However, problems and issues are linked to perceptions; their existence implies that there has been active cognitive knowledge production during the negotiation process. A problem's complexity, the various rationales operating within it and the ways in which they intertwine, only add to these difficulties, creating a further obstacle to agreement.

Negotiation on the substance of a potential agreement thus becomes a kind of communal fine-tuning of facts that are supposed to be scientific. Some parties, such as Russia or the US, tend to downplay the negative impact of climate changes. For instance, what the acceptable temperature range for global warming should be is quite subjective and thus highly negotiable.

For stakeholders such as governments, legitimizing a negotiation means formulating the problem in such a way that it looks serious but not so dramatic that it backfires if no agreement is reached at the end of the process. Information is omitted or included in the discussions according to strategic needs, or may simply be distorted.

Scientific data can be heavily instrumentalized through 1) what analytical categories are chosen to frame the issues to be discussed, 2) the selection of "relevant information," and 3) the criteria selected to rank the possible options. We find methodological biases common to decision-making procedures, whereby politicians making the initial decisions then look for scientific data only to back up what has been already decided based on other criteria.

Climate change issues are quite important as far as public opinion is concerned. Media coverage of any international conference on these issues feeds and develops public concerns. Some groups such as the Greens tend to overuse media to gain influence. This strategy may be counterproductive in terms of the effectiveness of the negotiation process because, as soon as commitments are made openly, it becomes less easy for any party to move from them than if the setting had been more discreet. There is also an amplification effect when pressure groups select what they consider as unacceptable and insist on denouncing the attitude of a party to the negotiation. Some groups also make abundant use of information leaks to criticize national positions.

In some settings, such as working groups, and on some issues, the negotiation may be subjected to a group-think effect, where each negotiator relies on the others and is unwilling to break the harmony of the group or to speak up against the general view. In draft-making sessions, for instance, passive disapproval of a proposition can be interpreted as a silent agreement.

With the shift in values away from the "polluter-pays" to the "wealthiest-countries-pay" principle, some countries tend to use the strategic ploy of resisting any move so that at the end they get funding to persuade them to agree. Such a system may obviously provide an incentive to disagree rather than to facilitate a consensus. At a certain stage of the process, especially for the chairperson, a way must be found to generate momentum. The point at which there is no more powerful opposing coalition, and at which the time has come to present a compromise must be realized. The Wall Street adage quoted by Sebenius (1993), according to which "time is the enemy of doable deals," may not necessarily apply in this case. As well as avoiding too strong a polarization in the positions, it is also important to feel when the moment may be ripe to propose a final package.

Outcome

The social/moral values governing any negotiation outcome can be defined as fairness. Establishing a regime to deal with climate change means taking measures that have economic and social consequences. Agreements have to abide by some principle of justice that will help to strengthen and stabilize them over time. The difficulty is that societies promote one principle at a certain stage of their evolution, and other principles at other stages. In fact, these principles may be mutually conflicting. Developing countries expect any new regime to be an answer to their basic social needs, whereas industrial countries tend to be more conservative, preferring the status quo but alongside efforts to prevent the situation from getting worse. Thus, conflicting principles of what constitutes fairness may make it more complicated for parties to reach an agreement.

The strategic shift from the polluter-pays to polluter-is-paid principle is designed to smooth negotiations, but this is not always understood by populations or the media. In some ways, such "smoothing" leads to challenges on strongly based moral grounds; people do not easily accept the idea of providing technical assistance and funds to those who, in their opinion, are behaving the worst. At best, for countries such as Brazil, China, India, Indonesia, and Mexico, there is always a strong temptation to be a "free rider", to scoop the benefits without incurring the costs. These countries continually point to the "historical debt" incurred by industrialized countries, which are responsible for around 80 per cent of the concentration of CO_2 in the atmosphere.

When the negotiation parties reach a possible formula, the question of its feasibility arises. Any compromise has to be politically acceptable for any elected government. Often, short-term efforts to achieve long-term effects lead to a strategic response being produced that simply buys time, effectively postponing any major decisions. By slowing down the process, some parties expect knowledge, politics, or technology will have changed enough to reduce the costs of whatever measures need to be taken.

Conclusion

If, as according to Benedick (1993), "politics is the art of taking good decisions on insufficient evidence," then climate negotiations provide the perfect example of the application of this maxim. There are plenty of obstacles along the way to making major joint decisions on climate change policy. Scientific knowledge and the actions to which it gives rise make sense only if they correspond to human needs. Science develops within society and is intended to serve some of its purposes, which is why a sociological view can highlight some of the stumbling blocks in climate change negotiations and help in analyzing them.

Climate change negotiations are impacted by the needs, apprehensions, and fears – real or imaginary, exaggerated or understated – of the various parties. For the time being, there are no scientific or economic paradigms for coping with the complexities of global climate change. In some ways, the Kyoto Protocol can be viewed as another victim of this bounded rationality. Based on the UN model, it literally invites excessive time and energy wasting on the conference's procedural aspects. For instance, endless discussions on the composition and mandate of working groups are held before these groups can start hammering out a draft paper. Formulas popular for a time and even somehow overused such as "sustainable development," "win–win situation," or "thinking global, acting local" are still waved around as slogans when substantive arguments are lacking in the discussions. Overall, the Kyoto Protocol addresses such a broad range of issues and involves so many parties that its effectiveness is always going to be less than optimal.

There is a contradiction between the nature of the negotiations, on the one hand, and their scope and reach, on the other. Of course, people find significance in a universal agreement hammered out in an international conference of all nations and stakeholders. But for results to be effective, small-scale meetings and agreements are far more useful. Note the substantive content of the 2006 Conference of the Parties serving as the Meeting of the Parties to the Kyoto Protocol (COP/MOP) which was based on workshops and seminars.

A more effective approach to climate negotiations would be to ensure that the basis on which decisions are made includes social values. For instance, naturally one wishes to live longer, but what would the ultimate cost be to the quality of life on Earth? The answer, though far from easy, cannot be sidestepped. For the time being, greater clarity and more realism are needed in dealing with these complex issues – and that will entail a much greater consideration of the basic human dimension of climate change negotiations, as well as a greater use of sociological approaches, if we want to come up with more effective answers.

References

Benedick, R.E. (1993) Perspectives of a Negotiation Practitioner, In Sjöstedt G. (ed.) *International Environmental Negotiation*. London: Sage. 219–243.

Faure, G.O. (1995) Conflict Formulation: Going Beyond Culture-Bound views, In Bunker, B. B. and Rubin, J.Z. (eds.) *Conflict, Cooperation, and Justice*. San Francisco: Jossey-Bass. 39–57.

Faure, G.O. and Rubin, J.Z. (1993) International Environmental Negotiation: Organizing Concepts and Questions, In Sjöstedt, G. (ed.) *International Environmental Negotiation*. London: Sage.17–26.

Sebenius, J.K. (1993) The Law of the Sea Conference: Lessons for Negotiations to Control Global Warming, In Sjöstedt, G. (ed.) *International Environmental Negotiation*. London: Sage.189–216.

Széll, P. (1993) Negotiations on the Ozone Layer, In Sjöstedt, G. (ed.) *International Environmental Negotiation*. London: Sage. 31–47.

Zartman, I.W. and Faure, G.O. (2005) *Escalation and Negotiation in International Conflicts*. Cambridge, UK: Cambridge University Press.

Part III
Stumbling Blocks

6 Defining a Politically Feasible Path for Future Climate Negotiations

Lessons from the EU–US divide over the Kyoto Protocol

Urs Steinar Brandt

Introduction

In December 2000, the United States (US) left the Kyoto Protocol on the grounds that it contained a number of fatal flaws (Chen 2003; Springer 2003; Springer and Varilek 2004).[1] To many, this decision was just one of the manifestations of an enduring conflict between the US and the European Union (EU) (see e.g. Christiansen 2003, Torvanger *et al.* 2004, Tjernshaugen 2005). This chapter seeks to infer the main lessons of this conflict so that the facilitation measures needed to avoid a similar situation in future climate negotiations can be shaped.

This paper argues that one (main) reason for the conflict between the EU and the US stems from a lack of political will be to accept the introduction of instruments that would accommodate both the interests of the EU and the US. While the EU favoured binding targets based on a responsibility based approach, and with only approved flexible mechanisms as supplement for national reduction initiatives, the US favoured full access to cheap reductions obligations and made its own ratification contingent on similar action from major developing countries.

We apply the concept of *political feasibility*, which means that the focus must be on policy measures that all countries can accept. This entails detecting the chief preferences of the main actors in the climate negotiations, such as what factors determine a country's willingness to engage in greenhouse gas (GHG) reductions above its business-as-usual level. The politically feasible solutions are those that move the process optimally in the direction of the overall targets.[2] The fairly general theoretical framework of political feasibility enables us to pinpoint the direction in which the negotiations should move so that both the current stumbling block between the US and the EU, and potential future stumbling blocks, can be overcome.

The overall target defined by the United Nations Framework Convention on Climate Change (UNFCCC) is to reach a "stabilization of GHG concentrations in the atmosphere at a level that would prevent dangerous anthropogenic interference with the climate system". It will be argued that, to achieve this target, three objectives must be met. First a number of short-run targets must be specified, which implies a succession of larger overall emission reductions. Second, participation in the Kyoto Protocol must successively be broadened. Finally, incentives must be provided for the development and implementation of renewable energy systems (RES). We propose that the first target is best achieved by instruments that make the overall abatement costs as low as possible. The second objective is best achieved by carefully chosen surplus allowances allocations, while the last one needs a specific agreement where an international cooperation can develop a Resolution on Climate Change.

The objectives of international climate policy

The objectives for the international climate negotiations were formulated in the UNFCCC at the "Earth Summit" in Rio in 1992. Article 2 states the overall objective as "stabilization of GHG concentrations in the atmosphere at a level that would prevent dangerous anthropogenic interference with the climate system." More specific targets were set five years later in Kyoto. According to the Kyoto Protocol, the objectives were to achieve the stated reduction targets and also to develop new cleaner technologies. The most important text for our purpose is found in Article 2, where it is stated, *inter alia*, that:

1. Each Party included in Annex I, in achieving its quantified emission limitation and reduction commitments under Article 3, in order to promote sustainable development, shall:

   a) Implement and/or further elaborate policies and measures in accordance with its national circumstances, such as:

      (i) Enhancement of energy efficiency in relevant sectors of the national economy;

      . . .

      (iv) Research on, and promotion, development and increased use of, new and renewable forms of energy, of carbon dioxide sequestration technologies and of advances and innovative environmentally sound technologies.

There are several interesting features in this text. Regarding the long-term target, only the damage side of the problem is mentioned, and hence, the costs of reaching that target are totally absent. In the Kyoto Protocol, there are two different levels of target, the one level being very specific short-run targets, and the other being a more diffuse objective of R&D on new environmentally friendly technologies.

The stands of the main parties

This section briefly describes the positions of the two main players in the climate change negotiations, the EU and the US, discusses the reasons for these positions, and evaluates the effect of these positions on the present state of the climate negotiations.

The "historic" development of strategies

At the first Conference of Parties (COP1) to the UNFCCC that took place in Berlin in 1995, the US in particular strongly advocated the idea of global CO_2 trade as a way of substantially reducing the costs of meeting emission reductions targets, because of the positive experiences the US has had with buying and selling environmental permits and in connection with the prospect of then being able to trade allowances applications (hot air).

On the other hand, the EU, not being too enthusiastic towards the use of flexible mechanisms, proposed coordinated policies and measures and the supplementarity conditions, meaning that flexible mechanisms should complement rather that replace domestic action (Christiansen 2003; Westkog 2002).[3] The Kyoto agreement included the use of flexible mechanisms, but the exact terms of how to operationalize this use were, however, postponed. The EU kept in a "quantitative interpretation of the supplementarity condition", and at COP6 at The Hague, the conflicting positions of the EU and mainly the US resulted in the US withdrawal from the Kyoto agreement (Brandt and Svendsen 2002).

Subsequently, the EU tried to keep the Kyoto agreement on track, even without the US. The Kyoto agreement was due to come into force once it was ratified by at least 55 parties representing at least 55 per cent of the total greenhouse gas emissions of Annex B countries in the year 1990. The first condition, that a minimum of 55 countries became party to the treaty, was fulfilled with Iceland's ratification on 23 May 2002. Hereafter, the main focus was to meet condition number two. The 2004 ratification by Russia was sufficient to reach the 55 per cent emission. To the surprise of many, the breakdown at The Hague was not the death of the Kyoto agreement, and in February 2005, the Kyoto Protocol formally came into force.

Two positions

Several papers have sought explanations for the divergence of positions between the EU and the US. Christiansen (2003) and Tjernshaugen (2005) argue that the divide can mainly be attributed to the following four points: 1) Difference in costs of participating (and a different perception of the costs), 2) different valuations of benefits and non-benefits, 3) different political values, and 4) historically, different contingencies.

The strategy of the US was focused not only on the costs of meeting the negotiated targets, but also on increasing the participation in the Kyoto Protocol (which might also have been due to a desire to minimize costs). The US signed the Kyoto Protocol under the presumption of access to cheap greenhouse gas reduction options (Brandt and Svendsen 2002). Nentjes and Woeldman (2000), Woeldman (2001) and Barrett (1998) argued that the Kyoto Protocol was negotiated in the expectation (at least for Russia and the US) of free trade, based on quotas. In the summer of 1997 the United States Senate protested against the so-called Berlin mandate, the foundation for the Kyoto Protocol.[4] With a vote of 95–0, it favoured a resolution that the US should not commit itself to sign an agreement that demanded the industrialized countries cut their emissions, without also demanding equally binding restrictions on the emissions of less developed countries, as this could harm the economy of the US. (This showed that the position of the US might be shaped, but not determined, by presidential preferences.) The arguments Bush used to leave the KP echoed this resolution.

The position of the EU exhibited elements of path-dependency. Brandt and Svendsen (2004) find that, historically, the EU has been a frontrunner in promoting renewable energy systems. The authors argue that the main reason for this can be traced back to the first oil crisis in 1973, when the oil price increased fourfold. In the 1960s and 1970s, Europe had huge imports of oil, whereas the US was self-sufficient. European dependency on oil imports meant that the price increase had a severe impact on the economies of European states, leading them to develop more new energy-efficient technologies. Two factors that contribute to the relatively tighter energy policy in Europe compared to that of the US are the level of energy taxation and geographical distances between home and work in Europe. Energy taxation is significantly higher in Europe than in the US (OECD 2002; Brandt and Svendsen 2006).[5] Therefore, European energy savings mean higher tax savings. This might have shaped the EU policy, in particular to use measures that increase the demand for renewable energy systems

However, genuine concerns about climate change issues at a higher political level in the EU than in the US could also have determined the EU position. The presence of green and left-wing parties in European parliaments probably shaped the greater EU focus on environmental and sustainability than on the costs of these policies. Tjernshaugen (2005) notes the determining role of proportional representation and coalition governments with, for example, green parties (and like-minded left and centre parties in countries such as Denmark and Norway) in Europe. Moreover, the much greater importance of private campaign funding in the US gives business lobbies, such as the oil lobby, a crucial advantage in policy determination.

Thus, a combination of potential first-mover advantages on the development of renewable energy systems, historically more regulated energy policies, and a greater desire to address the climate change issue at high levels in the EU are main reasons for the position of Europe and, ultimately, the EU.

A look at the arguments used to explain the positions of the two main parties to the climate change issue, the US and the EU, reveals two fundamental positions:

1. Strong focus on flexibility and cheap GHG reduction options (focus on short-run achievements)
2. Strong focus on sustainability and responsibility (focus on long-run achievements).

From the discussion later in this chapter, it will be obvious that the position of the US is strongly oriented toward position 1, while the EU position is more in accordance with position 2.

Proponents of the cost-driven approach argue that, by making the reduction costs as low as possible, it is easier to find acceptance of the targets by those who have to bear the costs of achieving them and hence, also easier to improve the prospects of tightening the targets in the future.

The main argument for the morally based approach (taking responsibility) is that, as the developed countries have the main responsibility for the present anthropogenic GHG stock in the atmosphere, and their economic development has been based mainly on burning fossil fuels, these countries must also take the lion's share of the reduction initiative. Moreover, proponents of this approach argue that making reductions a national responsibility creates incentives to intensify R&D in cleaner technologies, which in the medium term is a competitive advantage for countries, and in the long run also increase the prospects of being able to set more stringent future targets based on the diffusion of energy-efficient technologies.

Political feasibility and facilitation

In the previous section, the position of the EU and the US were explained in terms of historical, political and even accidental factors. In this section, a generalization of this reasoning is proposed. It might not appear to be an operational approach at first glance, in terms of being used as direct input to the actual negotiation process. However, it can serve as a tool to shape intuition and provide a better understanding of what determines the positions of the countries and, therefore, provide a theoretical framework for improving the quality of the guidelines for actual negotiations.

The principles behind political feasibility and facilitation

The position of any country is defined by numerous factors. In principle, it should be possible to derive a functional relationship between a country's GHG reduction level (compared to business as usual) and related factors.[6] By treating the reduction level as endogenous and related factors as exogenous, we can estimate this relationship.[7] For each state, the estimation would determine the weight attached to each of the factors country by country, and as these weights will vary between countries, would explain the differences in the positions of the nations concerned.

There might be numerous reasons why this approach is not feasible, (e.g. due to problems of measurement, the not very realistic assumption that none of the exogenous factors are correlated, and the problem that the degree of freedom is to small, due to too few observations), but the main gain from an analytical point of view is the way this approach structures our thinking. Having made the list of possible factors influencing the choice made by a country, the main task now is to identify key factors that mainly determine a country's decisions about a given problem. From this, it is possible to derive which dimensions a solution must incorporate in order to be acceptable for a country. If such solutions exist, they then define the politically feasible solutions.

Finding politically feasible solutions for future climate negotiations

Political feasibility starts with specifying and establishing the overall objectives. Given these targets, and given the politically reality (as defined in the section on political feasibility), how can an agreement that enables this target to be reached? As stated in section 2, the FCCC and the Kyoto Protocol lay down three overall objectives for global climate change policy. First, it is important to have short-run targets to keep the process going; second, it is necessarily to increase the number of countries participating in the agreements; and finally, a long-run objective with "deep" targets is needed.

A general lesson from economic policy theory is that it is necessary at least to employ the same number of different instruments as the number of objectives for which the overall policy is aiming. This idea is summarized in Figure 6.1, where the main objectives of future climate policy are stated together with the means by which these objectives are believed to be best met under the constraint of political feasibility: Targets must be set and instruments chosen such that the countries achieve these targets at lowest cost (not on the margin, but with respect to the overall costs) to keep the process going. To broaden participation, positive incentives should be given, particularly to less developed countries, and finally, to make the necessary deep changes in energy systems, incentives must be provided for the development and implementation of new cleaner technologies.

The next three sections aim to explain why this chapter arrives at the conclusion summarized in Figure 6.1, through a discussion of the instruments that can be used to meet these objectives.

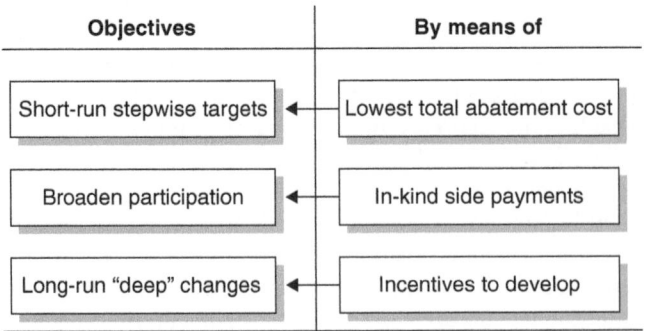

Figure 6.1 Several objectives need several different instruments

Short-run stepwise targets: Lowest possible total abatement costs

In this section it will be demonstrated on the basis of historical evidence that the cost issue is important, such that a treaty, to be politically feasible, must be designed in such a way that the overall targets are meet at a minimum cost (Newbery, 1990).[8]

Examples of the importance of the cost side

If we look at different instrument that have been proposed at the international level, not all cost-efficient policies are politically feasible, mainly because not all cost-efficient instruments minimize the direct costs from abatement. An example of this is the proposal of an EU-wide tax on CO_2 proposed in 1992. After the June 1992 conference in Rio establishing the UNFCCC, the European Commission launched a US\$ 15 tax (t/CO_2) directive proposal, but this proposal was opposed by several member states. Following a meeting on 5 October 1994 between the ministers of the environment of the EU, the Danish minister of the environment stated publicly that the proposed common CO_2 tax in the EU was dead. There are several reasons why the tax did not get approval. First, as a fiscal measure, it needed to be accepted unanimously by the EU countries. Second, the EU tax puts a large burden on firms in the EU, making them less competitive vis-à-vis rival firms from other countries not subject to taxation (like firms in the US). From the idea of a guiding tenet of the polluter-pays principle, the tax is optimal, but political reality (in terms of large resistance from major industrial groups) makes this proposal politically unfeasible.[9] Hereafter, the EU planned to auction tradable permits (TPS) which from the cost-side perspective resemble a tax, where the polluters still fully pay. Therefore, resistance from relevant stakeholders remained. In the end, the EU accepted a grandfathered TPS, where the relevant firms receive free permits for most of their current emissions. This kind of grandfathered TPS is not a fiscal measure but an environmental policy, and as such needs only a majority of supporting countries. Hence, the EU has moved from auctioned (equal to tax) TPS to a grandfathered system. The reasons for this being the preferred environmental instruments by industry, is recognized in several papers (e.g. Christiansen and Wetterstad 2003; Brandt and Svendsen 2003).

Another example is the EU proposal to make trading supplementary to national measures (Brandt and Svendsen 2002; Christiansen 2003). Brandt and Svendsen (2002) argue that the conflict between the EU and the US stems mainly from disagreement on the cost issue. The EU promoted three issues in The Hague. First, a 50 per cent national emission ceiling (the supplementarity principle), second, the use of carbon sinks, and third, an international market control system. As the US faces higher future reduction costs than the EU, it will be faced with considerably higher costs than those on which the negotiations in Kyoto were based.

It is not easy to say whether access to free trade and full access to sinks would have changed the US decision at The Hague. But it is obvious that cost issues were important for it. Thus, to encourage the US to join and participate in an international GHG emission-trading scheme, the EU must reconsider and acknowledge US claims for cheaper reduction options.

The connection between short- and long-run targets in negotiations

Without having secured a position that the short-run objectives are obtainable, it will probably also turn out that it will be impossible to achieve the long-run targets either. This is in particular true in situations where the obtainment of short-run targets is a prerequisite for progress with respect to achieving more stringent targets.

It is possible to identify situations where a high level of short-run political acceptability is a prerequisite for the long-run objectives to be achievable. Such a situation exists in cases where progress to reduce the international dispersed pollutant take place successively and where future reduction requirements are linked to past performances. The successive buildup is especially true for the negotiations over the climate change issue. Here, the first negotiations in Rio in 1992 stipulated stabilization of emissions, whereas the Kyoto agreement stipulated a total 5.2 per cent reduction by 2008–2012 compared with 1990 levels for most industrialized countries. A second round is expected to follow the Kyoto agreement after 2012, where even more stringent reduction levels are to be expected.

Considering the evidence above, it might not be unreasonable to assume that, once the negotiations break down, the process has to start all over again, resulting in a time delay. Such a situation is reproduced in Figure 6.2. The idea of this figure is that at any point in time there is an upper level of feasible achievements (regarding the level of total reduction possibilities), which is determined by, for example, the costs of compliance, the level of transnational trust, and so on. In the short run, cheap reduction options imply higher overall reduction, but high-cost options might in the long run give rise to higher reductions as high-cost instruments might spur technological progress. This is indicated in Figure 6.2 by the two upper lines that at some point in time will cross, indicating that the expensive early implementation has now implied the creation of new less polluting technologies, such that

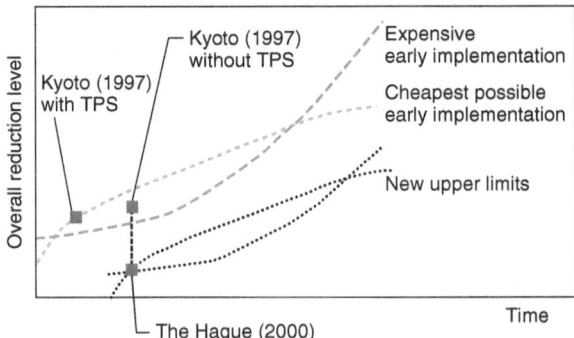

Figure 6.2 Heuristic picture of prospects for progress in climate negotiations

higher overall emissions reductions are now feasible compared with the case of cheap early implementation, but which has no significant technological progress.

However, if overly ambitious targets are stated, relative to the instruments that are supposed to be employed to achieve such targets, there will be a high risk of breakdown, resulting in a delay until new negotiations take place. Once a new agreement is renegotiated, the new upper level of feasible achievements is now at a lower level. The reason for this can be twofold, either because the level of trust between countries has diminished, or simply because it takes time before the process can be restarted. In this way, short-run choices have long-run implications.

As argued in Brandt and Svendsen (2002), the main reason for the collapse of the negotiations at The Hague were the increased costs of implementing the Kyoto Protocol, compared with the expectations that built the foundation for the Kyoto Protocol, because of restrictions on trade. This is illustrated in Figure 6.2 where the Kyoto Protocol without TPS has too high an overall reduction target, given the lack of cheap reduction options.

Broaden participation: In-kind side payments

Another issue that induced the US to leave the Kyoto Protocol was that the agreement excluded all the major developing countries. The inclusion of these countries together with a TPS would have reduced abatement costs considerably (Springer 2003).[10] In this section we examine the idea of surplus allowances allocations, which at the same time can broaden participation and lower overall reduction costs for those countries already joining the Kyoto Protocol.

According to Torvanger *et al.* (2004), one major challenge is to induce broader participation in climate policy, particularly the involvement of developing countries and the US. As clearly demonstrated in Figure 6.3, the inclusion of not only the US but also (major) developing countries is crucial for the achievement of the long run targets of the FCCC.

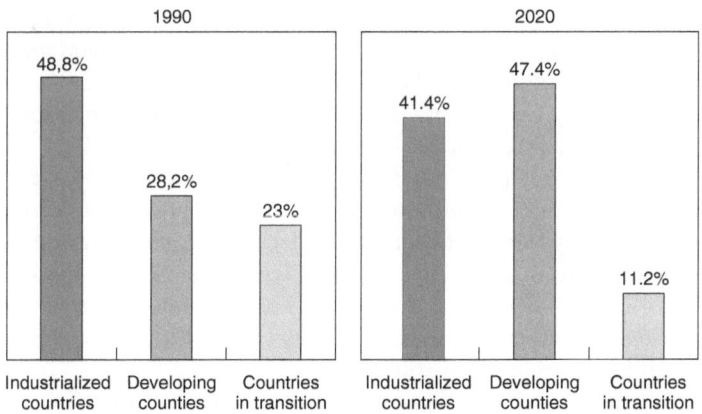

Figure 6.3 Change in relative CO_2 emissions 1990–2020

Given the idea of political feasibility, the most reasonable way to encourage countries to participate is to find a compromise that, in the best possible way, takes into account the individual countries' preferences. The problem with the developing countries is that constraints on carbon emissions also are constraints on economic growth. According to Kemfert and Kremers (2005), the participation of the developing regions would present the developed world with low cost opportunities for abatement. On the other hand, the economies of some developing regions, such as China and India, exhibit such fast growth that they are expected to be responsible for a significant part of future emissions during the next decade.

One compromise could be to use side payments to encourage participation. Such arrangements are, however, seldom seen: Mäler (1993: 27) mentions that: "it is somewhat surprising that instances where international environmental problems have been solved with the aid of side payments are very rare". He states several reasons for this observation. It goes against policy recommendations (the polluter-pays principle). The paying countries might be labelled weak negotiators, and it might create expectations of precedence. As with subsidies, it runs the risk of creating the wrong set of incentives. Although it could seem that side payments are not politically feasible, their relevance in promoting cooperation should not be ignored, in particular when countries differ greatly with respect to net gain or national income.

This brings about the notion of in-kind side payments: actions that imply the same type of changes in incentives as monetary side payments, but do not involve monetary transfers. There might be a good reason to search for less visible ways of payment to "support" countries to participate in an agreement. In Brandt and Svendsen (2005) it is argued that surplus allowances allocations (hot air) have been an important incentive to persuade Russia and other economies in transition (EIT) to participate in the Kyoto Protocol (see also Holtsmark 2003). "Hot air" means that a country's actual emissions level is lower than the number of allocated allowances. Under a TPS, where the allowances are allocated for free, the country can now sell such allowances without undertaking any abatement effort. In the Kyoto Protocol, hot air implies, for example, that Russia's actual emission level is lower than the number of grandfathered permits; such compensation can be seen as the price the international community had to pay in order to include, for example, Russia in the Kyoto agreement (Woeldman 2001: 9).[11]

The positive consequence of the inclusion of excess allowances is that it reduces the total cost of compliance and, therefore, in the best possible way supports the achievement of the stated environmental objectives. It, moreover, increases the political acceptability, by reducing the costs of compliance. It does this by lowering the permit price, and, hence, the shadow price of emission reductions (see Figure 6.8). Moreover, the presence of hot air does not erode the overall emissions reduction target (Brandt and Svensen 2005).[12]

The lesson from this is that hot air can be used as an "implicit side payment arrangement". It might be very difficult for countries that have experienced a large economic recession or countries where economic development is necessary to escape the "less developed groups of countries" to engage in costly abatement actions.[13] As the amount of hot air is linked to the size of the recession or alternatively, inversely related to economic growth, the larger the economic downturn or the less the economic growth, the more permits the country in question could sell, which in most cases would yield a higher amount of compensation for that country. The wealthier a country becomes, the fewer net transfers of resources will it receive. In this way, this proposal has a fairness component that resembles the fairness definition of Feinberg (1973), who defines fairness in the follow way: Equals should be treated equally and unequals unequally, in proportion to relevant similarities and differences.

In this way, the excess allowances are a mechanism to deal with questions of ethics. Many less developed countries have argued that the developed countries have the main responsibility for the present human-caused increase in the GHG levels. And moreover, these emissions have been a necessity as these countries develop. The less developed countries also have the right to get "unconstrained" access to cheap energy in order to develop.

However, to hand over surplus allowances allocations is neither politically uncontroversial nor easy to operationalize. The major problem is how to allocate the allowances. This question obviously deserves more attention. In principle, a route forward could be to apply a two-step approach, where the first step is used to either determine a "lowest acceptable permit price" or, alternatively, to specify an overall reduction target, and then let the parties bargain over the exact sharing rule in the second stage. The second effect of including hot air in a permit trading system is that it can be included as long as this implies a lowering of the permit price (as will be discussed later).

The conclusion is that including instruments that reduce the overall costs, such as surplus allowances allocations to broaden participation, also tends to reduce the marginal reduction costs (as the permit prices will fall) and, as a consequence, reduce the incentives to R&D and implementation of new and cleaner technologies. It might be impossible to find one instrument that, at the same time, can fulfill the objectives of broadening participation and provide the necessary incentives for R&D in cleaner technologies. Therefore, it is proposed in the penultimate section to make a separate agreement (treaty) on the development of new technologies.

Providing incentives for R&D in clean technology

Christiansen (2003: 352) states that "it is today widely recognized that technological changes and innovations are a key determinant for success or failure in climate policy, at least in the long run. The belief that technology can facilitate solutions to almost any problem, including climate change, is also shared across the USA" (Reuters 2005).[14]

This far, however, the cost of switching from fossil fuels to clean and renewable energy sources are at the present generally perceived as too high. Given the announcement of President Bush that controlling climate change must not cost one single American job, it seems impossible to increase the price of the conventional energy production to include all external costs (by for example a pigouvian tax), such that it is not likely that conventional energy production, at least in the US, will increase enough to make RES competitive in foreseeable time. Hence, an active strategy to develop new cleaner technologies in a way to make them competitive compared with conventional energy production is necessary. In this section we analyse the effort to provide incentives to develop R&D in cleaner technologies, in the efforts to make such technologies competitive.

Different policies to encourage development of new technologies

Article 2 of the Kyoto Protocol mentions explicitly the need for research on, promotion of development, and increased use of new and renewable forms of energy, of carbon dioxide sequestration technologies and of advances and innovation in environmentally sound technologies. The question is: how best to create incentives to develop new technologies, and how to increase the market penetration of renewable energy systems and trigger investment in new capacity?

One theoretical stand is to analyse which regulatory instruments yield the highest incentive to develop RES. The general lesson here is that the higher the shadow prices on emissions, the larger such incentives. The second stand is that direct incentives (or direct regulation) are the appropriate way to address this issue. Very generally, these two represent different ways of making RES competitive compared with conventional energy systems: first, by making conventional energy more costly by increasing the shadow price of emission and, second, by making RES cheaper by supporting its development in areas with a downward sloping learning curve.

Discussion of learning curve and differences in technologies

The evolution of production costs depends on the shape of the learning curve for the relevant technology (see e.g. Junginger *et al.* [2005], and references herein for learning curves for wind farms). The learning curve describes how unit costs of production change as the experience in using the technology increases. *Ex ante*, the exact shape of such a curve is uncertain. There are, however, some estimates and projections of learning curves. Hansen *et al.* (2003) estimated the learning curve for wind turbines in Denmark for 15 years (see Figure 6.4) comparing the actual with the price projected.

Although prices of energy produced by wind turbines have been falling due to "learning by doing", the projection of the trend presented in Figure 6.4, they still are not competitive (see Madsen *et al.* 2002; Sims *et al.* 2003).[15] The main reason for the downward sloping learning curve of the wind turbines is, according to Madsen *et al.* (2002), that the relationship between the size of a wind turbine and the costs/ kW is estimated and it is clearly negative.

However, the costs of producing energy by use of conventional energy systems could also change when exposed to greater pressure from competition. To get an

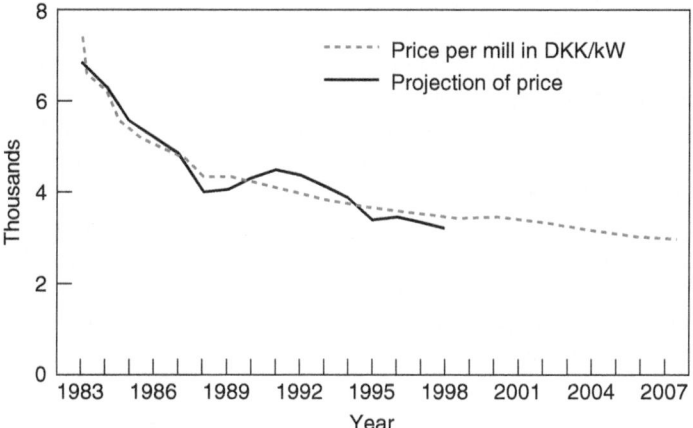

Figure 6.4 Price per mill DKK/kW, 1980 prices

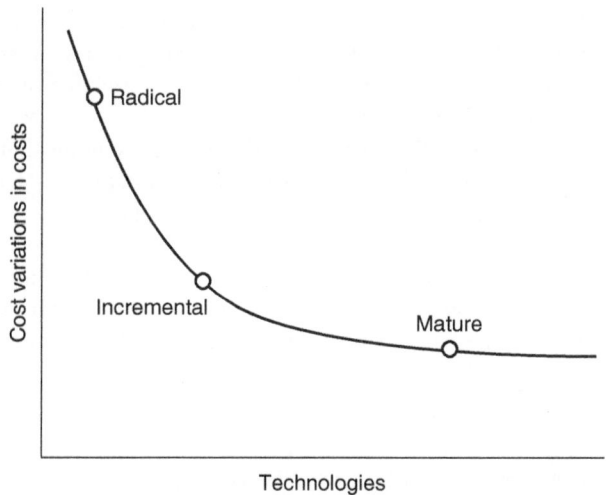

Figure 6.5 Position of technologies on learning curve[16]

idea of this, note that energy technologies reflect differences in costs and levels of development and can (as in Grübler *et al.* 1999)[17] be placed into three groups. The mature technology has received widespread usage and has well known specifications (e.g. combustion gas turbine, gas combined and conventional coal power plants).

Such technologies can be changed or improved under pressure from competition, but both the costs and the general level of energy efficiency is relative stable.

The incremental technologies have higher costs and exist in niche markets (e.g. biomass power plants, coal combustion cycle power plants, nuclear power plants and wind). They have the potential for higher efficiency and potential cost reductions if investment and development continue. The radical technologies are, by definition not widespread, but open to radical improvements in performance and costs (e.g. geothermal power plants, solar thermal power plants and PV solar) (see Grübler *et al.* 1999). Table 6.1 provides estimates of the costs of selected electricity-generating technologies.

This is important, as environmental targets change relative prices, and then also create incentives to make existing technologies more (energy) efficient. As seen from Figure 6.5, as the conventional energy-producing sector can be placed into the mature sector, costs of production will probably not change significantly as competition increases.

The dynamic cost-efficiency property, nationally and internationally

Much effort has been made in the area of environmental economics to disentangle what regulatory instrument provides the highest incentives to develop RES (an instrument that is the most dynamic incentive). Requate and Unold (2003) conclude that, in general, concerning national abatement policies, taxes provide more incentives that TPS in equilibrium.[18] Moreover, economic instruments in general provide higher incentives that direct regulation.

At the international level, things work differently. Assume that a number of countries have agreed on individual emissions reduction targets of an internationally dispersed pollutant like CO_2. Initially, let each country use a nationally cost-efficient way to achieve this target (e.g. a national emissions tax). Now let these countries engage in an international TPS. If a country has relatively high (low) marginal abatement costs, this country will be a buyer (seller) in this market, as the permit price will be below (above) the national tax level. This is illustrated in Figure 6.6.

Table 6.1 Cost of selected electricity-generating technologies[19]

Technology	US (1990)$ per KW installed capacity
Combustion gas turbine	200–700
Gas combined-cycle power plant	400–1150
Conventional coal power plant	1100–1650
Biomass power plant	950–2350
Coal combined-cycle power plant	1200–2000
Nuclear power plant	1500–2500
Wind	600–2800
Geothermal power plant	650–3950
Solar-thermal power plant	1600–3800
Solar PV	900–6150

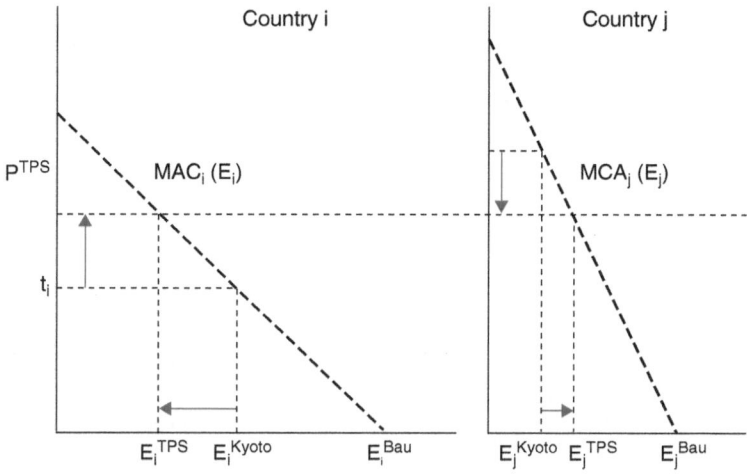

Figure 6.6 National tax vs. transnational TPS

In figure 6.7, $MAC_i(E_i)$ is the marginal abatement curve of country i, and measures the marginal costs of emission reduction compared to E_i^{BAU}, the business-as-usual emissions level in country i, while E_i^{Kyoto}, is the Kyoto emissions target and t_i is the tax on emission in country i. Finally, P^{TPS}, is the permit price in an international TPS. As country i is the low-cost country, it will receive higher incentives to develop R&D, while the high-cost country receives fewer incentives to develop RES in an international TPS compared to a national tax.

Table 6.2 presents a scenario of an estimate for the allowance price in the internal EU TPS. Note that all "old" countries experience a reduction in the shadow price on their emissions target in the TPS. Moreover, in the "new" EU countries, the emissions reduction requirement of the Kyoto Protocol is non-binding, implying a zero shadow price (the hot air issue). The result is that all of the "old" coun-

Table 6.2 CO_2 taxes under unilateral Actions and allowance price in an TPS (US$/t$CO_2$).[20]

Region	CO_2 taxes
Italy	29.2
France	19.1
Germany	19.0
UK	29.6
Scandinavia	16.5
Benelux counties	54.5
Southern Europe (except Italy)	16.5
Rest of EU	61.4
Assession countries	0.0
Allowance price in an ETS	12.9

Source: Klepper and Peterson (2004).

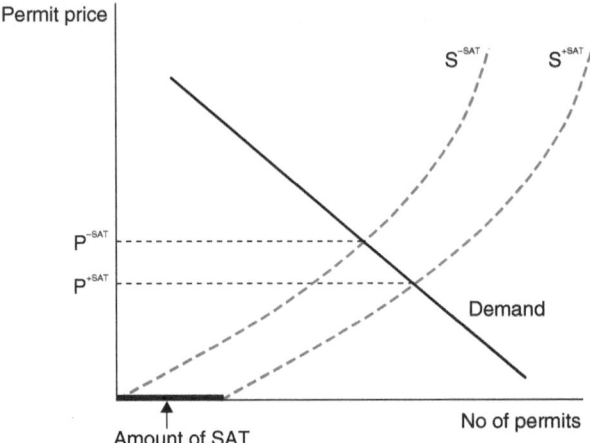

Figure 6.7 How surplus allowance trading affects permit price

tries are net buyers of allowances, and the assessing countries are net sellers. The allowance price, however, crucially depends on the availability of hot air.[21]

As also observed by Barreto and Kypreos (2004), the inclusion of surplus allowance trading will reduce the permit price even more.[22] In Figure 6.7, the last point is illustrated by a shift in the supply curve from S^{-SAT} to S^{+SAT} (where SAT means surplus allowance trading). The effect is that trade in permits increases and the equilibrium price is depressed.[23]

In addition to these general effects, both positive and negative "side effects" regarding the incentives for R&D in RES by use of an international TPS can be identified. If countries before entering the TPS did not use cost-efficient measures, the TPS will reduce incentives in most countries because of its cost efficiency property.

Moreover, positive effects of TPS also exist. Barreto and Kypreos (2004) note that because of actions of spill-overs of learning, imposing emission constraints on a given region may also affect the technology choice of other (unconstrained) regions. As a rule, the cheaper mitigation options brought about by the emissions trading mechanism produce a disincentive to deploy low-carbon technologies in permit-buying regions. But on the other hand, trade stimulates their penetration in (potentially) selling ones. This is particularly so in the case where non-Annex B regions join the trading system. Even in cases where the trade configuration strongly reduces the incentives for technology learning of low-carbon technologies in constrained Annex B regions (e.g. global emissions trade), the model still finds it cost-effective to stimulate early deployment if some global spillovers of learning provided by low-carbon technologies, although to a lower extent, are possible.

The conclusion is that, if surplus allowance allocations are allowed, a high-cost country that has not to date employed cost-efficient national measures will experience a drastic downturn in its marginal abatement cost of meeting its emissions target, and hence a large decrease in its incentives to develop and implement RES, which can only partially be outweighed by the deployment of RES to countries outside the TPS market.

Political feasibility and incentives to develop new technology

Creation of dynamic incentives by means of increasing the emitters shadow price on emissions means that new cleaner technologies are developed. Such developments reduce abatement costs and stimulate further progress. However, such thinking stems from national regulation theory, and does not consider resistance from the regulated entities or sovereignty issues in the international community. This way of thinking has been criticized in Barrett (2003: 392): "Like Montreal, Kyoto is meant to provide a 'pull' incentive for R&D. In capping emissions, Kyoto raises the cost of polluting, and so creates a demand for carbon-saving technologies, just as Montreal created a demand for CFC substitutions. The difference between the two situations … is that the cost of substituting for CFC was low. The costs of climate change mitigation will be much higher, and this matters. When the costs of supplying a global public good are high, the incentive not to participate is high and the burden on enforcement very great. If the treaty cannot support the burden, the result will be very weak incentives for the innovation and diffusion of new technologies."

As an example, if the purpose of the Kyoto Protocol was not only to secure the 5.2 per cent reduction in emissions of the Annex B countries, but also to promote incentives to develop new abatement technologies, then the best way to achieve these two objectives on the basis of the cost structure is by applying an approach that simultaneously yields low overall costs of compliance (to secure maximum political acceptability), and high marginal abatement costs of the last reduced unit (to secure maximum incentives to develop new and cleaner technologies). This reasoning is reproduced in Figure 6.8.

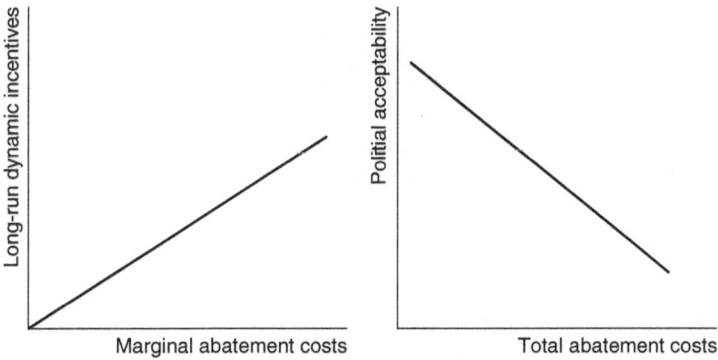

Figure 6.8 Level of reduction costs influences long-run progress

The problem with the EU policy around Kyoto was to address multiple objectives with only one instrument. It only focused on creating incentives to develop RES, but did not meet the short-run objectives, as illustrated in Figure 6.8. On the other hand, the inclusion of surplus allowance allocation will not meet the long-run targets.

The conclusion is that, in order to meet the targets, neither the US nor the EU original proposals will do, as illustrated in Figure 6.9. The TPS possible with surplus allowance allocation might be necessary to "keep the process moving in the right direction", but it will probably not provide the necessary incentives to met the long-run targets, while making the abatement effort too costly will stop progress.

Changing the design

From the discussion above, the hypothesis that the Kyoto Protocol will be unable to manage both short- and long-run objectives, is not easily rejected. Barrett (2003) questions whether the Kyoto Protocol has been structured correctly and whether a new design is needed to achieve progress.[24] Obviously, no single arrangement can deliver both the low-cost alternative and the high incentives to develop R&D. The at least as obvious answer is to think of different arrangements to deal with the different objectives. One possibility is to have more than one treaty, where one treaty concerns R&D.

The idea of an R&D protocol and a standards protocol is put forward by Barrett (2003). According to Barrett, the main advantage of such protocols is that they

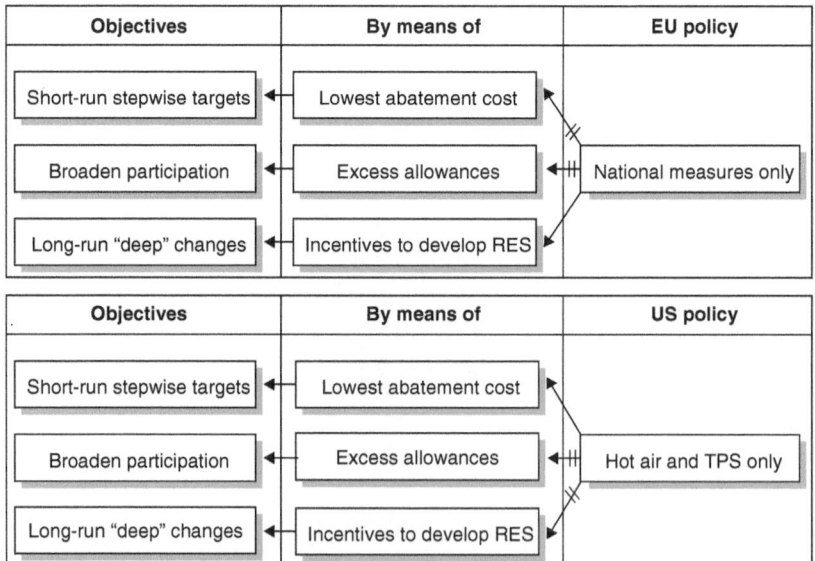

Figure 6.9 Traditional policy of EU and US

can be made self-enforcing.[25] One way of doing so is to make the financing of each country's contribution depend on an agreed total expenditure level, a share for each country determined by its circumstances (e.g. share of historic emissions) and the other countries' contribution (a system known in the literature as a matching mechanism). Such an arrangement can turn out to be self-enforcing in the context of technology development, if the abatement cost for all countries decreases sufficiently because of a positive relationship between financing, research effort and technology development.[26] In this way, a tipping point effect can be exploited, such that when a sufficient number of countries join the Protocol, the remaining will have an incentive to join (and no one has an incentive to leave the agreement).

As discussed in the previous section, there might be uncertainties attached to developing new technologies, learning curve effects might only materialize slowly, and the potential profitability of new technologies have a form of network externality inside it. To overcome such problems, a formal treaty is needed for coordination of R&D effort. An important result in Popp (2004) supports this claim. By making technology endogenous in an economic model of environmental policy (the DICE model), Popp shows that, while cost savings are significant, the effect of the induced innovated on emissions is small. The main reason, according to Popp, is the potentially crowding out of other R&D and market failures in R&D. Barrett (2003) addresses this last issue, as he points out that the coordination of R&D (e.g. to avoid the above) is the essential point of a treaty on technology.

For the US and the EU, this is a possible way of combining their efforts. While there is an ongoing disagreement over the Kyoto Protocol, there is by and large a common recognition that developments of new technologies are imperative for "real" progress (see Reuters 2005).[27]

However, technological progress will have an advantage when countries are willing to undertake real reductions, because technological progress and RES deployment typically have a network externality structure. A network externality exists when the more countries use a given technology, the larger the incentives are for the remaining countries to employ this technology as well. The reasons for this could be that when there are switching costs, buyers may be reluctant to purchase a technology that locks them in, making them orphans of a failed technology. Moreover, where there are substantial economics of scale, costs will be lower with technological specialization (Barrett 2003), as also seen in the case of wind turbines. Therefore, research might profit from the presence of countries that are willing to adapt a new technology (acting like a guinea pig). The effect could be a move down the learning curve, making the technology affordable/competitive for more countries, generating even more positive externalities.

Barrett claims that a protocol with standards will pave the way for R&D development; however, the Kyoto Protocol is now ratified, and, hence, the institutional set-up is in place. Moreover, the EU has launched the internal TPS successfully, and thus a "real cost" on GHG emissions is emerging, and it is obvious that this road should be pursued (see Tjernshaugen 2003).[28]

The idea expressed here is to have two different protocols, one a follow-up, building upon the Kyoto Protocol, with the same institutional set-up with caps

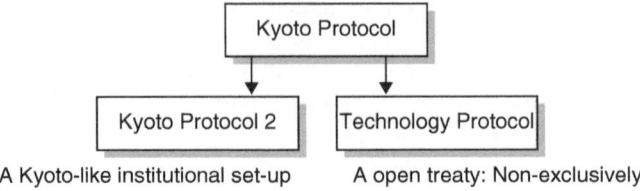

Figure 6.10 Splitting of Kyoto Protocol into two different treaties

on emissions, timetables and minimum ratification clauses (but possibly with a enlarged TPS), as illustrated in Figure 6.10. However, there should also be a parallel treaty on technology development, specifying each country's contribution depending on the other countries' total contribution, but open to any country.

Concluding Remarks

The analysis of this chapter builds on a very simple observation: in order to address *n* different objectives, at least *n* different policy instruments are needed. Figure 6.11 presents the main conclusion of this chapter.

In order to address the climate change issue, it is necessary to address three different objectives: to have continued stepwise progress, to successively broaden participation, and to end up with long-run deep targets. The policy instruments to achieve these objectives are proposed to be an unconstrained international TPS to provide low cost abatement opportunities, to apply surplus allowance allocations as "subsidies" for developing countries to cope with climate change without compromising their right to economic growth, and finally, a tailored international cooperation to develop RES and make them competitive.

Neither of these steps are easy to realize or uncontroversial. However, since the objectives seem to be unanimously agreed, the controversy will only be on how to achieve these objectives. While the grandfathered TPS is the only feasible cost

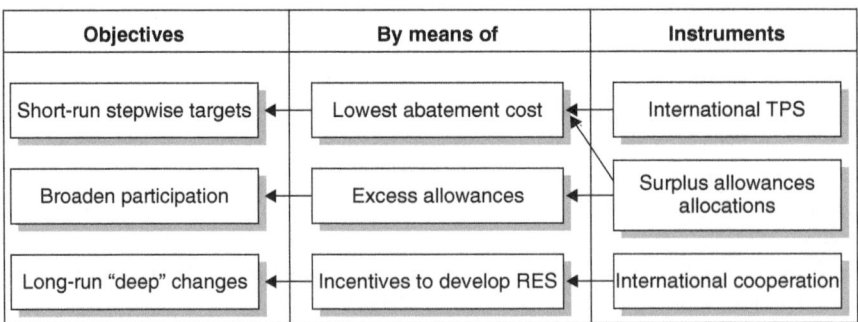

Figure 6.11 Politically feasible solution to climate change issue

efficient instrument that also provides the countries with cost abatement which is also a low cost, implementation of this seems unavoidable. There might, however, be other means of broadening participation, and there might be others ways to deal with the provision of incentives to develop RES. The main point is that these issues must be addressed in a politically feasible way, without compromising the other objectives.

Notes

1 America would have to reduce its production of carbon dioxide and the other gases that cause global warming by an estimated 30 per cent by 2012. President Bush said that would do too much harm to the American economy (BBC news, Sunday, 22 April 2001). The treaty excuses developing countries such as India and China from making cuts in greenhouse gases. Bush has said the mandatory pollution reductions would harm the US economy (*Washington Post* Staff Writer, Saturday June 2 2001: A06).

2 It is obvious that determining the overall targets and the possibility of achieving these targets cannot be treated independently. If, for a given set of overall targets, the set of politically feasible solutions is empty, the targets have to be changed, and if 'politically feasible solutions exists that can reach other (more ambitious) targets, the targets should be altered as well.

3 Westskog (2002) states that the main arguments put forward to support restrictions on trade are exertion of market power, transaction costs, the hot air issue and ethical concerns. However, Westskog (2002) largely rejects these concerns.

4 Where the parties to the FCCC (including US) agreed that the industrialized countries shall set emission limits, in specified time frames, such as 2005, 2010 and 2020, and that these should be part of an protocol, ready for signing at the end of 1997 (the birth of the Kyoto Protocol).

5 Brandt and Svendsen (2006) discussed why the EU has proposed that international measures should only be supplementary to national measures. They argue that also industrial interests could have shaped this position, e.g. in Denmark leading politicians mentioned that also increases competitiveness in renewable energy systems.

6 Country i's reduction level can be expected to be a function of this country's direct benefit from reduction (from reduced damage on own jurisdiction), its direct cost from abatement, how much the other countries reduce, the reputation (goodwill) earned internationally by acting (providing a public good), the level of expected reciprocation of other countries (positive/negative), anxiousness of losing sovereignty, the indirect gains in other issues like trade concessions, the prospect that domestic measures might spur technological advances, the preferences of the government of country i towards environmental issues, the moral aspect, personal matters (international negotiations involve face to face meetings).

7 The requirement is that countries make consistent decisions and that the institutional framework does not affect the countries' positions (that is, can be treated as exogenous as well).

8 In Newbery (1990) this view is formulated as follows: 'As economists, we have a duty to argue for cost-effective environmental policies. Inefficient policies not only achieve less than they should, but they also run the risk of alienating taxpayers and consumers'.

9 Great Britain was against common EU taxes in principle and wished to deal with the CO_2 problem on its own. Spain demanded the right to increase its CO_2 emission by 25 per cent because its industrial level was lower than that of other EU member states. Portugal and Greece argued against the CO_2 tax for the reasons that taxation could slow economic growth. Spain, Portugal, and Greece refused to accept further production costs until they reach an economic level similar to that of the more wealthy Northern EU members.

10 Springer (2003) gives a survey of 25 models of the market for TPS under the Kyoto Protocol. For global trading with CO_2, average permit price is estimated to 9 US\$/ton CO_2. For Annex B trading only, average permit price is estimated to 27 US\$/ton CO_2. In the absence of the US, aggregate permit demand from Annex B is, according the Springer (2003) more or less equal to hot air from former Soviet and Eastern Europe (depending on economic growth in Russia and other Annex B countries), with an estimated price in the range of 0–12US\$/ton CO_2.

11 Woeldman (2001: 9): "Eastern Europeans seem to consider the tradable hot air as a legitimate compensation for the emissions reductions induced by the economic decline which resulted from the deliberately established economic transition process".

12 Indeed, calculations undertaken in Brandt and Svendsen (2005) show that in case of Annex B, TPS where hot air trading is not allowed, and all Annex B countries are present in this market (including the US), then the overall reduction will be in the order of 11 per cent and not 5.2 per cent.

13 The example with the EU burden sharing rule also shows the necessity of tailoring environmental targets to meet other than purely environmental objectives.

14 In a speech in Brussels, Bush stressed the United States wants to address the problem of climate change. He has ruled out rejoining the Kyoto treaty, saying that to meet its requirements would cost millions of US jobs and billions of dollars. At his news conference, Bush said, "We care about the climate." "There's an opportunity now to work together to talk about new technologies that will help us both achieve a common objective, which is a better environment for generations to come," he said. (Source: Reuters: 17 February 2005 17:19:48 GMT).

15 The technological development has been stimulated both by the process and product innovations as the capacity of the individual mill has increased; see Madsen *et al.* (2002). For example, Sims *et al.* (2003) argue that wind based energy will not be competitive within a decade.

16 Mature technologies in widespread use have lower costs with lower variance; the costs of radical new technologies are higher and more variable. Variability of costs is also an indicator of the uncertainty of technology costs. Radical technologies are little tried.

17 See Grübler *et al.* (1999) for a very detailed discussion of the dynamics of energy technologies and a more thorough description of the different phases in the development of new technologies.

18 This conclusion, however, hinges on the assumption that the tax rate is not adjusted as new technology is introduced. If the tax rate is fully flexible, then the two instruments provide exactly the same incentives.

19 Values are taken from International Institute for Applied Systems Analysis (IIASA)'s comprehensive technology database (CO2DB). Data are for approximately ceteris paribus conditions (eg, coal plants include de-SO_x and de-NO_x equipment). Mature technologies in widespread use have lower costs with lower variance; the costs of radical new technologies are higher and more variable. Variability of costs is also an indicator of the uncertainty of technology costs. Radical technologies are little tried and their potentials for cost reductions are uncertain, and thus so are estimates of their costs.

20 Values are taken from IIASA's comprehensive technology database (CO2DB). Data are for approximately ceteris paribus conditions (eg, coal plants include de-SO_x and de-NO_x equipment). Mature technologies in widespread use have lower costs with lower variance; the costs of radical new technologies are higher and more variable. Variability of costs is also an indicator of the uncertainty of technology costs. Radical technologies are little tried and their potentials for cost reductions are uncertain, and thus so are estimates of their costs.

21 See e.g. Chen (2003), Springer (2003) and Springer and Varilek (2004).

22 Note that the issue of hot air and surplus allowance trading is only relevant in a TPS.

23 The marginal costs under free trade are equal to the permit price under free trade. Therefore, no country's marginal cost of abatement will ever be above the permit price

in equilibrium. It cannot be lower either, since then this country could reduce additional and sell with a gain. Hence, inclusion of hot air reduces the marginal costs of meeting the pre-defined emissions reduction targets.

24 It is, however, necessary to stress that the reason why Barrett (2003) claims that the Kyoto Protocol is wrongly designed is that it is not self-enforcing, whereas in this chapter it is argued that is wrongly designed in terms of lack of political feasibility. (The terms are, however, closely, related).

25 The self-enforcement property states the countries participate in an agreement only if the net benefit from participating is larger than unilaterally leave the agreement.

26 In the context of climate change, that there is such a sufficient relationship should, however, not been taken for granted.

27 Bush said, "We care about the climate." "There's an opportunity now to work together to talk about new technologies that will help us both achieve a common objective, which is a better environment for generations to come," he said. (17 February 2005 17:19:48 GMT, Source: Reuters).

28 According to Tjernshaugen (2003), it is doubtful whether UN conferences are particularly suitable to control agreements on technology. Moreover, such agreements do not demand any actions against the sources that pollute. In order to achieve this, other measures are needed, like emissions taxes or emissions permits. (Translated from Norwegian, Tjernshaugen 2003, p. 16).

References

Barreto, L. and Kypreos, S. (2004) Emissions Trading and Technology Deployment in an Energy-Systems "Bottom-Up" Model with Technological Learning, *European Journal of Operational Research*. 158. 243–261.

Barrett, S. (1998) Political Economy of the Kyoto Protocol, *Oxford Review of Economic Policy*, 14 (4). 20–39.

Barrett, S. (2003) Environment and Statecraft, the Strategy of Environmental Treaty-Making. Oxford: Oxford University Press.

Brandt, U.S. and Svendsen, G.T. (2002) Hot Air in Kyoto, Cold Air in The Hague- the Failure of Global Climate Negotiations, *Energy Policy*. 30. 1191–1199.

Brandt, U.S. and Svendsen, G.T. (2003) The Political Economy of Environmental Regulation in the EU, CO2 Trade, Grandfathering and Auction, *IME Working Paper*. 51 (03), Department of Environmental and Business Economics, University of Southern Denmark.

Brandt, U.S. and Svendsen, G.T. (2004) Fighting Windmills: The Coalition of Industrialists and Environmentalists in the Climate Change Issue, International Environmental Agreements: Politics, Law and Economics. Springer, 4 (4), December. 327–337.

Brandt, U.S. and Svendsen, G.T. (2005) Hot Air as Implicit Side Payment Arrangement: Could Hot Air Provision have saved the Kyoto Agreement, Forthcoming in *Climate Policy*.

Brandt, U.S. and Svendsen, G.T. (2006) Climate Change Negotiations and First-mover Advantages: the Case of the Wind Turbine Industry. *Energy Policy* 34. 1175–1184.

Chen, W. (2003) Carbon Quota Price and CDM Potentials after Marrakech. *Energy Policy*. 31 709–719.

Christiansen, A. C. (2003) Convergence or Divergence? Status and Prospects for the US Climate Strategy. *Climate Policy*. 3. 343–358.

Christiansen, A. C. and Wetterstad, J. (2003) The EU Frontrunner on Greenhouse Gas Emissions Trading: How did it Happen and will the EU Succeed? *Climate Policy*. 3. 3–18.

Feinberg, J. (1973) *Social Philosophy*. Englewood Cliffs, NJ: Prentice-Hall.

Grübler, A., Nakicenovic, A.N. and Victor, D.G. (1999) Dynamics of Energy Technologies and Global Change. *Energy Policy*. 27. 247–308.

Hansen, J. D., Jensen, C. and Madsen, E.S. (2003) The Establishment of the Danish Windmill Industry – Was it Worthwhile? *Review of World Economics*, 139. 324–347.

Holtsmark, B. (2003) Russian Behaviour in the Market for Permits under the Kyoto Protocol, *Climate Policy*. 3. 399–415.

Junginger, M., Faaij, A. and Turkenburg, W. C. (2005) Global Experience Curves for Wind Farms, *Energy Policy*. 33 (2). 133–150.

Kemfert, C. and Kremers, H. (2005) A Computable General Equilibrium Assessment of a Developing Country Joining an Annex B Emissions Permit Market. Forthcoming in *Environment and Development Economics*.

Madsen, Erik Strøjer, Jensen, Camilla and Hansen, Jørgen Drud (2002) Scale in Technology and Learning-by-Doing in the Windmill Industry. Working Papers 02–2, University of Aarhus, Aarhus School of Business, Department of Economics.

Mäler, K.G. (1993) Acid Rain Game II. *Beijer Discussion Paper Series*. 32. Beijer International Institute of Ecological Economics, Stockholm.

Nentjes, A., and Woeldman, E. (2000) The EU proposal on Supplementarity in International Climate Change Negotiations: Assessment and Alternatives. *ECOF Research Memorandum*. 2000 (28). Groningen, University of Groningen (RuG).

Newbery, D. M. (1990) Acid Rain. *Economic Policy*. 11. 299–346.

Organisation for Economic Co-operation and Development (OECD) (2002) Database on Environmentally Related Taxes, http://www1.oecd.org/scripts/-taxbase/queries.htm, accessed 6 July 2002.

Popp, D. (2004) ENTICE: Endogenous Technological Change in the DICE Model of Global Warming. *Journal of Environmental Economics and Management*. 48. 742–768.

Requate, T. and Unold, W. (2003) Environmental Policy Incentives to Adopt Advanced Abatement Technology: Will the True Ranking Please Stand Up? *European Economic Review*. 47. 125–146.

Sims, R.E.H, Rogner, H.-H. and Gregory, K. (2003) Carbon Emission and Mitigation Cost Comparisons between Fossil Fuel, Nuclear and Renewable Energy Resources for Electricity Generation. *Energy Policy*. 31. 1315–1326.

Springer, U. (2003) The Market for Tradable GHG Permits under the Kyoto Protocol: a Survey of Model Studies. *Energy Economics*, 25. 527–551.

Springer, U. and Varilek, M. (2004) Estimating the Price of Tradable Permits for Greenhouse Gas Emissions in 2008–2012. *Energy Policy*. 32. 611–621.

Torvanger, A., Twena, A., M. and Vevatne, J. (2004) Climate Policy beyond 2012: A Survey of Long-term Targets and Future Frameworks. *Cicero Report* 2004:02 Center for International Climate and Environmental Research, University of Oslo, Norway.

Tjernshaugen, A. (2005) United States Participation in Future Climate Agreements, an Assessment. *Cicero Policy Note* 2005: 01 (Available at http://www.cicero.uio.no).

Tjernshaugen (2003) Teknologisk alternativ til Kyoto, *Cicerone*, 6/2003. 16.

IEA (2001) World Energy Outlook 2001 Insights. Paris: International Energy Agency.

Westskog, H (2002) Why should Emissions Trading be Restricted. *Climate Policy*. 2. 97–103.

Woeldman, E. (2001) Limiting Emissions Trading: The EU Proposal on Supplementarity, Draft ECOF Research Memorandum, 2nd updated version, January 2001. Groningen, University of Groningen (RuG).

7 Between Two Giants

Lessons from the Russian Policy on the Kyoto Protocol

Vasily Sokolov

Introduction

The purpose of this chapter is to introduce the Russian national context in the international negotiations on climate change at the time of the 2009 Climate Conference in Copenhagen, in order to present clarification and sometimes explanation of the origins and scale of the stumbling blocks in the climate talks. Given this purpose, the chapter outlines the evolution of climate change issues on the national agenda, the interests and interactions of national actors involved in climate change management, and the manifestation of national processes in the international negotiations. Going beyond the limits of the position of Russian delegation in the climate talks, the chapter also gives the broader context of goals, tactics and constraints facing the negotiators.

In the management of global environmental risks, and climate change in particular, the Russian case highlights certain specific features which determine the position of the country, especially in the new international community which has emerged on the territory of the former Soviet Union. The case also displays a combination of high scientific interest in the global physical processes and low interest and capability to deal with such risks, on the part of social institutions inherited from the USSR.

The largest country in the world with visible geopolitical ambitions, and probably also with the largest regional differences, could not be ignored as a one of major players in the international negotiations and management of global environmental risks. The understanding of all deficiencies and also the positive sides of global risks management processes in Russia are important for the understanding of related trends in some other parts of the world.

The Russian participation in the Kyoto Protocol (KP) negotiations shows the dynamics of national interests in coping with the global climate problem, depending on the emergence or disappearance of specific opportunities linked to international regulations. The Russian case demonstrates also the deep interrelation of international negotiations with the changing position and interaction of national actors involved in the policy and goals formulation relevant to climate change. Finally, it draws attention to the existing interlinks between simultaneous negotiations and stumbling blocks, whose origins are often are found in areas outside the formal agenda of the negotiation.

The critical time for all these processes and their contradictions is the early 2000s, when Russia could determine the fate of the KP through its ratification or non-ratification, being in a position between two major blocs of international interests, led by the US and European Union, respectively. The usual tradition of complying with the negotiation processes suddenly faced a new situation, in which Russia could play its own "game" based on national interests and opportunities. "The freedom of choice" to support one of two major powers generated unprecedented domestic debate.

Background

Until its collapse in 1991, the USSR, the predecessor of Russia, played a special and visible role in the international efforts to manage global environmental risks. Despite the existence of strong ideological constraints, political isolationism, and economic self-sufficiency, the USSR clearly manifested interest in participating in the international efforts to combat global environmental risks and in creating the institutional, scientific and technological means for appropriate national actions.

Risk assessment efforts were based on rather well-developed scientific and monitoring networks inside the country and on linkages with international networks. The risks response actions were developed and transformed into national goals and strategy as a result of international obligations adopted with the signing of relevant conventions and agreements. Furthermore, Soviet science made a number of valuable contributions to the international understanding of global environmental risks.

A group of professionals from both the scientific community and the government played an exceptional role in the development of a global risks concept and in national policy formulation regarding risk prevention. Amongst them were M. Budyko, E. Fedorov, Y. Izrael, K. Kondratiev, G. Golitsyn and several others (Fedorov 1972; Budyko and Vinnikov 1976; Kondratiev and Moskalenko 1984; Golitsyn 1986; Izrael and Budyko 1987). For example, at the very beginning of the 1960s systematic observations were started which provided a decade later the first quantitative forecast for average temperature rise in the Northern Hemisphere (Inadvertent Climate Modification Report 1971). This forecast has been proven by real temperature measurements with data gathered throughout 1985. Considerable input had been made by Soviet scientists to the IPCC process.

The most common views on climate change impact among Russian scientists reflects higher expectations with regard to agricultural production, due to a longer vegetation growing period in Russia, counterbalanced by some uncertainties or even expected negative effects from adaptation to the adjustments in the agriculture structure and new agricultural plants selection resulting from climate changes (Golitsyn 1989).

However, scientific efforts were relatively ineffective in raising official or public concern about climate change. National implementation was successful only in these areas where environmental goals coincided with current economic goals. In all other areas, the implementation was retarded or even phased out due to funding shortages, technological constraints, and the absence of enforcement mechanisms. The major causes of this contradiction were the country's societal mechanisms. The highly centralized and bureaucratic state institutions were incapable of fully using and mastering scientific innovations and ideas. The decision-making process was isolated from broad public participation, and the immaturity of public organizations and the restricted role of the mass media meant that there was no strong connecting chain between experts and the public.

Explaining the evolution of the national management process

Several important factors should be taken into account when explaining the low concern about climate change risks in Russian/Soviet society until the time when the Kyoto Protocol entered the national debate. One factor was the uncertainty surrounding global risks; even in the scientific community, there were (and are) contradictory views on global changes and their effects. Second, there was a large gap between the natural and social sciences when it came to the management of global environmental risks. In the existing social sciences literature, one could at the time find a relatively low number references to climate change. Third, the dominance of economic development priorities meant that environmental goals had to wait – a view especially characteristic of the industrial actor and the government, facing more pressing problems of economic and social survival. This strategy of "waiting" for "matured" decisions in the international area determined much the Russian position in the international negotiations under consideration in our analysis.

Among a number of other factors at work, it is possible to isolate the monopolization of these issues by the state agency Hydromet (State Committee for Hydrometeorology and Environmental Monitoring) until the late 1980s, and the manner in which the discussion of these issues was nested within the Soviet government's broader foreign policy agenda. The evolution of efforts to understand and manage international atmospheric problems in Russia was shaped to a considerable degree by the manner in which responsibility for these issues was institutionalized. This model had clear implications for the post-Soviet period.

Hydromet, with long standing traditions and many prominent scientists (incorporated in the government structure), was nominated in 1972 as "lead agency" responsible for coordinating research on atmospheric pollution problems, environmental monitoring, and the USSR's national and international environmental policies. A single governmental body was the central and dominant player in the field, providing both membership and expertise for subsequent international negotiations on global environmental risks (Union of Soviet Socialist Republics 1986).

Hydromet concentrated both scientific and policymaking functions within its sphere of influence. It not only supervised and directed the scientific activities performed by an extensive network of specialized research institutions, but also sought to coordinate related scientific activities within the various institutions of the Academy of Sciences and universities. Hydromet was also responsible for coordination of the USSR's international activities – including both scientific discussions and intergovernmental negotiations. In effect, it provided a "full cycle" of risk management: not only the direction of scientific research and environmental monitoring, but also policy development, the formation of national goals and strategies, and coordination and control over the implementation of them.

All atmospheric risks were thus "processed" by a relatively small group of individuals who were members simultaneously of both scientific teams and the government. For example, Y. Izrael, the head of Hydromet from 1974 to 1991, remained closely involved in the direction of atmospheric research, co-authoring a number of important books and articles on the subject. The same individuals were also involved in international negotiations on global environmental risks. Therefore, the interaction of scientists with decision-makers was quite direct, since practically the same people were involved in the process of scientific research as well as in decision making and implementation.

Since the main directions and conclusions of scientific research were determined by a relatively small group of prominent senior scientists grouped in and around Hydromet, the manner in which the climate change issue was framed and assessed depended heavily upon these individuals' interests and expectations. This explains the delay in placing the issue on the national agenda; this was a function not only of scientific uncertainty, but also of an earlier formed opinion that global warming was likely to be a favourable phenomenon, at least with regard to some dimensions or impacts. In addition, the leaders of Hydromet sought to protect their bureaucratic control over the discussion of global atmospheric problems by depoliticizing them. They did so by framing the issues in purely scientific terms wherever possible. The nonscientific, policy-oriented dimensions of climate change were consequently brought onto the national agenda primarily through the country's participation in the negotiation on international environmental conventions. Only under the influence of international discussions did the Soviet debate over global warming move from abstract theoretical discussion to the consideration of practical policy options.

Hydromet's monopolization of international atmospheric issues came to an abrupt end during "the perestroika" with the creation of Goskompriroda (later Minpriroda). This step was the culmination of long-brewing dissatisfaction with the poor performance of the country's various and sundry regulatory agencies, and had been advocated for many years by the top national environmental specialists, including Izrael and his colleagues, who had assumed that any such centralization would take place under the aegis of Hydromet. In the event, Izrael and Hydromet were the biggest losers in the bureaucratic reshuffling, as Hydromet was stripped of all but its atmospheric monitoring functions, and responsibility for international environmental affairs was transferred to the new state committee.

Since the beginning of the 1990s, the Minpriroda became a major player in the management of climate change, participating at the same time as the governmental representative in international negotiations (Postanovlenie Soveta Ministrov 1989). It was also a period of rapid and far-reaching political change in the country. The result was a proliferation of new actors and channels of influence, ranging from the formation of local environmental organizations to a virtual explosion in transnational scientific and nongovernmental contacts. Minpriroda remained the most visible player in the formulation of the Russian approach to the management of global risks, but never again would it be able to monopolize those issues as Hydromet had.

The multisectoral interests within the climate change debate were brought also by the creation of the Interagency Commission on Climate Change, reformed by the Russian Government in April 1994 in response to the adoption of the UN Framework Convention on Climate Change (FCCC). Such a commission, comprising representatives from academia, government and industry, had large power in the decision-making process: appropriate resolutions were mandatory for the executive branch. The FCCC was ratified by the Russian Parliament on 25 October 25 1994 and signed by the President on 20 November 1994, making Russia one of the first signers of the Convention.

The transitions of national interests

Looking at the history of Russian/Soviet participation in the international environmental negotiations, two important conclusions can be drawn regarding the process. First, the political leadership's favourable attitude to participation in the appropriate negotiations can be explained by international political interests, but also by the fact that, for a long period of Russian environmental negotiations, the basic component and dominant focus was mainly professional academic-based. The political "supervision" of the process was mostly based on a "no-harm" policy, which meant evaluation of political and mainly economic impacts of new commitments resulting from the process, to check whether they could lead to disadvantages.

The Kyoto Protocol has dramatically changed this model, especially in the years when the protocol became a focal point for the national political debate. At least two critical areas can be outlined in this context.

The first was the emergence of potential economic benefits resulting from international regulation of climate change. The critical issues were the schemes of international emissions trading and joint implementation that were incorporated in the Protocol's provisions.

In face of this opportunity, the FCCC was taken out from the routine processing of international environmental issues for deeper consideration and finally given higher priority in the national agenda. New national actors with specific interests became involved in the national debate. The business sector and growing NGO groups joined the national discussions, and mass media started to reflect the issue as "a number one" on the environmental agenda.

The second critical factor emerged somewhat later, at the beginning of the year 2000, when Russia was in a position to take the responsible step to ratify the Kyoto Protocol, thereby determining its fate. However, preoccupied by internal problems of economic recovery, by the establishment of democratic structures, and by the need to formulate the new list of national priorities, Russia could hardly to take a leader role in the Kyoto process.

Still, after the formal withdrawal of the US from the Kyoto Protocol in 2001, Russia became a critical actor in the implementation process; being responsible for 17 per cent of the total greenhouse emissions, Russia made an important contribution to the 55 per cent of all emissions that were required to put the Protocol into operation. It was this unprecedented situation that enabled the Russian negotiators to determine the fate of the Kyoto process.

The ups and downs in the top-level political interest in the KP issue can be traced during the last few years. Soon after the US withdrawal from the Kyoto commitments in 2001, President Putin, hosting German Chancellor Schroeder, criticized the US decision, hence supporting the European position on the issue. Facing the fact that, for the first time in Russian political history, climate change was addressed at the highest level, some expected fast action regarding KP ratification.

However, the issue then practically disappeared from the top leadership' agenda, generating many doubts about Russia's actions, first of all in the EU, which expected solidarity with Russia on the climate issue.

From this time, the EU became a source of political pressure on Russia for the purpose of putting the KP into force. In March 2002 a special delegation of the European Union, including the ministers of environment for Greece and Italy and the EU Commissioner for the Environment came to meet the Russian Government and Parliament in an attempt to persuade them both to ratify the KP. In September 2002 a visit to Moscow of members of European Parliament had the same goal, as well as visits from other countries (e.g. Canada).

The outcome of these actions undertaken by EU seemed to be successful. At the World Summit on Sustainable Development (WSSD) in South Africa in 2002, the then Russian Prime Minister Kasyanov declared that Russia would ratify the Kyoto Protocol "in the nearest future". However, in reality the politicizing of the climate issue played a negative role in the Russian positioning on ratification of the Protocol. This was soon demonstrated at the World Climate Change Conference, specially convened in Moscow on September 2003. In his welcoming speech at the opening of the Conference President Putin said: "The Russian Government carefully considers the issue and all the complex problems linked to this problem. [A] [d]ecision will be made after the work is completed [...] of course, taking into account Russia's national interest" (Office of the Russian President 2003). The message was quite clear – Russia had some interest in joining the Protocol, for instance, to profit from regulations benefits, but not because of external pressure.

The transition of Russia's interests in the KP process was clearly connected to the critical issues described above, and led to a change from large expectations regarding the benefits from international regulations to a more realistic approach based on the assessment of the Kyoto commitments' impact on the country's development. In trying to counterbalance these contradictory interests, the political leadership left the climate issue to a public debate, which attracted many representatives from academia, business, government and non-government sectors.

For the first time in the history of Russian/Soviet international environmental activity, the Russian position started to be an important domestic political issue, which generated difficult discussions among national actors.

Three dimensions are usually taken into consideration in discussions on national interests in the Kyoto process – environmental, economic and political factors. In addition, the direct political pressure from the EU and the elevation of the climate issue on the international agenda generated suspicion among environmentalists because the estimations of the expected reduction of greenhouse gases showed only minor positive effects.

On the economic side, the benefits for Russian Federation from the Kyoto Protocol ratification were also considered to be illusionary. The major motivation for Russian participation in the Kyoto regime – emissions trading – had almost disappeared because of slow and unclear implementation mechanisms.

The last dimension of national interest – the political one – namely, to implement a sound international environmental policy, probably played a major role in the final decision on ratification taken in October 2004. However, the absenteeism of major global emitters – the US, first of all, but also China and India – always served as an argument that the political consequences of non-ratification would not be completely negative.

New arguments were brought into the discussion when the Russian political leadership in 2003 proclaimed an economic goal of doubling GNP in the coming ten years. According to calculations made by some members of the Academy of Sciences, the annual economic growth of 7.2 per cent (corresponding with the new political goal) would bring the country's emissions of CO_2 to the level of 1990 by the year 2009 (Russian Academy of Sciences 2004).

Two major results could emerge from this situation – 1) Russia could lose its potential for sales of emission quotas; and 2) economic growth would be burdened by the costs of emission reduction. Larger expectations for economic growth would face the problem at an earlier stage.[1]

As a result of the public debate, two major groups were formed. Both groups were trying to influence the governmental decision on KP ratification. An "anti-Kyoto" group was most actively supported by Y. Izrael, who found a new and strong partner – the President's advisor on economy, A. Illarionov, acting in opposition on some issues to the Ministry of Economic Development and International Trade, whose representatives favoured KP ratification. For Y. Izrael, full-member of Academy of Sciences and IPCC Board member, the debate was an opportunity to regain political visibility. His position was supported by some Russian academicians and politicians. On the President's initiative, the Scientific Council workshop within the Russian Academy of Sciences started its work in January 2004. Izrael and Illarionov were the most active members of this Council, explaining why the first outcome of the workshop exhibited a negative attitude to the Kyoto Protocol.

The major conclusions of the "anti-Kyoto" group were based on the following arguments: 1) KP has no scientific justification and does not meet the goals of FCCC; 2) global warming along with negative effects (permafrost melting) has some positive effects for a cold country like Russia (heating, biomass increase); 3) Russia has some serious economic constraints in meeting the first phase of KP; and 4) the carbon budget and flow need more investigation before any significant action can be taken at the international level (Yzrael 2002).

Among other arguments put forward by the "anti-Kyoto" group was Russia's CO_2 reductions during the 1990s had resulted in an industrial decline in Russia that was equal to 40 per cent of the increase in CO_2 emissions by other countries. These figures would show the Russian contribution to emission stabilization during the previous decade.

Sink credits were also in the focus of this group. Russia is among the biggest exporters of natural gas, contributing to the carbon intensity reduction of many European countries – this situation was not taken into account in the climate negotiations, furthermore, the Russian forests were not counted in the calculation of the total carbon balance. Finally, as a Northern country with some extreme regions, Russia needed to be treated separately in the process of goal formulation in the context of the climate talks.

The "pro-Kyoto" group was diversified, composed of representatives of academia, as well as of the governmental and non-governmental sectors. V. Danilov-Danilyan, former Minister for Minpriroda, G. Golitsyn, full member of the Academy of Sciences, A.Yablokov, former Advisor to the President on the Environment, and A. Kokorin from the World Wide Fund for Nature (WWF) were among the most visible in mass media reports.

The group's arguments had different directions: 1) Climate change has a visible anthropogenic genesis – humankind should be prepared for upcoming changes; 2) the greenhouse effect is a common hypothesis, but the processes of climate change are not quite determined; 3) natural anomalies resulting from climate change will not bring benefits to any country in the foreseeable future, including the Northern region; and 4) the costs of non-action could be much higher than projected costs of actions (Danilov-Danilyan 2003; Kokorin *et al.* 2004).

New actors, mainly business people and governmental economists, came to the forefront when Russia gained a potential key role in the KP process. A missed opportunity to trade quotas in case of non-ratification had raised some concern: "The science can debate the impact of greenhouse gases on climate warming for more 20 years, but we want to trade quotas now". These words by M. Delyagin, Director of the Institute for Globalization Research, were indicative of that kind of worry (Pas'ko 2004).

The other concern of the new actors in the climate debate was related to an attempt to link the KP issue to other international challenges facing Russia, or the need to have more investments in the energy sector; one particular issue was how to trade the ratification of the Kyoto Protocol with membership of the World Trade Organization (WTO).

The public debate was unexpectedly brought to an end by the decision made by the Russian Government and Office of the President to introduce the bill on KP ratification in October 2004 to the Russian Parliament. The bill was approved by both chambers in a very short time and was signed by the President on 5 November 2004. This was the final step in bringing the KP into force in 16 February 2005.

Climate negotiations: Searching for a niche

The evolution of the Russian positioning in the climate talks reflects changing priorities at the national level, as described above. Some elements of this evolution are of interest in an analysis of stumbling blocks and facilitation opportunities in those climate talks.

The Russian performance in the climate negotiation was never strongly active or aggressive, due to a relatively low priority given to this issue by Russian society. In addition, the limited circle of professionals participating in the international negotiations (mostly Hydromet and Minpriroda) did not represent the full spectrum of national interests – traditionally, the issue was considered to be an environmental issue with some economic and political implications. The role of Hydromet and Minpriroda, as main participants in the delegation, tended to cover mostly the scientific issues, while broad economic issues of economic were considered to be subsidiary in character.

The absence of clear political and economic goals in the international negotiations before Kyoto practically deprived the Russian delegation from the opportunity to have a "special niche" in these talks, allowing delegates to reflect on national interests. This situation partly explains the later association of the Russian delegation to a loosely aligned Umbrella Group including Australia, Canada, Japan, New Zealand, Russia, Iceland, Norway, and Ukraine.

Adaptation of KP at COP3 had put more light on the climate change issue in Russian society, bringing it in from the periphery of national priorities. The strategy of the Russian negotiators during the COP meeting in Kyoto was, however, based on inertia and caution inherited from previous experiences. Just a reminder – the Europeans suggested emissions be cut by 15 per cent. The Russian position was a quite modest 3 per cent goal. According to some analysts, the final resolution of 5.2 per cent meant big losses in Russian future benefits as a major seller of quotas, estimated to be in the range of US$ 50–70 billion (*Rossiiskaya Gazeta* 2001).

The successful result of determining the degree of stabilization of GHG emissions for 2008–2012 at the level of 1990 emissions can be considered a justified and well-argued position for the Russian negotiators. Dramatic decline in industrial activities, and consequently in emissions since 1990, made Russia the largest potential seller of GHG quotas, at the same time giving more flexibility in achieving emission reduction goals.

The reciprocal solidarity of major East and Central European countries during COP3 on fixing the 1990 level of emissions as a baseline has never emerged again during the climate talks because of the dramatic political transformations in the former USSR and Central Europe. That was the second reason for Russia for joining the Umbrella Group.

The major challenge and contradiction of Russian positioning in the climate talks came however from the divide that occurred between two major players – the

US and EU, at the time representing the world's two largest emitters. According to the International Energy Agency, at the end of 1999, energy related emissions in the United States and Europe amounted to 5,585 million tonnes (Mt) and 3,534 Mt, respectively. It can be noted that the main issues for the G77/China group had no real impact on the Russian position. For instance, Russia had a "low voice" in negotiating such issues as the Clean Development Mechanisms, Technology Transfer and use of international funding for climate change measures.

The hard choice of cooperating with one of two giants in climate politics was always considered in the context of two critical areas, described earlier, which formed the framework for Russian negotiators.

International emissions trading and joint implementation mechanisms allowed by protocol provisions represented a strong motivation for Russian engagement in the climate talks. This situation explains certain technical approaches proposed by the Russian delegation, particularly to consolidate all emissions from each Annex I country into an aggregate "carbon dioxide equivalent" and a "three gases basket". In other words, the "big bubble" approach could allow such a big emitter and seller of quotas as Russia to enter the world market of emission quotas without much interference in measures undertaken at the national level. From here came another proposal of the Russian delegation at COP3 – policy measures guidelines should be based on national efforts, allowing more flexibility, especially for a country with an economy in transition.

In some way this position corresponds well with the position of the US delegation, which favoured a target based on all GHGs, sources and sinks, flexibility, and no restrictions on emission trading, contradicting the EU emphasis on domestic measures which, according to the Russian delegation, led to constraints in the world emissions market.

The emerging divide between the European position and Russian interests became more evident at COP6. The talks in The Hague in 2000 floundered, owing to disagreement between the European Union and the US concerning the role of the so-called "flexibility mechanisms". The rules for emissions trading were not developed enough; an attempt to fill this gap in The Hague failed, mainly because European countries prioritized domestic efforts while the American delegation wanted a free choice approach; to cut emissions domestically or to buy quotas abroad. In fact the European restrictions on free trading of emissions could halve the benefits for Russia.

A correspondence of the Russian and the US positions is discernible also in their appeal for the meaningful participation of key developing countries in the climate talks. Particularly, at COP4 in Buenos Aires, the Russian delegation referred to an estimation demonstrating that, within the next 15 years, emissions from non-Annex I nations would exceed those of Annex I countries. At the same meeting the US stated that efforts by the developed countries would be insufficient for achieving the FCCC goals. For this reason, both Russia and the United States supported an Argentine voluntary commitment announced in Buenos Aires. The voluntary commitment of key developing countries was reciprocated by several developed countries in the Umbrella Group.

However, some American proposals contradicted Russian interests. For example, the US position on the legitimacy of emission trading before 2008 (the year in which national commitments were established) undermined the Russian interest in pursuing a favorable process of ratification. A turning point in the Russian position can be traced back to this time. In addition, the US withdrawal from the Kyoto Protocol drastically reduced prospects in Russia regarding economic benefits from emissions trading because a major buyer of quotas had disappeared from the global market.

These changed expectations explain the background to Russia's decision on KP ratification, motivated by predictions that two major players – the US and EU – would develop the common strategy in the future.

Looking at the future implementation phase, Russia has the potential to comply with the KP goals, despite the fact that this country is known as a country with an energy intensive economy, exceeding the energy use per unit of production by 9–10 times in comparison with some industrial economies. During the economic crisis of 1990–1998, the carbon intensity of the Russian GNP was increasing on average by 1.4–2.7 per cent per year, while economic growth in 1998–2003 showed an annual decrease of Russian carbon intensity of GNP by 4.3–5.1 per cent (Institut Ekonomicheskoko Analiza 2004). This rate of decrease was projected until the year 2012. Economic growth in Russia is associated with the improvement of energy efficiency.

Evaluating the potential for emissions growth, it should be noted that Russia does not expect population growth, and so heating will remain at the same level. Outdated heating equipment will gradually be modernized, which may lead to emissions reductions. The emissions in the oil and gas industries, constituting the major sector of the Russian economy, are quite limited. The heavy industries associated with large emissions have no growth prospects.

The major targets for Russia in future negotiations are conditioned by the need to take into account full carbon flows, particularly all sinks associated with forests and natural gas exports. The development of a workable system of emission trading and a joint implementation scheme is still of special interest in Russia.

Beyond ratification

The act of ratification made the climate change issue a practical enforcement question, despite the pressure from the "anti-Kyoto" group. In an interview reported in the *Izvestia* newspaper, the President's advisor A. Illarionov warned that the Kyoto provisions concerning enforcement would lead to a reduction of Russia's economic potential by 70–80 per cent. "The only way to reduce emissions is to stop the energy use", stated A. Illarionov. A Russian government plan emphasized energy efficiency measures along with a diversification of energy producers. It was expected that the efficiency of electricity production would increase by 8 per cent during the period 2004–2008. It was predicted that the number of independent natural gas producers would double by 2008, putting the so called natural monopoly "Gasprom" in a position to compete with an increasing number of producers.

In other words, the KP is expected to be used as an additional tool to achieve energy efficiency and economic restructuring goals.

In principle the KP targets add a new dimension to the fundamental dilemma for future energy development in Russia, similar to that of many other countries – to provide energy supplies through the expansion of energy systems or to regulate the energy demand through energy saving and increasing energy efficiency.

The bulk of legislative issues should be solved in the process of KP enforcement – mechanisms for state regulation of emissions and sinks, quotas for regions and individual enterprises, property rights regarding emission reductions, their certification and registration, and finally, the creation of a domestic market for emission reductions.

At international level, the major promises for Russia are still within emissions trade and joint implementation. The scale of such promises were, however, rescheduled to the level of specific industrial facilities and enterprises, as a resulting of a more pragmatic and practical approach. For example, the Danish Environmental Agency is investing around 20 million Euros in the modernization of two power stations in Russia (in the Khabarovsk and Orenburg regions) which will eventually allow a subtraction from Denmark's emissions account of 1.25 million tons of CO_2 per year. The major electricity producing company United Electric Systems of Russia announced recently its plans for 30 projects within the Kyoto provisions that will allow emissions cuts by 20 million tons per year. The "hot air" trade is still a quite elusive and non-practical approach, mainly because of the absence of major buyers on the world market.

The large differences in the estimates of Russia's potential in international emissions trade (from US$ 4 billion to US$ 250 million for 2012) and risks of sanctions for KP violations in the Russian heavy industries (iron and steel, non-ferrous, construction) bring new uncertainties about the future of the Kyoto process in Russia. The public debate around the Kyoto benefits and risks clearly shows that Russia's participation in the Kyoto talks after 2012 depends heavily on the progress and efficiency of the international negotiations and on the problem of achieving a compromise between two major players – the US and the EU.

Note

1 It should be noted that the other models showed that the emissions level of 1990 could be reached only after the year of 2020.

References

Budyko, M. and Vinnikov, K. (1976) Globalnoe poteplenie (Global warming), *Meteorologia i gidrologia* 7. 16–26.

Danilov-Danilyan, V. (ed.) (2003) Klimaticheskie izmenenia (Climatic change): vzglyad iz Rossii, Moscow: TEIS.

Fedorov, E. (1972) Vzaimodeisvie obschestva i prirody (Interaction between society and nature), in Izrael, Y. and Budyko, M. (eds.) *Antropogennoe izmenenie klimata*. Leningad: Gidrometeoizdat.

Golitsyn, G. (1986) Izmenenie klimata v **XX** i **XXI** stoletiyah (Climate change in **XX** and **XXI** centuries). *Izvestia* AN SSSR, t.22, 12. 1235–1249.

Golitsyn, G. (1989) Klimat i prioritety hoziastvovania (Climate and management priorities), *Kommunist*. 97–105.

Inadvertent Climate Modification Report (1971) Study of Man's Impact on Climate, Cambridge, MA: MIT Press.

Institut Ekonomicheskoko Analiza (Institute of Economic Analysis) (2004) Ekonomicheskie posledstviya vozmozhnoi ratifikatsii Rossiiskoi Federatsiei Kiotskogo protokola (The economic consequences of mutual ratification of the Kyoto Protocol for the Russian Federation), Moscow (unpublished report).

Izrael, Y. (2002) Kiotskii protocol – problemy ego ratifikatsii, *Meteorologia i gidrlogia*. 11. 7–13.

Izrael, Y. and Budyko, M. (eds.) (1987) Antropogennoe izmenenie klimata (Anthropogenic change of climate). Leningad: Gidrometeoizdat.

Kokorin, A., Gritsevich I. and Safonov G. (2004) Izmenenie klimata i Kiotskii protocol – realii i prakticheskie vozmozhnosti (Climate change and the Kyoto Protocol – the realities and practical features). Moscow: WWF-Russia.

Kondratiev, K. and Moskalenko, N. (1984) *Parnikovyi effect atmosfery i klimat* (The atmospheric green house effect and the climate). In *Itogi nauki i techniki*. Moscow: VINITI.

Pas'ko, G. (2004) Kiotskii Protocol: Vzglyad Pragmatika (The Kyoto Protocol: the view of the pragmatist), *Bellona: energia i klimat*, 26 February.

Postanovlenie Soveta Ministrov (Resolution of the Council of Ministers) (1989) O predotvraschenii otritsatel'nyh posledstvii izmenenia klimata dlya narodnogo khoziastva (On preventing the negative effects of climate change on the economic situation), 18 May, N 413.

Prezident Rossii (Office of the Russian President) (2003) Ofitsial'nyi sait (Official website), 29 September 2003.

Russian Academy of Sciences (2004) On Possible Anthropogenic Climate Change and on the Issue of Kyoto Protocol, Council Workshop, Press release, 14 May 2004.

Union of Soviet Socialist Republics (1986) Ob okhrane okruzha, ushche, sredy: Sbornik dokumentov partii i pravitelstva 1917–1985gg (Environmental protection: a collection of documents of the party and government 1917–1985), Moscow: Izdatelstvo politichesko literatury.

8 Leadership and Climate Talks

Historical Lessons in Agenda Setting[1]

Steinar Andresen

Introduction: Purpose and scope

One of the main purposes of this chapter is to draw lessons from past negotiations on climate change talks regarding the role of leadership in this area for the future. The chapter focuses on one complete cycle of the negotiation, starting with pre-negotiations in the 1980s and the post-negotiation that began after the entry into force of the 1997 Kyoto Protocol in 2005.

The year 2005 was a milestone for the climate negotiations, in the sense that the Kyoto Protocol came into force almost seven years after its adoption. Nevertheless, after COP10 in Buenos Aires (2004), frustration and pessimism came to characterize the situation. Progress was very limited, particularly in relation to the post-2012 period. Not only was the US unwilling to discuss this issue, so too were most members of the G77. None seemed willing to take on any specific commitment in the foreseeable future. After tough negotiations, the Parties were only able to agree on government experts arranging a Seminar in 2005 to discuss the future climate regime. Considering the thousands of negotiators and observers assembled in Buenos Aires, this did not augur well for the future of the climate regime. Looking at the situation at the present time, things have not changed much. Each COP meeting has produced some results like, for instance, the Action Plan at COP13 at Bali. However, optimism does not prevail because the scientific facts offer a different picture. The lack of optimism is determined by the huge discrepancy between how much emission reduction should and must be achieved to stabilize GHGs and the extent of the political willingness of countries – both developed and developing – to commit. There is still a long way to go before a viable, equitable, and sustainable solution to the climate problem can be attained. On a more positive note, there were efforts to bring the process forward in more exclusive forums, in high-level meetings between the EU and the US, as well as within the framework of the G8. The question was: Is there any way that the stalemate can be broken? Is there a magic "silver bullet" out there somewhere? In spite of the scientific, political and institutional energy mobilized by states, non-state actors and individuals for almost two decades of informal and formal negotiations, there did not seem to be any quick fixes. This situation, which remains with us, means therefore the departure point of this chapter is that there is reason to have modest expectations as to the future effectiveness of the climate regime.

The focus here is whether leadership can be one mechanism for bringing the process forward, as seen in a long-term perspective. To what extent, and how, has leadership made a difference in the period from the early days of agenda setting to the present? Unless we have an idea about this, we are not equipped to say anything meaningful about its possible role in the future. The chapter is organized in the following way. First, the concept of leadership will be discussed. In the second part, the historical role of leadership in the process is examined. In the third part, the possible future significance of leadership is discussed, before a brief conclusion is presented in section 4 of this chapter.

Leadership and its relation to the climate regime

We frequently hear policymakers claim that they play a leadership role in various areas of international politics, and that it is important for states to stand up as leaders on the international scene. While the concept is used without much precision in the media and among policymakers, in the scholarly debate, leadership is a far more complex as well as contested term. According to the *Blackwell Encyclopedia of Political Science*, leadership may be defined as "the power of one or a few individuals to induce a group to adopt a particular line of policy." According to the Dictionary of Political Analysis, leadership is what "enables an individual to shape the collective behavioral pattern of a group in a direction determined by his or her own values." Characteristic of these definitions is that they confine leadership to *individuals* and both stress that leadership is *relational*; it concerns the relationship between actors within a group, between leaders and followers. The above definitions are different in the sense that one underlines the significance of *power* in order to exert leadership; the other stresses the *values* of the leader.

Turning to the study of leadership in international relations, there is no consensus as to its significance or how it should be defined. While some analysts maintain that it is essential to foster agreement, others maintain leadership has little or no independent effect in international politics (Young 1991; Moravscik 1999). Second, while some maintain that leadership can only be meaningfully discussed in relation to individuals, others claim that it can also be associated with states (Young 1991; Underdal 1991). Third, while some analysts maintain that exerting leadership squares well with the application of power, others claim that the two should be separated. When power is introduced in the leadership arsenal, it becomes too similar to ordinary negotiation behavior (Malnes 1995).

Although disagreements relating to the concept and significance of leadership abound, all analysts agree that any empirical study of leadership is extremely complex, as distinguishing leadership from ordinary negotiation behavior is very demanding. Given the inherent complexities and methodological difficulties involved in studying leadership, the ambition of this paper is modest. Ideally, one would zoom in on a few specific events and, through careful process tracing, including interviews with key decision makers as well as independent observers, identify agents of leadership. Then, by counterfactual reasoning, one should attempt to establish whether leadership actually made a difference for the outcome produced (Skodvin 1999).

Be that as it may, leadership may be exerted through several mechanisms. Based on Underdal (1991, 1994) and Young (1991, 1998), four types of leadership mechanisms may be identified. An *intellectual leader* produces intellectual capital or generative systems of thought that shape the perspectives of those who participate in institutional bargaining (Young 1991). The *instrumental* or *entrepreneurial* leader often uses and amplifies the ideas of the intellectual leader in getting the ideas on the political agenda. He/she seeks to find the means to achieve common goals, and convince others about the (substantive) merits of the specific "diagnosis" offered or the "cure" prescribed. This type of leadership is a function of the actor's skill, energy and status (Underdal 1991). *Power-based* or *structural leadership* relies on the ability to deploy threats and promises, affecting the incentives of others to accept one's own terms. Such leadership will be in line with an actor's perceived self-interest. However, it has to go together with some notion of common interest to qualify as leadership, often difficult to decide in empirical analyses. Identifying *directional leadership* is associated with the setting of a good example or showing the way on how to deal with an issue (Gupta and Grubb 2000). This is often associated with the term "pusher." However, it takes more to be a leader than a pusher, as cheap and symbolic action does *not* qualify as leadership in this sense; some sacrifice has to be made to make it credible (Underdal 1991).

Different actors can exert different kinds of leadership. In our terminology, directional leadership is conducted by states or groups of states. A government, not an individual, may set an example on how to deal with a problem. At the other end of the continuum, intellectual leadership is usually associated with individuals. Although the money needed to fund research comes from government (or private sources), the ideas, concepts, and knowledge are created by individuals or groups of individuals. Instrumental leadership is also essentially exerted by individuals. While they will often need a formal role or position, the skill needed to be agenda setters, popularizers or to conduct creative bargaining will be vested in individuals. In power-based leadership, it will be individuals that act as creative leaders, but the links to the governments are normally close, especially if the stakes are perceived to be high. Such leadership is mainly linked to agents of powerful states.

What role can leadership be expected to play within the emerging climate regime? The climate change problem is an extremely "malign" problem: "climate change is characterized by high problem severity due to the broad scope, the long duration and the extreme complexity" (Børsting and Ferman 1997: 35). As a point of departure, it is therefore no surprise that progress within the climate negotiations is slow. Effectiveness can be expected to be low when extremely malign issues are dealt with internationally (Miles *et al.* 2002). Still, it is not only the nature of the problem that matters for the process of negotiations. Progress also depends upon the *problem-solving capacity* of the emerging climate regime. In simple terms, this can be conceived of as the three following components: the distribution of power between pushers and laggards, the institutional structure of the regime and leadership (Miles *et al.* 2002). A comprehensive survey of a high number of international regimes in this project has shown that the power dimension is by far the most important of the three, but leadership may also have an effect under certain conditions. Still, expectations should be modest as to what can be expected from leadership in the climate regime, considering the "malignancy" of the issue.

The development of the climate regime may be split into different phases: agenda setting, negotiations, and implementation (Young 1998). Agenda setting in the climate negotiations occurred from the late 1950s until the start of formal intergovernmental negotiations early in 1991. This stage can be subdivided into two phases. The first lasts until the Toronto Conference in 1988 where the idea of emission cuts first gained international currency. In the latter part of this phase (1988–1991), states and governments gradually moved in, paving the way for formal negotiations to start. The negotiation stage went from 1991 via the adoption of the Climate Convention in 1992, the adoption of the Kyoto Protocol in 1997, and ended up with the Protocol coming into force in 2005. Implementation has, to a modest extent, been carried out since the Climate Convention came into force in 1994 but can be expected to accelerate with the Kyoto Protocol in force.

To sum up, this paper assumes that leadership may have an effect, though expected to be of modest significance. All four types of leadership are discussed, and leadership is associated both with states (and the EU) and individuals. In part to narrow the focus and in part because of their significance in the making of the climate regime, the emphasis is on the US and the EU.

From agenda setting to negotiation: the role of leadership

Agenda setting 1: Mid-1950s to start of negotiations – individual leadership flourishes

The influence of increased greenhouse gas concentrations on the Earth's climate was the subject of sporadic scientific interest in the nineteenth century. However, it was in the 1950s that it became embedded with well-funded research programs. It required certain well placed individuals to get government funding and eventually establish institutions to push these research agendas. One key person was Roger Revelle, the director of the Scripps Institute of Oceanography. Together with his colleague Hans Suess, back in 1957 they issued the now famous warning: Human beings are now carrying out a large-scale geophysical experiment that may yield a far-reaching insight into the processes of determining weather and climate (Revelle and Suess 1957).

Revelle also played a key role in organizing the International Geophysical Year (IGY) of 1957 that gave a significant boost to climate-related research (Hart and Victor 1993). A decade later, as a professor at Harvard, Revelle inspired one of his students, Al Gore, to take up the issue of climate change. As a young Congressman, Gore held his first hearings in 1978 on the subject of global warming and Revelle was his first expert witness. Other key persons in this early period were Syukuro Manabe and Bert Bolin. No doubt, these were *intellectual* leaders, and some of them also played an *instrumental* role in increasing attention about this issue.

Climate research was also fostered by scientific networks under the aegis of the World Meteorological Organization (WMO) and the International Council of

Scientific Unions (ICSU). The role of the United Nations Environment Program (UNEP) was somewhat different, as it did not have the advantage of pre-existing ties to the scientific community. However, under its energetic Executive Director Mostafa Tolba, UNEP was able to sharpen the *policy* relevance of climate research by focusing on *societal impacts* of climate variability. Therefore, UNEP displayed considerable *instrumental* leadership from the late 1970s through to the mid-1980s, but UNEP was marginalized when nation states became fully engaged in the climate regime.

The Toronto Conference on the Changing Atmosphere in June 1988 was the high-water mark of the influence of environmental advocacy groups. Participation at Toronto, although at an individual level, was nevertheless broad and fairly elite. Thus, the political energy mobilized by the combination of activist scientists, green NGOs and policymakers in this final agenda-setting phase of the climate regime was impressive. However, key Toronto participants have later observed that many scientists were playing the role of "substitute policymakers," and it was politics not science that was, to an increasing extent, deciding the recommendations. Moreover, the target and timetable approach endorsed at Toronto that called for industrialized countries to cut their carbon dioxide emissions by 20 per cent relative to 1988 levels by 2005 was borrowed from other environmental agreements, most notably the agreement on the ozone regime. However, in terms of providing a focal point or a salient solution to the problem, the Toronto Conference does seem to have been a display of effective (if short-lived) collective instrumental leadership by activist experts and environmental advocacy groups.

During this phase it was the *interaction* between results emerging from decades of scientific research, creative individual intellectual and instrumental leadership, strong public demand for action, and a seeming visualization of the damaging effect of global warming that brought the issue to the international political agenda. In this phase, governments were essentially absent from the game. It was a period characterized by a search for quick diagnosis as well as a cure for the problem. However knowledgeable and creative, many of those active on the scene, in their rush to get the issue on the international agenda, underestimated the political and scientific complexity of the issue. Therefore, the cure they prescribed, based on the "ozone model" was neither necessarily very appropriate nor realistic (see Barrett 2003). This was clearly demonstrated in the next phase.

Agenda setting 2: Attempted leadership by small and "green" states (1988–1990)

One key development that took place shortly after the Toronto Conference was the establishment of the Intergovernmental Panel on Climate Change (IPCC) in November 1988. It was established around a small core of very reputed experts who were subsequently instrumental in attracting other high-caliber experts to the panel, thereby boosting its scientific credibility and political legitimacy. The long-standing intellectual leader, B. Bolin of Sweden, was elected as chair of the IPCC. Bolin's stature within the scientific community was critical in attracting

other reputed scientists to participate in the IPCC (Agrawala 1998b). He continued to play a critical leadership role over the next decade as chair of the IPCC.
The IPCC and key individuals like Bolin exhibited *instrumental leadership* through
their engagement of governments and their *intellectual leadership* through the Working Group 1 First Assessment Report.

Meanwhile, on the political front, public opinion continued to be green in the
OECD region, thereby helping sustain the demand for action and setting the stage
for a "green beauty contest." This period also marked the entry of the state actors,
but in most countries climate change was left to the environmental ministries,
thereby creating an impression that states were more ready to act than they later
proved to be. However, individual leadership no longer loomed large, as government actors moved in. Agenda setting continued but now it was essentially up
to states to come to grips with how to deal with the issue. Some small OECD
countries played an important role in this process by contributing to keeping the
issue "hot" by calling a number of international conferences. While innovative
approaches characterized some of the early conferences, many subsequently
turned into sterile ideological battles between "pushers" and "laggards." Some of
these countries also adopted ambitious unilateral emissions goals and introduced
CO_2 taxation, thereby underlining the importance of countries' *domestic* responsibilities. Thus, these "green" OECD countries may qualify as instrumental leaders
in part by continued international agenda setting, in part through *directional leadership* by demonstrating willingness and ability to deal with the issue.

However, not everybody was ready to follow, most notably the US. From the
end of the 1980s, US climate policy was taken over by strong ideologues in the
White House who cast doubt on the scientific message and, in contrast to most
states, emphasized the high costs of action to reduce greenhouse emissions. The
Chief of Staff in the White House, John Sununu, was the main architect behind
this policy and in firm control over the policies pursued by the executive branch.
The US advocated more research and a "no-regret" policy, instead of a "top-
down" target and time-table approach, which it found "impractical" as well as
infeasible (Agrawala and Andresen 1999). In pure instrumental terms, at least on
the domestic scene, he was a very effective leader. From our perspective he was
more of a negative leader in using the means at his disposal to achieve his goal,
which was "no cooperation," in contrast to our normative expectations from leaders as forgers of collective outcomes.

As Rio approached, the small "green" countries gradually became less visible
on the international scene while the EU and the US became the main actors.
Climate change was no longer a question of environment only, it was about to
become "high politics" and both actors were instrumental in polarizing the issue.
The EU had assumed a self-declared leadership role on the climate issue since
the late 1980s. The EU saw the issue in broader strategic terms, as it sensed a
leadership vacuum in the absence of strong US and Japanese climate policy positions. That is, the EU position was not necessarily only a reflection of concern for
an environmental problem, but perhaps equally important as a stepping stone to
standing as a strong and unified block on the world scene (Ringius 1999). The

main pusher on the international scene was the Environmental Commissioner of the Directorate-General of the Environment, Nuclear Safety, and Protection (DG XI), Ripa de Meana. The EU may have been perceived as a unified actor during the many clashes with the US, but there was no shared climate vision within the EU. In fact, most member states were quite reluctant to act on the issue. While de Meana was certainly an ardent pusher, it was doubtful whether he qualified as an instrumental leader, as he was better at playing to the gallery than trying to build workable compromises (Andresen and Wettestad 1990).

By 1990 both the scientific and political processes were about to become institutionalized and were under stronger governmental control. Nevertheless, intellectual and instrumental individual leadership was still exerted on the scientific scene. On the political scene, meanwhile, some small "green" states were attempting various forms of leadership, but their significance was reduced over time and the two political heavyweights, the US and the EU, dominated the scene. No doubt the EU was the pusher and the US was the main laggard, but instrumental leadership in terms of creating a basis for joint solutions was lacking.

Negotiation phase 1: The road to Rio – increasing polarization, less leadership

Most attention during the five sessions of the Intergovernmental Negotiating Committee (INC) leading to the Rio Conference was directed at the question of whether or not a specific legally binding emission target should be adopted. The focal point or "salient solution" became the stabilization of greenhouse gas emissions from developed countries relative to 1990 levels by the year 2000 – far less ambitious than the Toronto target adopted in 1988. Progress was slow because of the blocking role of the US until the last INC session just prior to Rio, where more informal and exclusive negotiation forums were established. In particular, bilateral negotiations between the US and the UK were critical in breaking the deadlock (Bodansky 1993). Although heavily criticized, this provided the foundation for the Climate Convention. Most OECD countries, as well as the green community, claimed that the Convention was basically flawed, as legally binding targets and timetables were not included. While the US and the UK paved the way for agreement on the crucial issue on commitments, the Chairman of the INC played a key role in ironing out the details of a Single Negotiating Text (SNT). This was done in a more exclusive meeting after formal negotiations had ended – in clear violation of INC procedures. Not least thanks to his effort, the most ardent blockers like Saudi Arabia did not raise procedural objections.

The Climate Convention was largely seen as a victory to the perceived main laggard, the US. However, the US was also the main architect behind the rather advanced institutional structure of the Convention (Bodansky, 1993). Nevertheless, the fact that the US got it their way reminded the US of the significance of power – and of veto politics. As to the EU, its self-imposed leadership role was tarnished because of strong internal disagreement over the question of CO_2 taxation, and the EU played no major role in Rio (Wettestad 2000). This paved the way for the compromise role played by the UK.

Overall, it is difficult to assign a leadership role to any of the participating states (and the EU). Identification of individual instrumental leadership is also elusive at this stage. The INC Chairman, Jean Ripert, may qualify, but it is uncertain what he had attained in the absence of the bilateral US/UK agreement.

Negotiation phase 2: The road to Kyoto – mostly traditional state bargaining

During the two COPs prior to the decisive COP3 in Kyoto, the EU and the US played varying roles. The EU and particularly Germany played a key role in making the Berlin Mandate possible, stressing the responsibility of the North to take a lead on the issue (Grubb *et al.* 1999; Gupta and Grubb 2000). The roles were reversed at COP2 when the EU again spent most of its time and energy in hammering out an internal burden-sharing agreement (Ringius 1999). In contrast, the most important reason for the adoption of the Geneva Declaration, underlining the need for legally binding commitments, was the result of the gradual change in the US position in a more "progressive" direction. COP2 has been described as a turning point in the climate negotiations because of the changed role of the US (Grubb *et al.* 1999: 53).

Prior to COP3 in Kyoto, the EU was finally be able to agree on a burden-sharing agreement, paving the way for an ambitious goal of 15 per cent reduction in greenhouse gas emissions by 2010.[2] Not only did this enable the EU to stand as a unified block, but the ambitious goal also gave it the moral upper hand in the run-up to Kyoto. Still, as no new ambitious goals had been adopted by the US or Japan, expectations as to what would come out of Kyoto were modest. Against this backdrop, most analysts saw the Kyoto Protocol as a significant step in the right direction through its combination of novel institutional procedures and fairly ambitious legally binding emission targets (Grubb *et al.* 1999). The Kyoto Protocol was a genuine compromise in the sense that "the EU got their numbers, the US got their institutions, Japan got prestige as a host, the JUSSCANNZ countries [Japan, US, Switzerand, Canada, Australia, Norway, and New Zealand] got their differentiation and the developing countries avoided commitments" (Andresen, 1998: 28). However, loopholes and ambiguities created problems subsequently.

In the final stages of negotiation up to Kyoto, traditional state bargaining in more or less closed settings was the name of the game. "One broad observation is that, for all the academic speculations about the decline of the nation state in the era of economic globalization, the Kyoto Protocol is very much an agreement struck by governments" (Grubb *et al.* 1999). Even though the Kyoto Protocol was a genuine compromise, there is little doubt which actor was most *influential* in bringing it about. As Grubb observes: "Within this panoply, US dominance is striking" (Grubb *et al.* 1999: 112). Moreover, he adds: "To discover the *source of most ideas* in the Kyoto Protocol, one needs only to read the US proposal of January 1997" (emphasis added). This especially applies to the flexible mechanisms, that is, the *means* by which the overall goal could be reached, essential to any US agreement to the Protocol. When comparing the US to the EU, according to Grubb *et al.* (1999: 112), "The coherence of the US administration contrasted with the unwieldy morass in EU decision-making in the negotiation process." Does this mean that the US stood forward as a leader in Kyoto, using an elegant combination of instrumental and power-based means? Some observers have reached this conclusion (Tangen 1999), while others will certainly disagree. For example, the environmental movement tended to see the novel institutions as no more than a creative way for the US to buy their way out of the problem, as the market mechanism would significantly reduce emission costs. In short, what some see as creative leadership, others see as a smart way to secure national interests. Leadership is easier to identify when there is consensus on what the best means are to reach a given goal, like the ozone regime. To the extent that the US was exerting leadership, this performance was considerably conditioned by the "two level game", in which the US negotiators were trying to balance domestic interests and pressure from other nation states (Putnam 1988). Considering that the US Congress would not ratify any climate agreement perceived to be costly, the US negotiators had to push strongly for the market-based mechanisms. US negotiators knew they had probably more opposition *domestically* than internationally, adding energy to search for innovative solutions. On the other hand, using Putnam's (1988) terminology, this gave the US a very small "win-set", giving the US negotiators considerable bargaining leverage.

Although the US played a dominant role in Kyoto, the significance of the EU should not be disregarded. The unity of the EU paved the way for the fact that emissions reductions was agreed by the Annex I nations. The EU may qualify for the role of directional leader by strengthening the commitments in line with the overall goal of the negotiations (Gupta and Grubb 2000). However, although unity may have increased its bargaining leverage, the effort to hold a unified front may have left less room for creativity and flexibility in the bargaining process (Sjølseth 1999).

In terms of individual instrumental leadership, two persons stand out as crucial in getting the Kyoto Protocol adopted: Vice President Al Gore and the Chairman of the Committee of the Whole (COW), Raoul Estrada-Oyela. In the buildup to the Kyoto summit, the latter played an important role by narrowing down and simplifying the final negotiation text. "This man has been brilliant. He has been able to create something out of nothing", observed John Prescott, the UK Deputy Prime Minister (*Financial Times*, 11 December 1997). Estrada was the only person with full insight into the final deliberations over different commitments allotted to the Annex 1 countries (Chasek 1997). He was also able "to declare a paragraph adopted by consensus when they had hardly finished their opposition" (*Financial Times*, 12 December 1997).

However, the work of Mr. Estrada would probably have been futile without the key role played by the US Vice President, Al Gore. A long-time advocate of policy responses to climate change in various key positions and since being elected Vice President, he and other key persons were gradually able to move the US climate policy more into line with the rest of the OECD countries. It is widely believed

that the mandate Gore gave the US delegation to demonstrate increased negotiation flexibility during his 16-hour visit to Kyoto was instrumental in breaking
the deadlock and paving the way for the subsequent agreement (Agrawala and
Andresen 1999).

The main effort of the EU was to push for an ambitious goal, not to engage in
detailed bargaining. That is, the EU has a high score on the directional leadership
dimension and a corresponding low score on the instrumental dimension. As the
US had the opposite score on these dimensions, they thereby in a sense *complemented* each other, which, with the help of key individuals, paved the way for the
Kyoto Protocol.

Post-negotiation: From COP3 to COP10 – Exit of the US–EU leadership?

The relative optimism created in Kyoto vanished quickly at COP4 in Buenos
Aires as it became clear that the Kyoto Protocol was full of "invisible brackets"
(Torvanger 1998). After the COP4 in Buenos Aires, the EU was again passive and
it was not able to table a single proposal (Tangen 1999). In contrast, the US stole
most of the attention by declaring that it would sign the Kyoto Protocol, although
this was mostly a symbolic gesture, considering the strong resistance to the Protocol in the Senate. The fact that Argentina agreed to take on voluntary commitments was seen as a major victory for the US.

The Buenos Aires Plan of Action set out the process for taking forward the provisions of the Protocol by COP6 (2000) in The Hague. COP6, however, ended in a
deadlock. For the first time a COP was suspended – until July 2001, when it recon-
vened in Bonn. Although controversies abounded, the most critical issue again
turned out to be between the US and the EU, this time over the role of carbon
sinks and whether and how countries should receive credits toward their emissions
reduction targets through the use of sinks such as forests. Most environmentalists,
as well as the EU, saw the US approach as close to cheating, a way of avoiding
painful domestic measures yet still get emission reduction credits. As a result of last
minute efforts to strike a deal, however, the distance between the US and the EU
was only 20 million tons of carbon emissions, a trifle compared to the six billion
tons of carbon that are emitted to the atmosphere every day. Looking back, this
may have been a window of opportunity that was closed for the foreseeable future,
as it appears that the US was very keen on striking an agreement in The Hague.
In fact, President Clinton himself was actively involved in trying to convince reluctant EU countries to accept an admirable last minute attempt to forge a compromise put forward by the British Deputy Prime Minister John Prescott (Agrawala
and Andresen 2001: 124). His energetic efforts represent attempted, but failed,
leadership. The Dutch Chairman of the Conference, Jan Pronk, also attempted to
forge agreement by his memo on the so-called crunch issues, but this also failed. It
appears that when polarization is strong, the room for leadership is limited.

The overall picture prior to the official US rejection of the Kyoto Protocol was
that neither the EU nor the US were able to stand up as strong leaders. Gupta and

van der Grijp (2000: 80) conclude that: "Should the US decide to take up the leadership role, it will be theirs for the taking." To date, it has refrained from taking this opportunity. Another conclusion regarding the role of the EU prior to the US de facto exit was: "high ambition–low performance" (Hovi *et al.* 2003).

With the perceived main stumbling block, the US, out of the picture since March 2001, the process toward ironing out differences between the remaining parties should have been a "piece of cake," but it soon became apparent that new laggards were emerging on the scene. However, for the EU, the US exit represented a window of opportunity in terms of finally realizing its leadership ambitions. In practice, the EU was now the only actor with vested interests as well as the political and economic potential to move the process forward. Neither developing countries nor economies in transition had the will or ability to do so. Japan has never seemed to have leadership ambitions in this process and the remaining OECD states are too small to make much of a difference.

The efforts by the EU and others to bring the US back on track failed. As a next step, the EU invested a great deal of political energy in rallying support behind the Protocol. Two strategies were applied; internally it aimed at setting a good example by heading for a swift EU ratification, externally by trying to persuade the reluctant "Gang of Four" (Australia, Canada, Japan, and Russia) – all previous allies of the US – to ratify the Protocol (Hovi *et al.* 2003). While the internal strategy was successful, the external strategy was time-consuming and less successful. After some time, Canada decided to ratify, although domestic opposition from the powerful fossil fuel lobby was considerable. After much hesitation, Japan also ratified, while Australia, with much the same argument as the US, rejected the Protocol. Thus Russia became the key player, as the Kyoto Protocol would not enter into force in the absence of Russian ratification. Mixed signals were given over many years before the Protocol was finally ratified during the fall of 2004 (Moe 2004). There is little doubt that there was strong lobbying both from the US and the EU regarding this issue, and this time the EU won the game, one of its rather rare victories in the climate battles with the US.

Clearly, the EU has had more of a leadership role after the US exit than before. Still, the EU mobilization of support for the Protocol essentially amounted to giving in to the demands of the "Gang of Four." Thus, while COP6 resulted in a "Kyoto Light", COP7 resulted in a "Kyoto Ultra Light" (Andresen 2001). In fact the EU now gave concessions on issues where it had previously refused to concede to US demands and the revised Protocol became very close to what the previous US administration had tried to achieve (Hovi *et al.* 2003). On the one hand, therefore, the Kyoto Protocol would probably not have entered into force in the absence of EU leadership. On the other hand, it was not able to secure the environmental integrity of the Kyoto Protocol. The indulgence of the EU in allowing the requirements of the "Gang of Four" to be accommodated suggests that the driving force behind this course of action were political benefits associated with leadership just as much as concern for the global environment (Hovi *et al.* 2003). The EU in 2005 launched a comprehensive emission trading scheme, a true market-based mechanism and a US "brain-child" that the EU previously opposed strongly.

Not much happened during the 2002–2004 COPs. Most of the time was used to negotiate decisions for implementing the Marrakesh Accords, not the least those associated with the operation of the Clean Development Mechanism (CDM). At COP10, to make sure that the house was in order for the Protocol's imminent entry into force, "the Parties gathered to complete the unfinished business from the Marrakesh Accords" (ENB Vol. 12 No 260, December 2004: 28). However important some of these issues may have been, they were of a more technical nature and received limited attention by most observers. In contrast, the question of future commitments after 2012 has been more hotly debated, as in Buenos Aires "it was crystal clear that some Parties are not ready to embark on post 2012 negotiations" (ENB Vol. 12 No 260, December 2004: 28). All that could be agreed was to arrange a *government* expert meeting for more informal discussions. The US was against discussing future commitments and so were the developing countries. The EU was not able to inject the necessary political energy into the process to move it forward.

Therefore, the "bottom line" after COP10 seemed clear, but somewhat depressing. A small minority of states, under EU leadership, wanted to build on the Kyoto architecture with strengthened commitments. Although the US, as we shall see later, has another approach, as of yet, no clear *alternative approach* emerged at this time. Thus, in a sense, the Kyoto Protocol remained "the only game in town." Still, as we know, the US was adamantly opposed to it and we do not really know the positions of most developing countries on a future regime. We do not know whether the Kyoto Protocol will become an appendix to the history of climate change or whether it will become an important stepping stone for the evolving climate regime. In that regard, it is important to bear in mind that the Kyoto Protocol, although ratified by 183 states, is a *"mini regime"* in terms of hard commitments. Some 130 states in the South have made no commitments, and it is not difficult to sign up to a document with no associated economic or political costs. It is also a fact that. among the Annex 1 states, very few states have made commitments that will be difficult and costly to meet (Andresen *et al.* 2002). This is not said to belittle the Kyoto Protocol, which was certainly a tough job and an important accomplishment by those who negotiated it. However, it is a fact that it is now that the "real" negotiations start, if we are able to deal with the global warming problem more effectively.

Future leadership

The EU and the US

Looking to the future, on the surface it seems that except for the US, the EU seems to be the only remaining serious leadership candidate. When we are concerned about the future, directional leadership seems to be the most important leadership mechanism. True, we do know that there will be demand and need for instrumental and power-based leadership as well, but we do not really have any way to predict when, how, and if it will take place. Moreover, instrumental power is mostly relevant in knotty negotiation situations and, as we have seen, its effect is often limited over time. What is needed now are initiatives and concrete action to increase the long-term effectiveness of the climate regime. Someone needs to show the way as to how emissions can be strongly reduced: directional leadership.

However, it may be that the leadership candidature of the EU is not quite as obvious as it may seem at first glance due to the *favorable circumstances* associated with EU climate policy. The backbone of the rather aggressive EU climate policies is the steep emission reductions in the UK and Germany for reasons unrelated to climate change (Gupta and Grubb 2000). Had this not been the case, there is little reason to believe that the EU would have been a strong pusher, as the specific climate policy measures adopted by the EU have not been very impressive. In fact it was recently reported that the GHG emissions had only been reduced by 1.7 per cent in the period 1990–2003, casting serious doubts on whether the EU target will be met (Euractiv.com, June 2005). This picture also reduces the role of the EU as a directional leader. Recall that according to Underdal (1991), "cheap and symbolic action does not qualify as directional leadership, some sacrifice has to be made to make it credible." The influx of new EU members from Eastern Europe will add to these favorable circumstances as the Economies in Transition (EIT) countries have also experienced strong emission reductions for reasons unrelated to climate policies. However, these "fortunate circumstances" for the EU will not last forever. They have so far paved the way for the very high political profile with much rhetoric, but we also need to see that the EU will not only "talk the talk but also walk the walk" (Agrawala and Andresen 2001).

In the future, therefore, the EU has a tough job to introduce new and "real" climate measures. The most obvious avenue in this regard is the EU (ETS) trading scheme. Compared with other similar trading schemes, the ETS GHG scheme is by far the biggest (Euractiv.com, 8 June 2005). This is a golden opportunity for the EU to show directional leadership by proving to the world that the system works and contributes to real emission reductions. It is premature to judge its effectiveness, but carbon allowances for delivery later in 2005 have increased sharply over time and approached 30 Euro (Point Carbon, July 2005; Euractic. Com, July, 2005). If this and trading schemes of other countries are linked and contribute to reducing emissions, this will also contribute to enhancing the legitimacy and prestige of the Kyoto Protocol approach. It has also been argued that the fundamental difficulty with the EU ETS is uncertainty about the Kyoto process and about the form and level of commitments beyond 2012 (Kanie 2005). We also know that present GHG trading emission schemes alone will not be able to sufficiently reduce emissions as envisioned in EU planning. The EU Commission was unusually pragmatic, as no new targets were set for the post-2012 phase when it presented its strategy paper in February 2005. It emphasized the need to bring more countries on board, the US and Australia as well as key developing countries like China, Brazil and India. In contrast, EU environmental ministers proposed a 15–30 per cent reduction in GHG emissions by 2020 and 60–80% by 2050 (!!) (Euractiv.com, 9 June 2005). This certainly underlines the high priority given by the EU to ambitious plans, but its practical significance remains elusive.

Turning then to the US, while the EU climate policy has to date been characterized by fortunate circumstances, the US has been in a different position. It is widely acknowledged that strong US climate policy measures are bound to be costly, although estimates vary strongly (Barrett 2003). As pointed out by senior analyst Christian Egenhofer of the Centre for European Policy Studies, the EU and the US have a fundamental difference of perception on climate and energy policy. While the EU uses mainly environmental lenses, the US frames it in a *security of supply* context (Euractive.com 9 June, 2005). The US is the largest energy consumer, producer, and net importer in the world and has the largest coal reserves (EIA 2004). As long as coal is a cheap and abundant source, the US is not likely to switch to other sources, and coal contains a much higher degree of carbon per unit of energy than oil and especially gas. The US economy is very energy intensive and the per capita energy use is more than the double of Western Europe (EIA 2004). Thus, there is not only an extremely strong fossil fuel lobby in the US, but people at large show little or no inclination to change their energy intensive behavior. The political feasibility of changing this at present seems to be non-existent. Still, as security and economy are highest on the US agenda, the strong dependence on imports of oil in particular and the recent very high prices may stimulate the search for alternative sources and stronger diversification (Bang *et al.* 2005; Froyn 2005). It should also be noted that the high US GHG emissions can be attributed to a much stronger economic growth as well as population growth compared to the EU. Still, US energy policy is very fossil-fuel-intensive, and the climate policy is but a small appendix of it.

With this caveat, surely the US has nothing to offer in terms of directional leadership? If we take a closer look at the roles of different key US actors, the picture is not necessary all that negative. First, the McCain–Lieberman Act was fairly narrowly defeated in the Senate (56–44). If such legislation is adopted, it will surpass the EU ETS both in terms of regulated sources as well as estimated value of annual allocation (EurActic.com, 8 June 2005). There have also been a number of other climate-related proposals in Congress, indicating that in the absence of leadership from the administration, Congress is becoming increasingly active. Second, there is a strong increase in climate policies at the level of states, adopting Kyoto-like approaches. Recently California pledged to cut emissions by 7 per cent compared with 1990 levels – and recall that California is the fifth biggest economy in the world. A majority of US states have developed, or are in the process of developing, mitigation policies (Christensen 2003). This "bottom-up" approach is in line with the approach that has been favored by varying US governments since the late 1980s, in contrast to the top-down approach by the EU. Historically US environmental policy has been developed at sub-federal level, creating a fragmented situation that has called for federal intervention to make more uniform regulations. Something similar might happen as regards US climate policies. Certain proactive US business companies cutting their GHG emissions strongly add to this impression of a significant bottom-up push for a more active climate policy in the US (Pew Centre).

Nevertheless, overall these policies can hardly be described as leadership; most of these actors are more like followers in the sense that they want a policy more in line with that of other OECD countries. What about the official climate policy of the George W. Bush administration: could we see any signs of leadership here? This question would probably either seem stupid or provocative to most observers. US emissions are increasing strongly, it refuses to discuss future commitments, and a bilateral and voluntary approach is the name of the game, not ambitious global emissions plans. Again, the picture is not all that simple, and science and technology are key components.

US scientists played a crucial role in getting this issue on the agenda and the US scientific effort has continued to be much stronger than that of other countries. Some see this as a part of a US strategy to emphasize science instead of introducing climate measures. This view may have some merit, but hardly any serious actor would claim that now we know enough about this issue to reduce our scientific efforts. Therefore, continued strong US scientific efforts, however motivated, is sorely needed to come to grips with this truly "malign" issue. Part of the scientific picture deals with technology. This is where the highest potential for US leadership is, as technology development has been emphasized much more strongly by the US than the EU or others. Technology is one main component of US climate change policy. A program of US$ 5.8 billion was adopted in 2005. This includes plans for methane capture from coal mining, zero- emission coal-firing plans, clean diesel and bio-energy programs. Other long-term research plans include hydrogen, nuclear energy, carbon sequestration as well as storage (EurActive.Com, 8 June 2005). Many have also tended to see this effort as an excuse for climate inaction. However, considering the need for some "real" reductions further down the road, this emphasis is sorely needed. The EU has advocated the "push" and "pull" technology strategy through the cap and trade system, "but you don't get breakthrough technologies in this way … with their long-term research program and massive investments the US has got something right there" (Egenhofer 2005). This indicates that the US has a directional leadership potential in terms of technology development (Kanie 2005). Although most US international climate policies are characterized by bilateralism, this is not the case in some of its technology initiatives. The US has taken the lead in "The Carbon Sequestration Leadership Forum" where both the EU as well as key developing countries participate (Bang *et al*. 2005). The inaugural meeting of the "International Partnership for Hydrogen Economy (IPHE) was held in Washington DC in November 2003, also with a limited attendance, but all key climate states, including the EU were there (Kanie 2005).

In short, structural, economic and political reasons contribute to explain how and why US climate policy differs from that of the EU. When these factors are considered, there is no doubt that the EU still has a more ambitious climate policy than the US, but the picture becomes more nuanced. Also, the US may have a directional leadership potential, particularly in terms of technology development.

The bigger picture

Although these two political heavyweights with directional leadership potential disagree on key issues, there have been various bilateral meetings between the two actors. The fact that such a dialog takes place is promising and indicates the presence of a pragmatism and openness that has not always characterized US–EU relations in this area. The fact that the then leader of the G8 group, the Prime Minister of the UK Tony Blair, moved global warming to the very top of the agenda of its 2005 meeting was also a positive at that time. As noted by the *Economist*, not a great friend of the Kyoto Protocol, "America should use Summit meetings to embrace carbon trading" (2005: 13). However, not much came of the 2005 meeting the G8 group. Still, the fact that it at the top of the international political agenda in contrast to *all* other international environmental problems contributed to keeping the pressure on the US administration and other reluctant actors. Recall also that the UK, due primarily to "fortunate circumstances," is in a good position to act as a directional leader. The UK is also in a favorable position to act as a broker between the US and the EU because of its long-standing close relations with the US.

The decision to invite major developing countries to discuss these matters was also a clever diplomatic move by Mr. Blair. The very idea of discussing global warming on a high level and in more exclusive context may be one way to bring the process forward. In the long run, the chance for progress among the main stakeholders is higher in more confined forums than in the highly politicized atmosphere of the global negotiations.

Still, to date, the EU and the US favor different approaches to this problem. The EU favors the Kyoto architecture with stronger commitments while the US relies on energy intensity targets and technology. Who will the developing countries follow regarding the post-2012 period? The South, as the US, sees climate change as a matter primarily of economy. Recall that the energy intensity approach was first elaborated by the US as a way to get the developing countries to take on commitments without sacrificing economic growth (Bang *et al.* 2005). In addition, the South sees the question in terms of equity, further strengthening their opposition toward traditional emission cut commitments. This indicates a stronger inclination by most G77 countries for support of the US approach in the future. In fact, there has been a tacit and informal "alliance" between the US and key developing countries at the last COP meetings, where the need for continued economic growth has been stresses (Jacob 2003). Moreover, there are severe doubts as to whether the EU *in practice* will support a climate policy that is contrary to economic growth – as limited growth is presently the biggest problem for the EU – in contrast to the US and China (*Economist*, 24 June 2005).

Finally, a few thoughts about the leadership potential of the non-state actors. As we recall, scientists were crucial in the agenda setting phase. However, over time, their significance became reduced because of the perceived high economic and political stakes involved. In short, there is a huge discrepancy between the scientific warnings and the political action, or lack thereof. The most important role of scientists is, typically, as agenda setters and suggesting how problems should be solved. Their influence tends to be reduced further down the road (Andresen *et al.* 2000). Although not able to exert a leadership role, the continued action of the IPCC is, of course, still needed to gain more precise knowledge of this scientifically incredibly complex problem. It is certainly also important that 11 scientific academies, including those of the US and China, India, and Brazil, called for urgent action against climate change prior to the G8 Summit in 2005. (EurActiv.com June 9).

As to the green NGOs, it has been maintained that they have lost influence and significance over the last few years: "The environmental movement's foundational concepts, its method for framing legislative proposals, and its very institutions are outmoded. Today environmentalism is just another special interest group" (*Economist*, 2005:11). This harsh statement does not come from the *Economist*, but is drawn from "The Death of Environmentalism", an influential essay published a few years back by "twos green with impeccable credentials (ibid.). The traditional green mantra of "mandate, regulate and litigate" is no longer valid (ibid.). Some will certainly find this criticism too harsh and one-sided, but to a large extent it also seems valid for climate change.

With a few notable exceptions in the US, the green movement has generally been critical of the market – and technology-based approaches to climate change that has gained ground since Kyoto. After the era of agenda setting, their influence on the process has been modest, unless they have been in alliance with strong states (Gulbrandsen and Andresen 2004). In short, unless the "greens" become more up to date with reality, we cannot expect much leadership here.

What about business and industry? It is well known that the fossil fuel industry, particularly in the US, has been highly influential in working against the Kyoto Protocol and strong commitments (Skjærseth and Skodvin 2003). This will surely continue. Significant parts of business and industry are likely to have the same position in the future and will be laggards, not leaders. However, this picture is no longer as simple as it used to be. A considerable portion of this sector, both in the US and in Europe, have taken on voluntary commitments, but with firm targets and timetables. Some of them have also succeeded in very substantial reductions of GHG emissions (Bang *et al.* 2005; Jacob 2003). In general business and industry, particularly in the US, are against the Kyoto Protocol, much for the same reasons as the US itself. However, an increasing number of businesses seem to be in favor of technology development as well as emissions trading. For example, in June 2005 the chief executives of some two dozen multinationals, including US firms like Ford and Hewlett Packard, asked the G8 to adopt a global carbon trading system (*Economist* 2005). In short, in the future, an increasing part of business and industry will probably have a directional leadership potential, simply because they find it to be in their long-term economic interest.

Summing up this section, the largest directional leadership potential is in the US technology approach, the EU trading approach, and among parts of business and industry. I am also quite confident that in the not-too-distant future the US will become a member of the trading system. As to the G77, they are more likely to "buy" the intensity approach rather the Kyoto approach as it can be combined more easily with continued economic growth. This direction toward more market and technology and fewer traditional top-down regulations will take place because the majority of actors will see this as the only way that is compatible with their broader political and economic interests. That is, the future of the climate regime will be decided not by environmentalists or environmental policy but by the broader lines of international economic and energy policies. The long- term trend is probably toward lower carbon intense economies, but it will take a very long time (many decades) before the climate regime is truly effective, if ever.

Conclusion

According to the findings of this chapter, leadership flourished in the first agenda-setting phase. In the absence of states in the process, there was plenty of room for *individual intellectual and instrumental leaders.* However, fortunate circumstances like high green demand from the public made their job easier. In the later agenda setting stage, some small "green" OECD states helped keep the issue on the political radar screen internationally and also adopted unilateral measures to fight global warming, thereby "showing others the way forward." But opponents of emissions reduction were also getting organized, and ideological clashes reduced the room for leadership and prevented progress.

The available evidence that *traditional state bargaining* was dominant throughout the period of negotiations before the Kyoto Protocol is covered in this chapter. However, both in Rio de Janeiro and Kyoto, effective instrumental leadership was conducted by key officials of the negotiation process negotiation as the deadlines approached, but their efforts may have been futile in the absence of a *positive interaction with representatives of key states.* That is, tacit or open alliances with *powerful states* may have been a prerequisite for exerting instrumental leadership. In the period leading up to Kyoto as well as after Kyoto, the EU has seemingly had a high score in terms of *directional leadership* through its ambitious emissions goals. However, its significance was somewhat reduced due to the "fortunate circumstances" surrounding the EU climate policy. Overall, the EU had a rather low score with regard to instrumental leadership. Until the US virtually withdrew from the Kyoto negotiation, they had the opposite profile, a high score in terms of instrumental leadership and a low score in terms of directional leadership. An important outcome from the pre-Kyoto talks is that a combination of EU directional leadership and US instrumental leadership contributed to bringing the process forward.

In short, in line with our expectation, considering the "malign nature" of the issue, the supply of leadership has been rather modest and its significance has also often proved to be rather short-lived. As to the future, it is argued that directional leadership is bound to be the most important. However, based on experiences to date, expectations should be modest as to the significance of leadership in bringing the process forward. It is argued, however, that the fact that climate is now discussed at a high level and more confined arenas is promising, and that some actors, like the EU and particularly the UK, have attempted to infuse the process with more political energy. It is essential, however, that the US and the key G77 countries are also willing to act. Although the US has long performed traditionally as a laggard, this position has recently started be modified. It can be argued that particularly its strong technology focus may bring the process forward, in combination with the EU trading system and a more proactive business approach.

However, the extent to which the climate regime will become more effective in a real sense depends primarily on the extent to which all key actors see it as being in their interest to work for a less carbon intensive global economy.

Note

1   Point three in this chapter builds on Andresen and Agrawala (2002) and Hovi *et al.* (2003).

References

Agrawala, S. (1998a) Context and Early Origins of the Intergovernmental Panel on Climate Change. *Climatic Change. 39* (4). 605–620.

Agrawala, S. (1998b) Structural and Process History of the Intergovernmental Panel on Climate Change. *Climatic Change. 39* (4). 621–642.

Agrawala, S. (1999a) Early Science-Policy Interactions Global Climate Change: Lessons from the Advisory Group on Greenhouse Gases. *Global Environmental Change. 9* (2). 157–169.

Agrawala, S. (1999b) Science Advisory Mechanisms in Multilateral Decision-Making: Three Models from the Global Climate Change Regime. Ph.D. dissertation, August. Princeton University, New Jersey, US.

Agrawala, S. and Andresen, S. (1999) Indispensability and Indefensibility? The United States in the Climate Treaty Negotiations. *Global Governance: A Review of Multilateralism and International Organizations.* 5 (4). 457–482.

Agrawala, S. and Andresen, S. (2001) US Climate Policy: Evolution and Future Prospects, In Agrawala, S. and Andresen, S. (eds.) *National Climate Policies: Evolution, Drivers, and Future Prospects.* Special Issue of *Energy & Environment* 2–3. 117–137.

Andresen, S., (1998) *The Development of the Climate Regime: Positions, Evaluation and Lessons.* FNI Report 3/98. Fridtjof Nansen Institute, Lysaker.

Andresen, S. and Wettestad, J. (1990) Climate Failure at the Bergen Conference? *International Challenges.* 10 (2).17–24.

Andresen, Steinar, Tora Skodvin, Arild Underdal and Jørgen Wettestad (2000) *Science and Politics in International Environmental Regimes. Between integrity and involvement.* Manchester: Manchester University Press.

Andresen, S. and Agrawala, S. (2002) Leaders, Pushers and Laggards in the Making of the Climate Regime. *Global Environmental Change.* 12(1). 41–51.

Bang, G., Tjernshaugen, A. and Andresen, S. (2005) Future US Climate Policy: International Re-engagement? *International Studies Perspectives.* 6 (2). 285–303.

Barrett, S. (2003) Environment and Statecraft: The Strategy of Environmental Treaty Making. New York, NY: Oxford University Press.

Bodansky, D. (1993) The United Nations Framework Convention on Climate Change: A Commentary. *Yale Journal of International Law. 18.* 451–558.

Børsting, G. and Ferman, G. (1997) Climate Change Turning Political, In Ferman, G. (ed.) *International Politics of Climate Change.* Oslo: Scandinavian University Press. 53–83.

Chasek, P. (1997) *Earth Negotiation Bulletin 12* (76), 13 December.

Christensen, A.C. (2003) Convergence or Divergence? Status and Prospects for US Climate Strategy, FNI Report 6/2003, The Lysaker: Fridtjof Nansen Institute.

Egenhofer, Christian, interview EurActiv (2005) Interview: CEPS-Researcher Christian Egenhofer to the climate policies of the EU and the US, available through http://46.4.92.154/de/nachhaltige-entwicklung/interview-ceps-forscher-christian-egenhofer-klimapolitiken-eu-usa/article-140356, last accessed 04 December 2012.

Froyn, Camilla Bretteville (2005) Decision Criteria, Scientific Uncertainty, and the Global Warming Controversy. In *Mitigation and Adaptation Strategies for Global Change*. 10. 183–211.

Grubb, M., Vrolijk, C. and Brack, D. (1999) *The Kyoto Protocol A Guide and Assessment*. London: Royal Institute of International Affairs.

Gulbrandsen, L. H. and Andresen, S. (2004) NGO Influence in the Implementation of the Kyoto Protocol: Compliance, Flexibility Mechanisms, and Sinks. *Global Environmental Politics*. 4(4). 54–75.

Gupta, J. and Grijp, N.M. van der (2000) Perceptions of the EU's Role. In Grubb, M. and Gupta, J. (eds.), *Climate Change and Leadership: A Sustainable Role for Europe*. Dordrecht: Kluwer Academic Publishers. 63–82.

Gupta, J. and Grubb, M. (eds.) (2000) *Climate Change and European Leadership*. Amsterdam: Kluwer Academic Publishers.

Hart, D.M. and Victor, D.G. (1993) Scientific Elites and the Making of US Policy for Climate Change Research, 1957–74, *Social Studies of Science* 23. 643–680.

Hovi, Skodvin and Andresen (2003) The Persistence of the Kyoto Protocol: Why other Annex 1 Countries Move on without the US. *Global Environmental Politics*. 3 (4). 1–24.

Jacob, T. (2003) Meeting Review: Reflections on Delhi, Climate Policy. 3.103–106.

Kanie, Norichika (2005) Mid-Long Term Target Setting and Challenges for its Internationalization: Reduction of Global GHG Emissions and Japan's Targets. Environmental Research Quarterly. 38. 84–92.

Malnes, R. (1995) "Leader" and "Entrepreneur" in International Negotiations: A Conceptual Analysis. *European Journal of International Relations*. 1 (1). 87–112.

Moe, A. (2004) Norge-Russland: Naboer gjenmom 1000 år (Norway-Russia: Neighbors for 1000 years), In Buchten, D., Dzjakson, J.P. and Nielsen, J.P. (eds.) Samarbeid om miljøet i nord (Environmental co-operation in the North). Oslo: Scandinavian Academic Press. 455–456. In Norwegian.

Moravscik, A. (1999) A New Statecraft? Supranational Entrepreneurs and International Cooperation. *International Organization. 53* (2). 267–306.

Oberthür, S. and Ott, H.E. (2000) *Das Kyoto-Protokoll. Internationale Klimapolitik für das 21. Jahrhundert*. Opladen: Leske + Budrich.

Putnam, R. (1988) Diplomacy and Domestic Politics: The Logic of Two-Level Games. *International Organization. 42*. 427–460.

Revelle, R. and Suess, H. (1957) Carbon Dioxide Exchange between Atmosphere and Oceans and the Question about an Increase of Atmospheric CO_2 during the Past Decades. *Tellus*. IX. 18–27.

Ringius, L. (1999) Differentiation, Leaders and Fairness: Negotiating climate commitments in the European Community. *International Negotiation 4*. 133–166.

Skjærseth, J.B. and Skodvin, T. (2003) *Climate Change and the Oil Industry: Common Problem, Varying Strategies*. Manchester: Manchester University Press.

Sjølseth, S. (1999) Explaining EU Climate Unity: Actors, Interests and Institutions. *FNI report 7/1999*. Lysaker: The Fridtjof Nansen Institute.

Skodvin, T. (1999) *Structure and Agent in the Scientific Diplomacy of Climate Change*. Dr Politics thesis, Department of Political Science, University of Oslo.

Tangen, K. (1999) About Negotiations and Leadership in Kyoto, *Internasjonal Politikk*. 1. 45–49 (in Norwegian).

Torvanger, Asbjørn (1998) Burden Sharing and Adaptation beyond Kyoto: A more Systematic Approach essential for Global Climate Policy Success. *Environment and Development Economics*. 3 (3). 406–409.

Underdal, A. (1991) Solving Collective Problems – Note on Three Models of Leadership. Challenges of a Changing World. *Festschrift to Willy Østreng.* Lysaker: The Fridtjof Nansen Institute. 139–153.

Underdal, A. (1994) Leadership Theory: Rediscovering the Arts of Management. *International Multilateral Negotiation: Approaches to the Management of Complexity.* 178–197.

Wettestad, J. (2000) The Complicated Development of EU Climate Policy, In Gupta, J. and Grubb, M. (eds.) *Climate Change and European Leadership.* Amsterdam: Kluwer Academic Publishers. 25–47.

Young, O., (1991) Political Leadership and Regime Formation: On the Development of Institutions in International Society. *International Organizations* 45 (3). 281–308.

Young, O.R., (1998) *Creating Regimes Arctic Accords and International Governance.* Ithaca, NY: Cornell University Press.

9 NGO Participation in the Global Climate Change Decision-making Process

A Key for Facilitating Climate Talks

Norichika Kanie

Introduction[1]

The Preamble to the United Nations Framework Convention on Climate Change (UNFCCC), which has been ratified by 192 countries including the United States (US),[2] acknowledges that "the widest possible cooperation by all countries and their participation in an effective and appropriate international response" is necessary for tackling "the global nature of climate change."

By the time of the first Conference of the Parties (COP1) after the entry into force of the Convention, it was becoming clear that many countries of the Organisation for Economic Co-operation and Development (OECD) were not on course to fulfill "an effective and appropriate response" and that commitments beyond 2000 were necessary to address climate change as a global and long-term issue (Grubb 1999; Oberthür and Ott 1999). Thus, the Berlin Mandate was agreed, under which the "Conference of Parties, having reviewed (the commitments) and concluded that these are not adequate, agrees to begin a process to enable it to take appropriate action for the period beyond 2000, including the strengthening of the commitments of Annex I parties." The negotiation which followed in 1997 agreed the text of the Kyoto Protocol, which included quantified GHG reduction objectives for Annex I countries for the period 2008–2012. The Kyoto Protocol entered into force on 16 February 2005.

However, in contrast to the UNFCCC, the Kyoto Protocol, which recalls the provisions of the Convention, has yet to achieve "the widest possible cooperation by all countries and their participation." The number of countries that have ratified is still lower than the number of parties to the Convention, and more importantly, the Kyoto Protocol lacks the participation by the largest emitter of greenhouse gas (GHG) in the world, the US. The *lack of global participation* on a global issue is the stumbling block that will be addressed in this chapter.

The approach to overcoming such an obstacle in the negotiating process is to facilitate and encourage participation in the negotiation. There are various solutions.

One way is to change the institutional framework to allow greater participation. The Kyoto Protocol includes only commitments covering the period 2008–2012, as no concrete commitments were made for the period beyond 2012. When the negotiation for further commitments started in 2005, as set out in the Article 3.9 of the Kyoto Protocol, this was seen as an opportunity to facilitate greater participation by means of institutional changes, even though doing so might somewhat hamper environmental integrity (NIES and IGES 2004; Bodansky 2004). To reduce GHG emissions, some nations preferred a country-based approach, others a sector-based approach. Some negotiation parties proposed that the issue should be dealt with by agreement among only the major players, but many others saw it as an opportunity for facilitating greater participation.

Another way is to direct facilitation activities directly to individual countries through diplomatic channels. This approach works occasionally when issue linkage is successfully made, as we saw in the Russian ratification process. It is widely recognized that Russian ratification of the Kyoto Protocol took place because it was linked to a submission tabled by the European Union in support of Russian membership of the World Trade Organization (WTO). Another case is that, since the US abandonment of the Kyoto Protocol, the Japanese and some European governments have used a variety of opportunities of bilateral (and some multilateral) talks to seek to convince the US to participate in the Kyoto regime. It is undoubtedly important for the US government to consider the perception of others. However, there was solid domestic opinion, within both the administration and the Senate, not to ratify Kyoto, and the US government does not normally change the course of its policy because of outside pressure. As historical institutionalism suggests, once a government program embarks upon a path, there is an internal tendency to persist in initial policy choices (Peters 1999). In the case of the US, the path can usually be altered only as a result of a great deal of domestic political pressure.

A third way of facilitating global participation is to support the increased participation of non-state actors, which facilitates not only global "participation" in a broader sense, or multi-stakeholder participation, but may also alter norms and domestic public opinions, and thus indirectly influence the position of a country. Given the growing role of nongovernmental organization (NGOs) in environmental policy processes and related networks across the globe, this mode of facilitation may have greater impacts in overcoming the stumbling block of constrained participation than one might at first think (Princen and Finger 1994; Potter 1996; Newell 2000). As Chasek (2001: 29) put it: NGOs are "increasingly serving as a catalyst" to initiate environmental negotiations. This chapter investigates how more NGO participation may help to facilitate the climate talks.

The chapter starts with a brief review of the formal decision-making structure of the UNFCCC and the status of NGOs in that structure. This is followed by a section identifying six modes of climate NGO engagement in the climate talks as (1) activist, (2) advisor, (3) observer, (4) legitimator, (5) implementor/monitor, and (6) in hybrid mode. The sixth, which is enabled by direct NGO member participation in the multilateral negotiation process in national delegations, will be further examined in the following section. The cases of Denmark, Switzerland and Canada are investigated more deeply – all have long had NGO members in their delegations to the UNFCCC COPs. It will be argued a coalition between state and non-state actor(s) would create new dynamics in the multilateral negotiation process, and thus also represent a potential for facilitating the climate talks. In the concluding section I will argue that organization of multi-stakeholder dialogs also facilitates participation.

UNFCCC Decision making process and nongovernmental actors

The existing decision-making procedure of the UNFCCC designates the Conference of Parties (COP) as the highest decision-making authority. COP meetings are normally held once a year, except in the case of COP6 where a further meeting (COP6 bis), took place because no agreement had been reached. COP reviews the implementation of the Convention and examines the commitments of parties in the light of 1) the Convention's objectives, 2) new scientific findings, and 3) experience gained in implementing climate change policies (Climate Change Secretariat 2002: 16). In principle, the Presidency of COP rotates among the five UN regions (Africa; Asia; Latin America and the Caribbean; Central, Eastern Europe and Western Europe; and Others), and often the country holding the Presidency offers to host the COP meeting in question. If there is no such offer, the session meets at the Secretariat in Bonn. COP decisions are taken in the plenary but negotiations on various individual issues usually take place in contact groups or other negotiation groups (see Figure 9.1).

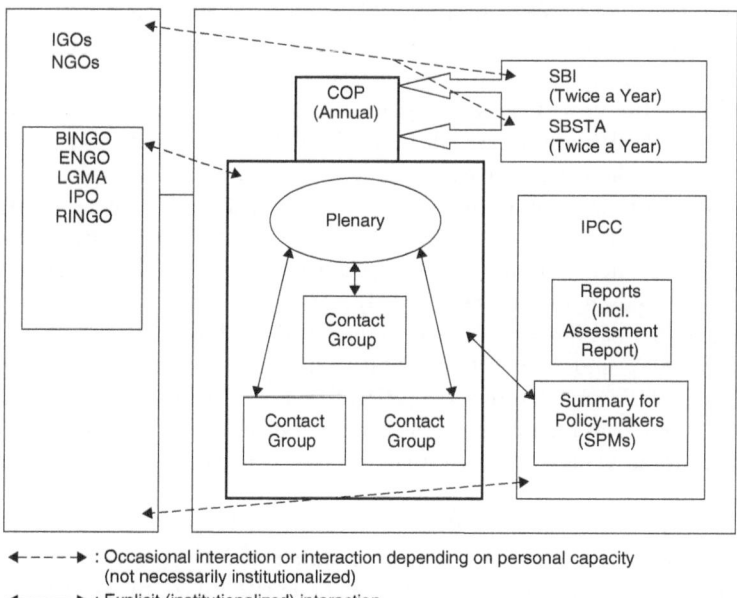

◀ - - - ▶ : Occasional interaction or interaction depending on personal capacity
          (not necessarily institutionalized)
◀——————▶ : Explicit (institutionalized) interaction

Figure 9.1 UNFCCC decision-making process and nongovernmental actors[3]

New scientific findings are expected to reach most of the policymakers (government officials) from the Intergovernmental Panel on Climate Change (IPCC), although in theory some basic information and research results may also come from other sources. Produced along with the IPCC Assessment Report, the best known IPCC publication appearing once every five years, are relatively short reports called executive summaries and summaries for policymakers (SPMs). These are probably the best source of scientific information for policymakers who do not usually have time to go through all the scientific reports and analysis. It should be noted, however, that SPMs are subject to line-by-line approval by IPCC working groups, and as the panel members usually consist of government officials and scientists, they are likely to be politically influenced (Kameyama 2004).

The FCCC has two permanent subsidiary bodies, the Subsidiary Body for Implementation (SBI) and the Subsidiary Body for Scientific and Technological Advice (SBSTA), which give advice to the COP, and usually meet twice a year. SBI deals with all matters related to the implementation of the FCCC. SBSTA deals with scientific, technological, and methodological matters. It plays a number of roles, including the important one of handling scientific information provided by experts such as the IPCC, and it also supplies policy-oriented information to them in return. In fact, it works closely with the IPCC and sometimes requests specific information or reports (Climate Change Secretariat 2002). Such a close relationship between a political body and the IPCC has created skepticism regarding the role and scientific credibility in terms of the IPCC being able to produce usable knowledge (Haas 2004).

Inside and outside the intergovernmental negotiation and international organizations, there is the NGO community. Through the Secretariat, NGOs are admitted to observe sessions at the conference venue, including the meetings of open-ended contact groups, and to observe intergovernmental negotiations.[4] In addition to acting as observers, a representative group of NGOs is invited to make statements to the COP under its agenda item "Statements by NGOs."[5] On some occasions, it is even allowed to make interventions. At COP9, for example, NGOs were given an opportunity to make interventions on two substantive agenda items in the COP plenary.[6] However, some high-level meetings and contact groups meetings such as the final negotiations of the compliance procedure were held behind closed doors (Gulbrandsen and Andersen 2004).

The number of NGOs registered to the COP sessions (from COP1 in 1995 to COP17 in 2012) and the number of NGO participants is shown in Table 9.1. The fourth row of Table 9.1 shows the ratio of total NGO participants as compared with the number of members of governmental delegation. With the exception of COP6 bis, held at shorter notice than the others because COP6 failed to reach an agreement (and, unusually, at mid-year), and of COP7, which was held right after 9/11, as well as after COP15 which witnessed a historic number of NGO participants, the record shows that the number of NGO participants exceeded the number of the governmental delegates. The interests of the NGOs in climate talks are thus not negligible. The UNFCCC secretariat recognizes that "NGOs play a pivotal role in the climate change intergovernmental process. Almost half

Table 9.1 NGOs in COP1–17

	COP1	COP3	COP4	COP6	COP6 bis	COP7	COP8	COP9
(1) Number of Delegation	869	2273	1430	2215	1819	2432	1468	1947
(2) Number of registered NGOs	165	236	148	275	219	194	168	267
(3) Number of registered NGO participants	979	3663	2357	3552	1587	1327	1858	2404
(4) Ratio ((3)/ (1)*100	112.7	161	164.8	160	87.3	54.6	126.6	123.5

	COP10	COP11	COP12	COP13	COP14	COP15	COP16	COP17
(1) Number of Delegation	2219	2809	2352	3516	3967	10591	5192	5413
(2) Number of registered NGOs	226	362	246	335	464	794	594	665
(3) Number of registered NGO participants	2888	5435	2533	4993	4463	12048	4560	4772
(4) Ratio ((3)/ (1)*100	130.1	193.5	107.7	142.0	112.5	113.8	87.8	88.2

Note: Total number of registered NGOs during COP2 and COP5 are not available on the official documents

of all registered participants of Convention bodies are from NGOs, with the share reaching 64 and 56 per cent in Kyoto and The Hague, respectively."[7]

NGO participants, or non-state actors, to UNFCCC climate talks are not limited to environmental groups. All participating organizations, except for governmental parties to the UNFCCC, are broadly speaking NGOs. Usually, however, intergovernmental organizations are acknowledged as IGOs (intergovernmental organizations, such as the OECD or UN agencies). In fact, the UN definition of an NGO in a resolution by the UN Economic and Social Council (ECOSOC) is "[a]ny international organization which is *not* established by intergovernmental agreements shall be considered as a nongovernmental organization" (Feld and Jordan 1994: 22). Currently over 985 NGOs and 67 IGOs have been admitted as observer organizations.[8]

NGOs form loose groups or "constituencies."[9] There are currently five constituencies acting as focal points for information exchange. Among them there are two with a relatively long history: the business and industry nongovernmental organizations (BINGOs) and the environmental nongovernmental organizations (ENGOs). Since COP1 there have been local government and municipal authorities (LGMAs), and since COP7 indigenous peoples organizations (IPOs). Most recently the research-oriented and independent organizations (RINGOs) have formed a constituency, some members of which are IPCC members.

In academic literature, NGOs in the climate talks may be defined differently. The Union of International Associations has set up a number of criteria to define international nongovernmental organizations (INGOs) that include aims, membership, governance, and financing.

The aims must be genuinely international in character and show intent to engage in activities in at least three states; membership must be drawn from individuals or collective entities of at least three states and must be open to any appropriately qualified individual or entity in the organization's area of operations; the constitution must provide for a permanent headquarters and make provisions for the members to periodically elect the governing body and officers; the headquarters and the officers should be rotated among the various member states at designated intervals; the voting procedure must be structured so as to prevent control of the organization by any one national group; and substantial budgetary contributions must come from sources in at least three states (Feld and Jordan 1994: 22).

According to this definition, many nongovernmental participants to the COP cannot be called NGOs, as the standards required are too high (e.g. that the funding source must be from at least three countries). Therefore, other organizations, have arisen – *pressure groups* – which seek to influence political decisions (Arts 1998; Willets 1982). Arts (1998) argues that pressure groups are composed of *sectional* and *promotional groups*. The sectional groups pursue the interests of a particular section of society, and include 1) economic groups (companies, commerce, trade, agriculture), 2) professional associations (doctors, nurses, lawyers, scientists), and 3) recreational groups (scouts). The promotional groups may be defined by general interests, including 1) welfare agencies (development associations), 2) religious organizations (World Council of Churches), 3) communal groups (indigenous groups, women's groups), 4) political parties, and 5) issue-specific groups (environmental organizations, peace groups). The main focus of this chapter is on the latter groups, particularly on the issue-specific groups on climate change. Other groups that come to the climate talks, such as indigenous peoples' groups (communal), can also be regarded as groups that have an interest in climate change. Some groups overlap, which means that some NGOs can belong to more than one group. The same can be said of NGOs interested in research, although some of these may overlap with professional associations.

Hereafter I will call the issue-specific groups concerned with climate change: climate NGOs. Business organizations, whose interests are in economic prosperity rather than facilitating climate talks per se, are not included in the group. The chapter's focus includes both activist and advisory NGOs, as long as they are climate-oriented (Newell 2000: 123; Gulbrandsen and Andersen 2004).

Modes of climate NGO engagement in climate talks

This chapter examines the best way for climate NGOs to break through the existing stumbling blocks in climate talks. With this in mind, we further conceptualize the current climate NGOs' modes of engagement in the climate talks which can be distinguished as: 1) activist, 2) advisor, 3) observer, 4) legitimator, 5) implementor/monitor, and 6) hybrid. NGOs may use one or more modes.

1. Activist
    Climate NGOs can engage in climate change negotiation process as activists, either by obtaining observer status to lobby negotiators *inside* the meeting, or by staying out of the institutional settings and protesting *outside* the venue. Many NGOs "involved in international meetings have also played a role in organizing protests that take place outside the conference halls during the meetings" (Fisher 2004: 179). For example, during COP3 in Kyoto, Japan, there was a 20,000-person demonstration (Rwomann 2000), and during COP6 in 2000 and COP6 bis in July 2001, demonstrations numbering 3,000 and 5,000, respectively (Fisher 2004: 184). Fisher (2004) argues that NGOs have a strong role in organizing protests; outside the conference venue, NGOs may draw attention of the negotiators inside to the need to shift the discussion to reflect public opinion or even block the negotiation itself. They can pressurize in their home countries and outside the conference through campaigning, rallying, direct actions, boycotts, and civil disobedience (Gulbrandsen and Andresen 2004).

    Inside the formal international negotiation structure, activist NGOs communicate the interests of their constituents to the representatives of governments and international institutions (Fisher 2004), putting pressure on negotiators, governments, and target groups, including industry, through lobbying and letters of protest. The UNFCCC secretariat has introduced web-based tools to "streamline the application procedures and to create an information resource with an electronic archive of side event presentations and reports," which helps climate NGOs to hold side events and gather information to enhance activities in this regard.[10]

    Generally speaking, long-term activities at home and through international networks (such as international campaigns) are more important than the activities at the conference venue, as the influence of activities relies on their main resource, that is, membership and public opinion. As Newell (2000: 129) put it, contact with diplomats at international meetings is only effective with popular backing.

2. Advisor
    NGOs with a strong intellectual base can engage in climate negotiations as advisors. They work closely with negotiators and government officials, and give policy recommendations and advice on legal, scientific, and technical matters (Gulbrandsen and Andresen 2004). Organizations such as the Center for International Environmental Law (CIEL) and the Foundation for International Environmental Law and Development (FIELD) are especially well known for giving advice to governments such as the Alliance of Small Island States (AOSIS) countries. They are said to "have counterbalanced the industrial lobby, worked closely with government officials and assumed

a new responsibility for the implications of their findings for policy options" (Chasek 2001: 28). Climate NGO engagement as advisors is not limited to giving advice to governmental officials but may also be directed to NGOs themselves or to both NGOs and, for example, BINGO members, to bridge the gap between the two. This does not influence the outcome of political negotiations in the short term, but can influence the long-term direction of climate talks, especially in an agenda-setting phase (Haas *et al.* 2004).

Advisor NGOs also hold side events and exhibits during the sessions, though which their research results are presented. They also attend workshops organized by the UNFCCC secretariat: among 14 workshops organized in 2003, ENGOs and RINGOs attended 10 each.[11] As the climate talks go into technical details, and given that most diplomats or policymakers change their position every few years, the role of advisory climate NGOs has grown in importance. Some advisory NGOs are even "qualified as intellectual leaders in the compliance system negotiations as a result of their ability to frame the compliance issue in a novel and constructive way" (Gulbrandsen and Andresen 2004; Young 1991). According to Chasek (2001: 231), NGOs "usually have better technical expertise than many governments and, thus, can assist and clarify the issues in the issue definition phase." During the negotiations themselves, governments can benefit from updated scientific, technical, and human-focused reports prepared by the nongovernmental community that shine new light on contentious issues.

3. Observer

Promoting transparency, while safeguarding negotiation effectiveness, is one of the UNFCCC secretariat's rationales in facilitating NGO participation. Observing the negotiation process remains a pivotal part of the NGO role. Fisher (2004) describes how international NGOs communicate the interests of their own constituents to the representatives of governments and international institutions inside the conference venue, and report the progress of the meetings to their members. In this way NGOs also increase the accountability of governmental delegations to the domestic constituencies. The most well-known observer climate NGO publication is *Earth Negotiations Bulletin* (ENV), published by the Canadian-based International Institute for Sustainable Development (IISD). Created in March 1992, the report covers every day of the negotiation and is usually distributed every morning; it has a readership of 34,000 worldwide, serving as one of the most reliable and comprehensive sources of information.[12] The newsletters of other NGOs such as the Climate Action Network (CAN) and the Kiko Network of Japan provide extensive coverage of the sessions and their own views on negotiations. Observing negotiations is of benefit to NGOs themselves. "By observing the negotiations, NGOs can obtain the information they need to follow the negotiation process, monitor the positions of governments, develop their own stance, and report back to their members" (Depledge 2005: 217).

Such observation activities, however, are not necessarily always welcomed by negotiators. Being observed by NGOs can prevent negotiators from speaking freely and also modify positions. Depledge (2005: 219) has shown how at Kyoto, Chair Estrada allowed parties privacy for bargaining and the negotiation took place behind closed doors; then, at the end of the deal making, he invoked transparency to place the strongest possible pressure on parties to compromise and not to block the agreements. In fact, the "absence of scrutiny from NGOs and the media is commonly viewed as pivotal to encouraging parties to speak more freely" (Depledge 2005: 219). When the negotiation goes into closed session, where NGOs cannot observe the negotiation and thus not implicitly pressurize negotiators, NGOs have to become "activists" and revert to traditional corridor diplomacy, lobbying negotiators and distributing documents during the sessions. This happens especially at an important phase of negotiation such as deal making in Kyoto or in the final negotiation on compliance issues (Gulbrandsen and Andresen 2004; Depledge 2005).

4. Legitimator

NGOs have important epistemic and legitimation functions in formulating international regimes (Hall and Biersteker 2002: 13). Especially on sustainable development–related issues, greater transparency and participation in the decision-making process may create greater stakeholder responsibility and government accountability. Thus, "the resulting agreements may be stronger than in cases where governments work in a vacuum" (Chasek 2001: 231). A unique aspect of climate change is that its impacts are most serious for the world's most vulnerable, whose voices are often neglected in global politics. Climate NGOs increase the legitimacy of negotiations by representing these vulnerable people. This is especially important in the case of climate change, where impacts take time to be recognized scientifically. NGOs associated with transnational social movements can pose "legitimate challenges to the existing international order" (Hall and Biersteker 2002: 13). Attention should be paid to the legitimator role of climate NGOs regarding climate change impacts.

5. Implementer/Monitor

An important and unique outcome of the 2002 World Summit on Sustainable Development (WSSD) in Johannesburg were the so-called partnerships for sustainable development – non-negotiated "type2 outcomes", based on General Assembly Resolution 56/226 encouraging "global commitment and *partnerships*, especially between Governments of the North and the South, on the one hand, and between Governments and major groups on the other", namely, voluntary multi-stakeholder initiatives aimed at implementing sustainable development that is complementary to the outcome of inter-governmental negotiation.[13] The guiding principle for the Partnerships for Sustainable Development states as follows:

> Partnerships should have a multi-stakeholder approach and preferably involve a range of significant actors in a given area of work. They can be arranged among any combination of partners, including governments, regional groups, local authorities, non-governmental actors, international institutions and private sector partners. All partners should be involved in

the development of a partnership from an early stage, so that it is genuinely participatory in approach. Yet as partnerships evolve, there should
be an opportunity for additional partners to join on an equal basis.[14]

NGOs are acknowledged "partners" in implementing sustainable development policies, and multilateral agreements on issues related to sustainable
development, including climate change.

Partnership initiatives registered to the UN Commission on Sustainable
Development (CSD), as mentioned above, are of a voluntary self-organizing
nature. In general, however, partnerships with NGOs in implementing policies are encouraged and important. In the realm of climate change, climate
NGOs are expected to be especially active in implementation of the Kyoto
Protocol and achieving UNFCCC objectives. The Kiko Network in Japan, for
example, works on environmental education and community-based climate
change prevention in the city of Kyoto to help achieve the 6 per cent emission
target of the Kyoto Protocol.[15] Similarly, the World Wide Fund for Nature
(WWF) work internationally in collaboration with electric power companies[16]
on a campaign called "PowerSwitch!" to encourage change in the source of
energy used from fossil fuel to clean energy.

Equally important is the monitoring of compliance activities that are "essential to prevent misuse of the Kyoto Protocol in general and the flexibility
mechanisms in particular" (Gulbrandsen and Andresen 2004: 69), for example, preventing the Clean Development Mechanism (CDM) being used to
finance coal-powered plants, and monitoring the quality of CDM sink projects and the use of the CDM in building nuclear power plants (Gulbrandsen
and Andresen 2004: 69). Comprehensive websites on the CDM and, such as
"CDM Watch" and "Sinks Watch", function as important monitoring tools
for NGOs.[17] Project investors will be sensitive about NGOs making any project shortcomings public, particularly if this might lead to an international campaign (Gulbrandsen and Andresen 2004: 69). Furthermore, although NGOs
are not allowed free access to the CDM Executive Board meetings, they still
can observe the meetings through live webcasts on the UNFCCC website,
which gives complete transparency. NGOs can also submit factual and technical information to two branches of the compliance committee, although
neither branch needs to consider the information in its deliberations.[18]

Exchanging views and interacting with each other on such mechanisms as
the CDM and sink projects can enrich the knowledge base for future projects
and even provide better ideas on evolutional institutional design change. An
efficient way of complementing information exchange and facilitating further participation is the holding of multi-stakeholder dialogs (MSD). After
the entry into force of the Kyoto Protocol, climate talks and policies entered
a new phase: implementation of the Kyoto protocol with agenda-setting for
post-Kyoto negotiation from 2005. This has meant a focus on future institutional innovation through experiential learning. In these global-level discussions multi-stakeholders demonstrate how to implement decisions, and

which tools, strategies, and partnerships are needed for sustainable policy and policymaking that encourage engagement and commitment (Hemmati 2002: 215). In fact, reviewing the implementation of the Convention and examining party commitments in the light of experience gained in implementing climate change policies is one of the objectives of COP (Climate Change Secretariat 2002: 16). Further discussion of MSD will be provided later.

6. Hybrid mode

An activist NGO can play the role of observer, and an implementer/monitor can also work as an advisor, depending on the situation. A single NGO can play a few modes of engagement even within the period of one conference.

Some members of governmental delegations to climate negotiations are also originally NGO members. This is a unique mode of NGO participation in the climate change decision-making process, changing the nature of global governance and, in particular, facilitating broader participation at both governmental and nongovernmental level. In the other modes of participation noted, there is a clear distinction between governmental organizations and nongovernmental organizations. However, in this mode, NGO members become governmental delegates to climate negotiations such as COP, and wear a red badge, while NGO members wear a yellow badge. This kind of erosion of government–NGO borders may change the whole dynamics of NGO participation in international negotiation process and decision-making procedures, and thus may lead to long-term facilitation of climate talks.

Since COP1 there have been climate NGO members in some government delegations, although it is hard to identify which organization is a climate NGO, as there are 189 parties to the UNFCCC and each country has different names for respective climate NGOs in its own country. Still, some internationally acknowledged NGOs such as Greenpeace, the WWF, and the Climate Action Network (CAN) have sent delegations to climate negotiations. Among the countries that have constantly included climate NGOs in their delegation are Canada, Denmark, Indonesia, Philippines, South Africa, and Switzerland, whereas countries such as Japan and the US usually do not have any NGOs in their delegations.[19]

The six modes of NGOs engagement in climate talks are summarized in Figure 9.2.

Hybrid mode of climate NGOs: Cases of Denmark, Switzerland, and Canada

For a closer look at the function of the hybrid mode of participation by climate NGOs, I have chosen the cases of Denmark, Switzerland, and Canada, as these countries tend to have climate NGO members in their delegations, and also because they represent different types of engagement. Interviews conducted at major climate negotiations, including workshops attended by delegates to the climate talks, although not comprehensive, show that these cases cover most of the important lessons to be learned for facilitating global participation.[20]

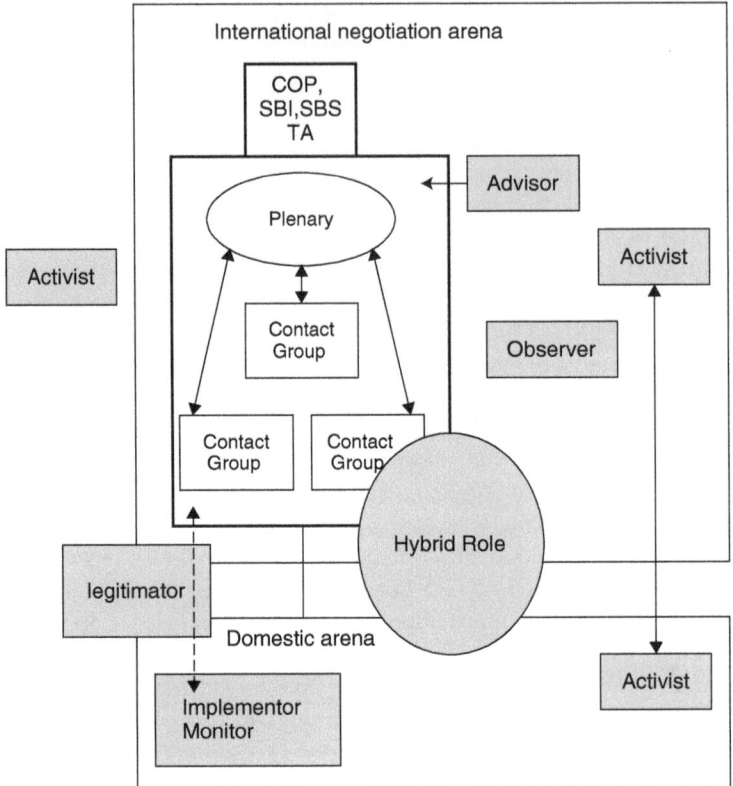

Figure 9.2 UNFCCC decision-making process and engagement of climate NGOs

The case of Denmark[21]

The Danish government started close collaboration with NGOs in the early 1990s in the process leading up to the UN Conference on Environment and Development (UNCED, Rio de Janeiro, 1992) on the grounds that sustainable development needs wider participation and transparency. The platform on the NGO side is the "Danish 92 Group", a coalition of 20 Danish NGOs that came together to work on environment and development to coordinate Danish NGO preparations for UNCED. Cooperation has continued since then, integrating environment- and development-related NGOs in the spirit of Rio Conference.[22] They have a secretariat in Copenhagen, and meet approximately once a month.

Delegates from the Danish NGO community have joined government delegations to climate change conferences since the 1993 International Negotiating Committee for a Framework Convention on Climate Change (INC), even before COP started. Usually there is one environment and one development delegation. Selection is usually decided by the 92 Group, which also sends delegations (in governmental delegations) to WTO meetings, among others, and their position paper is announced beforehand. This is not necessarily the same as the government position, but interestingly, up to 2001, almost 90 per cent of the positions coincided. If positions are different, a gentleman's agreement is made: before criticizing the position of the government in public, they raise the issue beforehand with the delegation, usually the delegation head. Danish NGOs also attend other international meetings, for instance, European Union meetings.

In those meetings, as very sensitive information can be heard, NGO delegates are sometimes asked to step out of the meeting, depending on the situation and the personality of delegates. There is another gentleman's agreement that NGO delegates do not disclose any information they hear in the meetings they attend until it appears from another sources.

Governments benefit from having a few NGOs in the delegation, as they can get to know the NGO position and developments, and this has a greater and more timely impact on the direction of the negotiation. Furthermore, government has an interest in NGOs acting on the basis of correct information, not on wrong information or rumors. In a sense, the delegates, even though from NGOs, are restricted by being part of the government delegation, and cannot speak freely as long as they have the red badge. However, they also have a higher chance of achieving their arguments if they are in the delegation. In Denmark's case, the position of the NGO and that of the Danish government – perceived generally as one of the most environmentally friendly governments – have been very similar, and this has made for close collaboration. At COP9, for example, a government speech referred to an NGO delegation's comment.

NGO delegates should thus have a good sense of their own broader views and of the prevailing political atmosphere, and be trusted by both sides, to act as a "thermometer" between the NGO and the government, and to interpret wisely what each other is saying. In fact, one of the Danish NGO delegations chose not to participate in COP as a government delegate when coordinator of an international NGO so that he had more freedom to speak. In short, NGO delegations' role, position, and effectiveness depend to a great extent on their personal capacity, trust, and vision.

The case of Switzerland and Canada

Switzerland has included usually two representatives from civil society in its delegation since COP1: one from environmental NGOs and one from development NGOs. Sometimes a representative from business and industry (mainly from a chamber of commerce or employers' confederation) as well as a researcher is also included. The NGO coordination office decides who to nominate, but the record shows that a representative from WWF Switzerland is often nominated.

Before attending a COP, representatives are invited to preparation meetings, and at COP they can attend delegation meetings every morning. The information gained is confidential, but there is no additional confidentiality agreement. Sometimes they also attend informal negotiation group as part of the Swiss delegation.

For this there is "no detailed mandate from the government about what to say in the negotiation."[23]

Governments expect NGO delegations to inform them of the position of the NGO community, namely, the Climate Action Network, and to play a bridging role between the government and the NGO. NGOs do not expect explicit information on the government's position. These measures ensure lack of conflict between the Swiss government and NGO positions.

The Canadian case is more or less similar to the other two, but one big difference is that delegates are bound by a confidentiality agreement and cannot disclose information obtained when in the governmental delegation. NGO representatives are not invited to a closed meeting. Canada had a negative experience at COP3 when an NGO delegate leaked information. At COP9 a NGO representative was a researcher and did not have a close working relationship with the government or the NGO community.

Lessons from the cases: Interactive diplomacy to facilitate greater participation

The following points are learnt from the cases:

- The hybrid mode of NGO participation depends on the strategy of the government and on how the government perceives NGOs. If the government wishes to collaborate closely with NGOs, there is a place for NGOs in a delegation, should the NGOs find it valuable. However, if there is close collaboration with NGOs, involvement in the delegation may be unnecessary (Gulbrandsen and Andresen 2004: 60–61). The Netherlands government, for example, has maintained a good relationship and close informal interactions with NGOs, and both sides are satisfied with the NGO not being in the delegation.[24]
- The hybrid mode of NGO participation also depends on the historical relationship between the government and NGOs in the given state. The nature of domestic decision making in Japan, for example, is top down, and NGO expertise and knowledge have not often been utilized in its external policy. Canada, on the other hand, works closely with NGOs in building international regimes such as for the "Ottawa process" for the Convention on the Prohibition of the Use, Stockpiling, Production and Transfer of Anti-Personnel Mines and on Their Destruction.
- Participation can, but does not necessarily mean, influence. Personal trust between the NGO representative and other governmental delegates will often determine the level of influence the NGO side is given.
- NGO delegates are likely to be restricted in repeating what they learn at a Conference in return for wearing the red badge. These restrictions can take the form of a written or a gentleman's agreement.
- When part of the delegation, NGOs will obtain more information giving them greater potential to develop their own agendas, but this again depends on their relationship with government delegates and the government itself.

- With NGOs in the delegation, the government gains closer access to NGO position and their possible actions. They can also ensure NGOs receive correct information to prevent misunderstandings.

The will and strategy of the government and of the NGOs will determine whether or not NGOs join a government delegation. However, it may not really matter whether NGOs are in the delegation or not as long NGO engagement in the process as a whole is satisfactory to each of the parties. Having NGOs in the delegation certainly has a symbolic significance and, since the 1992 UNCED, there are a growing number of countries with NGO members in the delegations, especially to the sessions of Commission on Sustainable Development (CSD), including WSSD, and to other social issues such as World Conference on Women. Moreover, growing attention has been paid to NGO relations with governments (Princen and Finger 1994; Newell 2000). Some NGO participation may be symbolic, having little practical significance, and NGOs may have limited access to the actual negotiation table or limited capacity to use the information gained. Nevertheless, lawyers from the British-based foundation FIELD which participated in intergovernmental negotiations as advisors to countries of AOSIS, gained much experience on the negotiation of legal issues such as compliance issues (Gulbrandsen and Andresen 2004: 60). Again, the important thing is not whether NGOs are in the delegation, but the relations between NGOs and the government concerned.

The hybrid mode of NGOs participation also has the potential to further erode the border between the government and climate NGOs. By being in the governmental delegation, the NGO member can go beyond being advisor and even become a (quasi-) negotiator as an accredited member of the government. More importantly, NGOs can bridge the information and position gap between the government and climate NGOs and might be able to create a new coalition through "interactive diplomacy" (Cooper *et al.* 2002). Let me explain this further.

Interactive leadership, as explained by Andrew F. Cooper *et al.* (2002), takes the form of joint management between state and non-state actors, based on the interaction between like-minded countries and the NGO community. For example, the case of the global ban on anti-personnel landmines, although not an environmental issue, shows that "non-hegemonic states and transnational social movements can achieve diplomatic ends by working in partnership" (Cooper 2000: 10). By developing a good working relationship with nongovernmental actors, a country or a group of countries can work together toward a common goal, which would have a great impact on mobilizing public opinion within and outside the conference venue, and then exert influence back on the negotiation outcome.

Climate NGO members in the delegation may be able to make a coalition of like-minded climate NGOs, using their network. What is important here is that NGO members should be in the governmental delegation, which means that like-minded countries (or a coalition of like-minded countries) are behind such a coalition. In a state-centric international multilateral negotiation process, governments make coalitions and NGOs utilize their networks (Zartman 1994; Chasek 2001) but, by having NGOs in the delegations, there may be a hybrid coalition between like-minded governments and NGOs. The government, even if it did not make NGOs part of the government delegation, should have maintained close collaboration with them; however, if the NGOs are part of the delegation, the collaboration would be tighter, and there would be a positive "multiplier effect" between NGOs and government (e.g. producing better ideas). Environmental ministries and ministers, for the most part, have a relatively weak position in the governmental hierarchy, but through international interactive coalition they can gain more normative backup, through NGOs as their "moral authority" (Hall and Biersteker 2002; Hall 1997).

In the long term there is also a possibility of expanding the membership of the coalition to, for example, like-minded local authorities or states (in the case of the US), which would have growing importance for climate policy implementation, including the Kyoto mechanisms. This kind of facilitation to create a new type of transnational, hybrid coalition can become a new type of interactive diplomacy that may even become a new type of leadership in the bottom-up multilateral negotiation process (Cooper *et al.* 2002). Climate NGOs from non-Kyoto parties such as the US and Australia also come to climate talks. Taking into account that those NGOs, such as Environmental Defense (ED) or World Resources Institute (WRI) in the US, also often work with the Federal Government (Gulbrandsen and Andresen 2004: 60–61), they too may facilitate global participation in the long run. States are not able to participate in the Kyoto regime by themselves, for example, but their policy or normative basis may be influenced, or informed at least, by the network developed though such a coalition.

Conclusion

To go back to the original research question, in order to overcome the stumbling block of the lack of global participation in tackling climate change, what can we do to facilitate the situation? At conclusion of this chapter, we should be reminded that there are two levels of facilitation.

One level is relatively short-term facilitation, directly addressing a non-party to the internationally negotiated agreement of the Kyoto Protocol (and its future form). To this end, traditional diplomatic efforts and stronger international leadership really does matter. Climate NGOs may exert influence as activists to apply international pressure to the non-Kyoto party. They may also apply pressure by observing and criticizing the international behavior of the non-party. Furthermore, they may also be able to create a new way to frame the negotiation (e.g. to suggest new issue linkages, to create a new institutional setting, and so on) through their advisory capacity. Climate NGOs have contributed and will contribute to changing the dynamics of the international negotiation process, but when it comes to influencing the decisions by a non-party that is only peripherally engaged in the negotiation process, their engagement is limited to performing as a "traditional" pressure group. Another level is long-term facilitation to enhance global participation in a longer timescale. This kind of participation includes not only the participation of states, but also broader societal participation that has normative impacts and will in turn influence decision making at state level. It is in this respect that the current variety of modes of engagement of climate NGOs to climate talks matters more for facilitating participation. In addition to the three modes mentioned above, the hybrid mode of engagement can blur the gap between governments and NGOs, and further facilitate wider participation by interactive diplomacy between coalitions of the like-minded in the government and climate NGOs. Equally important for long-term climate facilitation are the roles of NGOs as legitimator and implementor/monitor.

NGOs can "serve important epistemic and legitimation functions in formulating transnational policy decisions, regime rules, principles, and decision-making procedures", and therefore, their participation in itself legitimizes the participation of their country of origin in the long-run (Hall and Biersteker 2002). Moreover, implementation of CDM projects may also bring the development NGOs on board, so that the number of participants who work on climate issue will also grow to promote further long-term facilitation on participation, I propose here the organization of multi-stakeholder dialogues (MSD). This may complement the current institutional scarcity of interactive information exchange between the intergovernmental bodies and implementor and monitor NGOs (or even among implementor and monitor only), and facilitate further participation, while at the same time securing an efficient negotiation process. If this could include the issue of climate change impacts, it could also legitimate an intergovernmental process in terms of impact, as well as complementing the scientific evidence that has an impact on climate change. It has been pointed out that human resources and available budget means do not allow such a process to be developed. However, it could bring new ideas and practical realism into the deliberations and increase understanding between different positions.

To review the implementation of the Convention and examine the commitments of parties in light of the experience gained in implementing climate change policies is one of the objectives of the COP (Climate Change Secretariat 2002: 16). Participatory and transparent exchange of views improves the prospects for a more effective enforcement and compliance system (Haas 2004). As climate change is a long-term global issue, only broader participation, both in terms of parties and in terms of civil society, can create a sustainable solution. For this reason, NGOs are key to facilitating long-term climate talks.

Notes

1 Part of the research for this chapter is funded by the Grant-in-Aid for young Scientists (B) (KAKENHI-15710033).
2 As of 24 May 2004. See http://unfccc.int/essential_background/convention/status_ of_ratification/items/2631.php.
3 Details are simplified in this Figure. More detailed account of the institutions of the Convention and the future institutions of the Protocol can be found at Climate Change Secretariat (2002).
4 FCCC/SBI/2005/5.
5 FCCC/SBI/2005/5, para24.
6 FCCC/SBI/2005/5, para24.
7 FCCC/SBI/2005/5, para37.

8 http://unfccc.int/parties_and_observers/items/2704.php.
9 http://unfccc.int/resource/ngo/const.pdf.
10 FCCC/SBI/2004/5.
11 FCCC/SBI/2004/5. With regard to other constituencies, BINGOs attended eight workshops, IPOs attended two workshops, and LGMAs attended one workshop.
12 For more details, please see http://www.iisd.ca/.
13 307 Partnerships are listed as of 3 May 2005. See the updated information at the following website. http://webapps01.un.org/dsd/partnerships/public/browse.do.
14 http://www.un.org/esa/sustdev/partnerships/guiding_principles7june2002.pdf.
15 http://www.jca.apc.org/kikonet/english/index-e.html.
16 http://www.wwf.or.jp/lib/climate/powerpioneers.pdf.
17 Their homepages can be found at the following addresses. http://www.cdmwatch.org/index.php (CDM Watch), http://www.sinkswatch.org (Sinks Watch).
18 UNFCCC Decision 24/CP.7, VIII, 3, 4.
19 Japan included five NGO members in their delegation to WSSD, but not in UNFCCC COP. An exception is at COP6 when they included a youth delegate from an environmental NGO.
20 Interviews were conducted with the delegates of the following countries and NGOs: Argentina, Australia, Austria, Canada, Denmark, EC, Finland, Germany, Indonesia, Japan, Malta, Morocco, Netherlands, South Africa, Surinam, Sweden, Switzerland, and US; CAN Europe, Greenpeace, United Nations University, FIELD, WWF (China, Netherlands, Denmark, Switzerland), IISD, and members of IPCC.
21 The section on Denmark is drawn from my personal interview with Mr. Lars Jansen and Mr. John Nordbo during the COP9 and March 2004 in Copenhagen. I am grateful for their generous cooperation.
22 http://www.92grp.dk/inenglish/Default.htm.
23 Personal interview with Swiss NGO delegation at COP9, December 2003.
24 Interview with Mr. Yvo de Boer (Ministry of the Environment) and Mr. Sible Shone (WWF Netherlands). Also see Kanie (2003).

References

Arts, B. (1998) *The Political Influence of Global NGOs: Case Studies on the Climate and Biodiversity Conventions*. Utrecht: International Books.

Bodansky, D. (2004) *International Climate Efforts Beyond 2012: A Survey of Approaches*. Arlington, VA: Pew Center on Global Climate Change.

Chasek, P.S. (2001) *Earth Negotiations: Analyzing Thirty Years of Environmental Diplomacy*, Tokyo, New York and Paris: United Nations University Press.

Cooper, A.F., English, J. and Thakur, R. (eds.) (2002) *Enhancing Global Governance: Towards A New Diplomacy*. Tokyo, New York and Paris: United Nations University Press.

Depledge, J., (2005) *The Organization of Global Negotiations: Constructing the Climate Change Regime*. London & Sterling, VA: Earthscan.

Feld, W.J. and Jordan R.S. (1994) *International Organizations: A Comparative Approach*, 3rd edn, Westport, CT/London: Praeger.

Fisher, D.R. (2004) Civil Society Protest and Participation: Civic Engagement within the Multilateral Governance Regime. In Kanie, N. and Haas, P. (eds.) *Emerging Forces in Environmental Governance*. Tokyo, New York and Paris: United Nations University Press.

Giddens, A., (1984) *The Constitution of Society, Outline of the Theory of Structuration*. Cambridge: Polity Press.

Grubb, M. (1999) Global warming in an international context. In *Environmental Issues for the Gulf: Oil water and sustainable development*. P. Kassler. London: RIIA/Brookings.

Gulbrandsen, L.H. and Andresen, S. (2004) NGO Influence in the Implementation of the Kyoto Protocol: Compliance, Flexibility Mechanisms, and Sinks, *Global Environmental Politics*. 4 (4). 54–75.

Haas, P.M. (2004) When Does Power Listen to Truth? A Constructivist Approach to the Policy Process. *Journal of European Public Policy*. 11 (4. 569–592.

Haas, P.M. (2004) Addressing the Global Governance Deficit, *Global Environmental Politics*. 4 (4). 1–15.

Hall, R.B. (1997) Moral Authority as a Power Resource, *International Organization*. 51 (4), Autumn. 591–622.

Hall, R.B, and Biersteker, T.J. (2002) *The Emergence of Private Authority in Global Governance*. Cambridge: Cambridge University Press.

Hemmati, M. (2002) *Multi-stakeholder Processes for Governance and Sustainability: Beyond Deadlock and Conflict*. London: Earthscan.

Kameyama, Y. (2004) The IPCC: Its Roles in International Negotiation and Domestic Decision-making on Climate Change Policies, In Kanie, N. and Haas, P.M. (eds.) *Emerging Forces in Environmental Governance*. Tokyo and New York: United Nations University Press. 137–156.

Kanie, N. (2003) Leadership in Multilateral Negotiation and Domestic Policy: The Netherlands at the Kyoto Protocol Negotiation. *International Negotiation* 8. 339–365.

Kanie, N. and Haas, P.M. (2004) *Emerging Forces in Environmental Governance*, Tokyo and NY: United Nations University Press.

Newell, P. (2000) *Climate for Change: Non-state Actors and the Global Politics of the Greenhouse*. Cambridge: Cambridge University Press.

NIES and IGES (2004) *Framing Climate Protection Regime: Long-term Commitments and Institutional Options*. Tokyo.

Oberthür, Sebastian and Ott, Hermann E. (1999) *The Kyoto Protocol. International Climate Policy for the 21st Century*. Berlin/Heidelberg/New York: Springer Verlag.

Peters, G. (1999) *Institutional Theory in Political Science: The 'New Institutionalism*. London/New York: Pinter.

Potter, D. (1996) *NGOs and Environmental Policies: Asia and Africa*. London/Oregon: Frank Cass.

Princen, T. and Finger, M. (1994) *Environmental NGOs in World Politics: Linking the Local and the Global*. London/New York: Routledge.

Willets, P. (1982) *Pressure Groups in the Global System*. London: Frances Pinter.

Young, O.R. (1991) Political Leadership and Regime Formation: On the Development of Institutions in International Society. *International Organization*. 45 (3). 281–308.

Climate Change Secretariat (2002) A Guide to the Climate Change Convention Process: Preliminary 2nd edn. http://unfccc.int/resource/process/guideprocess-p.pdf.

Zartman, I.W. (ed.) (1994) International Multilateral Negotiations: Approaches to the Management of Complexity. San Francisco: Jossey Bass.

10 Institutional Capacity Building to Facilitate Climate Change Negotiations

A Need for New Thinking

Lisa van Well and Angela Churie Kallhauge

This chapter explores the concept of *institutional capacity* within the context of the international climate change negotiations, and suggests how capacity building or development could be facilitated in this regard. It attempts to do this by exploring how capacity is manifested within the negotiation processes in general and, specifically, how capacity is essential for facilitating the ongoing and long-term negotiations within the climate change regime.

Rather than drawing on the examples of traditional capacity-building activities targeting regions in developing countries, this chapter examines the institutional capacity of the climate change regime as a negotiating system. It then identifies several stumbling blocks with regard to institutional capacity and suggests options, including some examples of temporal interventions, for overcoming these to facilitate negotiations. This institutional perspective centers on how the structures and actors of the negotiations can make processes more efficient, equitable, coherent, and transparent, thus boosting the capacity of negotiating parties. Yet, although the study concerns institutional capacity of the climate change regime as an institution, it implicitly, and sometimes explicitly, crosses over into aspects of systemic and individual capacity. This is because the institution or regime, as a collection of norms, rules, principles and decision-making procedures, is composed of actors and their interests, capabilities, and expectations.

Multilateral environmental negotiations in general, and those on climate change in particular, are characterized by a high degree of complexity and horizontal issue linkages (Winham 1977; Zartman 1989, 1994). Thus, effective participation in the climate negotiations requires access to, and command of, the relevant scientific knowledge and negotiation skills, as well as knowledge of the unique processes of the ongoing climate change negotiation process. It also assumes a thorough understanding of national interests and preferences. Actors that are not able to mobilize the necessary expertise to match their interests in issues on the agenda often perceive the outcomes as being neither just nor fair, in the sense of not being representative of their countries' interests. In many cases, they accept the outcomes without sufficient analysis of the implications, a situation that often leads to poor implementation and follow up.

There is therefore a direct relationship between capacities to affect the nature of the negotiation process and those to comply with the provisions of the United

Nations Framework Convention on Climate Change (FCCC), the Kyoto Protocol, and other binding agreements. The ability to implement the Convention and the Kyoto Protocol's provisions depends on the extent to which they respond to the interests of the implementing countries. This in turn hinges on the extent to which Parties can articulate their interests in the process and make informed and balanced decisions. The question of the participants' capacity is thus a crucial factor in the development of any negotiated regime.

Another crucial element is the capacity of the climate change process as an institution to be responsive and flexible to the needs of parties so that complexity may be managed. This chapter addresses the capacity of the climate change regime, with a focus on institutional capacity to facilitate the negotiation phase of the regime building process. In this sense we diverge from the focus on capacity for implementing the commitments of multilateral environmental negotiations, to a more institutional focus on the capacity of the negotiation process itself. Many projects and programs in the field, as well as academic works and evaluations, have addressed capacity building in climate change negotiations in the sense of boosting the negotiations skills of various groups, in particular those from developing countries by training activities (cf. Churie *et al.* 2000). Our analysis recognizes these types of capacity building interventions at the individual or country level, but goes further, to distinguish and evaluate the institutional dimensions of capacity and capacity building.

Conceptualizing capacity

Discussions on capacity issues have been going on for several decades but only started taking on increasing importance in the late 1980s and 1990s (Morgan 1999). Donor agencies in general, and United Nations agencies in particular, have embraced the concept of capacity development in their work, which they have regarded as a necessary prerequisite to creating the conditions for effective and efficient use of development assistance. These efforts largely focus on technical assistance and human resource development and are often conceived within the framework of specified program or project goals. As a concept, it is an inherent part of many development assistance policies and is regarded an essential ingredient for the assurance of sustainable development.

The concept of capacity remains ambiguous, and the range of interpretations varies, depending on whether they are applied to the individual or to an entity or system. It also has different implications if viewed within a short- or long-term perspective, resulting in different prerequisites and strategies.

Various authors and organizations have conceptualized capacity in more general terms. For example, Fukuda-Parr *et al.* (2002) give a general definition of capacity as it relates to the actor as "the ability to perform functions, solve problems and set and achieve objectives" (cited in Willems and Baumert 2003: 5). The United Nations Development Programme (UNDP), on the other hand, provides a broader definition which refers to capacity as the ability of societies, individuals, or organizations to perform functions effectively, efficiently, and sustainably (UNDP

1997). This definition of capacity building allows for flexibility in the interpretation of the scope, nature, and elements of capacity building and alludes to different levels where capacities exist, thus conceiving capacity building as a process through which individuals organizations, institutions, and societies develop abilities, individually and collectively, to perform functions, solve problems, and set and achieve objectives (Morgan 1999). It implies an emphasis on the net sum of the effect of capacity development initiatives at all levels. This means that capacity building becomes an integral part of achieving specific goals and that it is a drawn-out series of events that is not limited to specific interventions.

Stumbling blocks in the process of climate change negotiation

The climate change issue exhibits all the typical characteristics constituting the issue of *complexity* of multilateral negotiations. The issue itself is multidisciplinary, cutting across the social, ecological, economic, and political spheres. Addressing climate change has required a multidimensional approach and has ramifications for all sectors of societies. While there is a relationship between addressing climate change and achieving sustainable development, little attention, especially in developing countries, is paid to the urgency of addressing climate change in the present time. This is because of other pressing priorities and also the lack of knowledge on the nature of the linkages that exist between climate change adaptation and mitigation and sustainable development. There is also little effort made in many developing countries, in particular, to disseminate information regarding the probable impacts of climate change.

In addition to the complexity of the issues, the climate change regime is characterized by a *range of actors*. While the negotiations are driven primarily by states, the roles of civil society, and in particular of the market systems, are key to identifying modalities for response measures. The negotiator has to play several roles beyond defending national interests and defining the framework for action. She is confronted with a range of cross-cutting and cross-sectoral issues that are continuously under development, whose resolution takes time, and where the impacts to be addressed are long lasting and effects of mitigation are not expected within the short term. Negotiators are therefore faced with a number of challenges. These challenges pertain to 1) *stakes and preferences*, 2) *analysis*, 3) *negotiations*, and 4) *collaboration*.

First and foremost, each negotiation party must be aware of its stakes and preferences. If these are not properly delineated prior to negotiations, this may give an unfair advantage to those parties that have clearly analyzed their negotiating positions in terms of their Best Alternatives to a Negotiated Agreement (BATNAs). However, understanding stakes and interests is primarily the responsibility of the negotiating country, in terms of providing negotiators with clear and coherent mandates.

It is also particularly important to understand stakes and preferences vis-à-vis other parties. This will help lubricate the negotiation processes with a higher degree of procedural fairness.

With regard to analysis, negotiators are expected to understand not only the scientific dimensions of the problem but also how these relate to the social, political, and economic systems. Understanding the interface between national interests and concerns – as well as the potential for responding to the issues and the global arena where the interests of negotiating partners, the global market, and potential for action at that level – is crucial. While most negotiators may not have the necessary scientific aptitude to master all the dimensions of the issue, they are expected to be able to identify legitimate sources and partners to collaborate with in generating and evaluating information necessary in the process. The degree of uncertainty in the issues requires the negotiator to act out of caution with a high degree of creativity, not only in the definition of policies and measures but also in how these will be communicated back to the constituents at the national level.

The institutional structure of the negotiations themselves also poses a major challenge to the actors. The proliferation of formal and informal meetings means that actors who are unable to participate in all of them have to prioritize the issues to focus on or identify allies to work with. The negotiator should also be familiar with the rules and procedures of the negotiating system and have a good command of the English language (as this is the language of choice in informal settings within negotiations under the UN).

Another stream of activity pertains to knowledge creation initiatives. A myriad of meetings and conferences to assess different aspects of the issues are continuously underway. While these may be organized aside from the formal negotiations, they contribute to enhancing understanding of issues, lobbying support from other sectors, developing new options for the negotiations and building relationships among different categories of actors. Collaboration is therefore necessary at all stages of the process. At the outset, it is the necessary ingredient for mustering support to initiate the process and define the key issues to be prioritized for attention. Within the negotiations, it is necessary to maintain a focus on collaborating with the other negotiators in working toward the goal set. In the climate change process, collaborative challenges also pertain to how other non-state actors can be mobilized to act to address climate change. Negotiators then not only represent the governments but also have to work to identify ways of involving the private sector, in particular. This means that incentives for action have to be built into the resulting agreements and translated into national policies. The negotiator also has the task of reconciling the national potential to provide such incentives with that being proposed at the global level. The capacity demands posed by the climate regime process are therefore immense, and cut across all the different phases from the agenda setting to the implementation.

Constituting capacity

When considering capacity, it is important to take into consideration the level and scope of its application. Three levels of capacity can thus be distinguished. Capacity development activities can be targeted at individuals (micro-level), organizations, or institutions (meso-level) or at the system as a whole (macro-level) (Forss and Venson 2002). The aggregation of activities at these different levels constitutes overall capacity development which can be assessed at the different levels, with the system level being where the multiplicative effect of organizational and individual capacity development would be most evident. According to UNDP (1997), the most typical entry point for capacity development is the institutional/organizational or entity level. This may take the form of structural reform to streamline procedures, planning, training, strengthening of the administration, increasing networking capabilities, among others. The systems level refers to the enabling environment under which organizational entities and individuals operate. It is at this level that the incentives and frameworks that would enhance their effectiveness and efficiency are determined. The capacity dimension here relates to policies, legal frameworks, resources, and process (UNDP 1997: 8) that are characterized by improvement of the functioning of the system through governance reform initiatives, decentralization of certain services, policy reforms, and restructuring of government institutions.

A major dimension of capacity is at the individual level (UNDP 1997: 10). This is the level where most activities are normally concentrated, and these often take the form of human resources and skills development through training and education programs with a specific goal that is related to the responsibilities of the individual or group of individuals. This level is the point where operational change can be made in organizations. Apart from initiatives to strengthen skills and knowledge, such programs increasingly focus on accountability, performance, values, ethics, incentives and security (UNDP 1997: 10), which can be interpreted as the application of the skills and knowledge and thus the result of enhanced capacity. In certain cases, capacity-building activities are not bound to the organization or entity but are specifically targeted at individuals without any particular association to an organization or entity. Such activities tend to have a shorter-term goal and, in many cases, effect. Even when activities are targeted at individuals, the broader context or systems level should be taken into consideration. This includes the structural limitations and opportunities that affect the ability of the individual to act, and the incentives available to ensure that the capacity is retained and appropriately utilized.

While these three levels of analysis are applicable within the context of the climate change regime, more attention has been paid to the capacities of the individual than to the institutional and systemic levels. This chapter focuses on the institutional capacity of the climate change regime and how this can be enhanced as a means of facilitating negotiation. That said, the chapter in no way ignores the other two levels of analysis. In fact it is nearly impossible to discuss the capacity of an institution without implicit and explicit reference to the systems level and the individual level. Institutions are "not only discrete organizations (e.g. government agencies) but also, more generally, sets of rules, processes or practices that prescribe behavioral roles for actors, constrain activity, and shape expectations (Keohane 1998, quoted in Willems and Baumert 2003: 11). Thus institutions are meaningless without the systemic contexts that shape them and the individuals or actors within the context. In this sense, the study of institutional capacity must include systemic and individual aspects. The individual is at the heart of the institutional and systemic contexts for capacity and capacity-building measures. Likewise, the system attributes frame the contexts for institutional and individual capacity analysis.

Regarding capacity development in a broader context would thus ensure that the net sum of abilities to act within a given context is maintained. It would distinguish capacity development as a goal from capacity development that aims at achieving a specific other goal, for example the assimilation of a particular technology. If society or a state is viewed as a system consisting of individuals and organizations or entities working towards a common goal, then capacity development should be targeted to the needs of the different levels within the system, that is the individual, the entity, and the broader system (UNDP 1997). Accounting for the broader context and the different levels reflects, to a certain degree, the potential for the multiplicative effect that actions at a particular level would have. This effect, while it may relate directly to the system characteristics, is also conditioned by a number of indirect factors in terms of the institutional frameworks, responsibility, and impact of added capacity, issues and goals.

Capacity and its development are not static phenomena but are dynamic in nature. Capacity development is therefore linked to change and management of change at the three levels (UNDP 1997: 13). Over time, the changes can either be incremental or lead to transformation of the particular system (Figure 10.1), the crux being the scale of the capacity and the goal for its development. Incremental changes tend to occur as the number of those involved increases. The multiplicative effect of capacity development of individuals is manifested in the increase in

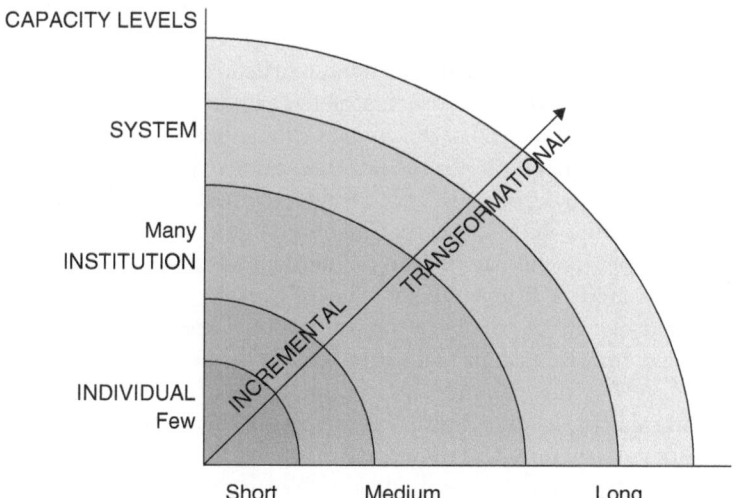

Figure 10.1 Incremental vs. transformational change

Source: Adapted from UNDP 1998, Capacity assessment and development in a systems and strategic management context. Technical Advisory Paper, vol. 3, p. 13.

the capacity of the entity, on condition that the individuals are retained within it. This, in turn, over time results in a net capacity increase within the system to a critical point, at which point longer lasting changes occur, thus a transformation.

Three components of capacity can be deciphered – capability, resources, and structural frameworks. Capabilities represent technical abilities, values, social (non-quantifiable) incentives, attitudes, cognition, values, cultural and social factors, relationships between different actors as well as the ability to undertake instrumental behavior to effect the desired change. Often, these factors are not considered directly in the design of capacity development activities. External interventions in identifying capacity needs may contribute to mustering the willingness to engage, but in some cases may be interpreted as impositions and thus have the opposite effect.

Resources, on the other hand, refer to the tangible assets available to support the actions being undertaken. These include financial, human, and technical resources. This is often the most apparent dimension of capacity and one that most interventions are directed to, as it is considered easiest to address. Provision of technical support, transfer of technical know-how and support through training programs, technology transfer and mobilization of financial resources characterize, to a large degree, capacity development activities.

The structural frameworks (Jänicke 2002: 5) for a particular actions being undertaken are of prime importance. These include the structures for the generation and use of knowledge, political and institutional frameworks, and economic and technological resources. The structural frameworks pertain to the institutional context of the negotiations and have two dimensions. One relates to the actual structures (physical setting, environment, procedures, rules and norms) which constitute the forum for the negotiations and include the setting of the negotiations, the resources available within this setting, the access and transparency of the organization, and the degree to which these structures empower negotiators to act. The other relates to the structures associated with the negotiation process and which are not directly involved in the actual negotiations. These include national institutions where preparations for the negotiations take place, and organizations within the research and epistemic community responsible for the generation of backstopping scientific information that is utilized in the process.

These three components can be seen as interdependent, in that they all have to be addressed in tandem to give the net effect of capacity. For example, the availability of resources without the necessary institutional frameworks or the willingness to act on them will result in a failure to achieve the ultimate capacity development goal. Likewise, the existence of a conducive environment characterized by awareness of capacity needs, political will and constructive attitudes, is incomplete if the institutional structures and resources are absent. It is often presumed that the resources for capacity building in individuals, particularly in the climate change regime, must be exogenous. However the focus on institutional capacity, which takes up the material as well as the nonmaterial resources, implies that nonmaterial resources may be drawn upon internally within the institution, often in the form of realizing previously unrecognized human potential.

The context: Capacity building under the UNFCCC and Kyoto Protocol

Capacity building is addressed in various articles of the UNFCCC, primarily as a means of achieving goals. The most explicit mention in the convention can be found under Article 9, which states that the Subsidiary Body for Scientific and Technological Advice shall provide advice on "ways and means of supporting endogenous capacity-building in developing countries." Within the Kyoto Protocol, Parties are committed to cooperating in, and promoting, "the strengthening of national capacity building" (Article 10e).[1] Capacity building cuts across many of the issues under consideration in the climate change process and has featured in several COP decisions. The issue of capacity building, however, was first considered as a separate agenda item only at COP5 (Bonn, October/November 1999) and has since remained an inherent feature, as the need to enhance the abilities of Parties is recognized in various articles.

During the fifth conference of the Parties in 1999, capacity building was considered for the first time as a separate agenda item. Two decisions, one on capacity building in developing countries and another on countries with economies in transition (EIT) were adopted. These decisions initiated a process to address capacity building in an integrated manner that takes into account existing capacity-building activities. It was through these decisions, that assessments of existing programs, their gaps and weaknesses, and the needs and priorities of the countries were initiated. They recognize that capacity building is a continuous process that is not limited to one sector or issue area but addresses all areas of relevance to climate change. It identifies the need for resources (financial, technological and know-how), structural frameworks (appropriate institutions), and capabilities (enabling environments). Annexed to the decision on capacity building for developing countries is a list of needs and priorities.

In 2001, the seventh conference of the Parties adopted frameworks for capacity building in developing countries and in countries with economies in transition. These frameworks set out the scope of capacity building and provide the basis for capacity-building actions related to the implementation of the Convention for participation in the Kyoto Protocol (in the case of EITs), and preparations for this participation (in the case of developing countries). They outline the guiding principles and approaches and acknowledge that there is no "one size fits all" formula for capacity building, noting that initiatives must be country-driven and focused on specific conditions, needs, and priorities.

While this is logical and expected, they do not explicitly address the crucial question of how the effective participation in the formulation and elaboration of the instruments would be achieved. One may argue that, while capacity development to participate fully in the negotiations is not directly alluded to in the decisions, the actions indirectly elaborated in the framework do address them. Building skills to assess country situations, enhancing the knowledge basis for decision making, and bolstering the capacity of national research and policy institutions, can all be seen as contributing toward providing a concrete foundation and enabling environment for participation in the global policy process. This feedback effect in the long term would enhance the ability of policymakers to make informed choices based on their national circumstances and on the state of know-how on specific problems. In the short term, however, the enhancement of the capacity to negotiate is not overt in these decisions, except where workshops and seminars to exchange knowledge on specific problems and to brainstorm on new avenues to pursue are provided for.

Technical language and scientific knowledge about the climate issue is not sufficient for effective negotiation. An enhanced capacity to negotiate effectively requires a combination of improved issue and process knowledge. Effective participation by the actors in the management of such complexity in an evolving negotiation process is a prerequisite for the development of instrumental strategies and tactics to serve the interests of an individual nation or coalition of states. The actor needs to learn how to master the use of complex knowledge at the negotiation table, with full consideration given to the constraints and opportunities of the process, which alter as the negotiation moves from agenda setting and negotiation on formulas to endgames and compliance problems. This way, the knowledge could be used effectively in framing issues to accommodate actor interests.

Facilitating institutional capacity of the climate change regime

Like the concept of capacity, institutional capacity defies tight definition in the absence of a more specific context delineating capacity "for what?" In our case, the object of the study is the process of climate change negotiations and ways to facilitate them. Thus the answer to this question is institutional capacity for the negotiations as a whole. Segnestam *et al.* (2002), as quoted in Willems and Baumert (2003: 10), conceptualizes institutional capacity as "a moving target. ... Today, institutional capacity often implies a broader focus of empowerment, social capital, and an enabling environment, as well as the culture, values and power relations that influence us."

Rather than conclusively applying a definition of a largely "undefinable" concept, we will rather look further into the institutional capacity of the climate change regime by employing three evaluative litmus tests: knowledge resources, relational resources, and mobilization capacity. This method of evaluating institutional capacity was developed in the past for a very different context by De Magalhaes *et al.* (2002: 54–58) in their study of urban governance and city center regeneration. It has also been used as a frame of reference for assessing the relationship between decentralization processes and sustainable development in local institutions in France, Sweden, and Russia (Veylon 2004). Still, the insights drawn from this framework for evaluation of institutional capacity can profitably be transposed to the level of international regimes with some modifications and augmentations for the different context.

Knowledge resources refer to flows of knowledge of various kinds among actors in a process, their frames of references, and how knowledge is processed within these frames (De Magalhaes *et al.* 2002: 55–56). Knowledge resources are incorporated into the structure of a system and thus incorporate contextual and cultural norms as well as the balance of power in the system.

Relational resources refer to the relational networks brought into the governance process by the stakeholders and other forms of social capital, including trust, fairness, and reciprocity (De Magalhaes *et al.* 2002: 56). Relational resources are focused on the agency or individuals and their identities. They include forms of partnerships, social networks, and channels of communication, including negotiations.

Mobilization capacity refers to the capacity of stakeholders to mobilize knowledge and relational resources to act collectively for a common goal, and thus implies the duality of agency and structure (De Magalhaes *et al.* 2002: 57–58). Mobilization capacity includes recognizing opportunity structures for change and is often stimulated by skilled "change agents."

Knowledge resources

Scientific knowledge and consensual knowledge

Knowledge generation and assimilation constitutes the most basic and obvious nonmaterial resource in evaluating capacity, be it individual, institutional, or systemic. In our analysis of the institutional capacity of the climate change negotiation process, the point of departure is the collective knowledge available to the delegates, both scientific and consensual. All delegates in principle have access to much of the scientific or "epistemic knowledge" produced by the Intergovernmental Panel on Climate Change (IPCC), the World Bank, and other groups. This knowledge comprises sophisticated climatology, geographic, economic, political, and other scientific models for predicting the environmental, social, and economic effects of climate change and the effects of mitigation or adaptation measures.

Yet this body of knowledge is so vast and often so complex that many parties with limited time, human, or financial resources find it difficult to make sense of the plethora or reports, analyses, and models available. Even briefer report extracts written specifically for policymakers may be beyond the scope of smaller delegations. To some extent this problem has been addressed through the efforts of many civil society organizations to aid in interpretation and assimilation of scientific knowledge, and this input into the negotiation system is an important resource.

The other type of knowledge that is vital to institutional capacity is "consensual knowledge." Consensual knowledge is an agreed-upon understanding of how an issue should be negotiated. This type of knowledge both builds on and is constrained by scientific knowledge, but is tempered by the collective values of negotiators. This is not to say that scientific knowledge cannot be consensual. The work of the IPCC is generally conceived of as a consensual effort (Alfsen and Skodvin 1998: 3), primarily because of its intergovernmental character. Consensual knowledge sets the conceptual framework for solutions implied from problem definition.

Consensual knowledge as a capacity resource is used as an institutional frame of reference in the negotiations; in fact, without such a frame, negotiations would not be forthcoming. But consensual knowledge accumulation also demands effort on the part of delegates. To negotiate effectively about consensual issues, delegates must have a good understanding and mandate of their own national and regional values and interests, both in absolute terms and relative to other delegates.

The processes of building knowledge are continual and cumulative. The greater an actor's capacity resources for assimilating knowledge are, the better-equipped a delegate will be to understand scientific knowledge and the better will be the prospect of creating consensual knowledge. However, transforming scientific knowledge to consensual knowledge is not without risks. As the character of knowledge changes in the process of negotiation, from scientific knowledge to consensual knowledge, important details and precision may be "negotiated away" by national interests. Young (2002: 189) asserts that, in one sense, this is inescapable: "All assessments, regardless of the care with which they are carried out, involved judgements that can never be strictly objective or unaffected by the beliefs and values of those that make them." A solution to this, Young proposes, might be to ensure that negotiating procedures are as transparent as possible, rather than being undercover bargaining on the part of self-interested actors (Young 2002: 189).

Constructing normative frames of reference

One structural element that encourages the self-interested actors to cooperate is the construction of common norms. Norms are defined as "collective expectations for the proper behavior of actors with a given identity" (Katzenstein 1996). Finnemore and Sikkink (1998) also note that norms "normally include standards for appropriate or proper behavior and by definition embody a quality of 'oughtness' and shared moral assessment". They prompt justifications for action and leave an extensive trail of communication between actors.

In the climate change regime, several norms operate as guiding devices for structure in the interests of the negotiating parties, and are the result of bargaining about consensual knowledge. Like consensual knowledge, they are a product of negotiation, but rather more institutionalized. An example of an institutionalized norm is the concept of "common but differentiated responsibilities," which has been lurking throughout the entire UNFCCC process, but became ingrained in the Delhi Declaration at COP8 in 2002. While this norm has little effect on the concrete attainment of the Kyoto Protocol targets, its acceptance sets the framework and thus eases further negotiations.

Relational resources

Forms of negotiation

But to what extent should the capacity needs that relate to the negotiations process be addressed through multilateral and/or bilateral channels? Arguably, it is the responsibility of the nation state to ensure that its interests are met within the processes. This means that the country should ensure that its negotiators and other representatives have a good understanding of what the country's interests are, what is acceptable or feasible within the country's contexts, which options they can propose or are ready to accept, and what support they may need in implementing actions to fulfill the objectives of the regime. Often many developing countries are unable to respond to all the demands placed upon them within the climate change and other processes. This limitation emanates from various forms of capacity limitation, not only in the preparation for the negotiations but also in the negotiating processes themselves.

While support to bolster the capacity to implement the outcomes of a negotiations process tends to be forthcoming, questions often arise regarding the appropriateness of supporting participation within the convention process. Often the discussion is limited to supporting the presence at the negotiating table, with little attention being given to enhancing the nature of that presence. Supporting the implementation is deemed appropriate, considering that the resulting agreement is based on consensus and, in turn, the interests of the supporting agent. But this is not as clear when considering negotiation capacity building.

As in all negotiations situation, two bargaining strategies exist: distributive bargaining or integrative bargaining (cf. Raiffa 1982; Lewicki and Litterer 1985). On the one hand, there are actors whose prime interest it is to "win" or advance their own positions. These actors see the negotiation process as a win or lose situation and engage in distributive bargaining to maximize their gains. On the other hand, there are those who view the process of negotiation as a cooperative effort where all involved will gain from the engagement. This form of bargaining, or integrative bargaining, means that the actors are focused on creating value within the process. The value can only be created if actors have, at the outset, an appreciation for and commitment to the need for collective action. This means that the participation of fellow actors, in these cases, negotiators, is necessary to ensure that value is created and distributed among all. It follows that it is in the interest of the strong actors to assist the weak actors to participate fully in the process to create value, which would be manifested in the nature of the outcomes and the perception of the outcomes (whether they are fair and just and representative of the actors' broader interests). Premised on this argument, capacity building as a tool to create value in a negotiations process is legitimized.

Whether various aspects of the climate change negotiations tend to be characterized by distributive bargaining or integrative bargaining depends not only on the strategies and mindsets, but perhaps more on the procedural aspects of the regime. Actors may be more likely to engage in integrative bargaining processes if they conceive of the negotiation setting or "rules of the game" as fair for all involved.

Capacity to negotiate fairly

International environmental agreements must be seen not only as effective, in terms of protecting the environment in a substantive outcome, but also as *fair* to attract participation, ease agreement among parties, and facilitate implementa-tion. Fairness concerns have especially permeated the climate change regime and the scholarly debate around it, particularly those dealing with how the burden of liability incurred from the damage of greenhouse warming should be borne in order to achieve a fair outcome, and by whom (cf. Agarwal and Narain 1991). However, a smaller number of studies address fairness concerns within the *process* of negotiation, and particularly with regard to the area of institutional capacity building.

In discussions of fairness, distinction is made between procedural fairness and justice of outcome. Justice is usually conceived of as an outcome depicting a con-sensual "common good," while fairness refers to the procedure by which the com-mon good is negotiated. In the climate change regime, the "common good" has been defined as sustainable development and one method of achieving this, in accordance to the Kyoto Protocol, is via a global reduction in greenhouse gas emissions. Albin (1999: 264) also makes a distinction between what is just and what is fair. She defines just agreements as agreements that are based on principles that the parties themselves have, of their own volition, agreed to honor. She states that fairness is the individual judgment of what is reasonable under the circum-stances, often in reference to how some principle of justice should be understood. Therefore "an outcome may be just in being in accordance with general distribu-tive principle, but unfair in how the principle is applied" (Albin 1999: 281). We examine here procedural fairness as a institutional facilitator of the climate change regime. The "rules of the game" are an essential structural element for achieving a just outcome and are an important relational resource for capacity.

As a guide to determining how processes of fairness ideally should be determined as collective decisions, John Rawls (1971) (re)introduced the theoretical concept of the veil of ignorance. Under this veil, which would screen out self-interest, actors or delegates would have only rudimentary information about the structure of the society in which they find themselves and other rational, self-interested, and equal agents. They have no awareness of structural or systemic conditions and do not understand their own interests and preferences in relation to this structure. In this "original position", any moral principles adopted will be generally ethical, as agents do not know where they themselves stand as a recipient of the fruits of justice, or pieces of the pie, if you will. This way the selection process cannot be influenced by self- interest and no-one's interests will be subordinated. On an indi-vidual level, agents will be under pressure to exercise caution in making rational distributive choices as, under the veil of ignorance, one could end up on the short end of the deal. It would thus be rational to try to "maximize what you would get if you wound up in the minimum, or worst-off, position" (the maximin strategy) (Kymlicka 1990: 65).

In Rawls' hypothetical original position, agents are endowed with equal duties, rights, power, and distribution of wealth. Yet this obviously holds neither in day-to-day life, nor in climate change negotiations, and Rawls' point in presenting the original position as a social contract argument is to summarize our notions of fairness and "help[s] us to extract their consequences" (Rawls 1971: 19–21). While the parties to the UNFCCC in principle enjoy equal status at the negotiation table, they are endowed with huge variations in wealth, power, and natural resources and capabilities. According to Rawls, this can be rectified under the difference principle. An inequality can be tolerated only if it improves the position of the worst-off. The important condition, however, is that everyone benefits from the transfer of assets. This has been addressed to some extent in the climate change negotiations in the form of differentiated, but common, responsibilities, and the range of greenhouse gas mitigation targets that allow for a relative improvement in the development standards of the least-developed countries.

In many respects, global climate change represents a classic commons problem. The economic costs of reducing greenhouse gas emissions are more than likely to outweigh the expected benefits for a party. It would thus be in each party's interest to have other parties follow the stipulations of the Kyoto Protocol, while free-riding themselves. Thus motivations to cut back would presumably be few and far between. Yet most countries have pledged to reduce greenhouse gas emissions under Kyoto and the distributive bargaining aspects of the debate are continually being addressed. Perhaps because agreement as to the scientific knowledge of the climate change debate has not wholly been transformed to consensual knowledge, the emphasis has been on the principles of bargaining.

Global climate change negotiations may be as close as we can get to a real-life example of Rawls' original position, where parties live under the veil of ignorance. Although a broad consensus has emerged from the scientific community that global climate change is "real, a necessary and continuing factor in our planetary history" (Mintzer and Leonard 1994: 14), the scientific evidence regarding the regional impacts of global change is still somewhat inconclusive. Some of these impacts may be positive in the short term; others may be devastating. And scientists are still uncertain as to the duration and scope of adjacent natural phenomena resulting from changing climatic patterns. Despite the plethora of global climatic models available, no one country may conclusively know how they will be affected, and few have the economic means for combating or adapting to these changes.

While negotiating parties obviously have some conception of the resources available to them to cope with climate change, they cannot be sure how hard they may be hit. The conception of the common good is not satisfactorily defined for these actors as consensual knowledge. As their stakes in the climate debate are still somewhat opaque, negotiating parties could be said to be operating under of a veil of ignorance. But theoretically, rational self-interest does not preclude justice under the veil of ignorance, as each party must decide on negotiation options that have the same consequences for themselves as they do for other parties (Kymlicka 1990: 64). Therefore, under a veil of partial ignorance, it would be advantageous for nations to proceed cautiously in constructing a process of distributive justice around climate issues.

Parties to the Kyoto Protocol obviously have a fairly good idea of their interests and stakes, and no party could be claimed to be operating under a veil of ignorance (although perhaps we would be better off negotiating under this veil!). Yet there is still wide variation in the structure of each party's capacity to negotiate in terms of knowing (and being mandated on) its own national interest, in assimilation of knowledge, and in understanding the current "rules of the game" in the climate change regime. This variation does exist, and those who assume to conclusively know their interests and stakes are at an advantage in terms of wielding greater power and influence under the negotiations.

As the "veil of ignorance" is an option only in theory, capacity development is needed to even out the playing field for parties that are at a relative disadvantage. Institutional capacity in terms of relational resources, such as fair negotiation processes, is essential for facilitating the negotiations. It is therefore important not only to increase the capabilities of the parties by attempts to understand one's own interests and stakes, but also the bargaining procedures should be structured so as to be as fair and transparent as possible to facilitate trust among Parties.

Networking and coalition building

As a relational resource, forms of networking and coalition building among the parties to the UNFCCC serve an important purpose in achieving procedural fairness. A negotiating structure that encourages processes of networking and cross-cutting coalition building among parties does much to boost the relational capacity of the regime as a whole.

For instance, the Group of 77, composed of 133 countries, is the largest developing country coalition operating under the UN. Within this group, other regional groups also operate as coalitions on specific regional and/or interest issues in the climate change process. The fact that the majority of the populations in developing countries either do not understand or do not care about climate change has fortunately not hindered developing countries from actively participating in the negotiations (Gómez-Echeverri 2000). Despite the fact that this group makes up the majority of the Parties to the Convention and Protocol, its contribution to the discourse does not match the level of interest one could expect, given the diversity of its members. As a unit, the group has been effective in providing analysis, synthesis, and political advice to a large group of developing countries which, as a result of their small and/or weak delegations, have had difficulties in keeping up with the pace of negotiations. However, the group has been less effective in uniting its members to counter the well-rehearsed positions presented by the wealthier Parties (Gómez-Echeverri 2000). They tend to limit their attention to discourses on specific issues, mainly those that relate to financing and technology transfer and with respect to ensuring that the developed country counterparts do not renege on their commitments in this regard. Invariably the main voices in the G77 have been those of the larger developing countries (Brazil, China, India, and Saudi Arabia) as well as that of the Alliance of Small Island Developing States (AOSIS).

Mobilization resources

Mobilization capacity is the ability of the institution and the individuals within it to bring about action to reach a specific goal. Knowledge resources and relational resources are important aspects of capacity and capacity building, but unless the two types of resources are integrated, the chances are that little will be accomplished. Indeed structural knowledge resources and actor-based relational resources constitute and reconstitute one another in a continual process, and there is a constant interplay between actors and their social structures (Giddens 1984). Structures are not static, but are constantly being redefined and interpreted by human actions, while at the same time, humans actions are patterned and constrained to some extent by the context of the social structure that we live in. This is what mobilization is about: the ability of actors, nations or delegates to recognize the structural framework under which they reside, and their efforts to change it. Mobilization capacity includes the skill to recognize endogenous untapped potential and the ability to generate and make good use of exogenous material resources.

Influence, leadership, and skilled change agents

Mobilization capacity is often stimulated by "skilled change agents" or strong leadership. Leadership is defined as the "asymmetrical relation of influence in which one actor guides or directs the behavior of others towards a certain goal over a period of time (Underdal 1994: 178). Influence in this regard refers to the successful use of power to produce an effect, without apparent exertion of force or direct exercise of command. While influence is the essence of actor interaction in negotiations, a distinction has to be made regarding the nature of the influence and thus the leadership. Underdal (1994) distinguishes two forms of leadership – *instrumental* and *coercive*. He defines instrumental leadership as that pertaining to finding means to achieve common goals. Coercive leadership refers to the imposition of an actor's preference on some other(s) or preventing others from doing so to others. According to Sjöstedt (1993), leaders tend to influence negotiations in all, or at least many, of the negotiations groups. Sjöstedt defines three qualifying conditions that define instrumental leadership. He notes that leadership constitutes calculated actions aimed at driving the process of negotiations in a certain or desired direction, that it exists more or less though the entire negotiations process or at least through the main developments, and that it is associated with the collective pursuit of some common good or joint purpose.

Under this definition, actors that demonstrate leadership qualities do not work only in their own interests, but have framed their interests within the broader context of the negotiation objectives. This also means that leadership in this sense is not exclusively conditioned by the "power" of the actor, and that the weakest actors (economically and politically) can provide leadership. The inability of a "strong" actor to provide instrumental leadership on a particular issue or set of issues can reflect a gap in their capacities, which may not be dependent on the availability of resources but resulting from a weakness of the capabilities. However, the net capacity deficits tend to be apparent in the case of many weaker actors (such as developing countries) which often suffer from a lack of resources and do not possess the necessary structural frameworks. It then follows that the capacity to negotiate is of concern to all actors with an interest in the fulfillment of the collective goal.

The demonstration of coercive leadership does not mean an absence of negotiation capacity. On the contrary, the capacities of actors playing the role of "brakers" in the process, and acting on the basis of their own interests, are determined by the extent to which they succeed in fulfilling their goals. This may not, however, contribute to the overall capacity of the process to fulfill the objectives set for it and, therefore, can be termed a negative process capacity.

Different actors have provided leadership within the climate change process in different ways. On the one hand, the expectation for leadership has been with regard to actions to mitigate climate change. Countries have looked to the big polluters to see if the momentum in the process can be increased if they set an example. On the other hand, there has been leadership relating to the formulation of the rules and modalities within the regime. This leadership is apparent at different points in the process as being provided by different actors. For example, the AOSIS, despite their limitations, small delegations, and extreme vulnerability to climate change, have provided instrumental leadership in that they have struggled to maintain the environmental integrity of the process by continuously reminding parties of the reality of the climate change impacts. Another example of instrumental leadership was provided by Brazil in the proposal that led to the formulation of the clean development mechanism of the Kyoto Protocol. Several other countries have provided leadership, but no one party has emerged as the overall leader within this process. It is this leadership, coupled with an indication of willingness to take on remedial measures, that creates the momentum to engage within the regime.

If an actor is described as being primarily passive on all issues, this signals a weakness in the actors' ability to participate in the negotiations, the absence of a clear mandate, or weak appreciation of the actor's own interests. A general reactive stance in most cases reflects a lack of mandate, even when the negotiator knows the general interest area. In some instances, the negotiators can take a proactive approach based on "gut feelings" or their own expert knowledge. Even when this happens, it still poses a difficulty in terms of communicating back to the constituents at the national level the outcomes of the negotiations and the need to implement them. Capability in the negotiation process therefore refers to the ability and skill required to effectively table an offer to other participants in the process, react to proposals advanced by other participants, and the ability of an actor to provide leadership in the process.

Temporal interventions for institutional capacity building

By focusing this analysis on knowledge and relational resources and the mobilization capacity of the climate change negotiation process, we examine how temporal interventions for institutional capacity of the climate change regime may move the capacity-building interventions from the well-developed individual level to the institutional level and, optimally, the systems level.

Knowledge resources, such as common frames of references and consensual knowledge, have long been acknowledged as imperative for negotiating processes.

Increasing *short-term* interventions in the agenda-setting stage of negotiations for attaining such knowledge and frameworks may, in the long run, be more conducive to attaining agreement. Transferable knowledge of this type is also the key to making negotiation processes more fair, and evening out the playing field for all parties involved. Greater transparency of procedures and structures is one starting point for this. *Short-term* interventions on the part of skilled agents may take the form of capacity-building workshops. Depending on the focus of the content of the workshops, different results can be achieved. Two examples where workshops were held in preparation for the sixth conference of the Parties to the UNFCCC in 2000 illustrate the different approaches taken. The first, targeted at African negotiations, was held in Dakar, Senegal, while the second, for Latin American and Caribbean negotiators, was held in Miami, US. These two workshops took different approaches to preparing the negotiators. The Dakar workshop focused mainly on substantive issues, with negotiation techniques being addressed through simulation games based on actual negotiating text. The Miami workshop, in contrast, addressed mainly negotiation skills with a particular focus on the process, its rules, procedures, and organization. This type of intervention aims to increase the knowledge resources of individual negotiations.

Medium-term interventions may not be targeted at a specific event but represent ongoing, but time-bound, actions such as coaching and mentoring. Coaching and mentoring over a period of time introduces new recruits to the process of negotiation and are effective in addressing the nuances of the process. As relational resources, networking and coalition building could greatly aid in facilitating fair processes of negotiation, but only if the focus is on the consensual aspects of integrative bargaining, rather than on networking to increase the share of the pie for specific groups in a distributive bargaining situation. Relationships with allies, understanding networking and coalition building, and the unwritten rules of protocol are best addressed in this way. Training programs can also be used as a means of building capacity in a longer time frame. The training of potential negotiators can start at an early stage. That way, participants gain an appreciation of the negotiation process as they watch the progress made over a particular period of time. This also allows for training in how to manage relational aspects of trust, knowledge, and strategy to ensure more fairness in the process. Medium-term activities tend to have incremental effects on the overall capacity threshold and, over time, can transform structures that affect the use and sustainability of the capacity. Within the negotiations context, these pertain mainly to the development of relational resources, especially those that play supporting roles to the process.

Long-term capacity activities are often pursued as a way of transforming structures for systemic change and providing the right environment under which capacities can be sufficiently utilized, not only to empower other delegates but to mobilize capacity from the individual level to the institutional level and eventually lead to transformation of the system. Institutionalized capacity development tends to be long-term. While individual negotiators can be influenced during the short and, to a certain degree, the medium term, the long term requires mechanisms to be put in place that will ultimately produce the necessary mobilization capacities. These pertain to interventions in education systems to incorporate the training necessary, intersectoral communication and cooperation to ensure that preparations are duly made, and a proactive culture to be nurtured at the country level. Such broad-based capacity building is not limited only to negotiating within specific regimes but results in better decision-making and governance processes. The responsibility for the initiation and continuity of such programs lies mainly with the recipient and depends on recognition of the broader capacity needs.

It is evident that many questions can be raised with respect to the logic underlying supporting capacity building for negotiations. This means that it is more acceptable to provide assistance in a situation where there is consensus and thus the bias toward more implementation-oriented capacity-building initiatives. Nevertheless, it does not mean that supporting capacity-building activities that strengthen the ability of parties to participate effectively in the process should be overlooked. On the contrary, the involvement of other parties in providing support to build negotiation capacity could contribute to enhancing the legitimacy of the process and the reputation of the other parties involved. It would also nurture a sense of trust and a willingness to explore options of mutual interest. The demonstration of willingness to recognize the limitation faced by certain parties creates an environment in which impasses exist and where past experiences in other forums can be overcome. However, the framing of such capacity-building initiatives, and the identification of the needs, has to be driven by the recipients. This would foster a sense of ownership and provide a concrete inroad and security for legitimate activities that would enhance the overall capacity of the process.

Institutional capacity building is thus an effective tool for managing the complexity of the negotiation process and the implementation of resulting instruments. By overcoming the constraints on actors' participation, it provides opportunities to actors to extend their potential and thus strengthen the foundation for stronger more effective regimes.

Note

1 UNFCCC 2003. Issues website <www.unfccc.int/ http://unfccc.int/issues/capbuild. html>, Updated, 10 April 2003, Accessed on 17 July 2003.

References

Agarwal, A. and Narain, S. (1991) *Global Warming in an Unequal World*. New Delhi: Centre for Science and Environment.

Albin, C., (1999) Justice, Fairness and Negotiation: Theory and Reality, In Berton, P. Kimura, H. and Zartman. I.W. (eds.) *International Negotiation: Actors, Structure/Process, Values*. New York: St Martin's Press.

Alfsen, K. A, and Skodvin, T. (1998) The Intergovernmental Panel on Climate Change (IPCC) and Scientific Consensus: How Scientists Come to Say what they Say about Climate Change. *Policy Note. CICERO*, Oslo.

Churie Kallhauge, A. and Gupta, J., (2000) *Preparing for COP6. Tiempo, Global Warming and the Third World*, Issue 36/37, September. London: IIED.

De Magalhaes, C., Healy, P., and Madanipour, A. (2002) Assessing Institutional Capacity for City Centre Regeneration: Newcastle's Grainger Town. *Urban Governance, Institutional Capacity and Social Milieux*. Farnham: Ashgate Publishing Ltd.

Finnemore, M. and Sikkink, K. (1998) International Norm Dynamics and Political Change. *Journal of International Organisation*. 52 (4). Special Issue on International Organisations at 50: Exploration and Contestation in the Study of World Politics.

Forss, K. and Venson, P. (2002) *An Evaluation of Capacity Building Efforts of UN Operational Activities in Zimbabwe 1980–1995*, Available through http://www.un.org/esa/coordination/Chpt8.pdf, last accessed November 2003.

Gomez-Echeverri, L. (ed.) (2000) *Climate Change and Development*. New Haven: Yale School of Forestry & Environmental Studies.

Jänicke, M. (2002) The Political System's Capacity for Environmental Policy: The Framework for Comparison, In Weidner, H.M. and Jänicke, M. (eds.) *Capacity Building in National Environmental Policy. A Comparative Study of 17 Countries*. Berlin: Springer Verlag.

Katzenstein, Peter J. (ed.) (1996) *The Culture of National Security: Norms and Identity in World Politics*. New York: Columbia University Press.

Kymlicka, Will (1990) *Contemporary Political Philosophy: An Introduction*. Oxford: Clarendon Press.

Lewicki, R. and Litterer, J. (1985) *Negotiation*. Homewood, IL, Ohio: Richard D. Irwin.

Mintzer, Irving M. and Leonard, J.A. (1994) *Negotiating Climate Change: The Inside Story of the Rio Convention*. Stockholm Environmental Institute. Cambridge: Cambridge University Press.

Morgan, P. (1999) *An Update on the Performance Monitoring of Capacity Development Programmes: What are we learning?* Paper presented at the meeting of the Development Assistance Committee (DAC), Informal Network on Institutional and Capacity Development, May. Ottawa.

Raiffa, H. (1982) *The Art and Science of Negotiation*. Cambridge: Belknap Press of Harvard University.

Rawls, J. (1971) *A Theory of Justice. Cambridge*. Cambridge: The Belknap Press of Harvard University.

Sjöstedt, G. (ed.) (1993) *International environmental negotiation*. London: SAGE.

UNDP (1997) *Capacity Development*, Technical Advisory Paper 2, Management Development and Governance Division Bureau for Development Policy. New York: United Nations Development Programme.

UNDP (1998) *Capacity Assessment and Development in a Systems and Strategic Management Context*. Technical Advisory Paper, No. 3, Management Development and Governance Division Bureau for Development Policy. New York: United Nations Development Programme.

Underdal, A. (1994) Leadership Theory: Rediscovering the Arts of Management, In Zartman, W. (ed.) *International Multilateral Negotiation*. San Francisco: Jossey-Bass. 178–197.

Willems, S. and Baumert, K. (2003) Climate Actions and Institutional Capacity: Current and Future Challenges. Paris: OECD.

Winham, G. (1977) Negotiation As a Management Process *World Politics*. 30. 87–114.

Young, O. (2002) *The Institutional Dimensions of Environmental Change: Fit, Interplay and Scale*. Cambridge, MA: MIT Press.

Veylon, B. (2004) Decentralization and Sustainable Development: Local Institutional Barriers and Opportunities for Sustainable Development, Master of Science Thesis 04–061, KTH Master Programme Series. Stockholm.

Zartman, W. (1989) *Ripe for Resolution*. New York: Oxford University Press.

Zartman, W. (ed.) (1994) *International Multilateral Negotiation: Approaches to the Management of Complexity*. San Francisco: Jossey Bass Publishers.

11 Stumbling Blocks in a Sectoral Approach

Addressing Global Warming through the Airline Industry

Lucas Bobes

Introduction

At present, questions concerning the airline industry are becoming increasingly integrated into the agenda of global negotiation on climate change. However, in the pre-Kyoto world, the situation was different. The airline industry was a new issue element in the complexity of the climate talks. One function of this chapter is to illustrate a situation where a new issue approaches the negotiation table in the process of building a climate regime.

This chapter also illustrates some stumbling blocks in pursuing sectoral approaches to an international climate policy framework. Establishing aggregate emission-reduction targets for different sectors is gaining more attention from policymakers, particularly those who wish to address the gaps of the Kyoto Protocol. Sector-based mitigation targets address the importance of cost-effectiveness as well as equity; that was noted in the Bali Action Plan. Mandatory cuts should be sector based, for instance, because of the huge variation of energy needs of different sectors. A sectoral agreement only provides for emission reductions in specific sectors, according to which numerical targets will be determined by stakeholders within the sector. However, as this chapter will discuss, a sectoral approach faces several stumbling blocks that should first be conceptualized in order to enable the formulation of strategic facilitation measures; this is valuable because the sectoral approach is usually an excellent complement to any international climate agreement.

The airline industry growth and its contribution to climate change

The contribution of aviation to global warming is significant; it has increased in the last years and is expected to continue increasing. According to the Intergovernmental Panel on Climate Change (IPCC), in 1992 aviation contributed by 3.5 per cent to the total radiative forcing[1] (IPCC 1999). In 2007, IPCC noted that aviation contributed about 2 per cent of globally produced CO_2 and accounted for 13 per cent of fossil fuels consumed by transport.

Aviation is a growing and high energy intensive industry. At the same time, civil aviation provides fundamental economic and social benefits: In 2003, more than 1.6

billion people used the world's airlines for business and leisure travel, air transport provided 28 million direct, indirect and induced jobs worldwide and more than 40 per cent of world trade of goods (by value) was carried by air (IATA 2003). While there was a decline of 9.3 per cent in passenger demand and 17.4 per cent in freight demand by June 2009, the first physical signs of economic recovery were anticipated at this point (IATA 2009). International passengers load factors stood at 71.2 per cent, down from the 74.5 per cent recorded in May 2008. Furthermore, air cargo shipped 35 per cent of the value of goods traded internationally (IATA 2009).

Since commercial aviation started operations in the 1950s, the industry has experienced continuous growth. It is worth analyzing this trend with more detail, both to better understand past expansion and also to provide solid foundations for what can be expected in the future. Therefore we will briefly review the trends in the passenger and freight sectors and the weight of these two segments compared with military aircraft operations.

From the 1960s until the end of the twentieth century, passenger traffic grew at a more or less steady annual rate of 9 per cent (IPCC 1999). One of the most commonly accepted scenarios for the period 2002–2022 depicts slower growth, at an annual rate of 4.3 per cent in passenger seats (Airbus 2003), as shown in Figure 11.1. This means that in this period of time, the number of available passenger seats will more than double.

Some of the most critical factors that favour this remarkable rise include increasing regional economic activity (especially in Asia), the rapid expansion of the so-called low cost carriers, greater personal freedom to travel, increased leisure time, globalization, EU enlargement, and so on.

The growth expectations for the freight sector are 4.9 per cent annually for the period 2002–2022 in terms of available capacity measured in tonnes (Airbus 2003) (see Figure 11.2), slightly higher than for passengers. This is due to some specific circumstances such as the transition to the market system of the former Soviet

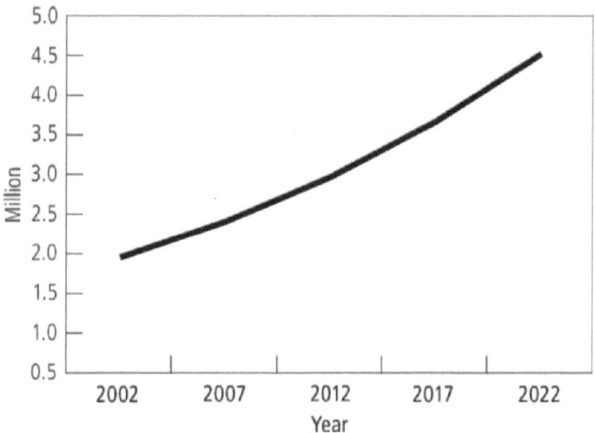

Figure 11.1 Passenger seats projected growth (Airbus 2003)

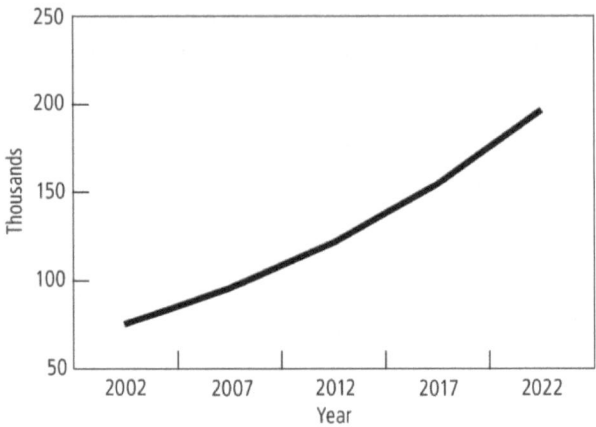

Figure 11.2 Freighter capacity growth in tons (Airbus 2003)

Union countries and the emerging Asian economies, as well as globalization. The freighter capacity is likely to multiply by three in the period 2002–2022.

It is important to take into consideration that, at this time, it is particularly difficult to produce reliable growth forecasts because the unprecedented events that have happened over the last three years – terrorist attacks, the SARS epidemic and wars in Iraq and Afghanistan – have strongly affected the way in which airlines operate their business. Enhanced security measures are, and will be, a more significant factor to consider. Furthermore, some governments are introducing new tax schemes for airline companies towards mitigating climate change. In 2009, US$ 6.9 billion were added by governments to the tax bill of airlines, when all other sectors were getting tax breaks (Bisignani 2009). Nevertheless, historically the airline industry has demonstrated noteworthy resilience to international crisis. This was the case, for example, in the previous Iraq conflict of 1991. Airbus considers that the crisis triggered by the 2001 September 11 attacks will delay growth by one and a half years (Airbus 2003).

Regarding military vs. commercial traffic, we have experienced a clear trend in favour of commercial traffic over the last 25 years, shown in Figure 11.3. The left hand side of the graph shows the split between the two types of flights, showing the fuel burned for each category.

Regarding the energy intensity of airplanes, the aircraft manufacturing industry has managed in recent decades to reduce significantly the energy intensity of transportation by air. Bisignani (2009) notes that the airline industry is committed to improving fuel efficiency by 1.5 per cent each year until 2020. Furthermore, he adds that IATA has set a target of 10 per cent alternative fuels by 2017. Some airlines have tested biofuels, making certification a reality by 2011. Figure 11.4 compares the energy-intensity of airplanes and passenger cars. The unit of comparison is the BTU (British Thermal Unit)[2] per passenger-mile. Obviously, it is

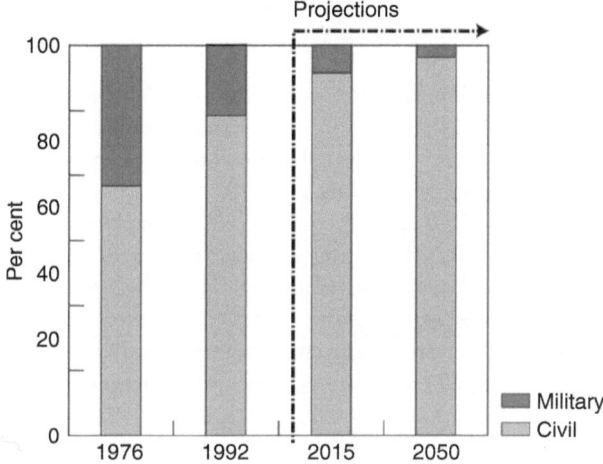

Figure 11.3 Civil vs. military aviation (IPCC 1999)

important to consider that the time required to produce emissions is much smaller for airplanes than for cars. At the same time, the infrastructure needed and related emissions are much greater for cars than for airplanes.

All in all, our conclusion regarding the airline industry is that anything but growth in traffic and emissions of greenhouse gases is improbable for the next two decades. The only circumstances that seem to dramatically bring air demand down are large catastrophes, which are difficult to predict by their nature and, of course, they are undesirable.

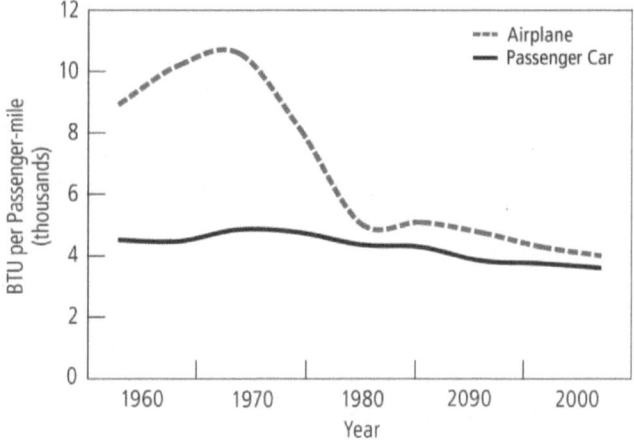

Figure 11.4 Energy intensity of airplanes and cars in BTUs[2] per passenger-mile (Bureau of Transportation Statistics 2004)

Regarding the impact of aviation over the climate, we have to say that understanding the contribution of the airline industry to global warming is complicated by a number of circumstances.

First of all, and unlike other type of local greenhouse gas releases, the effects of aircraft emissions depend on the altitude at which the airplanes fly, the temperature and the humidity. For these reasons, geographical point of emissions and time of the year are also principal variables that need to be examined when analysing greenhouse gas emissions from airplanes.

Second, there are various substances emitted by airplanes that contribute to global warming, principally CO_2, water vapour, NOx, formation of contrails and cirrus clouds, soot aerosol and sulphate aerosol (Bows *et al.* 2010).

From all these substances, emissions of CO_2 are often taken as a reference because they can be compared with emissions from other industries, since the greenhouse gas effect of CO_2 emissions is independent of point of discharge, unlike other types of airplane emissions. At the same time, CO_2 emissions can be reasonably easily quantified. It is estimated that CO_2 emissions represent about one third of the total global warming effect caused by airplanes (IPCC 1999).

Another substance emitted by airplanes, the Nitrogen Oxides (NOx), have a much shorter retention time in the atmosphere than CO_2. This short retention time doesn't allow uniform distribution across the globe and, for this reason, the point of emission becomes relevant; concentrations of NOx are higher in the zones affected by air traffic. This also means that the climatic effects of NOx have a more local or regional level than CO_2. Nitrogen Oxides affect the formation of ozone, which is a very powerful greenhouse gas. The intensity of this influence depends also on geographical point, time of the year and previous existing concentrations. NOx also contribute to the degradation of another greenhouse gas, methane, which has a long retention time in the atmosphere and consequently a global effect. Therefore, the understanding and measurement of the exact effects of NOx emissions from aviation is complex.

The water vapour emissions from airplanes have a significant effect in the lower stratosphere. Among other things, this means that aircraft flying at high altitudes, such as supersonic planes, play a more important role in global warming than those at lower elevations when it comes to water vapour emissions. Water vapour is emitted as a consequence of burning fossil fuel. The warm and humid water released by airplanes contrasts with the dryer and colder atmosphere, and the result is the formation of contrails as lines of clouds in the sky. If the weather conditions are relatively humid, these contrails don't vanish easily and instead they join other cirrus clouds naturally formed. In these cases, the warming effects of the contrails are more significant and difficult to isolate from the natural formation of cirrus clouds.

Moreover, as a result of the sulphur content of the aviation fuel and the incomplete fuel combustion, airplanes also emit sulphate and soot aerosol. The capacity to absorb and reflect radiation of these elements influences global warming. Sulphur aerosols have a cooling effect and soot aerosols have a warming effect. At the same time, aerosols facilitate the formation of cirrus clouds, since their particles help the water condensation process.

To the complexity and variety of substances emitted by airplanes, we have to add that there are some trade-offs in the climatic effects of these substances. For example, at high altitudes the emissions of CO_2 are minimized; nevertheless, the total climatic impact increases, due to the dominant counteracting effect of water emissions and ozone formation. Additionally, some of these effects are weather and geographically dependent and, for this reason, various global warming effects need to be considered individually and also in conjunction with the rest.

In the case of some particular global warming substances emitted by airplanes, like that of CO_2, the accuracy in determining their greenhouse effect is high, according to a significant part of the scientific community; nevertheless, the scientific uncertainties about the contribution of CO_2 emissions to global warming have been stressed by a International Civil Aviation Organization (ICAO) working paper presented by the Russian Federation (ICAO 2004b) in which it is suggested that changes in the concentration of CO_2 in the atmosphere are the effect of global changes in temperature, and not the cause of them. Surprisingly, the Russian Federation ratified the Kyoto Protocol only one and a half months later after presenting this argument. In other cases, such as the formation of cirrus clouds, the direct effects of aviation remain more uncertain and the margin of error in estimating them is very high. In the worst scenario, cirrus clouds could be the leading global warming effect from aircraft; on the other hand, the most optimistic estimations reveal that the formation of cirrus clouds by airplane activity could be negligible.

Apart from the scientific uncertainties described in the paragraphs above, a significant amount of information is needed to estimate the total climatic impact of aviation. Often, this information is expensive, difficult to obtain or simply unavailable. The following data can help determining the total climatic impact of aviation: amount of fuel consumption, type of engine, weight of airplane, flight duration, weather conditions, altitude and latitude.

Third, the climate effects and the future of civil supersonic aircraft fleet need further investigation; nevertheless it seems clear that the effects of supersonic aircraft are fundamentally different from those of subsonic airplanes. Supersonic airplanes have a cruising altitude of around 20,000 metres, as compared with the 9,000–12,000 metres of the subsonic planes. Since British Airways and Air France decided to stop commercial operation of the only civil supersonic aircraft in service, Concorde, the future for supersonic planes is uncertain. In any case, some major aircraft manufacturers (such as Boeing) have made attempts to approach the supersonic market. The Boeing Sonic Cruiser is an example of an effort to increase significantly the speed of regular airplanes. However, airlines have expressed higher interest in more economical planes rather than in supersonic ones, at least for the time being.

From the circumstances and complexities of understanding the global warming effect caused by the airline industry, we obtain our first and principal stumbling block in international negotiations for reducing aviation emissions. These complexities are preventing the international community from having a more advanced set of procedures, rules and penalties to control and minimize aviation emissions. Also, since the international community has judged it more easy to address other

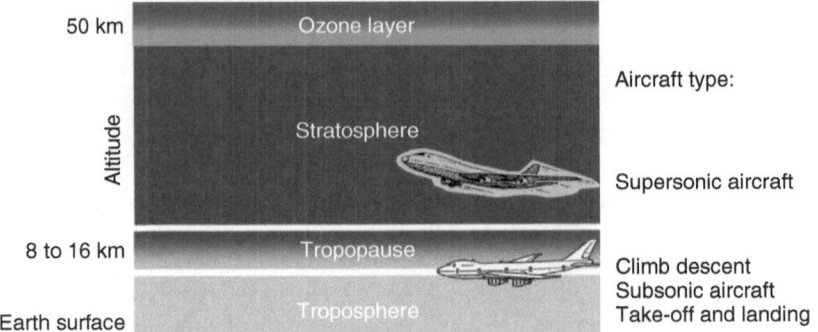

Figure 11.5 Aircraft flying altitudes and atmosphere layers

types of greenhouse gases releases, it has underestimated the potential benefits that an early action in aviation could have, both on the global warming effect as such, and also as a move towards a worldwide virtuous cycle to combat climate change provoked by humans. Following this line of reasoning, the stumbling block of the relatively small emissions from airplanes could potentially be capitalized upon to lead by example. In any case, under the scenario described earlier, the implementation of instruments to measure, control and reduce emissions from aviation becomes very relevant. In fact, the international community has already addressed these issues, although difficulties to arrive at concrete reduction commitments have not yet been overcome. In the following paragraphs we will review international negotiations aiming at including civil aviation in the climate regime.

Aviation emissions and the climate regime

Including aviation in the climate regime faces important stumbling blocks. On one side, the global nature of the climate change problem is common for all sources of greenhouse gases emissions, including airplanes, and this complicates the isolation of the aviation contribution to climate change and the quantification of the damage caused. Second, we have the specific characteristics of aviation emissions described in the previous section, which make the inclusion of aviation in the climate regime more complex. Finally, we have to bear in mind that it is estimated that 80 per cent of all air travel has an international nature; therefore it is difficult to assign responsibility for emissions of one flight to one particular country. On the other hand, progress in the scientific investigation of causes and consequences of climate change are increasingly putting pressure on high energy intensive industries like the airline sector. In the following section, we will review how these two contrasting elements – difficulty in reaching commitments and international pressure to reduce emissions – have been dealt by the international community and the current status of negotiations on the subject.

The signature in December 1997 of the Kyoto Protocol can be viewed as the first determined movement towards a carbon-constrained economy. From the industry point of view, this could be interpreted as a need to incorporate, sooner rather than later, the cost of emitting greenhouse gases into the business cycle. Article 2.2 of the Kyoto Protocol indicates that countries included in Annex I[3] should pursue limitation or reduction of emissions of greenhouse gases from aviation, and the ICAO received an specific mandate to work with the Parties of Annex I to achieve such objectives. Although Article 2.2 of the Kyoto Protocol is probably the major regulatory driver of any aviation emissions commitment to date, it faces at least two critical challenges: first, there is no time frame in which aviation emissions reductions should be achieved and, second, there is no quantification of such commitments.

However, it is not likely lead to weaker commitments from the aviation industry. The reason for this may be that emissions reductions have been linked or paired with energy efficiency measures (IATA 2001). The European Union has already adopted Directive 2003/87 introducing (as of January 2005) the largest emissions market in the world (Bobes 2004). Among other things, the entering into force of the EU Emissions Trading Market will mean that the EU subscribed to the Kyoto Protocol objectives even before Russian ratification was announced.

The coverage of the Kyoto Protocol regarding specific aviation emissions is limited to the inclusion of domestic flights emissions into the country's greenhouse gases emissions inventory. International flights, which account for approximately 80 per cent of civil aviation, bring a critical obstacle regarding the assignment of discharges, since there are several criteria that could potentially be applied: assignment according to point of origin or destination, nationality of the carrier, consideration of intermediate stop-overs of the flight, and so on. Given the difficulty in reaching agreement on the assignment of discharges produced by international flights, the Kyoto Protocol has left this crucial greenhouse gas source almost unaddressed. Also, with the United States being the largest market for domestic flights, and at the same time reluctant to ratify the Kyoto Protocol (KP), the inclusion of domestic flights in the climate regime through the KP is even less significant.

The European Union has addressed aviation emissions with different proposals that will be briefly discussed next, but the practical effects of these measures have not yet seen the light.

One of the first attempts has been the possible introduction of an aviation fuel tax (European Commission 2001). Since the Dutch EU Presidency of 1997, the EU has been discussing the possibility of introducing an aviation fuel tax for domestic and intra-EU flights. Aviation fuel is probably the cheapest fuel for any form of transportation. This fact has raised concern not only regarding the absence of internalization of environmental costs of aviation, but also on the subject of alteration of competition with other means of transportation.

The EU has worked together with the ICAO on the possible introduction of an aviation fuel tax and the main difficulties encountered have been the possible distortion of competition among airline companies, as well as economic distortion in general, since practices such as "tankering"[4] could be expanded, jeopardizing the benefits of such a measure and putting European carriers in a weaker competitive position.

The EU has also evaluated the implementation of an "en-route" charge. EUROCONTROL[5] suggests two alternatives: one revenue neutral option and one revenue-generating en-route charge.

In the case of the revenue neutral option, the most efficient airlines would get rebates funded by the least efficient carriers. A fundamental challenge for such initiative would be the development of an appropriate aircraft efficiency parameter. The revenue-neutral characteristic makes it more politically attractive but, at the same time, the environmental costs of air travel would remain outside the aviation business and, consequently, one of the major objectives of regulating aviation emissions would not be achieved. Also, the airlines of developing countries could potentially suffer all the costs, since their fleet are generally older and less environmentally friendly than those of developed countries.

The second alternative is a revenue-generating en-route charge. One possible way of a smooth introduction of these kind of measures would be to start with the implementation of a revenue neutral charge and progressively move towards the revenue generating charge. In any case, these "territory-based" approaches of the en-route charges are believed to lead to lower distortions of competition than the "fuel-based" structure described earlier (Dings 2004) and, for this reason, these measures would potentially be easier to introduce and be accepted by the several actors that have a say in the regulation and control of international aviation emissions.

A third way that the EU has tried to reach aviation emissions commitments has been the possible inclusion of the aviation sector in the EU Emissions Trading Scheme launched in 2005, but progress in this aspect has once more not happened at the desired speed. On one hand, the EU has not been effective in the communication to the general public on the benefits of the trading of emissions permits. The perception of such an instrument tends to be more in line with a form of right that industrialized countries get from developing countries to pollute, rather than a way to control and reduce emissions. The general public can be a very powerful actor in international negotiations to combat climate change, and international negotiators have not done enough to gain the general public's blessings and commitment.

As indicated above, Article 2.2 of the Kyoto Protocol requests Annex I Parties to pursue aviation emission reductions or limitations through the ICAO. Since domestic emissions are already accounted for in the assigned amounts of each country, it is generally understood that Article 2.2 refers to international aviation emissions.

In its efforts to comply with Article 2.2 mandate, the ICAO Executive Committee has recommended the adoption of a resolution to limit aviation emissions during the Kyoto First Commitment Period (2008–2012). Nevertheless, time has ran out to negotiate any specific commitment to be implemented so soon. In fact, negotiations for the Second Commitment Period (2013–2017) began in 2005.

In the thirty-fifth Session of the Assembly, the ICAO addressed aviation emissions and climate change from different angles. In the following paragraphs we will aim at describing the progress made and the barriers encountered.

First of all, market based approaches receive significant support and three possibilities are analysed: voluntary measures, emission levies and emissions trading. In the short term, ICAO recommends focusing the attention in CO_2 only, given the technical and scientific difficulties of expanding the market based mechanisms to other aviation emissions. At the same time, the ICAO admits that the lack of support from some countries to the Kyoto Protocol, as well as the delay in the entering into force of the agreement, have also delayed the process of introducing market based mechanisms to control aviation emissions. The discussion on whether or not it should be only the developed countries that carry the main burden, or if the developing countries should participate to some extent, has not yet been resolved.

The implementation of an emissions trading scheme for aviation has been recognized as a highly complex task, and further work on the subject is being carried out at the moment by external consultants at the request of the ICAO, which has expressed concern about the uncertainty of the results of the implementation of aviation emissions trading, although it recognizes that, theoretically, this market based system would maximize emissions reduction at the lowest possible cost. The debate on whether the aviation emissions trading should be integrated with other sectors, or operate in isolation, is still unfinished.

With respect to introduction of emissions levies, the ICAO has expressed preference for charges rather than for taxes. To date, the dichotomy between countries that accept charges versus those that don't has probably been the main reason for the slow progress on this subject. The discussion has also been centred on what should be the exact destination of the funds collected for such charges. The ICAO recommends a global aspiration of 2 per cent annual fuel efficiency of the international civil aviation in-service fleet. This has represented a cumulative improvement of 13 per cent in the short term (2010–2012), 26 per cent in the medium term (2013–2020) and about 60 per cent in the long term (2021–2050), from a 2005 base level (ICAO 2009). The ICAO judges the levies option as the least attractive, since the costs for some airlines could be critical and the environmental benefits very limited (ICAO 2004a).

From the aircraft manufacturing point of view, the industry is living in crucial times: two major manufacturers would have reasons to support progress of aviation emissions negotiations, since they are currently approaching the market with two new products based on revolutionary concepts. On one side, the Boeing 7E7, targeting the mid-sized sector, and on the other, the large Airbus 380. These new and more environmentally friendly airplanes would probably see their sales increased should airlines receive a mandate to improve their emission standards. Although significant progress has been made in the last decades in terms of fuel efficiency, no revolutionary technical concepts have been introduced in the airline fleets and some of the air fleet look quite similar today to how they did some decades ago. For example, the Boeing 747, which firstly appeared in the market in the 1960s, is still the principal aircraft for long distance and large capacity needs. The two major manufacturers mentioned, among others, could potentially capitalize on the emissions reductions commitments to foster a fleet renovation from airlines, promoting at the same time economic growth, technological development and employment. The ICAO pursues market-based mitigation measures through voluntary measures, emission charges and emission tradings (ICAO 2007). ICAO/CAEP furthermore developed a template to facilitate voluntary agreements and collects information for the purpose of information sharing among stakeholders. In addition, a new (draft) guidance document (ICAO Doc 9885) identifies a range of emission trading issues involved in including aviation in an open trading scheme (ICAO 2007).

Individual airlines have, in general, expressed concern and the desire to reduce the contribution of their businesses to global warming. The actions proposed are fundamentally of a voluntary nature and based on minimizing fuel use and optimizing operational measures. For example, British Airways is committed to a 30 per cent fuel efficiency improvement by 2010 on a 1990 baseline (British Airways 2004). Nevertheless, the airlines see any environmental objective as subordinated to the principal mission of satisfying air traffic demand, and consequently high cost options would most probably be rejected.

In the Table below we have aimed to briefly summarize the possible ways of achieving aviation emissions reductions and the advantages and disadvantages of each option.

The way forward

From Table 11.1 we can analyse what has been the progress to date and what are possible ways to move forward in order to achieve better results in aviation emissions and their climatic impacts. The first two options, namely, optimization of

Table 11.1 Options to achieve aviation emissions reductions: benefits and drawbacks

Description	Benefits	Drawbacks
Optimization of Fuel Burned (through engine improvement and airframe design)	Common interest with airline business. Relatively easy implementation	Progress depends on technical advances. Past improvements have proved insufficient
Optimization of Operational Measures	Common interest with airline business and economic development. No relevant disadvantages	Implies complicated agreement and difficult implementation
Implementation of Levies	They represent an incentive to minimize emissions and provide funds to mitigate impacts	Legal issues. Criteria for application of revenues obtained. Equity aspects. Sovereignty issues. Uncertain efficiency
Emissions trading	Minimization of costs to reduce emissions	Uncertain costs and effects. Possible complex implementation
Legal Limitation of Emissions	Environmental results guaranteed	Unrealistic approach. Emissions would not be limited at "any price"
Voluntary Measures	Ample room for manoeuvering	Uncertainty of results. "Goodwill" dependency

fuel burned and operational measures, have seen significant progress in the past decades; for example, airplanes produced today are about 70 per cent more fuel efficient per passenger kilometre than 40 years ago (IPCC 1999). Progress in air traffic control and management has also been impressive. Nevertheless the growth in the airline industry has counteracted the environmental benefits; and the emissions increase clearly exceeds the reductions achieved through new technologies and management systems. In conclusion, fuel and air traffic management optimization are necessary but not sufficient conditions to achieve the desired environmental results.

With respect to the next three options, namely, implementation of levies, emissions trading and legal limitation of emissions, it is reasonable to judge the progress achieved as unsatisfactory. The powerful economic interest of the air industry, the scientific complexity of aviation effects over the climate and the global complex negotiations are stumbling blocks that have so far jammed any chance of significant progress in addressing aviation emissions impact on the climate.

The implementation of levies faces opposition from the airline industry. Even the ICAO questions the efficiency of this measure, considering the drawbacks that would need to be faced. First, unlike other cases, it is difficult to quantify the damage produced by some specific emissions. As we saw in the first section of this chapter, the determination of the exact contribution of the airline industry to global warming is a challenge for the scientific community, and the quantification of the impacts and damages caused is even more difficult. It is therefore unrealistic to aim at fixing a levy that would be used to compensate the damage caused and also it would be very difficult to select a criterion by which to decide where the funds obtained should be applied. These uncertainties cannot provide a solid foundation for the implementation of the levies and the airlines fear that any measure of this type would be in part arbitrary and, consequently, could negatively affect fair competition between airlines or with other means of transportation. Other sources of possible distortions of competition could arise, such as the case where some countries adopt levies for their airlines and others don't. Also, any assessment of an emission charge produced over the high seas would require international agreement. In a case where charges were implemented, it is likely that a negative equity effect would occur, since the fleets of developing countries would most probably receive higher charges than those of the developed world.

The implementation of an emissions trading market requires agreement on the establishment of a limit to emissions. This is the way in which emissions rights would become scarce and therefore subject to trade. At the moment, the establishment of such a limit appears improbable in the short term, given the difficulties in reaching agreement and the uncertainties about which greenhouse gases should be included and how. Although relevant progress has happened in the emissions trading field over the last years, and several emissions trading schemes have been or will soon be launched, none of these plans include emissions from international aviation. In any case, there seems to be more support for an open emissions trading scheme in which the air sector could participate in wider schemes, benefiting from larger markets and probably lower prices. One major concern is the thin margin that the airlines have in reducing their emissions, which means that the air sector would most probably be located in the buyer side of the market. The implementation of an aviation emissions trading scheme has proved complex; since the ICAO recommended the development of an open emissions trading system for international aviation in 2001, the progress has been slow.

Another option could be to agree on broad and general "top-down" objectives and then see how the market and the industry can adapt to those goals. In my opinion, this is very much the scheme of the Kyoto Protocol, in which high level pledges and objectives have been set and approved first, and then each Party to the Protocol will work on achieving those objectives within the framework of its own commitments (Bobes 2004).

The Russian Federation ratification of the Kyoto Protocol and its subsequent entry into force proved that the KP approach is bringing some results, although we have to remember that it took seven years of intense negotiations to reach this result and that three of the top five emitting countries in the world are outside the KP emission reduction commitments, the US, China and India. Under these circumstances, it is relevant to evaluate what has been the opportunity cost for the ratification of the KP and what can be done in the future to accelerate commitments and further progress. Both for international aviation and for anthropogenic sources of greenhouse gases emissions in general, it is not realistic to set a quantified and unchangeable limit to emissions, particularly considering the close correlation of emissions with other factors as fundamental as population growth, economic development, welfare in general, and so on. Fernando Arlandis, Senior Manager Climate Change at PricewaterhouseCoopers expressed it very clearly. Referring to the fast economic growth that Spain has seen over the last years he said:

> If same growth trends continue, emissions will reach a 56% increase by 2010, compared to 1990, while our economy grows 66.4%. In other words, by all probabilities, Spain will not comply with the Kyoto commitments.
>
> (Point Carbon, February 04)

Mr Arlandis makes it very clear for the Spanish case: complying with the KP and maintaining a fast economic growth seems quite incompatible. Some similarities apply to other countries. These positive correlations between emissions of greenhouse gases and economic, population or welfare growth are, without any doubt, major stumbling blocks in today's negotiations on climate change in general and aviation in particular. So far, economic growth has historically implied increased emissions. Although technology and innovations have helped reducing that dependency, so far they have not been able to cancel it out. At the present stage of technological development, it is unrealistic to think that there is no trade-off between economic development and emissions. It will also be unrealistic to think that individual governments, which are currently the principal decision makers on the subject, will prefer to cut emissions at the expense of economic growth.

Mandating a fixed and unchangeable limit to emissions inevitably faces a hard opposition from many players, brings uncertainty into the economic system and, in many cases, discourages any kind of early voluntary action, therefore emissions should not be reduced at "any price". Additionally, considering how difficult it has been to come to a partial implementation of the KP, it doesn't seem likely that a global agreement limiting the international aviation emissions will be reached in the short or medium term. Under these circumstances, the role of other players in climate negotiations can be fundamental.

The work of the NGOs regarding aviation has paid special attention to the direct and indirect subsidies that aviation is receiving, given that these subsidies are an obstacle for the internalization of environmental costs of aviation and also a distortion to competition between different forms of transportation. Some organizations, such as Transport and the Environment,[6] are to some extent opposed to the ICAO recommendations in the sense that they favour unilateral European action in case significant international progress on aviation emissions control does not occur. Moreover, Transport and the Environment recommends the assumption of a reduction of aviation emissions of 5 per cent compared with the levels of 1990, the same as the Kyoto Protocol. Many other NGOs expressed concern about the current situation of the airline industry, insisting on the problems that have not yet been resolved such as the assignment of international emissions, implementation of levies for aviation, variable greenhouse gas effect of airplane emissions at different altitudes, delay of implementation of market based mechanisms to control and reduce emissions, and so on. Great efforts are being spent aiming at accelerating negotiations that would address these issues. But the barriers encountered have not yet been overcome.

Across the review made through this chapter, significant barriers for achieving sound results in addressing aviation emissions were found. Although in some cases the efforts to indicate the relevance of the problem are remarkable, agreements and actions are difficult to achieve and many fundamental players in the negotiations show extraordinary resilience against accepting emissions commitments. Under these circumstances, the pressure that civil society can execute in negotiations becomes fundamental and private companies should not underestimate their communication power to create and grow public awareness. In democratic countries, civil society mobilizations have many times demonstrated that many inflexible positions that seemed unchangeable have been diluted by the force of the voices of the public. This force can be critical in helping moving air emissions negotiations forward.

Environmental protection in general has achieved firm results when the efforts have been combined with a significant public support. For example, through communication campaigns and the positive reaction from the public, people perceive it as completely normal to spend higher amounts of money on products that protect the environment as compared with those products that don't offer that environmental protection guarantee. Following this line of reasoning, a positive answer from the public to an aviation emissions reduction possibility is quite likely. But what can the general public do to reduce aviation emissions? Obviously the clearest answer is to reduce travelling. This option, both in the business and leisure sectors, should not be underestimated. On the business side, many companies are promoting alternatives to travelling that are more economical, less time consuming, more friendly for the private life of the travellers and more environmentally friendly (such as the substitution of trips by the most advanced communication systems like video conferences). In the leisure sector, the explosion of demand for exotic destinations which occurred during the 1990s is now facing significant environmental consequences, both for transportation and also the local impacts at the destination; some travellers are already looking for closer destinations in order to avoid these effects. It is quite difficult to quantify these trends, but their presence is so obvious that it doesn't need to be proved with statistics. Since elimination or reduction of travelling is a rather radical option – sometimes it is not viable and it faces the opposition of many groups (airline, tourism and hotel industry, among others) – some flexible alternatives should be evaluated.

In recent years, according to the concept of the Flexibility Mechanisms[7] of the Kyoto Protocol, some individuals and organizations are opting for compensating emissions. Emission compensation consists of:

- Calculating the emissions corresponding to a certain activity (A), for example, travelling by airplane;
- Identifying a project (B) in which emissions can be potentially reduced at a low cost;
- Calculating what would be the cost of reducing the emissions produced in (A) at the price of reducing emissions in project (B);
- Funding emissions reductions in project (B) so as to compensate emissions produced in activity (A).

Many sources of emissions, like aviation, are insufficiently addressed by existing climatic policy instruments. Moreover, it is usually not possible or desirable to bring emissions down to the required levels. Compensation of emissions is a new mechanism that aims at covering these situations with the additional benefit of facilitating positive synergies in the climatic negotiations. By now many different offers for emissions compensation have arrived on the market (Sterk and Bunse 2004). Events with vast social impact such as the Football World Cup in Germany in 2006 are already opting for compensating for the emissions produced as a consequence of the trips related to the event.

In order to guarantee the environmental integrity of the mechanism, calculation of emissions should be done in a standard and certified manner. Also, the emission reduction projects should be adequately documented. Fortunately, the Kyoto Protocol Clean Development Mechanism[8] guidelines can be of great use for this purpose.

Although emissions compensation can be considered as an indirect mechanism of reduction, and therefore not so spectacular as a direct and dramatic drop of aviation emissions, I believe it has remarkable advantages that should not pass unnoticed.

Emissions compensation is voluntary and therefore is not in conflict with the interests of the airline industry; in fact it is quite the opposite since, by offering the possibility of compensating emissions, airlines would find a partial valve of escape for the immense pressure they are receiving to reduce the climatic impact of their activities.

New business opportunities are emerging since a new type of expertise is required in order to intermediate between the emissions emitter that funds reductions and the emissions reducer. The options to cover this task of intermediation are numerous.

Emitters in general are given a very realistic option to reduce their climatic impact without compromising their core activities and without jeopardizing the economics of their businesses.

More importantly, the general public will be able to actively participate and demonstrate its interest in the climate change. If public concern is significant, it can act as a hinge between the various parties involved in air transport emissions reduction negotiations, facilitating further commitments and agreements on a broader scale. Moreover, politicians and institutions involved in climate negotiations should reinforce their political statements with tangible efforts at their own levels. Compensation of travel emissions gives a bright opportunity to teach by example and help raise awareness of the problem of global warming among the general public.

The mechanism of emission compensation should nevertheless be carefully used, since compensation of emissions should not be understood as a substitute for emissions reductions but rather as a supplementary measure when reductions possibilities are not viable in specific cases. By definition, emissions compensation can only happen when it is possible to reduce emissions somewhere else, therefore compensation is an instrument to control or reduce emissions, not a solution in itself. If compensation is taken as the first option, emissions reductions would not take place and a possible rebound effect could happen at local levels, as the emissions could be increased in some places, as long as they are compensated somewhere else.

Additionally, emissions reductions – and not compensation – are normally linked to other environmental and social benefits, such as a decrease in traffic congestion or noise or the release of other pollutants associated with combustion of fossil fuels. If emissions reductions only happen in one place, the local environmental benefits would only happen in that place as well.

All in all, emission compensation can be viewed both as a second best option, and also as a first step to raise public awareness and interest in climate change.

Figure 11.6 illustrates how a compensation mechanism could work for the case of aviation. On the first instance, it would be reasonable to include only CO_2 emissions, given the difficulty in estimating other aviation emissions impacts over the climate. Technological advances in the airline reservation systems have significantly reduced the costs of booking travellers. As a consequence of the global spread of these computerized reservation systems, a few global reservation companies hold in their databases real time information about flight routes, schedules, type of airplanes and other relevant data that today are used for reservation and marketing purposes but which could potentially be used for estimation of

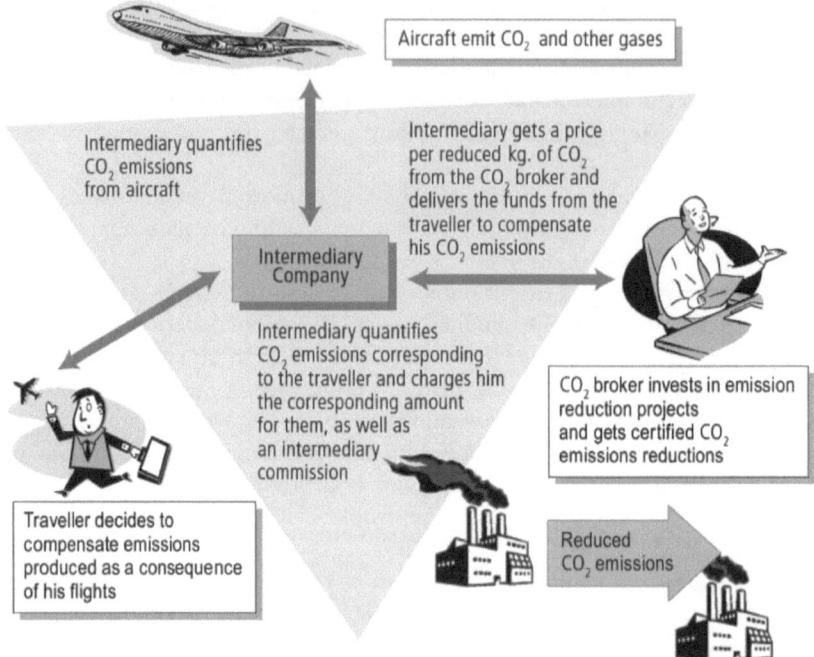

Figure 11.6 Compensation mechanism for aviation industry

emissions that correspond to one specific traveller. Other companies are specialized in identifying and carrying out emission reduction projects all over the world, given the increasing demand for certified emission reductions. An intermediate company could then put in contact the traveller and the emission reduction company (or CO_2 broker), so as to fix a price per reduced unit of CO_2. For every flight, the traveller would have the option of paying an additional amount of money in order to make his/her flight a carbon neutral flight. The idea has already been implemented by some companies such as Climate Care[9].

It can be assumed that there is still ample room for broadening the scope of the implementation by means of communication campaigns that could be sponsored by institutions involved in climate negotiations. Also, the success of this tool could be multiplied if the traveller is offered the carbon neutral flight option at the time of purchasing the ticket. This could be done by the reservation company introducing the information about emissions and prices, which is a realistically affordable project. The calculators used to estimate emissions could be substituted by a more sophisticated and precise *ad hoc* calculation.

If the idea is successful enough to reach a certain percentage of travellers that systematically compensate their air emissions, the negotiation scenario would be clearly improved and it would facilitate reaching commitments from the parties involved, since part of the objective would already be achieved by the voluntary measures implemented when travellers compensate their emissions.

Conclusion

In conclusion, international airline emissions have proved to be one of the most difficult and controversial elements of the climate talks, and their inclusion in the climate commitments should be facilitated by flexible options. This paper confirms voluntary commitments, and in particular compensation of aviation emissions, as one viable alternative to achieve emission reductions and to help move negotiations forward in order to include international air emissions in the climate regime.

These kind of projects cannot be expected to solve the aviation emissions reduction alone, but they will definitely contribute to moving in the right direction, as well as helping to shift the focus of international negotiations from the uncertainties and conflict of interests currently encountered to the solutions that can be implemented. Bisignani (2009) confirms that, even in recession, the aviation industry is aiming to achieve carbon-neutral growth by 2020. He notes that airlines are the first global industry to make such a bold commitment. The current stumbling blocks of international negotiations on aviation emissions would not be totally overcome, but they will become weaker and this would open opportunities for broader and more ambitious actions.

Notes

1 Radiative forcing is the change in the balance between radiation coming into the atmosphere and radiation going out.
2 BTU: The British Thermal Unit is a measure of the energy required to heat up a pound of water (0.454 kg) 1° Fahrenheit.
3 Annex I includes OECD (Organisation for Economic Co-operation and Development) countries plus countries whose economies are in transition to the market system. The latter are ex-Soviet Union countries.
4 Tankering is a term frequently used in the aviation sector that refers to the possible migration of airlines to airports outside the fuel-taxed zone or the action of taking untaxed fuel into the taxed area.
5 EUROCONTROL is the European Organisation for the Safety of Air Navigation. It currently has 34 member states and, among other responsibilities, EUROCONTROL collects air navigation charges on behalf of its member states.
6 The European Federation for Transport and the Environment is an umbrella organization for nongovernmental organizations that work in the field of transport and the environment, promoting environmentally sustainable transport in Europe. More information can be found at http://www.t-e.nu/.
7 CO_2 mixes more or less uniformly in the atmosphere. It is equivalent from a climatic impact standpoint to reduce emissions in one place or another. Kyoto Flexibility Mechanisms are based on this characteristic. Taking advantage of this substitutability property, the Kyoto Protocol and other International Markets for Greenhouse gases have developed mechanisms that are based on the trading of emissions rights or on the implementation of emissions reduction projects in those places where reducing emissions is less costly.

8 The Clean Development Mechanism of the Kyoto Protocol allows Annex I countries to implement sustainable development projects that reduce greenhouse gases emissions in Non-Annex I countries. The investing party receives Certified Emissions Reductions that can be used to meet its own emission targets. Annex I countries include developed countries and countries whose economies are in transition to the market system.

9 Some information including a CO_2 calculator providing the option of compensating aviation emissions can be found at www.climatecare.org.

References

Airbus (December 2003) *Global Market Forecast*. Toulouse Blagnac, France: Airbus.

Bisignani, G. (2009) *State of the Air Transport Industy*, Speech at the IATA Annual General Meeting and World Air Transport Summit, Kuala Lumpur, Malaysia, available through http://www.iata.org/pressroom/speeches/2009-06-08-01.htm, last accessed 19 July 2009.

Bobes, P.L. (2004) *Market Based Mechanisms for Achieving Kyoto Targets: Perspectives from Sweden and Spain*. Stockholm: Kungliga Tekniska Högskolan.

British Airways (2004) *2003/2004 Social and Environmental Report*. High Wycombe: Ethedo Press.

Bows, A., Manders, S., Randles, S. and Anderson, K. (2010) *Aviation Emissions in the Context of Climate Change: A Consumption-Production Approach*. Final Policy Report, Manchester, UK: Tyndall Centre for Climate Change Research.

Bureau of Transportation Statistics (2004) *National Transportation Statistics*. Washington, DC: US Government Printing Office.

Dings, J. (2004) *ICAO Assembly Oct 2004: Climate Disaster Coming Up!* Position paper, Brussels: T&E European Federation for Transport and the Environment.

European Commission (2001) *European Transport Policy for 2010: Time to Decide*. White Paper, Luxembourg: Office for Official Publications of the European Communities.

International Air Transport Association (2001) *Emissions Trading for Aviation. Workstream 3: Key findings and conclusions*. IATA, available through http://www.iata.org/soi/environment/emissionstradingstudy.htm, last accessed 4 February 2009.

International Air Transport Association (2003) *Fast Facts: the Air Transport Industry in Europe has United to Present its Key Facts and Figures*, available through http://www.iata.org/pressroom/industry_stats/2003-04-10-01.htm, last accessed June 2004).

International Air Transport Association (2009) Passenger Decline Stabilizes – Some Improvement in Freight, available through http://www.iata.org/pressroom/pr/2009-06-25-01.htm, last accessed 19 July 2009.

International Civil Aviation Organisation (ICAO) (2004a) Working Paper. 35th Assembly Session. *Aviation and Climate Change*. 2 August, ICAO.

International Civil Aviation Organisation) (ICAO) (2004b) Working Paper. 35th Assembly Session. *The Need for a Scientifically Substantiated Approach to the Introduction of Market-Based Measures to Limit or Reduce Aircraft Engine Emissions*. 8 September, ICAO.

Intergovernmental Panel on Climate Change (IPCC) (1999) *Aviation and the Global Atmosphere*. Cambridge, MA: Cambridge University Press.

Point Carbon (2004) *PwC Slams Spanish Kyoto Commitment*, 5th February, available through http://www.pointcarbon.com/article.php?articleID=3192, last accessed 9 February 2004.

Sterk, W. and Bunse, M. (2004) *Voluntary Compensation of Greenhouse Gas Emissions*. Wuppertal: Wuppertal Institute for Climate, Environment and Energy.

12 Overcoming Stumbling Blocks

Can the Intergovernmental Panel on Climate Change Deliver on Adaptation?*

Tora Skodvin

Introduction

One important issue contributing to the current deadlock in the climate negotiations is linked to the question of developing country participation. This is a tricky question. And it seems to be particularly tricky within a framework that almost exclusively focuses on *mitigation* of climate change, a feature that has hitherto characterized climate negotiations.

The European Union (EU) tried to raise the question of post-2012 commitments at the eighth Conference of the Parties to the Climate Convention (COP8) in New Delhi in October 2002, but met adamant resistance from developing countries. The G77 and China group refused to discuss measures that would imply greenhouse gas (GHG) emissions management by developing countries in the second commitment period, and advocated a major shift from the current emphasis on *mitigation* of climate change toward a stronger emphasis on *adaptation* to climate change. As pointed out by several observers, an increased emphasis on adaptation would be associated with a redistribution of capital from industry-restructuring efforts in the North to adaptation- and technology-development efforts in the countries commonly assumed to be most vulnerable to climate change: developing countries in the South (see, for instance, Jacob 2002). The result of the confrontations at COP8 was a deep division between the EU and developing countries (known together as the "Green Group") (Ott 2003).

The US, on the other hand, supported the position of the developing countries. Even though the US has traditionally been among the strongest supporters of developing-country commitments in an international climate regime, its position on this issue has shifted since the Bush administration's rejection of the Kyoto agreement. A main argument by the US government after George W. Bush took over the Presidency, and by the powerful US fossil fuel industry throughout the climate process, has been that even an effective Kyoto Protocol will not be sufficient to mitigate climate change. Thus, they have argued, it makes much more sense to channel resources for adaptation measures to those areas that are most vulnerable to climate change, while also investing in new technology to reduce greenhouse gas emissions in the long term. With this shift in the US position, we may be witnessing an emerging new alliance between the US and developing countries on this issue (see Jacob 2002, 2003). This impression is reinforced by the establishment of the Asia-Pacific Partnership on Clean Development and Climate, which is regarded as an innovative new effort to accelerate the development and deployment of clean energy technologies. The partnership, which includes Australia, Canada, China, India, Japan, Korea and the US,[1] is an agreement to work together and to work closely with private sector partners to meet goals for energy security, national air pollution reduction, and climate change in ways that promote sustainable economic growth and poverty reduction.

In the current situation, therefore, several observers have commented that the question of a stronger focus on adaptation seems to be difficult to bypass, for two main reasons. First, given the long atmospheric lifetimes of greenhouse gases, an exclusive reliance on mitigation is no longer an option. There *will* be changes in climate, which reinforces the need to readdress questions of equity in relation to climate change. Second, because of the developments we have witnessed in the negotiation process since the US withdrawal from the Kyoto agreement, a reframing of the climate problem towards a stronger and more committing emphasis on adaptation seems necessary to resolve the current deadlock in the negotiation process (see Ott 2003; Ott *et al.* 2004; Müller 2003). Moreover, given the position the US has signalled at these meetings, a stronger focus on adaptation also may hold a potential for reintegrating the US into a more active role in the international climate regime.[2]

A reframing of the climate problem toward a stronger emphasis on adaptation raises several important questions. One question is how adaptation measures should be funded. So far, this has been the main focus of the climate process. The development of financial mechanisms in this issue area was initiated at COP7 in Marrakesh, where three such funds were established.

An equally important question, however, is *what* a stronger emphasis on adaptation would imply: What is adaptation? What types of policy measures would be involved in an effort to enhance countries' capacity to adapt to climate change? In what regions are adaptation measures most critical? And, on what basis should these decisions be made? In short, one of the requirements of a reframing of the climate problem in this direction is that decisions can be based on a common foundation of state-of-the-art knowledge on adaptation.

While there is a growing body of knowledge on vulnerability and adaptation, there is also a growing recognition that there currently exists at least two different approaches to vulnerability and adaptation assessments (see O'Brien *et al.* 2004). The main differences between the two approaches lie in what constitutes the starting point for analysis and the extent to which the structural, social, and political dimensions of vulnerability and adaptation are included in the analysis. While the two interpretations are not incompatible, they seem to imply significantly different policy implications. Moreover, the different policy recommendations following from the two approaches may, to a varying extent, correspond to the agenda of developing countries in the climate issue, especially that related to sustainable development.

Currently, therefore, we have a situation where a stronger emphasis on adaptation is required but where there also exists two distinct scholarly interpretations of adaptation with significantly different policy implications. The focus of this chapter is the extent to which a knowledge base on adaptation is or can be provided by the Intergovernmental Panel on Climate Change (IPCC), which may also contribute to help negotiators overcome the deadlock in the negotiation process.

The IPCC is the scientific body of the climate regime, responsible for providing necessary assessments of state-of-the-art knowledge within fields and disciplines relevant to the climate change policymaking process. The IPCC is organized into three Working Groups (WGs), and vulnerability and adaptation assessments lie within the mandate of WGII. The fourth assessment report on climate change of the IPCC was published in 2007. A technical paper on climate change and water was released in June 2008.

In this chapter, I explore how vulnerability and adaptation is treated in IPCC reports, with a main focus on the distinction between the two interpretations of these concepts in the scholarly literature. The main research questions of the chapter are: What is the difference between the two interpretations of vulnerability and adaptation? What characterizes the IPCC approach to vulnerability and adaptation assessments? To what extent is the IPCC mode of operation appropriate for doing vulnerability and adaptation assessments according to both interpretations?

After a brief assessment of the status of adaptation in the current international climate regime in section 1, the two interpretations of the relationship between vulnerability and adaptation are presented and discussed in section 2, followed by a discussion in section 3 of the different policy implications the two interpretations generate. In section 4, attention is shifted toward the manner in which the IPCC has approached assessments of vulnerability and adaptation opportunities, with a main focus on the question of whether the IPCC mode of operation is appropriate for assessments of social vulnerability, which is addressed in section 5. In section 6, alternative institutional frameworks for social vulnerability assessments are discussed, before I conclude the analysis in section 7.

Adaptation and its current status in the international climate regime

Developing countries have criticised the current climate regime for its strong emphasis on mitigation. This is particularly pronounced in the Kyoto Protocol, which is almost exclusively focused on mitigation through reduction of GHG emissions. The dominant role of mitigation in the current climate regime implies that the question of equity in relation to climate change has also primarily been framed in terms of the distribution of costs associated with GHG emission reductions.

The equity dimension, however, is also closely linked to the distribution of climate impacts and adaptation costs. The impacts of climate change are not equally distributed across the globe. Some countries and regions are more exposed to climate impacts than others (Adger *et al.* 2003). Moreover, the costs associated with climate impacts will also be unequally distributed across the globe because of significant differences in levels of economic development. It is commonly assumed that developing countries will suffer more from climate change than industrialized countries because of both these factors (Adger *et al.* 2003). The longer the climate negotiations fail to develop effective means of mitigation, therefore, the more urgent the question of adaptation becomes, particularly for the developing countries which are most at risk.

In the United Nations Framework Convention on Climate Change (UNFCCC), this perspective on the equity dimension is acknowledged. Article 4.4 of the Climate Convention, for instance, states that "developed country Parties ... shall also assist the developing country Parties that are particularly vulnerable to the adverse effects of climate change in meeting costs of adaptation to those adverse effects." Similarly, Article 4.8 of the Convention states that consideration should be given to "meet the specific needs and concerns of developing country parties arising from the adverse effects of climate change."[3]

Financial mechanisms for adaptation measures were established at COP7 in Marrakesh in 2001, when three funds were established for this purpose: The Least Developed Countries (LDC) Fund, in which the Global Environment Facility (GEF) is requested to disburse funds for the preparation of National Adaptation Programmes of Action (NAPAs); the Special Climate Change Fund, also operated by GEF to finance adaptation, technology transfer, and mitigation measures; and the Kyoto Protocol Adaptation Fund to support adaptation projects and programmes in developing countries that are parties to the Protocol. The latter fund is to be financed by a levy on Clean Development Mechanism (CDM) projects (Huq and Burton 2003). Contributions to the LDC Fund and the Special Climate Change Fund are to be voluntary (Adger *et al.* 2003).

In relation to the establishment of the LDC Fund, and in recognition of Article 4.9 of the Climate Convention in which the special needs of LDCs are acknowledged, COP7 also instituted the NAPAs to "identify priority activities that respond to [LDCs'] urgent and immediate needs with regard to adaptation to climate change".[4] To date, this has implied that LDCs are invited to develop NAPAs according to a set of guidelines developed by the UNFCCC. An LDC Expert Group was also set up at COP7 to provide guidance and advice on the preparation and implementation strategy for NAPAs. The GEF, which constitutes the operating entity of the LDC Fund, is requested, "as a first step", to "provide funding from the LDC Fund to meet the agreed full cost of preparing the NAPAs, given that the NAPAs will help to build capacity for the preparation of national communications under Article 12, paragraph 1 of the Convention".[5]

Even though the institutionalization of NAPAs may represent a step toward a more active role by the UNFCCC on adaptation measures, the focus thus far has primarily been one of funding. The interventions by developing countries at recent COP meetings may imply that they are not satisfied by the mere establishment of some funds with voluntary contributions to take care of the costs associated with adaptation measures, and that they may demand more committed action on adaptation by industrialized countries. This is, for instance, reflected in India's proposal to develop an "Adaptation Protocol" to the UNFCCC (Müller 2003). More committed action on this issue area, however, would require a common foundation of state-of-the-art knowledge on adaptation.

Two approaches to vulnerability and adaptation

In the literature on vulnerability and adaptation, there is a growing recognition that there are two different conceptualizations of vulnerability with quite different implications for the development of adaptation policies and measures (see Kelly and Adger 2000; Burton *et al.* 2002; Adger *et al.* 2003; O'Brien *et al.* 2004). The first interpretation – often referred to as representative of a "first generation" of vulnerability analysis – sees vulnerability as the "end point" of analysis. This implies that vulnerability is understood as *net climate impact*. That is, vulnerability is what is left after we have "subtracted" a country or region's adaptive capacity from the physical effects this country or region is likely to experience from future climate change: Physical impacts of future climate change ÷ adaptive capacity = vulnerability (see Kelly and Adger 2000).

The second conceptualization, now increasingly emerging in scholarly literature on vulnerability and often referred to as representative of the "second generation" of vulnerability analysis, is the understanding of vulnerability as a "starting point" for analysis. According to this conceptualization, "vulnerability represents a *present* inability to cope with external pressures or changes," and climate change is just one of several causes of this external stress (O'Brien *et al.* 2004: 2, emphasis added). According to this understanding of vulnerability, the external pressures or changes are formally outside the definition of vulnerability. As vulnerability is always linked to a biophysical component, however, or to a "natural hazard," vulnerability and exposure remain inseparable (Kelly and Adger 2000: 327). The point here, however, is that vulnerability is seen as a current state caused by a set of structural, political, and social factors that determine a country or region's capacity to "anticipate, cope with, resist, and recover from the impact of a natural hazard" (Blaikie *et al.* 1994: 9; Kelly and Adger 2000: 327). In this sense, vulnerability by this interpretation "represents a potential *starting point* for any impact analysis" (Kelly and Adger 2000: 327).

Policy implications of the two interpretations

There are two main differences between the approaches. First, they differ in terms of what constitutes the starting point for analysis. While first-generation analyses start with exposure to "natural hazards" (like climate change), a group's current capacity to cope with external stress, and abrupt changes in general, constitutes the starting point for analysis in second-generation approaches. Second, the two approaches differ in terms of the extent to which structural, social, and political determinants of vulnerability and adaptive capacity are included in the analysis. While first-generation vulnerability assessments recognize the social dimensions of adaptive capacity, the analyses are nevertheless primarily focused on technical options for adaptation and hence technical measures to reduce vulnerability. Second-generation assessments, on the other hand, are often based on the premise that social, economic, and political factors represent crucial determinants of adaptive capacity and hence a group's vulnerability to external stress and abrupt changes. This interpretation, therefore, implies that in order to enhance a country's or region's adaptive capacity, measures have to be directed at its *current* vulnerability to natural hazards and external changes, including climate change. These differences thus imply that the two approaches require different types of knowledge input for conducting vulnerability and adaptation assessments and that they generate a significantly different output in terms of implications for adaptation policies and measures.

When vulnerability is defined in terms of net climate impact, the first step in a vulnerability assessment is to determine which physical climate impacts different countries are likely to experience in the future at national, regional, and local levels: Which climate futures should countries and regions be able to adapt to? The logic of the approach is that a mapping of regional effects of future climate change constitutes the foundation for the development of appropriate adaptation measures, which in turn will serve to reduce the country or region's vulnerability to adverse effects of climate change: "The assessment of vulnerability is the *end point* of a sequence of analyses beginning with projections of future emissions trends, moving on to the development of climate scenarios, thence to biophysical impact studies and the identification of adaptive options" (Kelly and Adger 2000: 327). Moreover, adaptation policies and measures following from this approach to vulnerability are largely technical in nature: they involve the building of houses, bridges, dams, irrigation systems, and so on, as appropriate, according to expected physical impacts of climate change in specific regions.

This has given rise to a large effort to develop models for assessing regional effects of future climate change. These assessments are characterized by uncertainties stemming from two main sources (Adger *et al.* 2003): First, scientists do not know accurately how the climate system might react to unprecedented emissions of GHGs and the possible feedback mechanisms such changes may generate. This uncertainty has been dealt with by developing ranges of estimates for key climate parameters, such as temperature, rainfall, and other factors. Predictions become more uncertain the further into the future they are extended, which is reflected in a widening of the ranges of the estimates (Adger *et al.* 2003). Moreover, climate model projections are based on Global Circulation Model (GCM) scenarios. While adaptation measures typically are local or site-specific, GCM scenarios only provide information at the global or (large) regional levels. Downscaling has been used to provide more detailed information, but is associated with a significant reduction in the accuracy of predictions (Burton *et al.* 2002: 152). Second, these predictions rest on assumptions about future emission trends, which in turn depend on assumptions about "unknowns," such as the rate of world population growth and development of non- or low-carbon technologies (Adger *et al.* 2003). What is more, assumptions have to be made, not only about *aggregate* levels of GHG emissions, but also about the *composition* of emitted gases. The composition of GHGs, and especially the relative share of short-lived and long-lived GHGs, can make a big difference for instance to the rate of change countries and regions are exposed to in the future (see Skodvin and Fuglestvedt 1997; Skodvin 1999).

Predictions about future climate impacts are therefore associated with significant levels of scientific uncertainty. The uncertainty increases, the further into the future we look and the more specific we want to be in terms of the geographical space the impacts concern. This implies that there is a significant risk that adaptation policies and measures that are implemented on the basis of such assessments do not contribute to reducing a country or region's climate vulnerability, simply because they are based on predictions of climate impacts that turn out to be false. At worst, adaptation policies developed and implemented on the basis of such assessments may even contribute to enhancing vulnerability. "Technological fixes" to reduce vulnerability "may make society more vulnerable in the long run because they may lead to increased investments and population concentration in locations subject to climate hazards" (O'Brien *et al.* 2004: 5).

Whereas the first-generation vulnerability assessments are based on predictions of future climate impacts, second-generation vulnerability assessments would to a larger extent focus on social and political factors that determine a country or region's current state of vulnerability. In relation to climate change, for instance, adaptation to climate variability is not something societies will have to begin doing when impacts of a human-induced climate change start taking effect; it is something societies are doing continually. Thus, even today, societies are more or less resilient to climate variability. An emerging research agenda with this point of departure focuses on the identification of generic determinants of resilience, including "the social capital of societies, the flexibility and innovation in the institutions of change, and the underlying health status and wellbeing of individuals and groups faced with the impacts of climate change" (Adger *et al.* 2003: 186; see also, Burton *et al.* 2002).

This approach has significantly different policy implications than a first-generation approach in the sense that policy responses are social rather than technical in nature. The approach generates policy measures directed toward fundamental causes of present-day vulnerability and include measures such as "poverty reduction, diversification of livelihoods, protection of common property resources, and strengthening of collective action" (O'Brien *et al.* 2004: 5). A basic assumption in this approach is that such measures would strengthen a society's *present* resilience and would thus also enhance its capacity to adapt to changing conditions in the future. Assessments of vulnerability and future adaptive capacity, therefore, have to start with knowledge about the present: "Understanding the present-day effects and response to climate variability at all levels of social organization is a prerequisite for studying the effects and responses to future climate change and for identifying the key determinants of successful adaptation in the future" (Adger *et al.* 2003; see also Burton *et al.* 2002). With this point of departure, vulnerability assessments constitute a basis for identifying adaptation capacity, including the identification of "opportunities and constraints to implementing specific adaptation policies" (O'Brien *et al.* 2004: 5).

In contrast to the static interpretation of vulnerability characteristic of the first-generation approach, this perspective emphasizes its dynamic element. Vulnerability is seen as a "dynamic entity, in a continuous state of flux as the biophysical and social processes that shape local conditions and ability to cope also change" (O'Brien *et al.* 2004: 3). In this sense, the approach "allows for adaptation to uncertainty" (O'Brien *et al.* 2004: 5).

The policy implications following from second-generation vulnerability studies are much more vaguely formulated than those arising from first-generation assessments. But both vulnerability and adaptation processes to climate change are likely to reinforce unequal economic structures (Adger *et al.* 2003). At the most general level, therefore, vulnerability is seen as closely linked to economic development, and a general "policy implication" of the approach is that vulnerability may be reduced by enhancing a country or region's level of economic development. Scholars of this approach also emphasize, however, that vulnerability to climate change "is not strictly synonymous with poverty" (Adger *et al.* 2003: 182; see also O'Brien 2000; O'Brien *et al.* 2004). For instance, pastoralists in the West African Sahel "have adapted to cope with rainfall decreases of 25–33 per cent in the twentieth century" (Adger *et al.* 2003: 181). Thus, although poverty is seen as a key driving force of vulnerability and a key constraint to individuals' coping strategy, "vulnerability to future climate change is likely to have distinct characteristics and create new vulnerabilities" (Adger *et al.* 2003: 182).

It has been maintained that the two approaches are distinguished by different assumptions about the causal relationship between vulnerability and adaptation (see O'Brien *et al.* 2004). Adaptation is a dynamic process, however, which implies that measures directed toward vulnerability at point T serve to enhance a region's adaptive capacity, which in its turn reduces vulnerability at point T_1. Further, although second-generation analysts emphasize that vulnerability is independent of the future impacts of climate change, it also seems clear that some knowledge of future exposure to climate impacts is necessary for assessing vulnerability even within the framework of this approach. The two approaches to vulnerability and adaptation are thus not incompatible. Rather, some combination of the two approaches is necessary as a knowledge base for the development of more committed adaptation policies within the framework of the UNFCCC. Second-generation scholars are thus not claiming that first-generation assessments of vulnerability are redundant, only that they urgently need to be supplemented by assessments of present-day resilience to climate variability. A crucial question, then, is whether the IPCC can contribute to delivering this type of assessment.

Vulnerability, adaptation, and the IPCC

Vulnerability and adaptation have been a key part of IPCC assessments since its First Assessment Report was issued in 1990, but especially since the provision of the *IPCC Technical Guidelines for Assessing Climate Change Impacts and Adaptations* (Carter *et al.* 1994). The main IPCC publications on vulnerability and adaptation since then include WGII contributions to the Second and Third IPCC

Assessment Reports (Watson *et al.* 1996; McCarthy *et al.* 2001) and a Special Report on impacts and vulnerability issued in 1997 (Watson *et al.* 1997).

In all its publications, the IPCC approach to adaptation has been "impacts-driven" (Burton *et al.* 2002). In the 1994 Technical Guidelines, for instance, a seven-step approach to impact and adaptation assessment is provided (often referred to as the "standard approach"), where the key element of the approach is step 4, selection of climate change scenarios: "By relying on climate change scenarios the standard approach directs attention to the impacts of future climate change and by default, away from current impacts and vulnerability" (Burton *et al.* 2002: 151). The approach rests on the assumption that adaptation responses are known, which is often not the case, and it does not include prescriptions for conducting assessments of the state of the system that will be impacted (for instance, in terms of present and future economic vulnerability), which may be equally or more important than projected climate change (Burton *et al.* 2002: 151).

In general, vulnerability plays a very minor role in the 1994 Technical Guidelines. While the term is defined,[6] it does not have a role in the prescribed methodology for impact and adaptation assessment. For instance, to the extent that present socio-economic conditions are included, they are included as baseline references against which to assess future *impacts* of climate change on socio-economic conditions, not as framework conditions for societies' capacity to *cope with* present and future climate variability. "In order to provide reference points with which to compare future projections, three types of 'baseline' conditions need to be specified: the climatological, environmental and socio-economic baselines" (Carter *et al.* 1994: vii; see also O'Brien *et al.* 2004). The assumed relationship between vulnerability and adaptation is reflected in the specification of objectives of adaptation strategies, defined, *inter alia*, as "the reduction of vulnerability" (Carter *et al.* 1994: x).

In the following reports, this conceptualization of vulnerability and its relationship to adaptation is further specified. In the 1997 Special Report, for instance, vulnerability is defined as a function of "the sensitivity of a system to changes in climate (the degree to which a system will respond to a given change in climate, including both beneficial and harmful effects) and the ability to adapt the system to changes in climate (the degree to which adjustments in practices, processes or structures can moderate or offset the potential for damage or take advantage of opportunities created, due to a given change in climate)" (Watson *et al.* 1997: 1).

The term "system" in this approach refers to ecosystems or sectors. In the 1997 Special Report, for instance, the approach is specified. "This report's assessment of regional vulnerability to climate change focuses on ecosystems, hydrology and water resources, food and fibre production, coastal systems, human settlements, human health, and other sectors or systems (including the climate system) important to 10 regions that encompass the Earth's land surface" (Watson *et al.* 1997: 2). The IPCC approach to adaptation, therefore, is mainly concerned with these systems' ability to adjust naturally or by human intervention, which then constitutes the basis for assessing their vulnerability to climate change. Two examples can illustrate the point, taken from the 1997 Special Report's assessment of the vulnerability of ecosystems, and hydrology and water resources.

With regard to the assessed vulnerability of ecosystems, the report concludes:

> Adaptation options for ecosystems are limited, and their effectiveness is
> uncertain. Options include establishment of corridors to assist the 'migration'
> of ecosystems, land-use management, plantings and restoration of degraded
> areas. Because of the projected rapid rate of change relative to the rate at
> which species can re-establish themselves, the isolation and fragmentation
> of many ecosystems, the existence of multiple stresses (e.g. land-use change,
> pollution) and limited adaptation options, ecosystems (especially forested sys-
> tems, montane systems and coral reefs) are vulnerable to climate change.
>
> (Watson *et al.* 1997: 3)

The assessed vulnerability of water systems is summarized in this manner:

> Various approaches are available to reduce the potential vulnerability of
> water systems to climate change. Options include pricing systems, water effi-
> ciency initiatives, engineering and structural improvements to water supply
> infrastructure, agriculture policies and urban planning/management. At the
> national/regional level, priorities include placing greater emphasis on inte-
> grated, cross-sectoral water resources management, using river basins as
> resource management units, and encouraging sound pricing and manage-
> ment practices. Given increasing demands, the prevalence and sensitivity of
> many simple water management systems to fluctuations in precipitation and
> runoff, and the considerable time and expense required to implement many
> adaptation measures, the water resources sector in many regions and coun-
> tries is vulnerable to potential changes in climate.
>
> (Watson *et al.* 1997: 3–4)

These examples illustrate very clearly that the main focus of the IPCC approach to
vulnerability and adaptation lies in *technical options* for adapting to climate impacts,
not *social, economic, and political constraints* to implementing these technical adaptation options. Social, economic, and political constraints to effective adaptation
policies are mentioned in the reports, however. On agriculture, for instance, the
1997 Special Report concludes:

> In regions where agriculture is well adapted to current climate variability
> and/or where market and institutional factors are in place to redistribute agri-
> cultural surpluses to make up for shortfalls, vulnerability to changes in climate
> means and extremes generally is low. However, in regions where agriculture
> is unable to cope with existing extremes, where markets and institutions to
> facilitate redistribution of deficits and surpluses are not in place, and/or where
> adaptation resources are limited, the vulnerability of the agricultural sector to
> climate change should be considered high.
>
> (Watson *et al.* 1997: 4)

Thus, while mentioned, these types of constraints – which are inextricably linked to second-generation interpretations of vulnerability as *cause* of, rather than *function* of, adaptive capacity – are not the main focus of IPCC assessments. It is not the vulnerability of populations or societies that is the main focus of the IPCC approach, but the vulnerability of systems and sectors such as these. The IPCC reports are largely silent with regard to the question of how social vulnerability can be assessed and addressed in policymaking.

This account of the IPCC approach to adaptation is representative also of later IPCC publications (WGII's contribution to the Third Assessment). The focus of the assessments is primarily on those climate impacts that can be expected in the future and (theoretically) available and largely technical options for adapting to these climate futures. The reports also reflect, however, an increasing difficulty in keeping the social dimension of vulnerability out of the equation. In WGII's contribution to the Third Assessment, for instance, it is recognized that "experience also demonstrates that there are constraints to achieving the full measure of potential adaptation" (IPCC 2001: 8). It is also recognized that "policies that lessen pressures on resources, improve management of environmental risks, and increase the welfare of the poorest members of society can simultaneously advance sustainable development and equity, enhance adaptive capacity, and reduce vulnerability to climate and other stresses" (IPCC 2001b: 8). The function of social factors as *determinants* of vulnerability and adaptive capacity, however, is not considered explicitly in the IPCC approach.

One implication of the impacts-driven and largely technical focus to vulnerability and adaptation characteristics of the IPCC approach, is that, while possible constraints to the effective implementation of adaptation options are recognized, they are not brought to bear on the assessment of vulnerability. Thus, as pointed out by Burton *et al.* (2002), even when it is recognized that not all of the theoretically available adaptation options will be effectively adopted and implemented by decision makers, assumptions about partial adaptation (for instance that 50 per cent of theoretically available adaptation options are implemented) "are not based on any knowledge or understanding of the adaptation process itself. There has been little or no consideration of the social and behavioural or other obstacles in the adaptation process" (Burton *et al.* 2002: 153).

This feature is not only characteristic of the IPCC approach. There are several impacts and vulnerability assessments that have been carried out, both by individual scholars and organizations, that are representative of an impacts-driven approach. The IPCC approach, therefore, might simply be seen as a reflection of the actual state of affairs in scholarly literature on vulnerability and adaptation. The IPCC is not a research institution. It does not, in principle, commission research or undertake research itself. The mandate of the IPCC is to give a fair and balanced assessment of published and peer-reviewed scholarly literature relevant to the topics covered by the IPCC reports in question. If a first-generation approach to vulnerability and adaptation dominates scholarly literature in this area, this approach will also necessarily dominate the IPCC approach to the same area. This is undoubtedly an important background factor for the current dominance of the impacts-driven approach in IPCC assessments. It is not the only explanation, however. This approach also seems to be a question of *framing*.

Ever since the problem of a human-induced climate change surfaced on the international agenda with the establishment of the IPCC in 1988, the problem has been framed as one of *mitigation*. The IPCC approach to adaptation studies lies completely within this mode of framing, in the sense that a definition of vulnerability as *net climate impact* "serves as a means of defining the extent of the climate problem and providing input into policy decisions regarding the cost of climate change versus costs related to greenhouse gas mitigation efforts" (O'Brien *et al.*, 2004: 1; see also Kelly and Adger 2000). This is also clear from the Foreword of the 1997 Special Report on impacts and vulnerability. "The report establishes a common base of information regarding the potential costs and benefits of climatic change, including the evaluation of uncertainties, to help the COP determine what adaptation and mitigation measures might be justified" (Watson *et al.* 1997: v).

This mode of framing the climate change problem in general and vulnerability and adaptation assessments in particular was not the only one possible, however. Since the early 1990s, there has been an emerging scholarly literature on social vulnerability to natural hazards (see Blaikie *et al.* 1994).[7] In this literature, the dominant focus is on the question of what makes peoples and societies vulnerable to natural hazards or disasters, whatever they might be. Thus, in contrast to first-generation interpretations of vulnerability to climate change, the "natural-hazards approach" divorces the concept of vulnerability from the precise nature of the natural hazard to which peoples and societies are more or less at risk. Rather than focusing on the precise nature of the natural hazard and potentially available technological coping strategies, this approach focuses on the social, economic, and political conditions that make people more or less vulnerable to natural hazards, in whatever form they may come. There is also a time dimension built into this approach, in the sense that "the more vulnerable groups are also those that find it hardest to reconstruct their livelihoods following disaster. They are therefore also more vulnerable to the effects of subsequent hazard events" (Blaikie *et al.* 1994: 9). While many of the natural hazards referred to in this literature do not have direct links to climate change (for instance, earthquakes and wars), others do (for instance, epidemics, famine, heavy rainfall, and mudslides, to mention but a few). Thus, the link to climate change as a "natural hazard", and hence the relevance of this approach to assessments of vulnerability and adaptation options to climate change, should not be all that difficult to recognize. This link, however, has not been made to any significant extent within the IPCC assessment process, despite a certain degree of cross-over between the "natural hazard" and climate change research communities.[8] One reason for this may lie in the framing of the climate problem as a problem of mitigation, which implies that the risk and vulnerability studies conducted by the "hazards community" were not considered relevant as basis for IPCC assessments on vulnerability and adaptation to climate change.

Can the IPCC conduct assessments of social vulnerability?

The importance of addressing questions of social vulnerability within the framework of the UNFCCC seems evident, particularly in light of the agenda of developing countries. Sokona *et al.* (2002), for instance, point out that a principal concern for developing countries on climate change is linked to the long-term goals of the UNFCCC on sustainable development. According to these authors, the focus on short-term goals in the Kyoto process has resulted in a "skewed" focus of the regime "towards minimising the burden of implementation on polluter industries and countries, instead of giving priority to the vulnerabilities of the communities and countries at greatest risk and disadvantage" (Sokona *et al.* 2002: 2).

To what extent, then, can the IPCC contribute to the provision of a common knowledge base on social vulnerability? What are the main features of the IPCC mode of operation and to what extent is this mode of operation appropriate for the development of a common knowledge base on social vulnerability?

In this section, I address two features of the IPCC mode of operation that are of particular relevance for the body's appropriateness for conducting assessments of social vulnerability:

a) The "top-down" approach characterizing current IPCC impacts, vulnerability, and adaptation assessments, and
b) The strongly politicized environment within which the IPCC operates and the resulting risk of a potentially paralysing politicization of the issue.

Supplementing the "top-down" approach of IPCC assessments

The IPCC's reliance on "fundamental science" such as biology and geophysics facilitates the development of common methodologies and tools and common standards of measurement. This approach to impacts, vulnerability, and adaptation assessments has been characterized as a "top-down" approach, a methodology that is maintained to be neither feasible nor desirable for assessments of social vulnerability:

> Effective adaptation policy has to be responsive to a wide variety of economic, social, political, and environmental circumstances. A different kind of creativity and ingenuity is required. It is, therefore, inappropriate to provide guidelines in a prescriptive style. What is required is a common framework of concepts, linked together in a flexible manner that helps in the design and organisation of research for adaptation policy to reduce vulnerability.
>
> (Burton *et al.* 2002: 154)

For decision makers, however, especially at the international level where vulnerability maps and assessments will constitute the foundation for the identification of "critical regions" and "hot spots" in the prioritization of policy measures (Kaspersson 2001), the comparability of assessments is crucial. The need to start any assessment of social vulnerability with an assessment of present states of vulnerability and adaptive capability does imply a stronger emphasis on "bottom-up" rather than "top-down" approaches. There are examples, however, of how this could be done in ways in which comparability is maintained and that, in principle (although perhaps not in practice), could be undertaken by the IPCC. The 2000 UNEP Country Studies is a good case in point (O'Brien 2000).

In the 2000 UNEP Country Studies, an objective to "test and apply the [IPCC] technical guidelines for assessing climate change impacts and adaptations" was combined with a focus on "current and projected environmental and socio-economic stresses" to expose the "context under which climate impacts must be considered" (O'Brien 2000: v). In this sense, the approach of the report represents a combination of first- and second-generation approaches to vulnerability and adaptation. This approach was developed with the aim of improving the methods for assessing climate change impacts. It is noted that current methods often ignore the needs and challenges facing developing countries, and that "data constraints, a lack of resources, different scientific traditions, and other priorities can render many of the existing impact methodologies ineffectual" (O'Brien 2000: v).

The country studies are conducted on the basis of a set of multiple methods. In particular, the approach combines "a mix of biophysical methods, economic models, empirical analogue studies, and expert judgements" (O'Brien 2000: 3). In contrast to the impacts and adaptation assessments conducted by the IPCC, for instance, these country studies include more qualitative data and analysis. This also implies that the design of the assessments has to be flexible to adjust to both the specific circumstances characterizing the regions subject to study and differences in the availability of local data. In the selection of the sectors that are subject to analysis in each country, for instance, the choice is based on assumptions about each sector's importance to the ecology or economy of the country (O'Brien 2000: 4). Thus, the sectors analysed in different countries may vary accordingly. The manner in which the sector studies were conducted, however, could also vary significantly because of variations in data availability, particularly historical climate data. Thus, the approach of the UNEP Country Studies shows that some variation in the design of the assessments is required. The comparability of different assessments is not necessarily compromised by this feature, as long as the "right" sectors are chosen as indicators of vulnerability. The introduction of qualitative data, however, implies that the assessments seem to depend on individual judgement to an extent that is unprecedented within the IPCC framework. The IPCC capability of handling such aspects, moreover, may be limited because of the politicized environment within which the IPCC operates.

The politicized environment of the IPCC assessment process

The IPCC is a rather unique construction in international cooperation. As a "scientific" body operating within the highly politicized framework of the negotiation process on climate change, the IPCC operates in the very difficult interface between science and politics.[9] One implication of this framework condition is a very alert awareness of the boundary line between science and politics, and a very rigid requirement (from both internal and external actors) not to move beyond the boundaries of science. In an institution such as the IPCC, that may very well be easier said than done. At most decision-making levels of the body, the majority of IPCC delegates are government officials operating on the basis of national instructions. To maintain the body's scientific authority, therefore, relatively rigid detailed time-consuming and inflexible rules of procedure have been incrementally developed during the years the IPCC has been in operation.[10]

Organizationally, the IPCC has three main decision-making levels: the "scientific core" of each WG, which is dominated by members recruited on the basis of their scientific merit; WG plenaries, where members participate in their competencies as national delegates, but where scientists also have significant decision-making power; and the full Panel where membership is entirely based on political nominations by national governments, and where scientists have a much more restricted decision-making power than in the WG plenaries.

The panel provides a set of reports differentiated according to the procedures by which they are endorsed by the Panel. The main IPCC publications include Assessment Reports, Summaries for Policymakers, Technical Papers, and Special Reports. The full Assessment Reports are the IPCC publications over which the politically nominated members of the panel have the least decision-making power. While Assessment Reports are accepted by the plenaries of the respective WGs for which the assessment is developed, they are not subjected to detailed discussions. Summaries for Policymakers are summaries of Assessment Reports and represent in some respects the opposite extreme of the full Assessment Reports in terms of being subjected to a line-by-line (often word-by-word) approval in the respective WG plenary for which the assessment is developed. Special Reports usually also constitute a bulk report supplemented with a Summary for Policymakers, and the endorsement procedure of Special Reports parallels that of the Assessment Reports. Technical Papers are reports on special issues based on the material of published Assessment Reports and where scientists are actually prohibited from using new material. In addition to these reports, the Panel provides a Synthesis Report of the assessments, which is a summary (synthesis) of the reports of all the three WGs that make up an assessment endorsed by the full panel plenary in a so-called "paragraph-by-paragraph" endorsement procedure.[11]

These various rules of procedure for the panel's endorsement of its publications imply that a report's journey from the "scientific core" of the organization to a published report is a cumbersome one that requires careful navigation on the boundary between science and politics.

Claims that the Panel is overstepping the boundaries of science have become a frequently used strategy to influence the substantive content of the Panel's publications in the desired (and, paradoxically as it may seem, usually politically motivated) directions. There are several examples in the Panel's history that illustrate this trend, but a good example in this context is its endorsement of the 1994 Technical Guidelines, which took place at a WGII plenary session in Nairobi in 1994.[12] On this occasion, the question of whether the panel could accept references to "adverse effects of climate change" without accompanying references to "positive effects of climate change" spurred an almost two-day long discussion. The argument of the parties which required insertion of the phrase "and positive" every time that the "adverse effects of climate change" was mentioned in the report, was that this was a "scientific" report which should adopt neutral language. While that may be true, it was also quite clear that the demands of these parties were politically motivated.[13]

Assessments of social vulnerability are likely to involve many difficult concepts that may generate this type of discussion. This can be illustrated by a Table showing "selected characteristics of vulnerability and their assessment" (from Adger and Kelly 1999, cited in Kelly and Adger 2000: 331).

Table 12.1 illustrates some of the very difficult concepts that may be involved in assessments of social vulnerability, concepts with political connotations that may be difficult to disentangle from their scientific meanings, especially within a politicized environment such as the IPCC. While some of these factors can be quantified, such as poverty and marginalization, others cannot be quantified in meaningful ways. The difficulties in quantifying some of these factors may reinforce the risk of politicization inherent in the IPCC framework (Kelly and Adger 2000).

There are several examples in the IPCC's history of the paralysing effect a politicization of the issues under discussion can have. And even though the Panel may be able to disentangle and resolve such controversies, the result would be compromise text that might undermine the scientific authority of the assessment or jeopardize its applicability to policymakers. Thus, the proximity to politics of the issues involved in assessments of social vulnerability may represent a serious obstacle to the IPCC's provision of such assessments. While an impacts-driven approach to adaptation also may involve value-related issues, this is an approach

Table 12.1 Selected characteristics of vulnerability and their assessment

Vulnerability indicator	Proxy for	Mechanism for translation into vulnerability
Poverty	Marginalization	Narrowing of coping and resistance strategies Less diversified and restricted entitlements Lack of empowerment
Inequality	Degree of collective responsibility, informal and formal insurance and underlying social welfare function	*Direct*: concentration of available resources in smaller population affecting collective entitlements *Indirect*: inequality to poverty links as a cause of entitlement concentration
Institutional adaptation	Architecture of entitlements determines resilience Institutions as conduits for collective perceptions of vulnerability Endogenous political institutions constrain or enable adaptation	Responsiveness, evolution and adaptability of all institutional structures

that often is referred to as "scientific" and thus more appropriate for the mode of operation of the IPCC:

> [T]he inherently political nature of issues such as property rights that affect levels of vulnerability cannot be ignored. Arguably, it has been the impossibility of engaging with the more political aspects of the issue of vulnerability that has led the Intergovernmental Panel on Climate Change, operating as it must within a framework of consensus decision-making, to focus on the biophysical aspects of the subject. Analysing biophysical impacts is not, of course, a value-free activity, but problematic areas may not be as obvious.
>
> (Kelly and Adger 2000: 329)

Alternative institutional frameworks?

There may be alternative ways of incorporating social vulnerability assessments within the UNFCCC framework. In January 2004 the UNFCCC Secretariat issued a *Compendium on methods and tools to evaluate impacts of, and vulnerability and adaptation to, climate change*.[14] This report is a response to the request to the Convention Secretariat by COP3 "to continue its work on the synthesis and dissemination of information on environmentally sound technologies and know-how conducive to mitigating, and adapting to, climate change; for example, by accelerating the development of methodologies for adaptation technologies, in particular decision tools to evaluate alternative adaptation strategies" (Decision 9/CP.3, cited in UNFCCC 2004: 1–2). The compendium is an update of a similar 1999 report. It is neither an assessment of social vulnerability nor a methodology for doing assessments of social vulnerability; it is a compilation of the decision tools for impacts, vulnerability, and adaptation assessment that have been used in various guidelines, policy framework reports, and assessments by a number of organizations and bodies. The compendium is worth mentioning, however, because of its explicit recognition of the distinction between first- and second-generation approaches to vulnerability and adaptation:

> The more recent emphasis on current climate variability, and current vulnerability and adaptation, has been associated with more sophisticated approaches to socio-economic scenarios, to stakeholder participation, to adaptation policies and measures, and to the assessment and strengthening of adaptation capacity. These changes are reflected in the content and structure of this updated version of the compendium, making it more relevant to today's needs.
>
> (UNFCCC 2004: 1–3)

Similarly, in the specified rationale for NAPAs it is emphasized that:

> the NAPA takes into account existing coping strategies at the grassroots level, and builds upon that to identify priority activities, rather than focusing on

scenario-based modelling to assess future vulnerability and long-term policy at state level. In the NAPA process, prominence is given to community-level input as an important source of information, recognizing that grassroots communities are the main stakeholders.[15]

The message from second-generation vulnerability scholars thus seems to have been well received in the UNFCCC. Given the potential obstacles to the provision of full IPCC assessments of social vulnerability, future COPs might wish to consider transferring this issue to other bodies that are less cumbersome and less in the limelight within the UNFCCC institutional framework. This could also facilitate a commissioning of the actual assessments of social vulnerability to other expert bodies than the IPCC that operate within less politicized environments and decision-making procedures.

Conclusion

In the climate issue, developing countries' primary concern is sustainable development:

> It is time now to refocus on the longer-term objectives of the UNFCCC, particularly on its stated goals regarding sustainable development. ... [D]eveloping country concerns, which had always been marginal to the thrust of the UNFCCC, have become even more marginalised in recent COPs as the focus has been concentrated on getting the Northern countries ... to accede to the Kyoto Protocol. This has happened at the cost of sidestepping, if not outright ignoring, Southern priorities.
>
> (Sokona *et al.* 2002: 1–2)

One important mechanism whereby sustainable development can be reintegrated in the international climate regime is through a stronger focus on vulnerability and adaptation.

Currently there exist two distinct conceptualizations of vulnerability with markedly different implications for adaptation policies and measures. The impacts-driven approach pursued by the IPCC leads to a focus on "technology and transfer of technology, rather than on development" (O'Brien *et al.* 2004: 12). To address adaptation capacity and needs, an understanding of the more fundamental causes of vulnerability is required. The current IPCC assessments of future climate impacts thus need to be supplemented by assessments of social vulnerability. As pointed out by O'Brien *et al.* (2004: 13), this seems to constitute an essential basis for the development of adaptation policies, to prevent the risk of "treating the symptoms of vulnerability instead of its causes."

There are elements in the IPCC mode of operation, however, that may represent obstacles to its provision of assessments of social vulnerability. In particular, the "soft science" input of qualitative data and assessments, combined with the politicized environment and decision-making procedures within which the IPCC operates, may seem to represent particular challenges for the provision of assessments of social vulnerability by the IPCC. This politicized environment implies a strong need to draw a very clear line between science and politics. This distinction may be difficult to maintain in assessments of social vulnerability, which in turn may result in long and difficult discussions in the WG and a counterproductive politicization of the issue. Thus, the question of whether assessments of social vulnerability can be transferred to other institutional frameworks within the UNFCCC needs to be addressed.

Even with the provision of social vulnerability assessments, however, difficult obstacles to more committed adaptation policies remain. In contrast to impact-driven assessments, which generate relatively clear technological options for adaptation policies, the policy implications of social vulnerability assessments are likely to be vaguer. This implies that their conversion into explicit policy commitments, as demanded by developing countries, also is likely to be more difficult. Within the context of climate change, moreover, an additional potential problem is the need to distinguish between developmental problems in general and adaptation to climate change in particular. As pointed out by Kelly and Adger (2000: 191):

> the Kyoto Protocol, and related mechanisms around the international agreements on climate change, has authority only to focus on environmental impacts and adaptation provoked by a narrowly defined human-induced climate change. Hence there is a fundamental dilemma at the heart of international action on this issue – the need for reductionist identification of the 'climate'-related part of global social and economic trends, versus the desire to see all climate change as another important dimension of global environmental threats to development.

In the current international climate regime, the climate problem is framed as an environmental problem – a problem of pollution. Developing countries – and the US – however, do not only see this problem in environmental terms. To them, climate change and efforts to abate climate change are as much an *economic* as an environmental problem. An important motivation for the US defection from the Kyoto agreement was the economic implications of its Kyoto commitments for the national economy, especially as what they perceive to be major economic competitors in an international market, large developing countries such as China and India, are exempt from regulatory commitments. Developing countries are concerned with the developmental dimensions of climate change, both as cause and consequence of climate change policies.

This is also reflected in different perceptions of the link between mitigation and adaptation. The current perception is that the costs of future climate change (expected to be high) can be avoided by mitigation of climate change through reduction of GHGs. Mitigation of future climate change can also be achieved, however, by pursuing a less carbon-intensive economic development in developing countries. Thus, a shift in the framing of the climate problem to allow for a stronger emphasis of its economic dimensions can represent a potential pathway whereby mitigation and adaptation policies can be combined.

Different perceptions of the climate problem itself thus seem to be a root cause of the current situation in the climate negotiations. Further progress in the negotiations, and a shift in the emphases between mitigation and adaptation, may thus depend very heavily on the development of common terms within which the problem is understood and addressed. To break the current deadlock, it seems essential that different perceptions are addressed and resolved. Assessments of social vulnerability could represent an important step in that direction.

Notes

* Acknowledgements: I want to thank Karen O'Brien, Jon Hovi, Dag Harald Claes and Espen Moe for useful comments to an earlier draft of this chapter.
1 See, for instance, Reuters Environmental News Service, Planet Ark, 2 November 2005: "Six-Nation Climate Change Meeting set for Mid-Jan" at http://www.planetark. com/ See also US Department of State Fact Sheets, "President Bush and the Asia-Pacific Partnership on Clean Development", 27 July 2005, and "Vision Statement of Australia, China, India, Japan, the Republic of Korea, and the US for a New Asia-Pacific Partnership on Clean Development and Climate", 28 July 2005. Source: http:// www.state.gov/.
2 It should be noted that currently it is difficult to assess whether US support of developing countries on this issue is rhetoric to trouble the waters for Kyoto Parties, or whether this actually is an issue where the US could be willing to take on some sort of commitments.
3 The text of the Climate Convention may be found at http://unfccc.int/essential_ background/convention/background/items/2853.php Accessed: 8 December 2004
4 UNFCCC: "National Adaptation Programmes of Action (NAPAs)". Source: http:// www.unfccc.int/adaptation/napas/items/2679.php Accessed: 8 December 2004.
5 UNFCCC Decision 27/CP.7. Source: FCCC/CP/2001/13/Add.4, 21 January 2002, available at http://unfccc.int/essential_background/convention/items/2627.php Accessed: 8 December 2004.
6 "Vulnerability is the degree to which an exposure unit ['the activity, group, region or resource exposed to significant climatic variations'] is disrupted or adversely affected as a result of climatic effects" (Carter *et al.* 1994: 3). The low significance ascribed to this term is reflected in that this definition only appears in the bulk report, not the Executive Summary or the Summary for Policymakers.
7 Kaspersson links this approach to even earlier works, notably "Amartya Sen's (1981) classic study of famine in Bengal" (2001: 2).
8 Personal communication with Karen O'Brien.
9 This section is based on previous work by the author, published in Skodvin (2000).
10 IPCC Rules of Procedure can be found at: http://www.ipcc.ch/about/procd.htm Accessed: 16 December 2004.
11 Synthesis Reports very often focus on a specific subject or a set of policy-relevant questions. The Synthesis Report to the Second IPCC Assessment Report, for instance, focused on the scientific basis for interpreting the ultimate objective of the UNFCCC to prevent a "dangerous anthropogenic interference with the climate system" (article 2). The Synthesis Report of the Third Assessment Report was structured around nine "policy-relevant, but not policy-prescriptive, questions". See: http://www.ipcc.ch/ Accessed: 16 December 2004.
12 The Second Plenary Session of (the reorganised) Working Group II, Nairobi, November 1994.

13 A detailed account of this discussion is provided in Skodvin (2000: 162–164).
14 Source: http://www.unfccc.int/ Accessed: 8 December 2004.
15 UNFCCC: "National Adaptation Programmes of Action (NAPAs)". Source: http://www.unfccc.int/adaptation/napas/items/2679.php Accessed: 8 December 2004.

References

Adger, W. N. and Kelly, P. M. (1999) Social Vulnerability to Climate Change and the Architecture of Entitlements. *Mitigation and Adaptation Strategies for Global Change.* 4. 253–266.

Adger, W. N., Huq, S., Brown, K., Conwaya, D. and Hulmea, M. (2003) Adaptation to Climate Change in the Developing World. *Progress in Development Studies.* 3 (3). 179–195.

Blaikie, P., Cannon, T., David, I. and Wisner, B. (1994) *At Risk: Natural Hazards, People's Vulnerability, and Disasters.* London and New York: Routledge.

Burton, I., Huq, S., Lim, B., Pilifosova, O. and Schipper, E.L. (2002) From Impacts Assessment to Adaptation Priorities: The Shaping of Adaptation Policy. *Climate Policy.* 2.145–159.

Carter, T. R., Payy, M., Nishioka, S., Harasawa Hideo, Christ, R., Epstein, P., Jodha, N.S., Stakhiv, E. And Scheraga, J. (1994) *IPCC Technical Guidelines for Assessing Climate Change Impacts and Adaptations.* Oxford, UK/Tsukuba, Japan: University College and Center for Global Environmental Research.

Huq, S. and Burton, I. (2003) Funding Adaptation to Climate Change: What, Who and How to Fund? London: IIED. *Sustainable Development Options.* Source: www.iied.org.

IPCC (2001) *Climate Change 2001: Impacts, Adaptation and Vulnerability.* Summary for Policymakers of the Working Group II Report. Geneva: Intergovernmental Panel on Climate Change.

Jacob, T., (2002) Reflections on Delhi – The Rise of Economic Realities? Internal Memo after COP8, DuPont Global Affairs, October 2002.

Jacob, T. (2003) Meeting Review: Reflections on Delhi, *Climate Policy.* 3.103–106.

Kaspersson, R. (2001) Vulnerability and Global Environmental Change. *IHDP Update, Newsletter of the IHDP.* (2). 2–3.

Kelly, P. M. and Adger, W. N. (2000) Theory and Practice in Assessing Vulnerability to Climate Change and Facilitating Adaptation. *Climatic Change.* 47. 325–352.

McCarthy, J. J., Canziani, O.F., Leary, N.A., Dokken, D.J. and White, K.S. (2001) *Climate Change 2001: Impacts, Adaptation and Vulnerability.* Contribution of Working Group II to the Third IPCC Assessment Report. Cambridge, UK: Cambridge University Press.

Müller, B. (2003) Framing Future Commitments: A Pilot Study on the Evolution of the UNFCCC Greenhouse Gas Mitigation Regime. Executive Summary, EV 32, June 2003. Oxford Institute for Energy Studies.

O'Brien, K. (2000) Developing Strategies for Climate Change: The UNEP Country Studies on Climate Change Impacts and Adaptations Assessment. *CICERO Report,* 2002. Oslo: *CICERO.* 2.

O'Brien, K., Eriksen, S., Schjolden, A. and Nygaard, L.P. (2004) What's in a Word? Conflicting Interpretations of Vulnerability in Climate Change Research, Working Paper 2004:04. Oslo: *CICERO.*

Ott, H. E. (2003) Global Climate. *Yearbook of International Environmental Law.* 13 (2002). Oxford: Oxford University Press.

Ott, H. E., Brounds, B. and Winkler, H. (2004) South-North Dialogue on Equity in the Greenhouse: A Proposal for an Adequate and Equitable Global Climate Agreement, May 2004: *Deutsche Gesellschaft für Technische Zusammenarbeit (GTZ) GmbH.*

Skodvin, T. (1999) Making Climate Change Negotiable: The Development of the Global Warming Potential index, Working Paper, 1999: 09. Oslo: *CICERO*.

Skodvin, T. (2000) *Structure and Agent in the Scientific Diplomacy of Climate Change*. Dordrecht: Kluwer Academic Publishers.

Skodvin, T. and Fuglestvedt, J. S. (1997) A Comprehensive Approach to Climate Change: Political and Scientific Considerations. *Ambio*. 26 (6). 351–358.

Sokona, Y., Najam, A. and Huq, S. (2002) Climate Change and Sustainable Development: Views from the South. IIED. *Sustainable Development Opinion*. London, Source: www.iied.org.

UNFCCC (2004) Compendium on Methods and Tools to Evaluate Impacts of, and Vulnerability and Adaptation to, Climate Change. Final Draft Report, January 2004: *UNFCCC Secretariat*. Source: http://www.unfccc.int/.

Watson, R. T., Noble, I.R., Bolin, B., Ravindranath, N.H., Verardo, D.J. and Dokken, D.J. (1996) *Climate Change 1995: Impacts, Adaptation and Mitigation of Climate Change*, Working Group II contribution of the Second IPCC Assessment Report. Cambridge, UK: Cambridge University Press.

Watson, R. T., Zinyowera, M., Moss, R. and Dokken, D. (1997) *The Regional Impacts of Climate Change: An Assessment of Vulnerability*. Working Group II Special Report: Summary for Policymakers. Geneva: Intergovernmental Panel on Climate Change.

13 Common but Differentiated Responsibilities

The North–South Divide in the Climate Change Negotiations

Ariel Macaspac Penetrante

Introduction

Although the climate change negotiations have a mainly scientific basis[1] (see Lieberman *et al.* 2007; Alley *et al.* 2007; Weaver 2004), it is the political preconditions affecting actors, structures, processes, issues, and outcomes that, for the most part, determine how climate change is addressed (Penetrante 2010, 2012).

In February 2007 the Intergovernmental Panel on Climate Change (IPCC) released its Fourth Assessment Report. This not only stresses the role of anthropogenic activities as major contributors to global climate change (IPCC 2007) but also the urgency of formulating policies to address these. The current climate change negotiations demonstrate various stumbling blocks to obtaining an equitable and sustainable agreement on climate change. At the international level, in particular, climate negotiations must address changing political conditions resulting from, for example, an increasing number of national and international stakeholders, changing mindsets, growing agendas on certain topics, and evolving expectations regarding outcomes. How do these influence the negotiation process? How can they be conceptualized and dealt with?

The transformation of the climate change issue from "low" to "high politics" in recent years is very obvious in the growing public attention to climate change.[2] This has been partly due to the "discovery" of the inter-linkages between the climate change issue and other "high politics" issues such as trade, economics, and security. Policymakers have become increasingly aware that climate change undermines the sustainability of human wellbeing (Cowie 2007). It is now seen as a phenomenon that affects national security and has the potential to destabilize states, if not properly addressed. Of course, this transformation of climate change to the high political sphere could be seen as desirable, in that it attracts the attention of policymakers and the public. However, it also leads to the securitization of climate change and therefore to an increase in its complexity. Thus, securitization and complexity become the basis of climate change decision and policymaking, rather than objective scientific knowledge. "Technical formulas" based on scientific expertise are increasingly being subordinated to "political formulas," which have always been at the root of the climate talks agenda. In the same manner, climate issues have intruded in other areas. Depledge (2006) describes the situation in which climate politics have crossed environmental boundaries. These linkages have created self-enforcing dynamics, further increasing the complexity of climate change negotiations.

The climate change context is complex and fraught with uncertainty (see Adger *et al.* 2003; Depledge 2005; Hamal 2010; Marland 2006; Dimento and Doughman 2007). For example, there is considerable uncertainty regarding emission estimates. Scientific projections regarding the effects of greenhouse gas (GHG) emissions on climate are uncertain, as they cannot be estimated with complete accuracy. The emissions timescale is uncertain – should GHG emission measurements begin with the start of the industrial revolution in the eighteenth century[3] or from the year 1990, frequently used by scientists as the baseline year for estimation of GHGs? This choice over baseline (e.g. 1990 or 2000) has, for example, inherent notions of justice and fairness. Industrialized countries would probably find inclusion of the industrial revolution within the time frame unfair, on the basis that present generations should not be held accountable for the actions of their predecessors. However, developing countries, anticipating unavoidable emissions as they pursue industrialization goals, would find it unfair for the past not to be included in total emissions. They would argue that the developed countries have contaminated the environment and should pay the price for the resulting climate change. According to one commentator,[4] developed countries have consumed three-quarters of the pizza and are now demanding that the developing states not only limit themselves to the consuming the remaining quarter but also share that quarter with developed countries. Adding to the uncertainty, the climate change scenarios used by scientists as a basis for modeling climate change are too general at present to be applicable to a specific locality. Downscaling them to provide more localized information brings a significant reduction in the accuracy of predictions (Burton *et al.* 2002: 152).

The complexity, multiple causation, and uncertainties inherent in the climate change negotiation process mean that a theoretical, context-based understanding of the interaction of multiple variables within a specific framework and structure is needed. Only an analysis of this type – a systems analysis – can provide an adequate response to the self-driving dynamics that constitute the climate change bargaining and decision-making process.

The North–South Divide

The climate change negotiations seem to have resuscitated the North–South divide[5] in the international system, something which had come to be seen as outdated (Nigel 1986; Robinson *et al.* 2000).[6] In 1980 the Brandt Report – Common Crisis North–South[7] – stressed the growing "Third World" problem and the increasing divergence between the North and the South. This paved the way for more studies on the North–South divide. The North–South divide is the socio-economic and political division between developed ("North") and developing ("South") countries (see Hayes and Smith, 1993; Zartman 1987; White 1993). It represents the development gap between countries as described by the Human Development Index (HDI) and exemplified by Walter Rostow's model of development (2000). This paper highlights the problems brought about by the social evolutionary feature of this model, which assumes that the North "invented" development and that the "South" should follow the same path. In other words, the experiences of the North are supposed to serve as a yardstick for formulating development policies, which implies the need to assert "development principles" determined by the North. This paper argues that development can be removed from its "Western-centric" orientation and instead conceptualized contextually, with socio-cultural and local specificities being properly considered.

For the purposes of this paper, the North consists of the member states of the Organisation for Economic Co-operation and Development (OECD) and the so-called "economies in transition," that is, former Warsaw Pact countries in Eastern Europe.

The contesting notions of justice and fairness, which can also be found in areas such as trade and global economics, have sharpened the relevance of the North–South divide in the climate change negotiations. The boundary between the North and the South has been primarily determined by positions about who should shoulder the costs of climate change mitigation and adaptation measures. On the one hand, the North pursues a "forward-looking" notion of justice and fairness, maintaining that present generations should not be punished for "crimes" they did not commit and for "crimes" that were not crimes, as they were carried out by earlier generations (Caney 2009). Furthermore, the North argues, it would be unfair for present and future generations to have to carry these costs and unfair, too, that mandatory emission cuts would be required by developed countries under the Kyoto Protocol and not by major emitters in the developing world, such as Brazil, China, and India (see Schelling 1995). Developed countries, on the other hand, argue that mandatory cuts should apply to all countries, regardless of the past, especially as some developing countries have rapidly increasing emissions and therefore should adopt substantial mitigation targets of their own (Buchner and Lehman 2005: 45ff.; Gardiner 2004: 38ff.).

The South follows a "backward-looking" notion of justice and fairness, according to which present generations in developed countries have benefited from the decisions and actions of their forebears and should therefore carry the associated costs of climate change (La Viña 1997: 174; Muller 2009). Furthermore, as the North "owes" the South the environmental space it has contaminated, there should be a scheme of compensation for the developing countries. As La Viña (1997: 65) notes, historical data show that industrialized counties of the North have utilized more than half of the world's fossil fuels over the past 120 years. Many developing countries find it unacceptable that their traditional agricultural practices are blamed for the increase in methane emissions. As these practices support the subsistence of billions of people, they should be distinguished from energy-wasteful agriculture, animal husbandry, and industry in the North, which reflect "luxurious needs" and also contribute substantially to methane emissions. Furthermore, technology transfers and other developmental measures should not be based on goodwill, but rather seen as compensation payable by the North to the South for the historical damage to the environment. The South now also demands that climate change should not compete with the conventional development assistance provided by industrialized countries, especially to the least developed countries, as this will lead to unfair distribution. However, it is important to find where the boundaries of such commitments lie in order to avoid a "blank check" situation. The question of justice and fairness in the distribution of global warming "burdens" has been described as "a tricky one practically, philosophically, and politically" (Grubb *et al.* 1992: 306). Justice and fairness refer to apportionment of the costs of adaptation, mitigation, and/or compensation, which is contentious in both developed and developing countries.

Growing involvement of the South in the climate talks

The complexity of climate-change-related issues has increased recognition of the importance of developing countries – usually bit players in international negotiations – in the climate talks process. For instance, several developed countries have made the participation of developing states, particularly those with emerging economies such as Brazil, China, and India a precondition for their own participation. This has paved way for developing countries to become more important players. President George W. Bush announced during his term in office that the United States would not return to the climate change negotiation table unless developing countries also formally accepted the same or "comparable" responsibilities as countries in the North. The US Special Envoy on Climate Change, Ambassador Todd Stern, (2009) argues, "China and the developing world's current emissions are tomorrow's historic responsibility" and notes that developing nations are on track to produce more than 80 per cent of the growth in carbon emissions during the next several decades. According to Stern, China is now the second-largest economy and second-largest trading power after the United States. In 1992 when China signed the UNFCCC, it emitted 2.5 Gt of CO_2, half of the US total. China currently emits more than 7 Gt of CO_2 per year, surpassing the US as the world's largest emitter. Stern also notes that China's emissions under business-as-usual assumptions would alone be big enough to put the world on track to global concentrations of 540 ppm of CO_2 and a 2.7°C temperature increase.

Thus, excluding developing countries from climate talks is no longer an option. The lack of binding commitments agreed on mitigation at COP15 in Copenhagen has been somewhat compensated for by developing countries becoming considerably more involved in the mitigation negotiations than they had been at earlier COP and COP/MOP meetings. This process development may help a more comprehensive mitigation agreement to be passed in the future.

Developing countries have found ways to empower themselves and actively engage in climate change negotiations, help formulate the climate talks agenda, and also influence the negotiation process. The empowerment of developing countries has come about partly through the organizational settings available within the international system. The establishment in 1964 of the "negotiation vehicle" called the Group of 77 (G77, with currently 133 member states) has contributed to the increase in negotiation capacity of the developing countries. In 1972 the UN

Conference on the Human Environment in Stockholm counselled the formulation of environmental policies that would not hamper development. Under the 1972 Stockholm Declaration of Principles (Cowie 2007: 392–395), developed countries are encouraged to help developing countries carry out environmental management and establish environmental safeguards The United Nations Conference on the Environment and Development in Rio de Janeiro in 1992 established a process for negotiating further measures to help developing countries and introduced mechanisms for transferring finance and technology to developing countries (Cowie 2007: 400).

The question is whether there will be any spill-over of this new-found sense of identity and "self-esteem" from the climate change context to other contexts such as nuclear proliferation. Certainly, the spill-over seems to have mainly occurred the other way around – from trade and global economics to climate change issues. As Yu III (2008: 1) states: "the increasing share and influence of developing countries in global economic affairs – both in terms of shaping global economic policy and in terms of actual weight in the global economy … has been clearly evident in both the WTO and UNCTAD."

Methodologies – New "Old Paradigm"

This chapter seeks further understanding of the complexities, uncertainties, and multiple causation issues in the climate talks that have led to changes in organizational patterns in the international system and introduced new settings. These changes include:

1. new and unaccustomed coalitions – for example, the G77 entering the climate change realm; and
2. the revival of the old notion of the North-South divide (a non-organizational setting) – said, like the East–West rift of the Cold War era, to be "deceased" (van Evera 1990) and "discredited" (Sebenius 1991: 87).

Both the new organizational patterns and the notion of the North–South divide, determined by contesting notions of justice and fairness, define the identity, decisions, and actions of states. However, it should be noted that the new "old paradigm" of the North–South divide is still more or less limited to the climate change context. The extent to which there will be further spill-over into other realms such as education, migration, and energy remains to be seen.

The first part of this chapter will analyze the organizational and non-organizational settings contributing to developing countries' identity-building process, how this identity can be conceptualized, and how it can be understood in the climate change context. The second and third part of the chapter focus on an analysis of the structural features of the climate change negotiations and the implied coalition building established through the identity-building process. It will ask several questions: What are the forces behind the coalition building in the context of climate change? What dynamics are created through the new organizational setting? As

COP15 in Copenhagen showed that leadership in the climate talks seems to be missing, is the hardening of the "front lines" between coalitions preventing any meaningful leadership from emerging? How does structure influence the decisions and behavior of states? The fourth part of the chapter discusses technology transfer, capacity building, and equity in relation to developing countries.

The approach here is not reductionist in the sense of segregating a problem into its constituent elements and analyzing them separately, but is concerned rather with an integrated analysis of its quantitative and qualitative aspects. Limiting the analysis to simple interactions among actors would not be enough to shed light on the complexity of the climate change issue. The process, structure, and outcome can be analyzed only at the system level, where the system's complexity and multidimensionality demonstrate various patterns of interaction and self-driving dynamics (Cavallo 1979: 24, 33; Kindler and Kiss 1984: 1). This approach allows the inherent dilemmas and paradoxes – and thus the stumbling blocks – in the climate change negotiations to be identified.

Structural Features of the Climate Change Negotiations

An analysis of the relationships among actors (agency) and structure, particularly in the political context of climate change negotiations, reveals the presence of consensual relationships (see Adler 1997; Carlsnaes1992; Dessler 1989; Keohane 1989; Wendt 1987) in which political effects, outcomes, and events can be explained in terms of structural or contextual factors (Hay 2006: 2). Patterns of political behavior observed, for instance, on the part of states within the international system cannot simply be explained in terms of, say, actor logic, but must be looked at from the perspective of overall system logic. In the climate change negotiations, the dominance of the rational choice, for instance, in formulating policies based on costs and benefits, could be seen as undermining the validity of structuralism. However, as Hay (2006: 103) argues, even rational choice implies structuralism, even though the most basic assumption upon which rational choice theory is premised is that actors are egoistic and self-regarding utility maximizers who behave rationally in pursuit of their preferences. Such preferences are relative, or they should be understood as a response to other actors, which means that they can only be explained in a social context. George Tsebelis (1990: 4) points out the paradox between optimal adaptation of actors to an institutional environment and the assumption that these actors have perfect knowledge of their environment. It is, in fact, the prevailing institutions (the rules of the game) that determine the behavior of the actors and this, in turn, produces political or social outcomes.

This paper notes that the structures created by the institutional environment are merely subjective constructions: interpretations by the actors of the "realities," using their own rational system as a reference point. While each actor can have its own rational choice system, this may be different from the rational choice system of other actors. The rational choice system of each actor is determined by various learning processes and experiences leading to subjective priorities for decisions and behavior. However, the subjective reality of the actors is, in fact, the objective realities of their environment. Given that actors tend to subordinate structures through their own cognitive "interpretation process," actors are recognized as having greater sovereignty over structures. Nevertheless, structure does have an influence on the decisions and behavior of actors – however, not to the extent that actors lose sovereignty over and control of the process. The concept of "new institutionalism" reflects this sovereignty over structure, emphasizing institutional settings as merely mediating and constraining factors in the realization of outcomes (Hay 2006: 105). This concept stresses:

i)   the ordering of social and political relations in and through the operation of institutional (structural) constraints such as established practices, processes, and tendencies (Pierson 2000: 264);
ii) the normalization of institutional functions through shared codes, rules, and conventions imposing value systems that may constrain behavior (Brinton and Nee 1998: part I);
iii) definition of the logic of appropriate behavior in anticipation of sanctions (March and Olsen 1984, 1989); and
iv) reliance upon existing institutional templates (DiMaggio and Powell 1991).

If actors use appropriate behavior in anticipation of sanctions, then they are "controlled" by and "subordinated" to the structure. Structure thus may be a constraining factor in decision making; however, it should be noted that this will occur for as long as the structure (in the form of institutions and conventions) is considered legitimate. Structure can be challenged by actors and modified according to the will of an actor. The effect of structural constraints is avoidance of contingencies[8] (lack of information about a specific behavior on the part of another actor; see Parsons 1960; Luhmann 2004: 13–14) through social communication enabled by structural elements. Institutions should be defined in terms of rules, norms, and conventions (Hall 1986: 6; March and Olsen 1984, 1989) which are themselves (evolving) outcomes of a (dynamic) bargaining process. How structure influences decisions and how behavior influences conduct is context- (or structure-) dependent, not in the sense that actors are dictated to by structures in which they have to behave in a particular manner in a given context, but in the sense that *decisions and actions become habitual*. The new institutionalism thus tends to replace rational choice theory's "logic of calculus" with "logic of appropriateness" (Hay 2006: 106).

The climate change bargaining table: The burden of burden sharing

Climate change offers a unique and more or less unprecedented bargaining table. In addition to the multi-causation and inter-linkages of issues, the bargaining table is inclusionary, enabling developing countries, NGOs, and the private sector to influence the negotiation process. Furthermore, the asymmetry in information, resources, and even the negotiation capacity of the actors, leads to coalition building. Are coalitions in the climate negotiations different in nature from coalitions in other realms? Negotiations under uncertainties contend with different dynamics; for instance, dependence on the scientific community which contributes models and scenarios to illustrate uncertainties is an important factor here. Climate change is also unique in the sense that negotiations usually aim to get the biggest portion of the pie. However, in the climate change context, actors are negotiating to get the smallest portion of the pie, as the pie represents costs or burdens. As the climate change negotiation is all about burden sharing, the bargaining process has different components from other bargaining processes. Furthermore, although there is a consensus about the goal or about the desired outcome of the negotiation (to stabilize the GHG concentrations in the atmosphere or to maintain the 2°C threshold), the procedural setting (i.e. how to achieve the outcome) is a major stumbling block in the negotiation process. Finding an "acceptable" procedure (e.g. voting procedure) is made more difficult by systemic conditions which will be analyzed in this section. A descriptive analysis of the "climate change bargaining table" will help understanding the context of climate change and the ways in which it influences the course of the negotiation process and its outcome. The following features of the "climate change bargaining table" as a system show that negotiation, both in and of itself, has become a major stumbling block to addressing climate change.

1. Multiple Causation

Multiple causation is a stumbling block to addressing climate change. The concept of multiple causation is based on the understanding that there may be various possible causes of a given effect, any one of which may be a sufficient but not necessary condition – or a necessary but not sufficient condition – for it to occur. A specific policy can lead to "products," which require other "products" to complement them. This policy may also lead to a proliferation of "substitutes." In one way or another, policies will have impacts on other sectors that were not intentionally targeted, which may lead to transparency and accountability problems. Theodorson and Theodorson (1979) explain another type of multiple causation, in which the given effect may occur in the absence of all but one of the possible sufficient (but not necessary) causes and, conversely, in the wake of some but not all the necessary but not sufficient causes. In short, the uncertainty over the exact causes of global warming, and how global warming itself leads to land scarcity, food insecurity, migration and health problems, is a question that cannot be answered without referring to other (socio-political) issues. Food insecurity is not only caused by global warming, but also by poor infrastructure or poor policies. Politically, adaptation measures will require a detailed description of the causal relationships among the various factors involved in order, for instance, to legitimize assistance – where developed nations may be disinclined to extend adaptation assistance to corrupt authoritarian regimes. The diffuse nature of climate change brought about by multiple causation leads to simultaneous bargaining (Carraro and Siniscalco 1997; Folmer *et al.* 1993; Alesina *et al.* 2001) on different issues, which, *inter alia*, further complicates agenda setting, increases the number of stakeholders, and requires coordination of a larger number of stakeholders.

2. Horizontal interlinkages and cross-cutting of issues

Climate change negotiations involve issues such as mandatory emission cuts, technology transfer, capacity building, flexible mechanisms (e.g. Clean Development Mechanism), food security, mass migration (particularly the so-called environmental refugees), trade (e.g. emission trading), project-based mechanisms, land use, land use change and forestry, governance (inter-agency coordination of national government units), human rights, health problems, indigenous people, gender-specific issues, and so on. This results in problems in agenda setting and issue clarification in the negotiations, which may lead to outcomes that are too broad to actually adequately address the climate change problem. For example, there could be a perception that COP15 was more about development than about climate change. Even though some cause-and-effect relationships can be shown in environmental problems, these are never strictly linear but part of a complex web of interactions (Malabed 2001). An approach based on interlinkages further recognizes the complexities inherent in ecosystem dynamics and their interface with the equally complex social, economic, and political dynamics inherent in human development and governance, particularly policies, laws, and institutions. To address the interlinkages and cross-cutting nature of these issues, a dynamic understanding and flexible problem-solving orientation are needed. The coordination of actions across governmental agencies (environment, energy, trade, economics, finance agencies) and non-governmental agencies supporting the government is of importance. This blurs institutional boundaries and integrates the needs of the different interests at the national level. This is sure to improve the bargaining capacity of state actors at the international level, not to mention the negotiation structure and agenda.

3. Multi-party setting – Multilateralism vs. bilateralism

Bargaining in the context of climate change is bargaining in a multi-party setting that thus requires a multilateral approach to negotiation. "Bargaining offers" are not only made to all state actors simultaneously at the international level, but concessions also need to be negotiated with stakeholders at the national level. As shown by COP15 in Copenhagen, multilateralism is not the approach preferred by major countries such as China and the United States. However, bilateralism is perceived by several non-major players as an attempt to marginalize them from the negotiation system, and this perception does not contribute to building trust in the system. Smaller developed countries, particularly those from the European Union, prefer a multilateral approach, as they have conditioned their resources for this type of setting, for instance, by distributing tasks among themselves. How can diverging settings be accommodated within the negotiation system? Another converging cleavage is that of international policy determined by a national mindset due to its legitimacy deriving from the national constituents. Because international negotiation is two-level (international and national), it results in "glocalized" policies that are intended to serve national interests. Partnerships between a national government, NGOs, and other civil society groups, research institutions, and policy think tanks, provide countries with a more diverse bargaining capacity at the international level. Furthermore, governments find themselves communicating with other countries in the region before going to the international level. Links between institutions and organizations operating at different spatial scales are required. Regional partnerships are necessary as, for instance, some water basins cut across national boundaries and involve multiple users and stakeholders. At the international level, as discussed earlier, the inclusive nature of the climate change context allows broader state participation, particularly by developing countries. With regard to the climate change context, Pruitt and Carnevale (1993) note the more likely increase in competition in the bargaining process in a multiple party setting. If an actor in assertive mode is confronted by demands from different sides and levels, it will have to defend its own position more strongly. The climate change negotiation confirms the notion that no single actor or institution has a monopoly on ideas, issues, agendas, or bargaining power (Zartman 1994; Zartman and Crump 2003; Hampson 1995; Susskind and Crump 2008), which further contributes to the complexity of the climate change context, already made complex enough through scientific uncertainties. The multiple party setting of the bargaining table, in short, requires the understanding that negotiation per se represents a major stumbling block in the climate change negotiations. This argues for a more theoretical understanding of the interaction of multiple parties in the bargaining and decision-making processes.

4. Asymmetries

Although the climate change context is an inclusive negotiation process and participation is the first step in influencing that process, players are frequently confronted by asymmetries in their bargaining strengths due to asymmetries in resources and capacity. Asymmetries in bargaining strength imply stronger parties having greater influence on the procedures and on the outcome, with coercion through economic, political, and military resources possibly being an element of the bargaining process (Miller 1995: 60). However, it should be noted that not only "having more than the others" but also "having less than the others" can also be a source of bargaining strength (see Larson 2003: 145), particularly in the climate change negotiations. These should not only be understood in terms of the possession of resources and capacity but as also as the possession of "negative" power, namely, the capacity to freeze, delay, or even veto achieving an objective (Hardy 1985), with a state dictating the speed of the negotiation because it lacks capacity to cope with the negotiation issues. A state may also legitimize its positions through deficits in resources and capacity illustrating a normative source of power in the context. Asymmetries in negotiation capacity, as Carraro and Sgobbi (2008: 1490) argue, lead to differing preferences regarding the negotiation setting. A weak player (high discount rate/high uncertainty) prefers to negotiate simultaneously, a strong player (low discount rate) bargains sequentially to signal his bargaining strength. Asymmetry is therefore ambivalent in its nature: on the one hand a facilitation mechanism ("having something to offer" strengthens negotiation capacity) and, on the other hand, is also a stumbling block as it can obstruct the bargaining process.

5. *Outcome – Procedural justice and burden sharing*

The climate change negotiation involves the combination of negotiating in absolute and relative terms regarding the outcome. The climate change negotiations focus primarily on the processes through which decisions will be made and how the burdens (costs) of mitigating and adapting to climate change will be distributed. The climate change context involves a variety of notions of justice such as procedural justice (fairness in dispute resolution and resource-allocation processes, e.g. participation of least developed states in all parallel meetings), distributive justice (fairness in the distribution of rights and resources, e.g. emission rights due to development and in the distribution of the adaptation fund), and retributive justice (fairness in the rectification of wrong, e.g. compensation to vulnerable countries) (see Solum 2004; Bone 2003; Dworkin 1985; Rawls 1971). The variety of notions of justice involved influences the situation at the bargaining table, in the sense that they are to a significant degree instrumentalized by actors as a source of bargaining power. However, the different notions of justice to some extent compete with each other, for instance, in agenda formulation. Which notion of justice is then to be prioritized on the agenda? In addition, the focus on procedural justice may have some negative implications in terms of effectiveness of the potential outcome, as effectiveness is not always synonymous with appropriateness.

Integrative summary – Bargaining about climate change

Understanding the five above-mentioned features of the situation at the climate change bargaining table will enable decision and policymakers to formulate plausible, prioritized negotiation strategies. Bargaining, as introduced by Rubinstein (1982), describes the process through which negotiating agents try to reach an agreement with players making offers and counter-offers over the terms of the agreement. The climate change context, however, requires the expansion of the simple offers/counter-offers relationship. Multiple causation, interlinkage, and cross-cutting of issues lead to simultaneous bargaining (Carraro and Siniscalco 1997; Folmer *et al.* 1993; Alesina *et al.* 2001) which then enlarges the zone of possible agreement (ZOPA). Whether this really does increase the likelihood of reaching an agreement (Carraro and Sgobbi 2008: 149) is still an open question. What should be also be considered is the extent to which parties can formulate strategies that will converge to a feasible solution (Carraro and Sgobbi 2008: 1502). The involvement of multiple and cross-cutting issues implies an increase in the number of stakeholders needing to be integrated into the negotiation process, which will eventually change the course of the negotiation. Because of these asymmetries in resources and capacities, new mechanisms are introduced, such as coalition building and the promotion of flexible mechanisms such as carbon trading. The multilateralism at the climate change negotiation table further leads to an increase in the number of interests and positions in the negotiation process, which could obstruct the pace of negotiation. Furthermore, diverging notions of justice and fairness can be identified in the climate change negotiation which, in turn, offers the possibility of sustainable and equitable outcomes due to the inclusive and integrative nature of the process. To sum up, as Raiffa (1982) argues, it is always more efficient to bundle issues in the negotiation procedure, so as to exploit the trade-offs among different issues. However, he notes that complexity considerations should be accommodated in formulating negotiation procedures such as proceeding issue by issue. Even though multiple causation, interlinkages, and cross-cutting of issues may cause asymmetries, multiparty diverging notions of justice, inefficiencies, and delay in finding an equilibrium, according to Raiffa's argument, there is a higher likelihood that the climate change problem will be addressed properly through bundling. Finally, the creativity in finding feasible solutions is limited by the concept of state sovereignty. According to La Viña (1997: 2), national boundaries are irrelevant in dealing with the climate change problem; however, there seems to be no alternative to using the "national lens" in negotiations. As Harrison and Bryner (2004: 329) conclude: "uncertain problems are especially difficult to handle in the anarchic international system because the nature of the problems must be negotiated through contending national preferences and interpretations." Climate change is thus an international phenomenon addressed at the national level. This discrepancy is a major stumbling block in formulating effective policies to address it.

Coalition through identity in the climate change negotiations

The coalition between the G77 and China (G77 + China) in the climate change negotiations is a spill-over from the trade and economic to the climate change realm caused by interlinkages and cross-cutting issues. Climate change negotiations came to involve the very same justice and fairness issues perceived in trade negotiations, thus reviving the coalition in the climate talks. This part of the chapter illustrates how the notion of justice and fairness constructed an identity which then paved way for the establishment of this coalition. The identity of the "South" determines the negotiation positions as well as the negotiation strategies of the collectivity influencing the negotiating structure (e.g. agenda setting, negotiation arena). As Najam (1994, 1995), Sauvant (1981), and Weiss (1986) claim, the developing countries do, in fact, negotiate as a collective in most global negotiations and particularly in global environmental negotiations.

Non-organizational setting: Identity building through injustice and unfairness

Identity reflects the existence of a collective group, both in terms of how the group sees itself and how it is perceived by others (see Meyer 2002). Identity defines the frame of reference for the "rationality system" of a collectivity, influencing the motivations and actions of both the individual members and the collectivity as a group. Identity should be regarded as an entrepreneur (seller) of social norms and, therefore, is to be understood as a dynamic and evolving process of inclusion and exclusion (see Jenkins 1997: 44f.; Barth 1996: 300). This chapter argues that identity can also be constructed on the basis of real or imagined experiences of marginalization. Identity building is therefore primarily an adaptation process and thus a positive strategy for acquiring political benefits. The South in the context of the climate change negotiations is a collectivity of diverse countries which seem to be too heterogeneous (e.g. Bangladesh, China, Saudi Arabia) to be included in one collectivity.

In the climate change context, the logic of primordial ties and kinship (stateship) connections[9] (see Shils 1957; Geertz 1963) among states is not the driving force behind the construction of the South's identity, but rather their common real or imagine experience of marginalization. The self-definition of the South, as the South Commission (1990: 1) states, is a definition of exclusion: these countries believe that they have been "bypassed" and view themselves as existing "on the periphery." Identity is ascribed through memories (see Lowenthal 1985; Alexander 2004: 1) and experiences of injustice and unfairness which lead to an internal status-evaluation process and a common understanding among members. This paves the way for consolidation of a collective identity. The South's identity is a collective "decision" to represent "social pain" (injustices and unfairness combined with vulnerability) (Alexander 2004: 1). Hence, as Victor Roudometof (2006: 7) argues, memory of events is closely connected to the collective representations produced by contemporary policies. The South's identity is therefore a process defined by social interactions implying that the South identity is only explicable in a social context and should therefore be understood as an open bargaining process between self-perception and the image defined by the social partner (Meyer 2002: 41), known also as the "North." The South as a collectivity then identifies the "others," and this event can lead to the further politicization of the differences, as threats coming from each side can be perceived, leading to further securitization of issues. Moreover, identity is an open bargaining process both among members of the collective group and with the social partner (Meyer 2002: 35–38). To sum up, the construction of identity occurs when members (in this case, countries) wish to symbolize their closeness to one another, whereas the closeness is established through the "collective fear of the future" (Lemarchand 1994: 27), the latter serving as the source of political mobilization of the collectivity. Mobilization requires the consolidation of identity and serves as an instrument of resistance. Identity with its organizational functionality is engaged to ensure protection from the dominance of the other group, driven by memories of marginalization.

Nevertheless, there are arguments against the viability of the concept of the South as a collectivity. Harris (1986) announced the end of the Third World as economic reality and ideological representation in his book, *The End of the Third World: Newly Industrialized Countries and the Decline of an Ideology*. The Third World[10] notion is argued to be obsolete because of a "global manufacturing system" where no clear identification can be made between rich and poor, have-and-have-nots, industrialized and non-industrialized (Harris 1986: 200–202), making the North-South divide obsolete. The spatial restructuring of the last 20–30 years has eliminated the structural divide between first and third worlds (Burbach and Robinson 1999: 27–28). The issue of internal heterogeneity questions the ability of the developing countries to come up with a united voice, "masking major dissimilarities among (developing) countries" (Head 1991: 71) and missing the point that interests of states do not always coincide with the conventional view of whether a Southern or Northern position is called for (La Viña 1997: 172).

Notwithstanding, Williams (1993: 10) states that internal heterogeneity may pose a "problem of management," but should not be taken as a "sign of irrelevance or disintegration." A significant question is how diverse developed countries are among themselves. Compared to the developing countries, developed countries have regular forums such as the G8, OECD, EU, and to some extent the World Economic Forum. Therefore, any "problem of management" is not an agent-based deficiency, but rather an institution-based condition. As Bernarditas Muller, G77 and China coordinator for the climate change negotiations, (interviewed in June 2009 in Geneva), stated: developing countries have limited possibilities of conducting meetings to internally discuss climate change issues, unlike the developed countries, which use regular forums to internally discuss climate change issues prior to actual negotiations.

In the climate change context, the North–South divide is more than a development gap, as it includes justice and fairness issues. As Sabri-Abdalla (1980: 40) articulates, the root cause of Southern solidarity has been the notion of "dependence." As long as the "perception of dependence remains intact, so will the South." To sum up, many developing countries do not fully trust the developed countries and feel coerced into making sacrifices to support their "materialistic" values and lifestyle.

The organizational face of the South, the Group of 77 + China will be discussed in the next section. Furthermore, it will be considered whether there are other alternative organizational faces of the developing countries aside from this grouping.

Organizational setting: Coalition building through common interests

As discussed earlier, the features of the climate change negotiation table, particularly multilateralism,[11] involve the establishment of coalitions which represents a self-driving dynamic of the bargaining process. Multilateral negotiations offer not only the possibility but also the necessity of coalition building (see Hampson and Hart 1995; Polzer *et al.* 1995; Raiffa 1982; Zartman 1994). In the climate change negotiations, coalitions are a product of strategic calculation on the part of actors to achieve the least portion of the "cost pie." Coalition is a temporary alliance of groups to achieve a common purpose or engage in joint activity (Yarn 1991: 81). They are also a response to asymmetries, enabling weaker parties to combine their resources and become more powerful than when acting alone. In the climate change context, developing countries feel they are in the weaker position in the international system not because of their poverty, but because of their "poverty of influence" (South Commission 1990). Developing countries strive for a say in the political decisions that affect them (Krasner 1985; Thomas 1987). As Najam (2005: 128) claims, the politics of the South is the "diplomacy of influence, not the politics of power." However, coalitions require a minimum level of trust among members which can be partly guaranteed by having similar values, interests, and goals. In the climate change context, the common interest of influencing seems to be more intense than the diversity in the G77 + China. Coalition building is, therefore, the "primary mechanism through which disempowered parties can develop their power base and thereby better defend their interests" (see Watkins and Rosegrant 2001).

The G77 + China coalition

The G77 + China coalition in the climate change negotiations is a process coalition in which broad-based interests are coordinated over a long-term period. The coalition and the resources available to its members are not exclusive to the climate change realm. The Group of 77 (G77) was established on 15 June 1964 by 77 developing countries through the "Joint Declaration of the Seventy-Seven Countries" signed at the end of the first session of the United Nations Conference on Trade and Development (UNCTAD) in Geneva. Beginning with the first Ministerial Meeting of the Group of 77 in Algiers in 1967, which adopted the Charter of Algiers, a permanent institutional structure gradually developed. As the largest Third World coalition in the United Nations with 133 member countries, the G77 allows the developing world to articulate and promote its collective economic interests and enhance its joint negotiating capacity on all major international economic issues in the United Nations system and promote economic and technical cooperation among developing countries. As described above, the interlinkages of the climate change with other issues such as economics and trade make it possible and practicable for developing countries to use the G77 as a negotiation vehicle. The G77 + China coalition for the climate change negotiations is a remarkable and, to some extent, unprecedented coalition of states. The G77 with 133 member countries is the largest coalition within the UN framework. The coalition represents Asia, South America, and Africa, leading to its apparent recognition as a significant force in the negotiations. The G77 + China coalition has identified the continued widening of the gap between developed and developing countries and this inequality and inequity seem to have entered the realm of the climate change negotiations. Climate change should be understood through a bigger picture also called "glocalization" (global + local), whereas global issues have varying local implications (Abatzoglou *et al.* 2007). For some elements, such as temperature or precipitation, large regional changes may occur that are not in accordance with the global mean change (McCarthy *et al.* 2001: 938). In this regard, quite a wide variety of local impacts of climate change can be observed. As Skodvin (Chapter 12) argues, the use of downscaling is necessary to provide more detailed information; however, downscaling is associated with a significant reduction in the accuracy of predictions (Burton *et al.* 2002: 153). This chapter argues that downscaling of issues is a significant feature of the climate change negotiations and will have implications for the dynamics of coalition building within the climate change negotiations.

Climate change is not only seen as impeding development through its implications for food security, water supply, and health; the developing countries also perceive inequity and other disadvantageous elements in the proposed and implemented mitigation and adaptation measures to confront climate change. As described earlier, the "common fears" of the developing countries defines the identity of the collectivity of the South. The climate change context further exemplifies "some historical and structural inequities" (Yu III 2008: vi) which have been identified in trade and global economics.

Issue Coalitions of Developing Countries

New coalitions emerge in the course of the climate negotiations. These new coalitions are usually issue based and may cross the North–South divide such as the Environmental Integrity Group, which was the first coalition involving both developed and developing countries. Issue coalitions such as the Coalition of Rainforest Nations, Comision Centroamericana de Ambiente y Desarollo (CCAD), and the Bolivarian Alliance for America (ALBA) played a significant role in including various issues in the international negotiation process.

A major issue coalition of developing countries is the Alliance of Small Island States (AOSIS), which is a coalition of small island and low-lying coastal countries and functions primarily as an *ad hoc* lobby and negotiating voice for Small Island Developing States (SIDS) in the global climate change negotiations. Similar to other issue coalitions, although AOSIS is mostly composed of developing countries,[12] this coalition is not seen as an alternative to the G77 + China coalition. Even though Palau withdrew from the G77 + China coalition in 2004, having decided that it could best pursue its environmental interests through AOSIS, 27 of its 43 members and observers are still members. Two AOSIS members are EU members and the rest are non-G77 developing countries. AOSIS is not an alternative, but rather a complementary negotiation vehicle for the 27 developing countries. The Least Developed Countries (LDCs), of which the Maldives, for instance, is a member, is another. However, LDCs are not to be regarded as a *de jure* coalition, as LDC is merely a categorization used within the United Nations system to refer to countries with the Human Development Index ratings of all countries in the world. The LDCs are, nevertheless, a *de facto* group with members having a common identity. The multiple membership of similar coalitions is, to a significant extent, enabled by the nature of the different interlinkages of climate change negotiations.

As Stevenson *et al.* (1985: 261) point out, coalitions typically lack a formal structure. The AOSIS, for instance, has no formal charter, budget, or secretariat. Nevertheless, AOSIS enriches the G77 + China, as it provides further issue-focused understanding of the climate change context. AOSIS is among others (such as

OPEC and BASIC – Brazil, South Africa, India, and China) an issue coalition in the climate change negotiations, with member states maintaining membership of other issue and process coalitions.

Technology transfer, capacity building, and equity – The North–South divide continues

To understand the relevance of the North–South divide in the climate change negotiations, the issue of technology transfer will be taken as an analytical framework to showcase diverging positions between developed and developing countries; it can thus illustrate the gap between North and South, particularly in areas relevant to sustainable development. This chapter introduces the following nexus to point out contesting perspectives: 1) technology transfer–climate change negotiations; 2) technology transfer–capacity-building; 3) technology transfer–equity.

Technology transfer – Climate change negotiations

Achieving the objective set out in Article 2 of the UNFCCC will require technological innovation and the rapid and widespread transfer and implementation of technologies. Article 4.5 of the UNFCCC states that developed country Parties and other developed Parties included in Annex II "shall take all practical steps to promote, facilitate and finance, as appropriate, the transfer of, or access to, environmentally sound technologies and know-how to other Parties, particularly developing country Parties, to enable them to implement the provisions of the Convention."[13] Technology transfer is seen as one way of reducing the costs of climate change mitigation (see Nagashima and Dellink 2008). A system for innovation in technology in every sector plays a significant role in controlling climate change costs and, as argued by traditional neoclassical models, can increase factor productivity (see Arrow *et al.* 1961; Uzawa 1965; Peck and Teisberg 1994; Nordhaus 1994). Furthermore, technology is seen as having the potential to increase economic growth (Romer 1990; Griliches 1992). Technology transfer has changed from being a means of achieving emission abatement to becoming an issue on which states negotiate. The disappearance of the boundary between means and ends contributes to greater complexity and can be understood as a self-driving mechanism of the climate change negotiations. Nevertheless, technology transfer as an instrument and as an end may stabilize agreements, as it increases the so-called "Zones of Possible Agreements," embodying desired payoffs because of spill-overs to other realms aside from emissions reduction and adaptation. Furthermore, as La Viña (1997: 107) notes, technology transfer is an incentive for most developing countries to participate in the negotiation process.

Technology transfer – Capacity building

Technology transfer in areas such as biotechnology and energy not only increases the capacity of developing countries to mitigate and adapt to climate change, but also provides spill-over benefits in other areas such as securing food supply, reducing migration, and preventing social unrest, as well as assisting developing countries pursue sustainable development. This raises a problem for developed countries in terms of justifying possible unemployment increases in their own countries due to "subsidies" to developing countries. Will it be seen as legitimate to "give the butcher the knife" to cut their own prosperity? Or is it wrong to think of prosperity as absolute rather than relative? Furthermore, how can one be sure that capacity building does not include military capacity building? Developing countries could also legitimately ask whether technology transfers from developed countries really are intended to build capacity, or in fact to increase dependency. As Tandon (2008: 1) argues, technology transfer, like aid, sounds positive because it is associated with "development," solidarity, and humanitarian causes. As Tandon (2008: 3) continues, "the moment 'dependence' is added to aid (and technology transfer), it loses its lustre." In transferring its technology a donor nation inevitably transfers its own ways of thinking and doing, its own institutions and values. According to Tiles and Oberdiek (1995: 2) these interact profoundly but unpredictably with the way of life of the recipient nation. A conflict arises if the transferred technology was not conceptualized to take account of the complex economic, political, sociocultural and infrastructural contexts of technological decision making (see Mowery and Rosenberg 1989: Chapter 11). Technologies can destroy certain values and make others virtually impossible to fulfill (Tiles and Oberdiek 1995: 57). Another problem concerns for example the development of biofuels, which could lead to competition over land for food and potentially the need for developing countries affected by this to import food.

Technology transfer – Equity

The challenge of global warming should be met in partnership between developed and developing countries. The longstanding argument about responsibility for global environmental action for all nations was mentioned above. The World Climate Conference of 1990 recognized that "the principle of equity and the common but differentiated responsibility should form the basis of any global response to climate change" and that "developed countries must take the lead." The Conference also proposed that contributions to the stabilization effort should be equitably differentiated according to countries' responsibilities and their level of development.[14] However, according to Slinn (1990: 78), developing countries believe that equity "entitles them to rights of indemnity in respect of the cost of remedial measures." This notion is to a significant extent politically unacceptable to developed countries, paradoxically for the same principle of equity. Although it is recognized that developed countries should assist developing countries in meeting the goals of the UNFCCC, however, it is not equitable that developing countries pursue a "business-as-usual" approach regarding emission cuts. The "common but differentiated" responsibility has been interpreted to mean that developed countries will initially undertake commitments (as through the Kyoto Protocol) and that commitments by developing countries and by economies in transition should be consistent with their economic and developmental needs (La Viña 1997: 106). Developing countries such as Brazil, China, and India are accused of hiding under the mantle of the G77, rejecting further commitments of emissions reduction, and ignoring the initial agreement that commitments should be consistent with their economic status. As La Viña (1997: 108) further argues, "the decisions on global warming will not be taken for the purpose of achieving equity" but rather to pursue self-interest.

Developing countries still question whether developed countries have already materialized their own commitments. Because technology transfer is difficult to quantify, there can be no absolutely reliable documentation about how many climate-relevant technologies are transferred annually (IPCC 2000: 4). Common pathways include government assistance programs, direct purchases, licensing, foreign direct investment, joint ventures, cooperative research arrangements and co-production agreements, education and training, and government direct investment. One criticism from the developing countries is that the technology transfer scheme is competing for resources initially intended for the Official Development Assistance (ODA). Governments of the developed countries have pledged to put aside 0.7 per cent of their gross national income (GNI) for aid. The 0.7 principle was adopted by the United Nations in October 1970. As Tandon (2008: 2) argues, this promise remains unfulfilled and some developed countries "make use of creative conceptual and accounting tricks to boost their ODA figures." Furthermore, he argues, aid is combined with conditions to the developing countries which increase their dependence on the developed countries. In his 2008 book, *Ending Aid Dependence*, he sets out examples of how aid has been abused to pursue national interests and how, under current mechanisms, aid has only increased the dependency of developing countries. The technology transfer mechanism under the UNFCCC should not be added as ODA. For several developing countries, particularly the Least Developed Countries (LDCs), which are highly dependent on ODA, technology transfer schemes mean the reduction of financial allotments for their own development programs.

Another contentious area of the technology transfer scheme is the issue of patents and proprietary rights. Developed countries argue that most technologies were developed in the private sector and are protected by intellectual property rights. While proprietary rights are a key element in the commercial development of technology, commercially transferring technology to developing counties will mean additional financial burdens on developing countries. Furthermore, it is regarded as unfair that "developed countries give developing countries money to buy technology from developed countries and count this as technology transfer or OAD" (interview Bernarditas Müller 2009). This scheme is nothing other than subsidies to the economies of the developed countries. Winner (1993: 289) introduced three "maxims" about technology that can be applied to the climate change context. First, no innovation without representation: those who are most affected by a technology often have the least to say about its development and use. In the technology transfer scheme, developing countries should have the choice as to which technology will be used and where it will be applied. Second, no technology (engineering) without political deliberation: as most applied science and engineering occurs without any political deliberation, technology transfer needs public debate, public inspection, and political discussion. Technology transfer should not exclude democratic procedures. Third, no means without ends: technology carries with it moral and political responsibilities, and therefore accountability should not be forgotten. Climate-relevant projects, such as using biomass for biofuels which had implications for the corn price in Mexico and food security in the Philippines, should be held accountable. Technology transfer should be complemented by developmental policies which make use of indigenous knowledge.

Conclusion

The analytical description of the climate change bargaining table enables the in-depth analysis of the interdependence between structure, process, and actors. The qualities of the bargaining table (multiple causation, interlinkages and cross-cutting of issues, multiparty setting, asymmetries, and dominance of the justice and fairness issue in the formulation of an outcome) determine the conditions available in this context. For instance, multilateral negotiations encourage the formation of coalitions, either to establish a leverage against a stronger actor or to gain a significant amount of leverage within the coalition in which coalitions unintentionally moderate the extreme positions of some members of the coalition. Although the G77 + China coalition, as complemented by the AOSIS and to some extent by the LDC, is internally a heterogeneous group with diverse member states interests, the common interest of minimizing the vulnerability of its members to the perceived domination of developed states in the international regime serves as the "cohesive glue" encouraging complementary efforts among member states. Coalition building, although seen as an obstructing factor in the international climate talks, can also be considered as a facilitating mechanism, empowering developing states through expanded participation and ensuring that the negotiation outcomes are equitable, resilient, and effective.

Considering that the North–South divide implies a major stumbling block in reaching agreements to confront climate change, it seems necessary to entertain the idea of moving away from the North–South paradigm as La Viña (1997: 189) proposes, arguing that there is a need to ask whether this paradigm should be reconsidered or even rejected. The new paradigm should not be a stumbling block, but should rather represent a facilitating mechanism. Whether it is UNCTAD, UNCED, GATT or the climate talks, the negotiations, according to La Viña (1997), always begin with political rhetoric about the inequality in the international economy, historical responsibility, and the need for radical restructuring of the international political and economic system, and usually end up with no satisfactory resolutions of the core issues that have been raised. He adds, the North–South paradigm often leads to deadlocks or unsatisfactory compromises, and this dichotomy does not offer a promising way to deal with climate change. However, it should be noted that the North–South divide is an implication, as La Viña himself recognizes, of the primacy of states and the concept of nation-states in international negotiations. As this paper identifies, the gap is created by a global problem confronted through the national lens. Nevertheless, it may not be feasible, under the present condition of the absence of a world government and the supremacy of the notion of sovereignty of states, to find an alternative. As Zartman (1987: 6) points out, the structural explanation is given in terms of power, either as relative positions of the parties or as the relative ability of the parties to make their options prevail. From the structural point of view, there can be no other paradigm aside from the North–South one because there will always be power (measured not only in resources but also in influence) imbalances that create and maintain the North–South divide. The North–South divide is a social reality and, to some extent, a social imperative.

Recognizing this, coping with it, identifying the positive connotations behind the North–South divide, can be a more practical approach. Coping with the divide means promoting further dialogs between the developed and developing countries. Furthermore, finding inclusive mechanisms such as editorial discipline in documents, "equal opportunity" in international employment policies (for example, "equally qualified candidates from developing countries will be preferred"), correcting historical errors (Rawls 1971), using geographic rotations of leadership in international regimes can favor "global integration." Noting that these mechanisms are already common practices in the international system, it makes sense to ask why these mechanisms are not working.

The North–South divide: Opportunities waiting to be discovered

The North–South divide in the climate change negotiation offers opportunities to push forward the negotiation process. Negotiation is about trust building. The promises made by developed countries of billions of US dollars to support climate change policies in developing countries are not legally binding. If these pledges bring the experience of reliability (and therefore an increase in the credibility of the donors), then these pledges could enhance trust between developed and developing countries. Trust is the key to the increased willingness of some developing countries to accept some binding commitments in future climate talks.

Furthermore, opportunities may be found in looking for a more sustainable solution to the common problem because of the wider participation allowed by the divide. The divide was, for instance, responsible for several organizational settings, such as the G77. The North–South divide represents one of the necessary "rooms of constructive debate" where interests hidden behind positions can be expressed. It offers an overview of the structural conditions in the international system, which countries should consider as their point of departure in negotiating their interests. For instance, the North–South divide enables states to establish which coalitions should be allowed to negotiate, encouraging coalitions to be established on the basis of common interests, which make the negotiation process less uncertain. The North–South divide reflects a basic function of a multilateral process which requires the building of a like-minded community of negotiating parties.

Challenges of the North-South Divide

The North–South divide has been one of the "front lines" in the climate change negotiations where two identities meet. The construction of identity along this line reflects the interests (or stakes) of members. Although it should be noted that there can be other front lines apart from the North–South divide, the divide was the most dominant one in the last few COP meetings. Several challenges brought by the North–South divide can be found in the COP15 negotiations. Conflict arises not through the incompatibility of goals. In the last decades, agreements have been made that have conditioned the goal-building process. The Copenhagen Accord is an integral part of this process, where the 2°C goal was codified. *Consensual knowledge has been achieved.* Conflict comes from the incompatibility of behavior, in other words, the instruments that should be used to achieve the desired outcome. The North–South divide highlights the differences in behavior due to different national conditions. The challenge that this behavioral incompatibility offers can be addressed through an in-depth analysis of the needs to be met, so that behavior can be changed. Any reframing of behavior requires the management of knowledge to ensure flexibility and adaptability, a quick response, and the strategic coordination of appropriate tools. Considering the North–South divide as the point of departure in analysis need offers the opportunity of a focused yet comprehensive approach to confronting climate change.

The North–South divide is a not a "disease" that needs curing. It hurts, but it also shows where actions are needed. The common perception that the process has been delayed by this division is misleading. The North–South divide shows the need to find a more effective and inclusive approach to administering the knowledge base for negotiation.

Notes

1 Oreske (2007) notes that there is a scientific consensus on climate change, with all but tiny handful of climate scientists convinced that earth's climate is heating up and that human activities are a significant cause, see Oreske, N. (2007) *The Scientific Consensus on Climate Change: How Do We Know We're Not Wrong?* Furthermore, in January 2001, the United Nations Intergovernmental Panel on Climate Change (IPCC) released a report stating that there is now new and stronger evidence that most of the climate warming observed over the the last 50 years is attributable to human activities, sending strong signals to governments that urgent action is needed. Therefore, the climate change negotiations do not follow a top-down approach where governments set the agenda for negotiations, but rather a bottom-up where scientific communities and NGOs set up the agenda setting process.

2 DiMento and Doughman (2007) mention in the book Climate *Change* that the movie *The Day After Tomorrow* grossing nearly half a billion dollars in 2004 and the 2006 documentary film *An Inconvenient Truth* which was produced by and featuring former Vice President Al Gore have increased public concerns about climate change. Surveys conducted before and after the release of *The Day After Tomorrow* indicated a significant effect on climate-change risk perceptions, conceptual frameworks, policy positions and voting intentions of those who saw the movie, see Dimento, J. and Doughman, P. (2007) *Climate Change. What it Means for Us, our Children, and Our Grandchildren.* The MIT Press, Cambridge, MA, and London, UK, foreword.

 3 See Fagan, B. (2008) for a detailed historical description of global warming.
4–5 For further reading on *third worldism*, see Malley, R. (1999), "The Third World Moment", in *Current History*, November.
 6 Furthermore, Hardt and Negri (2000), Held *et al.* (1999), Hardt and Negri (2000) and Hoogvelt (1997) have argued as well that because of the elimination of the clear separation along the developmental threshold between the North and the South, the concept is to be regarded as outdated.
 7 The Brandt report was led by West Germany's Chancellor Willy Brandt. The report came from a commission of 18 politicians of international repute.
 8 Double contigencies is a sociological in the "structural functionalism" and in the "systems theory" represented by Talcott Parsons and Niklas Luhmann. The concept refers to a situation where individuals find themselves in a social interaction. Each of them is confronted with the lack of information about a specific behavior of the other (called "contigencies"). The social interaction involving actors without knowledge of the sense of the actions of the others will be difficult to manage. Institutions and conventions are constructed to prevent such contigencies.
 9 Of course, primordial ties are to a limited extent applicable to the climate change context. This paper argues that identity-building, according to concept of primordial ties, is developed when elements which are "given" such as language, kinship and racial ties are referred to as fundamental bases for political attachment. In the climate change context, developing countries based their decision to take the South identity not on given elements such as common culture or language, but rather on something that is "taken" such as the common experiences of injustice.
10 For the purpose of this paper, the Third World which comes from the economic lens and the South which comes from the geographical lens are considered the same entities.
11 In this paper, multilateralism is referred to multiple party setting.
12 Two members, Malta and Cyprus, are also member states of the European Union.
13 Cited from IPCC Special Report, Methodological and Technical Issues in Technology Transfer, p. 3.
14 See paragraphs 5 and 11 of the Ministerial Declaration of the Second World Climate Conference, Geneva, Switzerland, 7 November 1990, reprinted in *International Law and Climate Change*, p. 356.

References

Abatzoglou, J., DiMento, J., Doughman, P. and Nespor, S. (2007) Climate-Change Effects: Global and Local Views, In Dimento, J. and Doughman, P. (eds) *Climate Change. What it Means for Us, our Children, and Our Grandchildren*. Cambridge, MA and London: The MIT Press. 45–64.

Adger, W.N., Huq, S., Brown, K., Conway, D., Hulme, M. (2003) Adaptation on Climate Change in the Developing World. *Progress in Development Studies*. 3 (3).179–195.

Adler, E. (1997) Seizing the Middle Ground: Constructivism in World Politics. *European Journal of International Relations*. 3. 319–363.

Alesina, A., Angeloni, I. and Etro, F. (2001) The Political Economy of Unions, NBER working paper.

Alexander, J. (2004) Toward a Theory of Cultural Trauma, in Alexander, J., Eyerman, R., Giesen, B. (eds.) *Cultural Trauma and Collective Identity*. Berkeley, CA: University of California Press.

Arrow, K.J., Chenery, H.B., Minhas, B.S., and Solow, R.M. (1961) Capital-labor Substitution and Economic Efficiency. *The Review of Economics and Statistics*. 43 (3). 225–250.

Bone, R. (2003) Agreeing to Fair Process: The Problem with Contractarian Theories of Procedural Fairness. *Boston University Law Review*. 83. 485–552.

Brinton, M.C. and Nee, V. (eds.) (1998) *The New Institutionalism in Sociology*. New York: Russell Sage Foundation.

Buchner, B. and Lehmann, J. (2005) Equity Principles to Enhance the Effectiveness of Climate Policy: An Economic and Legal Perspective, In Bothe, M. and Rehbinder, E. (eds.) *Climate Change Policy*. Utrecht: Eleven International Publishing. 45–72.

Burbach, Roger and Robinson, William I. (1999) The Fin De Siecle Debate: Globalization as Epochal Shift. *Science & Society*. 63 (1) (Spring). 10–39.

Burton, I., Huq, S., Lim, B., Pilifosova, O. and Schipper, E.L. (2002) From Impacts Assessment to Adaptation Priorities: The Shaping of Adaptation Policy. *Climate Policy*. 2. 145–159.

Caney, S. (2009) Human Rights, Responsibilities and Climate Change, In Beitz, C. and Goodin, R. (eds.) *Global Basic Rights*. Oxford: Oxford University Press.

Carraro, C. and Sgobbi, A. (2008) Modelling Negotiated Decision Making in Environmental and Natural Resource Management: A Multilateral, Multiple Issues, Non-cooperative Bargaining Model with Uncertainty, ScienceDirect – *Automatica*. 44 (6). 1488–1503.

Carraro, C. and Siniscalco, D. (1997) R&D Cooperation and the Stability of International Environmetal Agreements, In Carraro, C. (ed.) *International Environmental Negotiations: Strategic Policy Issues*. Cheltenham: E.Elgar. 71–96.

Carlsnaes, W. (1992) The Agent-Structure Problem in Foreign Policy Analysis. *International Studies Quarterly*, 36: pp. 245–270.

Cavallo, R. E. (ed.) (1979) *Systems Research Movement: Characteristics, Accomplishments, and Current Development*. General Systems Bulletin Special Issue – Summer, IX (3) (A report sponsored by the Society for General Systems Research).

Cowie, J. (2007) *Climate Change. Biological and Human Aspects*. Cambridge, UK: Cambridge University Press.

Depledge, J. (2005) *The Organization of Global Negotiations: Constructing the Climate Change Regime*. London/Sterling, VA: Earthscan.

Depledge, J. (2006) The Opposite of Learning: Ossification in the Climate Change Regime. *Global Environmental Politics*. 6 (1). 1–22.

Dessler, D. (1989). What's at stake in the agent-structure debate? *International Organization*, 43 (3): 441–473.

DiMagio, P.J. and Powell, W.W. (eds.) (1991) *The New Institutionalism in Organizational Analysis*. Chicago, IL: University of Chicago Press.1–38.

Dimento, J. and Doughman, P. (2007) *Climate Change. What it Means for Us, our Children, and Our Grandchildren*. Cambridge, Massachusetts and London, England: The MIT Press.

Dworkin, D. (1985) *Principle, Policy, Procedure in a Matter of Principle*. Cambridge, MA: Harvard University Press.

Fagan, B. (2008) *The Great Warming. Climate Change and the Rise and Fall of Civilizations*. New York: Bloomsbury Press.

Folmer, H., van Mouche, P. and Ragland, S.E. (1993) Interconnected Games and International Environmental Problems. *Environmental Resource Economics*. 3. 313–335.

Gardiner, S. (2004) Ethics and Climate Change. *Ethics*. 114 (3). 555–600.

Geertz, C. (1963), The Integrative Revolution: Primordial Sentiments and Civil Politics in the New States, In Apter, D. (ed.) *Old Societies and New States*. New York: The Free Press of Glencoe. 47–76.

Griliches, Z. (1992) The Search for R&D Spillovers. *The Scandinavian Journal of Economics*. 94. 29–47.

Grubb, M., Sebenius, J., Magalhaes, A. and Subak, S (1992) *Sharing the Burden. Confronting Climate Change: Risks, Implications and Responses*. Cambridge, UK: Cambridge University Press.

Hall, P.A. (1986) *Governing the Economy: The Politics of State Intervention in Britain and France*. New York: Oxford University Press.

Hamal, K. (2010) Reporting GHG Emissions: Change in Uncertainty and Its Relevance for Detection of Emission Changes, IIASA Interim Report IR-10-003 (July 2010), available through http://webarchive.iiasa.ac.at/Admin/PUB/Documents/IR-10-003.pdf, last accessed 4 December 2012.

Hampson, F.O. (1995) *Multilateral Negotiations: Lessons from Arms Control, Trade, and the Environment*. Baltimore: The John Hopkins University Press.

Hampson, F. O. and Hart, M. (1995) *Multilateral Negotiations: Lessons from Arms Control, Trade, and the Environment*. Baltimore: The John Hopkins University Press.

Harris, N. (1986) *The End of the Third World. The Newly Industrializing Countries and the Decline of an Ideology*. Harmondsworth: Penguin.

Harrison, N.E. and Bryner, G.C. (2004) Toward Theory, In Harrison, N.E. and Bryner, G.C. (eds.) *Science and Politics in the International Environment*. Landham, MD: Rowman & Little Publishers. 327–350.

Hardt, M. and Negri, A. (2000) *Empire*. Cambridge, MA: Harvard University Press.

Hardy, C. (1985) *Understanding Organizations*. London: Penguin Business.

Hay, C. (2006) Political Ontology, In Goodin, R.E. and Tilly, C. (eds.) *The Oxford Handbook of Contextual Political Analysis*. Oxford: Oxford University Press. 101–115.

Hayes, P. and Smith, K. (eds.) (1993) *The Global Greenhouse Regime. Who Pays? Science, Economics and North-South Politics in the Climate Change Convention*. New York/London: United Nations University Press and Earthscan Publications.

Head, I. L. (1991) *On a Hinge of History: The Mutual Vulnerability of South and North*. Toronto: University of Toronto Press.

Held, P., McGrew, A., Goldblatt, D. and Perraton, J. (1999) *Global Transformation, Politics, Economics and Culture*. Stanford, CA: Stanford University Press.

Hoogvelt, A. (1997) *Globalization and the Postcolonial World: The New Political Economy of Development*. Baltimore: The John Hopkins University Press.

Intergovernmental Panel on Climate Change (IPCC) (2000) Methodological and Technological Issues in Technology Transfer. IPCC Special Report. Summary for Policymakers, available through http://www.ipcc.ch/ipccreports/sres/tectran/index.php?idp=0, last accessed 28 July 2009.

Intergovernmental Panel on Climate Change (IPCC) (2007) Climate Change 2007: The physical science basis: summary for policymakers. A Report of the Working Group 1 of the IPCC, available through http://www.ipcc.ch/pdf/assessment-report/ar4/wg1/ar4-wg1-spm.pdf, last accessed 21 July 2009.

Keohane, R.O. (1989) *International Institutions and State Power*. Boulder, CO: Westview Press.

Kindler, J. and Kiss, I. (1984) Future Methodology Based on Past Assumptions? In Tomlinson, R. and Kiss, I. (eds.) *Rethinking the Process of Operational Research and Systems Analysis*, Frontiers of Operational Research and Applied Systems Analysis. 2. Oxford and London: Pergamon Press.1–17.

Krasner, S.D. (1985) *Structural Conflict: The Third World Against Global Liberalism*, Berkeley, CA: University of California Press.

Larson, M.J. (2003) Low-Power Contributions in Multilateral Negotiations: A Framework Analysis. *Negotiation Journal. 19* (2). 133–149.

Lieberman, D., Jonas, M., Winiwarter, W., Nahorski, Z. and Nilsson, S. (2007) Accounting for Climate Change: Introduction, In Lieberman, D., Jonas, W., Nahorski, Z. and Nilsson, S. (eds.) *Accounting for Climate Change. Uncertainty in Greenhouse Gas Inventories*. Dordrecht: Springer.

Lowenthal, D. (1985) *The Past is a Foreign Country*. Cambridge, UK: Cambridge University Press.

Luhmann, Niklas (2001) *Soziale Systeme: Grundriß einer allgemeinen Theorie* 11. Auflage (1. Auflage 1984). Frankfurt am Main: Suhrkamp.

Malabed, R. N., (2001). *Ecosystem Approach and Interlinkages: A Socio-Ecological Approach to Natural and Human Ecosystems*. Discussion Paper Series 2001–005. Tokyo: United Nations University.

Malley, R. (1999) The Third Worldlist Moment. *Current History*. November, 98 (631). 359–369.

March, J. G. and Olsen, J. P. (1984) The New Institutionalism: Organizational Factors in Political Life. *American Political Science Review*. 78. 734–749.

March, J. G. and Olsen, J. P. (1989) *Rediscovering Institutions*. New York: Free Press.

Marland, G. (2006) The Human Component of the Carbon Cycle. Testimony before the Committee on Government Reform, Subcommittee on Energy and Resources, US House of Representatives, 27 September.

McCarthy, J., Osvaldo, F., Canziani, N., Leary, N., Dokken, D. and White, K. (eds.) (2001) *Climate Change 2001: Impacts, Adaptation and Vulnerability*. Intergovernmental Panel on Climate Change. Cambridge, UK: Cambridge University Press.

Meyer, Thomas (2002) *Identitätspolitik. Vom Missbrauch kultureller Unterschiede*, Frankfurt am Main: Suhrkamp Verlag (SV).

Miller, M. (1995) *The Third World in Global Environmental Politics*. Boulder, CO: Lynne Rienner.

Mowery, D.C. and Rosenberg, N. (1989) Technology and the Pursuit of Economic Growth. Cambridge: Cambridge University Press.

Muller, B. (2009) Interview on June 26 at the South Centre, Geneva.

Nagashima, M. and Dellink, R. (2008) Technology Spillovers and Stability of International Climate Coalitions, *International Environmental Agreements: Politics, Law and Economics*. 4 (8), December. 343–365.

Najam, A. (1994) *The Case for a South Secretariat in International Environmental Negotiation* (Program on Negotiation Working Paper 94–98), Cambridge, MA: Harvard Law School.

Najam, A. (1995) An Environmental Negotiation Strategy for the South, *International Environmental Affairs*. 7 (3). 249–287.

Najam, A. (2005) A Tale of Three Cities: Developing Countries in Global Environmental Negotiations, In Kallhaugem A.C., Sjöstedt, G. and Corell, E. (eds.) *Global Challenges. Furthering the Multilateral Processes for Sustainable Development*. Sheffield, UK: Greenleaf Publishing. 124–143.

Nigel, H. (1986) *The End of the Third World. Newly Industrializing Countries and the Decline of an Ideology*. Harmondsworth, Middlesex, UK: Penguin Books.

Nordhaus, W.D. (1994) *Managing the Global Commons: The Economics of Climate Change*. Cambridge, MA: MIT Press.

Oreske, N. (2007) The Scientific Consensus on Climate Change: How Do We Know We're Not Wrong? In Dimento, J. and Doughman, P. (eds.) *Climate Change. What it Means for Us, our Children, and Our Grandchildren*. Cambridge, MA and London: The MIT Press. 65–99.

Parsons, Talcott (1960) *Structure and Process in Modern Society*. Glencoe: Free Press.

Peck, S.C. and Teisberg, T.J. (1984) Optimal Carbon Emissions Trajectories When Damages Depend on the Rate or Level of Global Warming. *Climate Change*. 28 (3). 289–314.

Penetrante, A. (2010) Politics of equity and justice in climate change negotiations in North–South relations. In Brauch, H.G., Spring, U.O., Mesjasz, C., Grin, J., Kameri-Mbote, P., Chourou, B., Dunay, P. and Birkmann, J. (eds.) *Coping with global environmental change, disasters and security – Threats, challenges, vulnerabilities and risks. Series on human and environ-mental security and peace.* 5. Berlin, Heidelberg/New York: Hexagon Springer-Verlag. 1355–1366.

Pierson, P. (2000) The Limits of Design: Explaining Institutional Origins and Change, *Governance.* 13 (4). 475–499.

Polzer, J. T., Mannix, E. A. and Neale, M. A. (1995) Multiparty Negotiation in its Social Context, In Kramer, R. and Messick, D. (eds.) *Negotiation as a Social Process: New Trends in Theory and Research.* Thousand Oaks: Sage. 123–142.

Raiffa, H. (1982) *The Art and Science of Negotiation.* Cambridge, MA: Harvard University Press.

Rawls, J. (1971) *A Theory of Justice.* Cambridge, MA: The Belknap Press of Harvard University Press.

Robinson, William I. and Harris, Jerry (2000) Towards a Global Ruling Class: Globalization and the Transnational Capitalist Class. *Science and Society.* 64 (1). 11–54.

Romer, P.M. (1990) Endogenous Technological Change, *The Journal of Political Economy.* 98 (5). 71–102.

Roudometof, V. (2006) *Collective Memory, National Identity, and Ethnic Conflict. Greece, Bulgaria, and the Macedonian Question.* Westport, Connecticut: Praeger.

Rostow, W. (2000) The Five Stages of Economic Growth. A Summary, In Corbridge, S.(ed.) *Development. Critical Concepts in the Social Sciences.* London/New York: Doctrines of Development. 105–116.

Rubinstein, A. (1982) Perfect Equilibrium in a Bargaining Model. *Econometrica.* 50. 97–109.

Sabri-Abdalla, I. (1980) Heterogeneity and Differentiation: The End of the Third World? In Haq, K. (ed.) *Dialogue for a New Order.* New York: Pergamon Press. 22–24.

Sauvant, K.P. (1981) *The Group of 77: Evolution, Structure, Organization.* New York: Oceana Publications.

Schelling, T. (1995) Intergenerational Discounting, *Energy Policy.* 23 (4/5).395–401.

Sebenius, J.K. (1991) Negotiating a Regime to Control Global Warming, In Matthews, J.T. (ed.) *Greenhouse Warming: Negotiating a Global Regime.* Washington, D.C.: World Resources Institute.

Shils, E. A. (1957) Primordial, Personal, Sacred and Civil Ties. *British Journal of Sociology.* 8. 130–45.

Slinn, P. (1990) Development Issues: The International Law of Development and Global Climate Change, In Churchill, R. and Freestone, D. (eds.) *International Law and Global Climate Change.* London/Dordrecht: Graham and Trotman/ Martinus Nijhoff.

Solum, L. (2004) Procedural Justice. *Southern California Law Review.* 78. 181–322.

South Commission (1990) The Challenge to the South: The Report of the South Commission. Oxford, UK: Oxford University Press.

Stern, T. (2009) Todd Stern on China and the Global Climate Challenge, June 3, Speech at the Center for American Progress, available through http://www.americanprogress. org/events/2009/06/av/stern_remarks.pdf, last accessed 28 July 2009.

Stevenson, W.B., Pierce, J.L. and Porter, L.W. (1985) The Concept of Coalition in Organizational Theory and Research. *Academy of Management Review.* 10 (2). 256–268.

Susskind, L.E. and Crump, L. (2008) *Multiparty Negotiations.* London: Four-Volume Set, SAGE Publications.

Tandon, Y. (2008) *Ending Aid Dependence.* Cape Town, Dakar, Nairobi and Oxford: Fahamu Books/ South Centre Publications (Geneva).

Thomas, C. (1987) *In Search of Security: The Third World in International Relations.* Boulder, CO: Rienner.

Theodorson, G., and Theodorson, A. (1979) *A Modern Dictionary of Sociology*. New York: Barnes and Noble.

Tiles, M. and Oberdiek, H. (1995) *Living in a Technological Culture. Human Tools and Human Values*. London/New York: Routledge.

Tsebelis, G. (1990) *Nested Games. Rational Choice in Comparative Politics*. Berkeley/Los Angeles/ Oxford: University of California Press.

Uzawa, H. (1965) Optimum Technical Change in an Aggregative Model of Economic Growth. *International Economic Review*. 6 (1). 18–31.

La Viña, A.G.M. (1997) *Climate Change and Developing Countries: Negotiating a Global Regime*. Quezon City, Philippines: Institute of International Legal Studies, University of the Philippines Law Center.

Watkins, M. and Rosegrant, S. (2001) Building Coalitions, In Watkins, M. and Rosegrant, S. (eds.) *Breakthrough International Negotiation: How Great Negotiators Transformed the World's Toughest Post-Cold War Conflicts*. San Francisco: Jossey-Bass Publishers. 211–227.

Weaver, A. (2004) The Science of Climate Change, In Coward, H and Weaver, A. (eds.) *Hard Choices: Climate Change in Canada*. The Centre for Studies in Religion and Society, Waterloo, Ontario: Wilfrid Laurier University Press. 13–43.

Weiss, T.G. (1986) International Secretariat or Servant of the G77, In Pitt, D. and Weiss, T.G. (eds.) *The Nature of United Nations Bureaucracies*. Boulder, CO,:Westview Press. 84–108.

Wendt, A. (1987) The Agent-Structure Problem in International Relations Theory, *International Organization*. 41 (3). 335–70.

White, R. (1993) *North, South, and the Environmental Crisis*. Toronto: University of Toronto Press.

Williams, M. (1993) Re-articulating the Third World Coalition: The Role of the Environmental Agenda. *Third World Quarterly*. 14 (1). 7–29.

Winner, L. (1993) Artifacts / Ideas and Political Culture, In Teich, A. (ed.) *Technology and the Future*, 6th edn. New York: St. Martin's Press.

Yarn, D. H. (1991) *The Dictionary of Conflict Resolution*. San Francisco: Jossey-Bass Publishers.

Yu III, Vicente Paolo, B. (2008) *Unity in Diversity: Governance Adaptation in Multilateral Trade Institutions Through South-South Coalition-Building*. Research Papers 17, Geneva: South Centre.

Zartman, W.I. (ed.) (1987) *Positive Sum. Improving North-South Negotiations*. New Brunswick, US/Oxford, UK: Transaction Books.

Zartman, W. I. (1994) Two's Company and More's a Crowd: The Complexities of Multilateral Negotiation, In Zartman, W.I. (ed.) *International multilateral negotiation: Approaches to the management of complexity*. San Francisco: Jossey-Bass. 1–10.

Zartman, W. I. and Crump, L. (2003) Multilateral Negotiation and the Management of Complexity, *International Negotiation*. 8. 1–5.

14 Developing a Legal Toolkit

Institutional Options to Remove Stumbling Blocks in the Climate Change Negotiations[1]

Dirk Hanschel

Introduction

The aim of this study is to systematically analyze the effectiveness of the institutional toolkit used in the climate regime, and to discuss possible improvements and innovations that can be pursued by negotiators striving for the environmentally best regulatory solution to be achieved under the circumstances of the current bargaining situation (so-called "environmental advocates"). Effectiveness is measured by the extent to which the forms of institutionalization constituting this toolkit *do not* help to remove stumbling blocks in the negotiations or present stumbling blocks themselves (low effectiveness), or help to remove stumbling blocks and to facilitate negotiations, leading to a more effective result (high effectiveness) (Hanschel 2005: 11ff.).

The institutional design of the climate regime follows a particular approach that has been applied in many other environmental treaties before. In order to balance the interest of states in preserving their national sovereignty and the need to protect important common goods, new forms of international environmental lawmaking have been designed that focus on post-agreement negotiations (Riedel 1998: 28).[2] Instead of creating only single legal instruments that deal with the issues in an all-encompassing way, international regimes have been established that consist of two or more instruments tackling the regulatory problem in a step-by-step approach. Parties to these regimes are thus constantly and simultaneously pursuing the aims of implementation, improvement, and innovation of the existing regulatory framework, as the former Executive Secretary of the climate regime, Michael Zammit Cutajar, has pointed out (Wittneben *et al.* 2006: 90). While this analysis focuses on climate negotiations, conclusions will also, where feasible, be drawn from the regimes on long-range transboundary air pollution and ozone depletion.[3] All three of these regimes are in many ways typical of international environmental law making: They deal with transboundary problems that can only be solved by a joint effort of the international community; they were all hampered by an initially low consensus and a substantial degree of scientific uncertainty about the issue; and they responded to this challenge by establishing continuous forums of negotiation in the form of international regimes.

For the purpose of this analysis, such regimes are defined as treaty-centered normative systems with an integrated norm-enforcement and development structure for the solution of transboundary problems (Hanschel 2003a: 54). In the initial phase of negotiations, environmental problems are typically characterized by substantial scientific uncertainty and by a low consensus regarding how they should be addressed. Accordingly, international environmental regimes are composed of dynamic and flexible forms of institutionalization that can largely be divided into various institutions, as well as into different regime enforcement and development procedures. These procedures may be described as negotiation-oriented, as they set the framework for dynamic conflict management relying on cooperative processes of joint decision making instead of adversarial or even confrontational approaches.

These negotiation-oriented forms of institutionalization, in particular the regime development procedures, form the basis of the so-called framework-protocol approach: First a framework convention containing extensive development procedures, but only very limited substantive rules, is adopted. These procedures then promote the creation of protocols filling the regulatory gap (Hanschel 2003a: 260). The first phase sets up a legislative framework, within which a scientific, economic, and fairness-related discourse may take place (Aman 1993: 3; Poitras 1999: 104f.). The results of this discourse are then integrated into the second phase leading to the agreement on concrete and substantial obligations (Aman 1993: 3; Poitras 1999: 105). This two-step approach, which may also be called a cascade model, is typical of international environmental law-making (Riedel 1986: 164f.; Susskind and Ozawa 1992: 144ff.; Malanczuk 1995: 992).

In the course of this analysis, the prominent debate on whether institutions matter in international relations will be applied to environmental regimes (Hasenclever *et al.* 1997: 3ff; Young 1989: 216ff.; Bernauer 1995: 355). The focal point of discussion has gradually shifted from the question as to *whether* institutions matter to the question of *how* they matter (Young and Levy 1999: 1ff.).[4] International negotiations are dominated by aspects of the bargaining situation, such as power, interests, preferences, the structure of the regulatory problem and the bargaining skills of the negotiators. It will be suggested, however, that an adequate institutional design may also affect the outcome of negotiations. This analysis aims to show that the negotiation-oriented forms of institutionalization set up in the climate regime are generally able to play a facilitating role, although there are several stumbling blocks that need to be removed to make this institutional toolkit more effective. Some of the suggestions made here are of a rather tentative character and may prove not to be entirely feasible in practice; others may well deserve implementation in the further regime building process.

The limited effectiveness of the climate regime

It may appear strange at first sight as to why the ozone and the climate change regimes have taken such different routes.[5] While the ozone regime (composed of the Vienna Convention on the Protection of the Ozone Layer [Vienna Convention] and the Montreal Protocol on Substances that Deplete the Ozone Layer [Montreal Protocol]) led to the achievement of an almost complete ban of all ozone-depleting substances and is thus widely considered a big success story (Chasek 2003: 187ff.), the climate negotiations provide quite a different record so far (Hanschel 2005: 16ff). The United Nations Framework Convention on Climate Change (UNFCCC) contains rather vague provisions, and the effectiveness of the Kyoto Protocol (KP) was severely hampered by its modest targets for only a number of industrialized countries, as well as by the lack of ratification in the United States and Australia. Even after the ratifications of Canada, Japan, and Russia allowed the Kyoto Protocol to enter into force in 2005 and Australia finally ratified the treaty in December 2007, the progress made has still been limited, as the reduction targets contained in this treaty are rather modest and only concern industrialized countries.[6] The first Conference of the Parties serving as the Meeting of the Parties to the Kyoto Protocol (COP/MOP1) passed more than 40 decisions relating to the so-called Marrakesh Accords, to the strengthening of the so-called flexible mechanisms, the compliance mechanism and adaptation.[7] After intense negotiations, COP/MOP1 furthermore agreed to "initiate a process to consider further commitments for Parties included in Annex I for the period beyond 2012 in accordance with Article 3, paragraph 9, of the Protocol", and to this end it established an open-ended *ad hoc* working group (AWG-KP), meeting for the first time in May 2006 in Bonn (Wittneben *et al.* 2006: 91f.).[8] While this may be considered a success, it merely constituted a first step towards tightening the obligations and, in spite of the Roadmap adopted in Bali in 2007 and subsequent conferences in Poznan in 2008 and Copenhagen in 2009, some major emitters have still not accepted substantial binding commitments (Morgenstern 2009: 235ff.). Furthermore, many countries still have a long way to go in terms of an effective implementation of the existing targets. This is why adaptation measures are becoming more and more important.[9] However, they cannot be a substitute for effective mitigation action.

In spite of the review process under Article 9 KP and the dialogue on long-term cooperative action under the UNFCCC (AWG-LCA)[10], the climate regime has not yet produced a sound solution to the regulatory problem (Brouns and Langrock 2006: 3ff.; Barrett 2006: 550ff.; Sterk *et al.* 2007: 1ff.; Morgenstern 2009: 235ff.). On a world scale carbon dioxide emissions will further increase, especially if one considers the quickly rising emissions of countries in economic transition such as China, India, Brazil, or South Africa. The overall 5 per cent target agreed upon in Kyoto is, by a long way, not enough to tackle the problem effectively.[11] At recent climate change summits, no substantial progress was made in terms of the future regime design (Sterk *et al.* 2007: 1, 3ff.; Morgenstern 2009: 235ff).[12] In spite of high expectations, even the 2009 Copenhagen Conference merely ended with a non-binding Accord (not even endorsed by the COP), which provides for a pledge-and-review system allowing parties to voluntarily enter their commitments.[13] This is quite a difficult situation, especially in the light of the Fourth Assessment Report of the Intergovernmental Panel on Climate Change (IPCC), clearly stating that "warming of the climate system is unequivocal, as is now evident from observations of increases in global average air and ocean temperatures, widespread melting of snow and ice, and rising global average sea level."[14]

The Characteristics of the Bargaining Situation

A very important factor for success in the ozone regime was a powerful progressive coalition called the "Toronto Group" (composed of the United States, Canada, and the Nordic countries) that was able to encounter and finally overcome the initial opposition by the European Community.[15] Ozone depletion is caused by a limited amount of man-made substances (Thoms 2003: 823). Therefore, the US chemical industry managed to develop and produce substitutes at reasonable costs, which allowed the United States to obtain a leadership role in the process (Parson 1993: 41; Oberthür 1997: 78ff.). Furthermore, the discovery of the ozone hole helped to produce undefeatable scientific results, making it easier to convince recalcitrant states to agree on far-reaching regulations (Parson 1993: 60; Oberthür 1997: 79).

By contrast, the nature of the climate change problem is far more complex and concerns a much wider range of industries (Thoms 2003: 823ff.; Pierrehumbert 2006: 574 ff.). It can thus be classified as a "malign" problem which is character-ized by "high uncertainty and little consensus, low saliency and sharp asymmetries in vulnerability, affecting core economic activities and including intricate com-petitiveness issues, and with little entrepreneurial capacity to smooth and reduce all these complications" (Wettestad 1999: 18). Real substitutes or other simple solutions are not at hand. Thus, the problem is apparently much more compli-cated and expensive to solve than the problem of ozone depletion.[16] This again reinforces the North–South conflict between industrialized states and developing countries, making it very hard to get the latter on board without substantial com-pensation payments (Hanschel 2005: 16). These reasons caused the United States, a crucial player in these negotiations, not to ratify the Kyoto Protocol, which it has described as being "fatally flawed".[17] At several Conferences of the Parties (COPs), the country presented an international program involving joint implementation projects with other industrialized and with developing countries. It proposed a reduction of greenhouse gas intensity (i.e. the "rate of greenhouse gas emissions per unit of Gross Domestic Product") by 18 per cent until 2012, which was a step forward, but which also meant that emissions in absolute terms may still rise if the economy is growing (Knox 2004: 145 f.).[18] Recently, after a change of govern-ment policy, the United States has suggested an emission reduction in the range of 17 per cent by 2020, based on 2005 levels, in conformity with expected national legislation.[19] China proposed a 40–45 per cent reduction of carbon dioxide emis-sions in the same period.[20] However, these and other pledges under the Copen-hagen Accord are of a voluntary nature only,[21] unless they are fortified in other documents.[22] The Fourth IPCC Assessment Report indicates that very ambitious global mitigation action needs to be taken in the near future to avoid irreversible consequences.[23] Whether this will be achieved, even in the light of a 20–30 per cent reduction target by the European Union based on 1990 levels,[24] is questionable.

The gap between the immediate and substantial measures called for by scientists, and the action taken within and outside the framework of the Kyoto Protocol, is to a large extent due to the high costs involved in terms of the impact perceived on the national economy, short-term political impact, and influence on strategic partnerships. Some issues have simply been too hot to handle up to now, but – as already COP/MOP1 has shown – the climate regime is making progress, although subsequent conferences have unfortunately not fulfilled the high expectations (Sterk *et al.* 2007; Morgenstern 2009). At least the scientific dispute concerning anthropogenic interference with the climate has, by and large, been resolved and shifted to the question as to how strong this interference is and how serious its consequences are – although some fundamental skeptics may still continue to exist.[25] Furthermore, studies such as the important review by Sir Nicholas Stern have made it clearer than ever before that the costs of inactivity will most probably be much higher than effective joint efforts to combat climate change now.[26]

Stumbling Blocks and the Institutional Design

Although the bargaining situation as outlined earlier is a very influential factor, the institutional set-up also plays a major role in blocking or promoting international negotiations in the climate change regime. While negotiation-oriented forms of institutionalization have often appeared to have a positive impact on the negotiations, there are also major stumbling blocks in the regime which need to be addressed.

One major institutional problem has already been solved: The requirements of the Kyoto Protocol concerning its entry into force, as laid down in Art. 25 (1), were rather difficult to fulfill, especially as the United States, which is responsible for a substantial amount of the relevant overall emissions, and several other industrialized states initially decided not to ratify.[27] Ratifications by Canada, Japan and Russia, which came after various concessions had been granted to these countries in subsequent negotiations, allowed the treaty to finally enter into force (Barrett 2006: 549).

The biggest remaining challenges in the climate regime are to make it work in practice, to enhance its credibility, to achieve a greater adherence to the Kyoto Protocol including adoption by all states, and to set up and implement rules for the second commitment period after 2012. Lack of substantial consensus, as well as the high complexity of the negotiation process, are probably the biggest stumbling blocks in this regard. Art. 3 (9) and 9 KP provide some guidance concerning the further development of the climate regime (Wittneben *et al.* 2006: 90ff.). However, these norms are open-ended and leave room for speculation on possible alternative structures, which can be used in a creative way, but which can also be abused in order to derail the whole process instead of focusing on the solution.[28]

A further stumbling block is the inter-institutional relationship between the permanent treaty bodies (in particular the COP, the COP/MOP, and the subsidiary bodies) and the contact groups which are smaller issue-specific governmental negotiation and drafting forums set up in an *ad hoc* way to deal with issues referred to them.[29] Furthermore, the lack of equal representation in the contact groups may be quite detrimental to the success of negotiation rounds. Finally, this analysis suggests that independent experts might be integrated at a higher level in the negotiations. It will be argued that the failure to do so could be one reason why negotiations within contact groups are sometimes not as effective as they might be.

The consensus requirement in the COP may be considered as another obstacle for the successful further development of the regime rules. Parties should strive to replace it by qualified majority voting as a last resort, although this will obviously be very difficult to achieve (Wittneben *et al.* 2006: 99). Finally, in spite of important decisions taken by COP/MOP1 relating to the compliance mechanism,[30] the procedures on compliance control and on compliance assistance, as well as the flexibility mechanisms, require further fine-tuning, while the actual dispute settlement procedure has, up to now, turned out to be of little practical value.[31]

Thoms identifies further potential stumbling blocks in the UNFCCC, such as the fact that it deals with a basket of several gases instead of focusing on carbon dioxide alone, or the fact that "those who could potentially profit from international regulation" were not separated "from those who would not" (Thoms 2003: 848). However, because of scientific uncertainty and the global character of the problem, it was wise to address all issues from the very outset and to include countries with divergent interests already in the framework convention, in order to establish a global platform for negotiations and discussions (Hanschel 2005: 16ff.).

1. Institutions

    The striking feature of institutions in the climate regime, as in other international environmental regimes, is their cooperative working mode and the constant revision process, which allows them to "learn" (Brown Weiss 1997: 302; Haas *et al.* 1993: 422ff.; Sand 1990: 36). If they are skillfully designed, they may contribute considerably to the enforcement and the further development of the substantive obligations (Birnie and Boyle 1992: 160f.). Institutions may *inter alia* be classified according to their composition, size, and competences (Lang 1995: 206). Another valid criterion is the degree of institutionalization that may be determined by the decision procedure applied within the institution (unanimity, consensus procedures, various forms of majority decisions, etc.), the subject matter of the decisions (substantive rule making or rather technical/procedural matters), the decision-making forum (i.e. government representatives or expert panels), the mode of decision making (negotiation or confrontation) and the resulting degree of legal obligation (binding "hard law" as opposed to mere recommendations or other kinds of "soft law") (Hanschel 2003a: 214).

    a) Main organs

        A basic dividing line can be drawn between main and subsidiary regime bodies. Environmental regimes usually set up one main treaty body, which is often (as in the climate regime) the COP in the framework convention and the COP/MOP in the protocol. These organs are typically composed of government representatives and fulfill legislative, executive, and quasi-judicial functions, so that the regimes in which they operate rank between international organizations and *ad hoc* state conferences (Ott 1998: 86, 271). In the process of decision making, negotiations and the creation of a consensus are stressed; sometimes, however, qualified majority decisions may be taken (Ott 1998: 214). The COPs operate as forums for political guidance and establish a strong link to the preferences of the national governments, thus enabling states to ensure a permanent influence on the further regime development (Hanschel 2003a: 215). At the same time, they may present a focal point for an active exchange of ideas and arguments.

The COP is – as in the ozone regime – the central political steering body within the climate regime, with far-reaching and very detailed competences pertaining to the enforcement and advancement of the substantive obligations, as well as the development of the institutional structure (Art. 7 UNFCCC) (Werksman 1996a: 105). To some extent, it has managed to avoid stagnation and to set in motion a dynamic process leading to more effective implementation and a continued process of rule-making. While the earlier regime on long-range transboundary air pollution only provides for an Executive Body with rather limited competences, the COP set up under the ozone regime is equipped with much wider powers, allowing the regime to develop autonomously (Chasek 2003: 187ff.). This model has (with some caveats) by and large been transferred to the climate regime. On the basis of its mandate, the COP developed a differentiated monitoring system. The existing reporting obligations led to the creation of an extended data basis. This procedure was flanked by softer procedures, such as duties of cooperation and renegotiation procedures (*pacta de negotiando, de contrahendo*) that may pave the way for facilitated amendment procedures similar to those already created under the ozone regime (Hanschel 2003a: 127ff.). Based on the model of the COP, the Kyoto Protocol then established the COP/MOP which, at its first session in 2005, passed a number of crucial decisions that had been drafted by the COP in the preceding years (Wittneben *et al.* 2006: 93ff.). In the light of current post-2012 negotiations, it still remains to be seen whether this institution will live up to the high expectations raised by its very successful model in the ozone regime, the Meeting of the Parties to the Montreal Protocol.

b) Subsidiary organs

The climate regime is marked by a tightly woven web of different institutions driving ahead the process of negotiations and the solution of the regulatory problem.[32] The regime institutionalizes an extensive and multifold dialogue between scientists and politicians, as well as between state and non-state actors. Similarly to the ozone regime, a large number of subsidiary organs are interlinked and thus contribute substantially to the enforcement and further development of the regime rules. These institutions and the corresponding procedures create a complex negotiation-oriented regulatory system, allowing for a constant revision of regulated substances and contributing considerably to the success of the regime (Hanschel 2003a: 127ff.). They furthermore facilitate fruitful forms of forum shopping (i.e. if one institution is blocked, another one may still function).

First and foremost, secretariats play an important role in many international environmental regimes (Brown Weiss 1999: 1571). Their main task is to collect, channel, and distribute pieces of information, thus acting as an interface between the states and the regime bodies (Wettestad 1999: 235; Brown Weiss 1999: 1571f.). Their administrative capacity, however, largely depends on their funding, which is often rather weak (Wettestad 1999: 235). Furthermore, a secretariat can only operate effectively if member states provide it with the data necessary to establish the extent to which they have complied with their obligations (Wettestad 1999: 199). The secretariat of the climate regime in Bonn generally takes a very active stance in the organization of the COPs, the dissemination of information, the coordination of the various regime bodies, and the actual negotiations.[33] Furthermore, the parties to the UNFCCC set up a Subsidiary Body for Implementation (SBI), which was followed by the creation of the Compliance Committee (CC) under the Kyoto Protocol.[34] Moreover, there are important technical and scientific advisory bodies, in particular the Subsidiary Body for Scientific and Technological Advice (SBSTA) and the IPCC (Werksman 1996a: 384ff.).[35]

Nongovernmental organizations (NGOs), which are abundant in climate negotiations, fulfill functions of a public conscience and watchdog (Susskind and Ozawa 1992: 154; Giorgetti 1998: 136ff.; Morgan 2006: 1). While they are usually not formally integrated into the negotiation process, they often operate *de facto* as subsidiary regime bodies. They convey new scientific and technical information as well as public opinion to the actors and point out alternative policy options, so that their reports may constitute an important basis for decisions on the implementation and amendment of existing standards (Susskind and Ozawa 1992: 154, 158; Brown Weiss 1993: 693f.). Moreover, NGOs afford important switchboard services between the regime bodies and the public, and they may pass on observations and decisions by the regime bodies to the media, so that recalcitrant states can be put in the pillory (Giorgetti 1998: 137; Wolfrum 1999: 52).[36] Finally, they may influence the negotiation process by forming alliances with key actors (Andresen and Gulbrandsen 2005: 178).

The Secretariat, SBSTA, SBI, the IPCC and other subsidiary bodies provide the necessary administrative and scientific expertise to promote the further negotiation process in the climate regime. In doing so, they are supported by NGOs that use independent sources of information and exert pressure on the parties. In particular, the scientific bodies play an important role, as they help to reduce the high level of scientific uncertainty. Scientific information serves as an important basis for argumentation in favor or against further regulatory measures (Werksman 1996b: 59). Privileged access to such information may improve the bargaining power of an actor. Therefore, it is crucial to have a transparent and objective system of information gathering in place. As the Fourth Assessment Report has demonstrated again, the findings of the IPCC are generally of a highly authoritative character, because this institution was created for the purpose of providing sophisticated scientific input for the climate negotiations and is closely connected with the regime.[37] Current allegations concerning the improper handling of data in scientific reports show that it is crucial for the IPCC to maintain that status.[38] A loss of reputation and credibility might play into the hands of those wanting to create additional stumbling blocks in the process. The IPCC needs to be firmly supported by the COP, and it requires substantial resources in order to make sure that it remains the central point of reference to the negotiators in all scientific questions (Pierrehumbert 2006: 591). The decision of the United States to set up and fund its own climate change scientific program must not result in a competing system which is administered by one state only instead of the international community of states as a whole.[39] One may hope that all countries will rely on the research results compiled by the IPCC and consider their own research programs merely as an additional source of information.[40] The Copenhagen Accord of 2009, which refers back to the IPCC findings, justifies that hope.[41]

While the scientific bodies in the climate regime provide important pieces of information, they are mostly composed of government representatives, specialized in the relevant fields (Bryce 1999: 385). Government influence may ensure equal representation of interests and thus help to get all states on board when taking a decision or publishing a report. It may also help to translate scientific results into a language that can be "digested" by the political decision makers. Hence, the IPCC does not conduct research itself, but selects and compiles relevant research results by external researchers. Its task is to "assess on a comprehensive, objective, open and transparent basis the scientific, technical and socio-economic information relevant to understanding the scientific basis of risk of human-induced climate change, its potential impacts and options for adaptation and mitigation".[42] The members of the IPCC are likely to gain a certain degree of autonomy from their home government after being affiliated with that institution. Furthermore, the review of scientific evidence by the IPCC is generally carried out in a very sophisticated way.[43] Nevertheless, while it could be difficult to agree on its composition, it might be wise to try to set up an issue-specific, high-ranking, and influential permanent forum of independent experts outside the secretariat, which can discuss the science of climate change without having to pay regard to any national constraints.[44]

This relates to another major stumbling block: The interaction of the COP, SBI, SBSTA, and the contact groups shows some deficits, which raises the question of effective delegation between these institutions. As the COP, SBI, and SBSTA are composed of government representatives, delegation of an issue to a contact group is based on a political decision, even though this act of delegation may in fact have been prepared to some extent by the secretariat. Furthermore, the fact that various regime bodies are authorized to delegate an issue to a contact group causes multiple problems. At the COPs, there is a repeated intense discussion about the exact mandates of individual contact groups, which often overlap. For example, delegates debate the question of whether certain issues fall within the scope of SBI or SBSTA. What is achieved in one contact group often depends on the results pending in another one, which makes it difficult to establish helpful issue linkages. Especially when several contact groups convene simultaneously, lack of information and uncertainty are serious stumbling blocks. This creates numerous options to block the process, for example, by postponing the session until another contact group (where the same blockade strategy might be applied) has presented its results.

It is not realistic to hope for a change of strategic behavior on the part of parties in the COP, SBSTA, or SBI, and an overlap of issues would not completely be remedied by independent experts, either. Still, it could be helpful to let such experts advise the secretariat in drafting a plan on how the issues can best be subdivided and timetabled alongside the conference schedule. This will often mean that contact groups should not convene simultaneously, but consecutively. In order to gain further time for this procedure, the plenary part of the high-level segment, sometimes stretched by rather general statements by the parties, might be shortened by strictly enforcing a time limit for each speaker. Furthermore, experts could serve as advisors to the chairs of contact groups and provide information on parallel negotiations in other contact groups. One may assume that the secretariat already takes care of these issues to a certain extent, and there are clearly limits of feasibility. Still, further steps in this direction might be useful.

Another stumbling block is the voting procedure in the COP which is dominated by consensus decisions. From the beginning of the negotiations, there has been continuing disagreement on the rule of procedure Nr. 42, which deals with this question.[45] This is a major stumbling block in the negotiations, as binding majority decisions are excluded, although they could help to spur the decision-making process (Wittneben *et al.* 2006: 99). However, chances to alter the consensus requirement are rather low, as parties have an essential interest in not being voted down. In some cases a "consensus minus one" decision can be achieved, but only a very strong chairperson can do this, and certainly not under all circumstances.[46]

Moreover, there is a problem of unequal representation in the contact groups. Industrialized countries that have the necessary resources may send representatives to all contact groups at the same time, thus gaining access to the full amount of information, and they may process the information accordingly. By contrast, some of the developing countries, in particular the least developed countries, can often send only one or two delegate(s) and therefore have to select which forum they consider to be the most important. One might argue that, in any contact group, all states are at least represented by one member of their own bargaining coalition, such as the so-called Group of 77 (G77 + China), the Alliance of Small Island States (AOSIS), and so on.[47] However, the interests in such a coalition are often less homogenous than they appear at first sight, which can be illustrated by the fact that Bangladesh, China, India, and Saudi Arabia are all in the same coalition (Wittneben *et al.* 2006: 99ff.). Thus, joint decisions of such a group often require cumbersome internal pre-negotiations resulting in rather weak and inflexible decisions. Representatives carrying such a decision into the contact group may be paralyzed by a very low margin of discretion, making it impossible to react flexibly to sudden changes in the dynamic process of negotiations. If they exceed that margin, the decision taken in the contact group will often not be accepted by the coalition as a whole, which will cause severe problems in the COP that has to adopt this decision. Therefore, representation of individual countries by only one actor of the bargaining coalition is often an insufficient substitute.

Some decisions at COP meetings have been taken in a contact group without being fully consented to by all delegations, which may lead to further discussions and a blockade situation in the COP on the very last day of negotiations. If all states were represented in every contact group, a clear rule of not going backwards in the COP could be established, but this would mean that the contact groups, which by definition tend to be small and less than fully representative, could not function anymore. The effectiveness of the contact groups is strongly determined by their small size, which allows for a more informal and matter-of-fact atmosphere and a quicker decision-making process. If all states participated in the contact group, it would just function like another COP. Therefore, it will remain necessary for some countries to be represented by others in such a group. But without real agreement within sometimes rather loosely knit coalitions such as G77 + China, it would be more helpful to send more than just one or two representatives of that coalition into the contact group. The alternative would be to restructure the coalitions to make them more homogenous. While the secretariat might set up criteria such as a comparable per capita income, vulnerability, and so on, as conditions for membership, such an intervention does not appear feasible, as the coalitions are based on autonomous decisions of the parties and establishing criteria would probably raise a debate as to their equitableness.

Another problem lies in the fact that the expertise of the chairs and co-chairs of the contact groups may vary. They are usually appointed by the President or SB Chair on the basis of proposals by the parties and the secretariat. However, apparently not all of them have the same level of training and experience and the necessary authority to conduct the negotiations in an equally effective way. Furthermore, the chairs are in the ambiguous situation of being country representatives, while at the same time having to operate as independent and neutral chairpersons, which can cause friction, but is a common practice at many international conferences and difficult to alter.

As a possible facilitation strategy vis-à-vis these stumbling blocks, the COP could decide on a procedure regarding negotiations within the contact groups. This procedure should contain provisions about equal representation of all coalitions and include the necessary financial support for adequate capacity-building measures. Furthermore, there might be stricter rules on how the chairs of the contact groups are appointed and who qualifies for such a position. These rules, however, have to leave some margin of discretion so that the necessary flexibility of negotiations is ensured. Moreover, one has to admit that difficult negotiations most of the time result from a difficult bargaining situation rather than from individual mistakes by the chairpersons.

2.   Negotiation-oriented enforcement procedures

A main problem of international environmental law in general and the climate regime in particular is its inherent compliance deficit (Bothe 1996: 13). This problem has been addressed frequently without a proper solution being found.[48] Since the traditional mechanisms of norm enforcement only work to a very limited extent, the main instruments of implementation are cooperative negotiation-oriented enforcement procedures (Ott 1998: 259; Ehrmann 2000: 34ff.). This so-called sunshine approach or manager-model is based on a combination of monitoring, regular reporting, access to information, participation of NGOs, among othes, and relies on the assumption that states do not like to be put in the pillory, but want to preserve their good reputation within the international community (Chayes and Handler Chayes 1995: 3, 22ff.; Brown Weiss 1997: 299). A main target of this approach is to achieve transparency and thus responsibility, as states often disguise their true interests and merely pay lip service to their obligations (Chayes and Handler Chayes 1995: 22ff.).

a)   Compliance control

The main instruments of enforcement are mechanisms for compliance control, which appear to be inspired considerably by international human rights instruments (Brown Weiss 1997: 299; Wolfrum 1999: 36; Riedel 1996: 95ff.; Hanschel 2003b). In international environmental law, direct sanctions are rarely ever applied (Brown Weiss 1997: 298; Wettestad 1999: 237). Indirect sanctions or negative incentives may result from the withdrawal of advantages, but in most cases positive incentives are used (Brown Weiss 1997: 299, 303; Marauhn 1996: 719). The final decision about the consequences of non-compliance is often taken by a political organ such as the COP in the climate regime, which is left with a rather wide margin of discretion (Marauhn 1996: 719). The crucial advantage of negotiation-oriented means of compliance control is that states are more willing to accept a regulatory scheme that does not interfere with their sovereignty (Werksman 1996a: 102; Chayes and Handler Chayes 1993: 205). Accordingly, Chayes and Handler Chayes assume that "indirect mechanisms that induce compliance but do not hinge on the rewriting of international law are more likely to produce results more quickly and reliably" (Chayes and Handler Chayes 1993: 205).

Compliance control is brought about by a number of institutions and procedures cooperating in a coherent system (Victor 1998: 677). As in many other international environmental regimes, the enforcement system in the climate regime consists of an implementation review mechanism and a non-compliance procedure (Marauhn 1996: 698f.; Wolfrum 1999: 36; Ehrman 2000: 476f.). According to Art. 12 UNFCCC, state parties have to communicate information relating to implementation, which is then considered by the SBI (Art. 10 UNFCCC). Pursuant to Art. 4 (2) (b), UNFCCC state parties periodically submit information on their policies and measures regarding the mitigation of climate change. The COP has progressively developed methods of monitoring these data on the basis of its mandate in Art. 4 (3) UNFCCC.[49] Art. 5 and 7 KP contain far-reaching reporting obligations. Furthermore, guidelines were adopted under Art. 8 (4) KP which provide for in-depth reviews of communications by so-called Expert Review Teams (ERTs).[50] The ERTs, whose members are selected on the basis of expertise and geographical distribution, shall "provide a thorough and comprehensive technical assessment of all aspects of the implementation by a Party of the Kyoto Protocol [...] [and] conduct technical reviews to provide information expeditiously to the COP/MOP and the Compliance Committee".[51] The CC consists of members both from industrialized and developing countries and is composed of a facilitative branch and an enforcement branch (Knox 2004: 144; Wittneben *et al.* 2006: 94f.). The facilitative branch advises parties in a cooperative fashion in order to improve compliance, while the enforcement branch determines "whether an Annex I party is not in compliance with its emission requirements and its eligibility requirements for participating in the flexibility mechanisms" which leads to certain consequences (Knox 2004: 144). At COP6.5 in Bonn, these consequences were stated, namely the deduction of 1.3 tons from the party´s assigned amount for the next commitment period for every ton of emissions above the respective commitment, the requirement to develop a compliance action plan, and finally the suspension of eligibility for the emission trading system (Knox 2004: 144). At COP7 in Marrakesh the compliance mechanisms were further elaborated, including questions of access, investigation, and decision making (Wittneben *et al.* 2006: 93ff.). By comparison, the enforcement of the obligations in the ozone regime is guaranteed by an implementation committee and a corresponding procedure that became quite effective after 1992 and developed its force, as the examples of Russia and other states in transition to a market economy have shown (Hanschel 2003a: 103ff.).[52] Against the background of that experience, COP/MOP1 in Montreal passed the former draft decisions relating to the CC in the climate regime, and named its first members.[53] COP/MOP2 in Nairobi then adopted the rules of procedure of the CC and invited parties to further contribute to its funding.[54] However, in order to set up the compliance mechanism in a legally binding form, Art. 18 KP appears to demand a formal amendment to the Protocol, which according to Art. 20 (4) and 20 (5) KP, requires instruments of acceptance by the member states, although there has been a vigorous debate on this question at COP/MOP1 (Wittneben *et al.* 2006: 94). A long-term perspective will show whether the CC can take a similarly active stance as the implementation committee in the ozone regime. The effectiveness and necessary legitimacy of its work, especially with regard to the enforcement branch, would increase through a binding treaty amendment (even if the ratification requirement may be cumbersome and cause a delay).[55]

The constructive dialogue between the CC and the states, which mainly relies on carrots rather than sticks, will probably remain the central focus, as states are not ready to compromise further on their sovereignty. The reaction on cases of non-compliance generally correlates to the non-adversarial and cooperative character of the non-compliance procedure (Marauhn 1996: 718). However, this reaction needs to take into account the specific reasons for non-compliance (Brown Weiss 1997: 302; Werksman 1996a: 102f.). Mitchell thus distinguishes "non-compliance as a preference, non-compliance due to incapacity, non-compliance due to inadvertence" (Mitchell 1996: 11ff.). Chayes and Handler Chayes rightly claim that "the tendency is to winnow out reasonably justifiable or unintended failures to fulfill commitments – those that comport with a good-faith compliance standard – and to identify and isolate the few cases of egregious and willful violation" (Chayes and Handler Chayes 1993: 204f.). In cases of non-compliance due to a lack of capability, a more supportive reaction (especially by way of compliance assistance)[56] is needed than in cases of intended non-compliance, where at least mildly adversarial elements such as shaming (putting states in the pillory) or indirect sanctions such as the forfeiture of certain advantages in the climate regime (i.e. participation in the flexible mechanisms, eligibility for compliance assistance, accounting with regard to emission tons, etc. [MacFaul 2005: 5]) are necessary.[57] Stronger sanctions, however, are not feasible as the parties would probably not agree on them. Due to the complexity of the climate problem and the vagueness of some of the regime rules, it is quite difficult for a state to fully comply and for the CC to establish cases of non-compliance or of a violation of the treaty provisions. Thus, just as the implementation committee in the ozone regime, the CC is likely to initiate its enforcement mechanisms only when parties exceed an "acceptable" degree of non-compliance, which in turn needs to be defined by the CC itself.[58] Verification was one of the most disputed issues at COP15 in Copenhagen. China, in particular, preferred to rely on national measures, which was not acceptable to the European Union and the United States. The Copenhagen Accord after all relies on verification, because of the pledge and review system, there is no real obligation as long as the Accord (including the pledges made) is not formally endorsed by the COP and, potentially, ratified by the parties.[59]

b) Compliance assistance

As stated above, one major reason for non-compliance with international environmental obligations is the lack of administrative, technological, and financial capacities, so that procedures of compliance assistance as a means of facilitation play a central role here (Lang 1996a: 97). But no matter whether employed as a reaction to non-compliance or as preventive measures, the receipt of technological, financial, and other forms of aid may help developing countries to fulfill their cost-intensive obligations, and may encourage them towards a more active participation in the regime (Gündling 1996: 796ff.). In international environmental law, compliance assistance has in many instances proven to be a more effective means to ensure compliance than sanctioning (Gündling 1996: 801f.). The Rio Conference of 1992 showed that, in the long-run, only a constructive dialogue with developing countries may lead to an effective solution of global environmental problems (Gündling 1996: 802ff.). Furthermore, compliance assistance may spur the further development of a regime. The receiving states are not only enabled to comply better with the existing obligations, but they may also be more likely to agree with the tightening of the regulatory framework (Hanschel 2003a: 238). However, measures of compliance assistance require a considerable degree of consensus between the donor states, as they produce not only benefits, but also substantial costs. Aspects of fairness, such as an equitable distribution of costs and an equal right of all countries to economic development, often play an important role here (Gündling 1996: 797ff.).

The ozone regime procedures concerning the transfer of financial resources and technology made an important contribution to the enforcement of the obligations (O'Connell 1992: 300). Transfer mechanisms including various forms of capacity building were equally considered as crucial requirements for the success of the UNFCCC, because both mitigation and adaptation measures are relatively expensive (Churchill 1991: 173). Hence, probably most of the developing countries could not afford to take the necessary measures without any support by the regime institutions. According to Art. 4 (3) UNFCCC, the industrialized states have to bear the agreed full costs for the provision of data. Furthermore, they have to pay for the agreed full incremental costs for the mitigation and adaptation measures, as far as they are adequate.[60] Art. 4 (4) UNFCCC stipulates further support to the particularly vulnerable countries. In Art. 4 (5), UNFCCC states provide for a transfer of technology. As part of the Marrakesh Accords, COP7 set up a "framework for meaningful and effective actions" in order to enhance implementation of the latter provision; this framework comprises "technology needs and needs assessments", "technology information", "enabling environments", "capacity building", and "mechanisms for technology transfer".[61] Art. 11 UNFCCC establishes a financial mechanism, which is refined in the Kyoto Protocol. Thus, Article 11 KP requires the industrialized states to provide new and additional resources for the developing countries. Finally, the Facilitative Branch of the CC plays an important role, by being "responsible for providing advice and facilitation to Parties in implementing the Protocol, and for promoting compliance by Parties with their commitments under the Protocol, taking into account the principles of common, but differentiated, responsibilities and respective capabilities [...]".[62]

Participation of the developing countries is crucial for the regime, because in the long run there can be no effective solution of the climate change problem without having them on board. Some economies in transition, such as China, India, Brazil, or South Africa, are developing so rapidly that a dramatic increase in emissions has occurred and is still continuing. Meanwhile, China has even surpassed the United States as the world's largest emitter.[63] This is why many argue that such countries should now adopt substantial mitigation targets of their own. Conversely, developing countries claim that historically climate change is a result of Northern industrialization, so that industrialized states should take the lead in the climate regime.[64] Notwithstanding the exact obligations of developing countries, measures of compliance assistance should be promoted, as they have proven to constitute important instruments bridging the economic and technological gaps between the parties (Bodansky 1992: 64). Even where no mitigation targets apply, compliance assistance may still help countries to fulfill their reporting obligations. The reporting guidelines developed by the IPCC help to structure and thus facilitate the reporting, but often technological and financial problems in collecting the necessary data remain.[65]

In summary, the measures of compliance assistance under the climate regime are quite substantial, but not yet sufficient. For example, the mechanisms of financial assistance around the Global Environment Facility (GEF) need further fine-tuning. There have been intense discussions at various COPs on how to improve the existing mechanisms, which cannot be dealt with extensively at this point. Suffice to say that two of the three financial mechanisms provided in the Marrakesh Accords, namely the Adaptation Fund (AF) and the Special Climate Change Fund (SCCF), "continue to be under heated debate" even after the successful Montreal round in 2005 (Wittneben *et al.* 2006: 96; Ott 2001: 211ff.). While in Nairobi some progress was made on the funding of the AF and the activities to be sponsored under the "Buenos Aires Programme of Work on Adaptation and Response Measures" during the first two years, parties still could not agree on the exact institutional set-up of the AF (Sterk *et al.* 2007: 9). A satisfactory replenishment of the various funds in the climate regime, including also the Least Developed Countries Fund (LDCF), and transparent rules concerning their operation are necessary in order to maintain a truly effective system of compliance assistance. The expansion of the existing financial support mechanisms was one of the few partial successes of the Copenhagen Accord, where developed countries agreed to provide US$ 30 billion between 2010 and 2012 and "set a goal of mobilizing jointly 100 billion dollars a year by 2020 to address the needs of developing countries." In institutional terms a "High Level Panel will be established under the guidance of and accountable to the Conference of the Parties" and a "Copenhagen Green Climate Fund shall be established as an operating entity of the financial mechanism of the Convention".[66] It remains to be seen whether this non-binding understanding will become binding and/or will be implemented in practice.[67] Equally, technology transfer remains an important and controversial issue on the agenda, with the negotiations in Nairobi focusing on the future mandate of the Expert Group on Technology Transfer (EGTT) and its potential replacement by a Technology Transfer and Development Board and a Multilateral Technology Acquisition Fund (Sterk et.al, 2007: 9). Due to divergent views on this issue, the EGTT mandate was extended and negotiations were postponed.[68] In Bali a set of actions was decided upon, serving to strengthen the implementation of the technology transfer framework.[69] The latest step is the Technology Mechanism set up by the Copenhagen Accord "to accelerate technology development and transfer in support of action on adaptation and mitigation", which follows a decentralized approach and is "based on national circumstances and priorities".[70] Again, this does not constitute a binding commitment, and the question remains to what extent this will be connected to the international framework on technology transfer.[71]

c)   Flexibility mechanisms

In order to facilitate implementation of the mitigation targets, three mechanisms, namely, International Emission Trading (IET), the Clean Development Mechanism (CDM), and Joint Implementation (JI) were created in the climate regime (Freestone 2005: 11ff.; Hovi *et al.* 2005: 6ff.; Sands 2003: 372ff.). All these mechanisms are designed to make implementation of regulatory measures more cost-effective (i.e. efficient) by institutionalizing a flexible cooperation between the member states (Hanschel 2003a: 238). Therefore, they were named "flexibility mechanisms" (Hovi *et al.* 2005: 6ff.; Cullet 1999: 168ff.; Yamin 1998: 113ff.). They operate on the basis that climate change is a truly global problem so that, in terms of environmental effectiveness, it does not matter where in the world the necessary emission reductions are achieved.

Joint implementation in a more general sense can be found in the regime on long-range transboundary air pollution, the ozone and the climate regime, and comprises all activities implemented jointly between member states in order to fulfill their targets (Mason 1995: 296ff.; Gillespie 2005: 296ff.). According to Art. 6 UNFCCC, JI encompasses joint activities between Annex I parties which are credited by so-called emission reduction units (ERUs) (Gillespie 2005: 299ff.). CDM (Art. 12 KP) widens this concept by granting credits called certified emission reductions (CERs) for projects undertaken in non-Annex I countries (Davies 1998: 457ff.; Werksman 1998: 147ff.).[72] As this mechanism also serves as an instrument of compliance assistance, there are some restrictions taking account of the structural inequality between industrialized states and developing countries (Heintz *et al.* 1994: 169; Gillespie 2005: 306ff.). As the last step of flexible implementation, IET (Art. 17 KP) allows Annex I countries to acquire assigned amount units (AAUs) by trading emissions among them in order to fulfill their obligations under Annex B of the Kyoto Protocol (French 1998: 235ff.; Frischmann 2001: 463ff.; Evans 2004: 167ff.).[73]

While all three mechanisms became operational with the formal adoption of the Marrakesh Accords by COP/MOP1 in Montreal in 2005, further elaboration and fine-tuning, in particular concerning their interaction, is required (Wittneben *et al.* 2006: 95).[74] Moreover, the interpretation of terms such as the "additionality requirement" in the CDM or the "supplementarity" of IET has led to heated discussions (Wittneben *et al.* 2006: 95; Gillespie 2005: 313). According to Mason, the success of flexibility mechanisms largely depends on the existence of opportunities for cost-saving and of incentives for structural change, so that "joint implementation must ... combine the flexibility required for the former with the rules and institutions necessary for the latter" (Mason 1995: 299). The concept of flexible implementation has encountered the great difficulty of transferring economic theories into practical, acceptable and effective international legal obligations and agreements (Loske and Oberthür 1994: 45; Bodansky 1993: 520f.; Sands 1995: 178). However, according to many studies, market-oriented incentive-based mechanisms are more easily implemented than command-and-control approaches, since they are better suited to balance costs and benefits between the contracting parties (Aman 1995: 460ff.; Dudeck and Palmisano 1988: 220). At the same time, such flexibility mechanisms must not lead to an (often criticized) complete sellout of obligations or a mere dealing with "hot air" (Sterk *et al.* 2007: 7ff.). Finally, these mechanisms have created "enhanced transaction costs associated with verification and review" to be tackled

in a satisfactory way by the compliance control system described above
(Hovi *et al.* 2005: 7).

As the Kyoto Protocol and its "rulebook," namely, the Marrakesh
Accords, only became effective a couple of years ago, and the first commitment period is still running, it is yet too early to make a final judgment
on the effectiveness of the flexibility mechanisms. However, when considering the high economic costs that are involved in climate protection
and the global nature of the regulatory problem, the idea of cost-effective mitigation measures is very convincing, and the flexible mechanisms
have started to turn out as an important institutional facilitation measure
based on economic incentives. COP/MOP1 has taken important steps
towards the realization of this idea: It finally launched JI and set up its
governing body, and it helped to secure funding of the CDM in 2006–
2007,[75] although COP/MOP2 urged parties to provide further contributions for the operation of the newly established JI Supervisory Committee
(JISC).[76] All in all, chances are rather high that further stumbling blocks
concerning the flexible mechanisms, such as the asserted lack of long-
term legal certainty (in particular with regard to the second commitment
period) and of effective linkages with existing regional systems, will be
removed in the near future. More experience, as it is currently gained
through the emission trading scheme set up by the European Union (EU-
ETS) which serves as an important laboratory in this field, will be most
precious in deciding which improvements and adjustments need to be
made in the system in order to make it even more effective.[77]

d)   Dispute settlement

Many environmental regimes contain both procedures of compliance
control and dispute settlement, which may be distinguished by their concrete design and their aim (Sands 1996: 101). Dispute settlement basically presupposes a conflict between two states and the participation of
an impartial third party, thus a triangular relation, while compliance
control rather presents a bipolar relation between an individual state and
an implementation body (Marauhn 1996: 721). However, this distinction
becomes blurred to some extent when one considers the fact that submissions to the CC can also be made by "any Party with respect to another
Party".[78] Nevertheless, the aims of establishing compliance and of solving
a dispute between two parties remain distinguishable.

Traditional dispute settlement procedures are rightly considered as
an anachronism, as they often serve to put all the blame on one state
(Victor 1996: 84f.; Werksman 1996a: 102; Sands 1996: 70ff.). Therefore,
they are hardly ever used in international environmental regimes (Werksman 1996a: 102; Victor 1996: 85). By contrast, modern environmental
dispute settlement procedures are usually negotiation-oriented (Rubin
1993: 275ff.; Lang 1996b: 695). Thus, according to Art. 14 UNFCCC
and Art. 19 KP, the dispute settlement procedure in the climate regime
contains three elements: negotiations and other peaceful means, followed

by a referral of the case to the International Court of Justice, and finally a conciliation committee (Hanschel 2003a: 144f.).

In theory, dispute settlement procedures can be useful in every regime, as they may clarify the problem, enhance motivation for agreement and offer creative solutions and means for their implementation (Rubin 1993: 287ff.). However, even the negotiation-oriented dispute settlement procedures in the ozone regime, the climate regime, and in the regime on long-range transboundary air pollution were either never or hardly ever used and provide several stumbling blocks (Ott 1996: 737). In spite of their cooperative character, they often produce a public reaction that may be harmful to the reputation of a government. Furthermore, the results of such a procedure in an individual case are difficult to calculate (Hanschel 2003a: 241). States obviously prefer to solve disputes by way of informal, low-key, and confidential consultations during the official conferences. As long as the disputes are solved here (which is certainly not always the case), a formal dispute settlement procedure may appear futile. However, its existence does not harm the negotiation process, either, and may still become useful in the future. In order to make dispute settlement more attractive, the element of negotiation should be emphasized even more, and negotiations should – at least in the initial phase – be conducted in a confidential way. A further option would be to allow other actors, such as NGOs, to initiate this procedure. However, it appears quite unlikely that member states will accept such a move. The bulk of non-compliance cases should be dealt with by the CC, which may reduce the need to make the dispute settlement procedure operational. Still, in order to allow for the possibility of forum shopping, it might be useful to try and modify the existing dispute settlement procedures along the lines indicated above.

3.　　Negotiation-oriented regime development procedures

a)　General duties of cooperation
　　Long-term cooperation is at the very heart of international environmental regime-building (Raustiala and Victor 1998: 693). Since, in the initial phase of negotiations, actors lack precise information about the actual scope of the problem, about possible solutions to the problem and about the views of other actors on the issue, they need to establish a process of continued discussion (Werksman 1996b: 58ff.). Some authors consider duties of consultation and cooperation to be rather doubtful obligations (Heusel 1991: 244ff.). However, while these duties are often rather vague as to their content, they are legally binding parts of international environmental treaties (Ipsen 1999: 95ff.). As they do not interfere with the sovereignty interests of the member states, they usually require only very little consensus between the parties. They may thus form the basis of further-reaching procedures of regime development and help to build up public pressure. Therefore, one way to raise the chances of getting more states

to ratify the Kyoto Protocol is to "launch informational campaigns" in order to convince the actors and the public that the problem exists and that more substantial rules are necessary to tackle it effectively (Thoms 2003: 857ff.). Duties of cooperation in the climate regime are contained in Art. 4 (1), 5 and 6 UNFCCC as well as Art. 2 (1) and 10 KP and concern research cooperation, joint development of technology, exchange of information and support of public opinion.

Duties of cooperation should also entail measures of capacity-building other than those concerning compliance assistance already mentioned. One important aspect here is the specific training of diplomats of any country asking for support in this field. This is a very controversial issue: While further training might help a country to represent its interests in a better way, it equally means that the country can block negotiations more effectively, if it wants to do so. Equity concerns demand that training measures should definitely be carried out because, as a matter of principle, states should be enabled to effectively represent their own legal interests, no matter what those interests are. But, even in terms of regime effectiveness, such measures may be helpful, although they should not be biased in any way. As long as training programs do not only focus on bargaining skills, but also on information about climate change and the regime, this may by itself underline the necessity to take some further action. Conversely, delegates who are not sufficiently trained are more likely to be manipulated. They may even take irrational decisions, because they feel insecure but want to take their own stance in the process. Negotiations by well-trained diplomats are likely to have a higher level of sophistication and thus to produce good results. However, there is a thin line between good training and subtle manipulation, which must not be crossed.

b) Renegotiation procedures

Duties to continue negotiations on the already existing rules are useful in all treaties that by themselves are not sufficient to fully solve the regulatory problem, be it due to a lack of the necessary consensus or because the problem and its possible solution have not yet been fully revealed. In both cases, renegotiation procedures contribute to the maintenance of the process of negotiation and rule-making, thus helping to prevent the existence of a singular "sleeping treaty" which does not solve the problem (Ott 1998: 250). Such renegotiation procedures are laid down in Art. 4 (2) d and 7 (2) UNFCCC and require the Parties to review, in regular intervals, the existing material and institutional rules. Art. 9 KP puts this into more concrete terms by calling for regular sessions of the regime bodies in order to carry out this review, in particular concerning the reduction targets. Art. 3 (9) KP even requires the COP/MOP to "initiate the consideration of ... commitments (for subsequent periods) seven years before the end of the first commitment period". This obligation to renegotiate led to the establishment of the rather hesitantly framed "Seminar of

Governmental Experts (SOGE)" at COP10, which, although merely convening "talks about talks about talks" concerning action under both legal instruments, proved to be rather effective as an informal round (Ott *et al.* 2005: 90; Wittneben *et al.* 2006: 93). A further and more robust renegotiation procedure was then passed by COP/MOP1, which created the Ad Hoc Working Group on Further Commitments for Annex I Parties under the Kyoto Protocol (AWG-KP) in order to negotiate further targets for the second commitment period to be completed "in time to ensure that there is no gap between the first and the second commitment period". (Wittneben *et al.* 2006: 92).[79] At COP/MOP2, the parties confined themselves to stating that the Kyoto Protocol "has initiated important action", but some elements "could be further elaborated upon" and implementation could be further enhanced".[80] Although parties failed to agree on a clear timetable, there was still enough time to negotiate an agreement on post-2012 commitments within the mandate of the AWG (Sterk *et al.* 2007: 4ff.).[81] Unfortunately, only limited progress has been achieved on subsequent meetings, either, so that time is getting short.[82]

The dual track approach pursued by COP11 President Stéphane Dion implied that, simultaneously, a quite informal low-key "dialogue on long-term cooperative action to address climate change by enhancing implementation of the Convention" was initiated (Wittneben *et al.* 2006: 92).[83] While this informal process carefully avoided the term "negotiation" and seemed to lose some of its momentum at COP12, it remained important as an overarching forum allowing for the integration of the developing countries and industrialized countries (such as the United States) into the discussion on substantial mitigation targets (Sterk *et al.* 2007: 3ff.; Morgenstern 2009: 237). At COP13 in Bali, the informal review under the UNFCCC was strengthened through the Bali Action Plan which launched "a comprehensive process to enable the full, effective and sustained implementation of the Convention through long-term cooperative action".[84] The aim of the Ad Hoc Working Group on Long-Term Cooperative Action under the Convention (AWG-LCA) was to achieve a decision at COP15 which, however, failed: The Copenhagen Accord was only taken note of by the COP, but not formally endorsed, mainly due to protest from developing countries, including economies in transition, that have begun to defend their interests more vigorously in the negotiations. Still, one may expect negotiations within the UNFCCC framework to continue in the future.

The above-mentioned procedures may be qualified as *pacta de negotiando* or *pacta de contrahendo* (Heusel 1991: 243f.; Beyerlin 1976: 407ff.), obliging the states to renegotiate in "good faith as properly to be understood," to keep up "the negotiations over a period appropriate to the circumstances, to show an "awareness of the interest of the other party," and to embark on a "persevering quest for an acceptable compromise."[85] While there are different views on the distinction between both terms, basically the *pactum de contrahendo* requires states to try to reach consensus concerning a specific agreement, whereas a *pactum de negotiando* only obliges them to take up serious negotiations with a view towards reaching consensus (Seidl-Hohenfeldern 2000: 65; Ipsen, 1999: 98). Due to the rather open-ended character of the renegotiation procedures in the climate regime, they should rather be qualified as *pacta de negotiando*.

Translated into economic theory, environmental regimes may also be considered as "incomplete contracts" (Ayres and Gertner 1992: 730; Cooter and Ulen 1997: 3ff., 161ff.). While lawyers consider a treaty to be incomplete when the contained obligations are not fully specified, economists imply by this term that the obligations do not provide optimum economic efficiency, that is, "contracts that fail to fully realize the potential gains from trade in all states of the world" (Ayres and Gertner 1992: 730). Incomplete information and the inability to foresee the future are typical reasons for incompleteness (Richter and Furubotn 1996: 247ff.).

Both from a legal and from an economic point of view, the setting up of renegotiation procedures appears necessary to complete the regime-building process in order to achieve a legally effective and economically efficient outcome. Procedures of renegotiation have thus proven to facilitate negotiations, by pushing them further ahead, guiding them into the right direction and allowing for the necessary regime amendments (Hanschel 2003a: 253f.). In the climate regime, these procedures have turned out to be rather successful soft tools, able to smooth away or circumvent existing stumbling blocks resulting from divergent party interests by the pre-structuring of future regime development through formula compromises. Even though recent progress has been scarce and parties have been cautious enough to phrase the mandates of institutions, such as the AWG, in a careful and often hesitant fashion, it appears plausible that, in the long run, the dynamic character of these renegotiation procedures will have a more positive effect on the outcome.

While renegotiation procedures have been skillfully designed in the climate regime, the current dual track approach has also turned out to be a stumbling block to the extent that it has increased complexity of negotiations, as illustrated in the run-up to COP15 and at the conference itself. The advantage is that progress can be made on both tracks, independently of each other. Nevertheless, the comparison with the regimes on long-range transboundary air pollution and on ozone depletion teaches that, in the long term, it is wise to focus regime development mainly on the protocol(s), not on the convention. When discussing future design options, this should be kept in mind.

c) Facilitated amendment procedures
Facilitated amendment procedures have, in many cases of international environmental rule-making, turned out to be the central element of treaty development (Wettestad 1999: 198; Poitras 1999: 104). The facilitated amendment options may relate to the treaty itself, and to annexes to the

treaty, as well as to new treaties that are connected to the initial one (Wettestad 1999: 198). International environmental negotiations are characterized by a compromise between national sovereignty and transnational interdependence (Benedick 1993: 229). Facilitated amendment procedures try to reconcile these two aspects and finely balance those elements that maintain and those that limit national sovereignty, thus promoting a dynamic advancement of the existing obligations and securing a permanent acceptance by the member states (Benedick 1993: 229). Ideally, facilitated amendment procedures should be backed up by renegotiation procedures and by general duties of cooperation, since these mechanisms require a much lower substantial consensus between the parties as to the solution of the problem and thus present the ideal first step towards the agreement on stronger development procedures (Hanschel 2003a: 258).

While facilitated amendment procedures usually spur further negotiations, they may also hamper them, because, as a matter of principle, the majority could simply impose the desired changes upon the minority. However, if – as in the ozone regime – qualified majority decisions are only provided as a last resort, as a mere option on the horizon, negotiations are usually promoted, since this may help to pressurize parties into achieving consensus even on the more controversial issues (Hanschel 2003a: 259). As Wettestad (1999: 259) puts it: "Although ... several regimes opened up opportunities of majority decisions, consensual decisions were the general order of the day."

Facilitated amendment procedures can basically be recommended for all regimes that depend on a dynamic development of the existing rules in order to react to changing technical and scientific knowledge (Hanschel 2003a: 257). However, if these procedures lead to decisions and measures that are ultra vires, that is, beyond the scope of the initial treaty, they may easily run the danger of losing the necessary acceptance by state parties. Non-consented decision procedures or practices may thus encounter massive problems when it comes to implementation (Chayes and Handler Chayes 1993: 183; Haas *et al.* 1993: 24). In order to avoid disputes on a large scale, decision processes should be formulated in a very clear and unambiguous fashion and take into account aspects of fairness (Young 1993: 259). Furthermore, the decision-making process should include non-state actors in order to allow for a greater creativity (Poitras 1999: 106). Poitras rightly states that "by limiting the proposal of amendments to contracting parties, the amendment process is narrowing its potential to generate new ideas" (Poitras 1999: 106).

The regime on long-range transboundary air pollution, which contains first steps in this direction, has served as a kind of "regulatory laboratory" for the ozone and the climate regime and has provided important experiences concerning the functioning of such a dynamic procedure (Hanschel 2003a: 257). The ozone regime displays the most advanced facilitated amendment procedure, which has contributed immensely to regime effectiveness (Werksman 1996b: 61; Palmer 1992: 274ff.; Hanschel 2003a:120ff.): Direct amendments to the text of the ozone treaties are only binding for the member states ratifying them. But the regulatory content of the regime may be indirectly amended (adjusted) by changing the annexes to the protocol according to the procedures laid down in Art. 2 (9) of the Montreal Protocol: These annexes do not merely contain technical rules, but also a list of controlled substances. Due to a referral in the Montreal Protocol, these annexes are part and parcel of the treaty itself. As a last resort, changes may be brought about by qualified majority decisions (Art. 2 [9] [c]) and do not require ratification in order to be valid. Instead they automatically enter into force six months after the decision was taken and are binding upon all parties (Art. 2 [9] [d]). This procedure has been described as revolutionary, since it deviates from the principle expressed in Art. 39 and 40 of the Vienna Convention on the Law of Treaties (VCLT), according to which binding amendments usually require consent by all parties (Beyerlin 2000: 46; Ott 1998: 55ff., 155ff.; Werksman 1996b: 619).

Unfortunately, this very dynamic facilitated amendment procedure does not exist in the climate regime, where the rule of consent still dominates the amendment provisions (Wittneben *et al.* 2006: 99). Apparently, states were not as convinced as in the ozone regime that such a dynamic procedure was really necessary to tackle the regulatory problem, or they simply did not want to embark on such a dynamic process again (Werksman 1996b: 62). According to Art. 15 (3) UNFCCC and Art. 20 KP, the decisions on formal amendments require a three-quarters majority, but the amendments only enter into force for those parties ratifying them (Art. 15 [4] UNFCCC and Art. 20 [4] KP). A similar procedure applies to annexes to the UNFCCC or the Kyoto Protocol (Art. 16 [2] UNFCCC, Art. 21 [4] KP): Majority decisions are possible as a last resort, but the parties may opt out in order not to be bound by them (Art. 16 [3] UNFCCC, Art. 21 [5] KP).

This amendment procedure is too weak and should be modeled more closely according to the ozone regime. A lot will depend on the question to what extent states are willing to agree upon a further transfer of sovereignty rights, including majority decisions binding all parties as a last resort. While it is correct to assume that consensus is a very important resource in a negotiation process, it is equally important to have the option to raise the pressure when necessary. This model proved to be quite successful in the ozone regime (as only as a distant threat on the horizon) and should therefore be promoted in the climate negotiations, as well. However, there are grave doubts about whether states will be willing to submit to such a procedure, since at least some may fear that this will cause a dynamic development similar to the ozone regime, which may be hard to control. While a dynamic institutional design may thus facilitate negotiations in one regime, this very fact can mean that the same device will be very difficult to achieve in another one. Nevertheless, parties should strive for such an innovation in the climate regime as an important instrument of long-term facilitation with regard to existing stumbling-blocks.

4. The framework-protocol approach and future prospects

The framework-protocol approach may set in motion an extended bargaining process leading to the agreement on the lowest common denominator and to weak enforcement procedures (Susskind 1994: 32; Kelly 1997: 481f.; Brown Weiss 1996: 276). While the framework agreement is usually widely accepted, many states may abstain from signing or ratifying the protocol, hence severely reducing the effectiveness of the regime (Gollnisch 1995: 89ff.). The alternative would be to focus on the advancement of the initial convention, instead of creating new instruments which may only blur the problem (Poitras 1999: 109). If state parties to a framework convention simply do not want to set up and use effective development procedures and avoid creating an effective protocol, the framework protocol approach may become a mere face-saving exercise fulfilling an alibi function while lowering the pressure for further rule-making. Moreover, positive developments are not always spurred by the regime itself, but often by external factors such as changing preferences or scientific developments (although these factors may in turn be influenced by the institutional set-up). These are clearly the limits of institutional effectiveness.[86]

However, if there is a minimum of goodwill of the parties to advance the initial regulatory framework, practice shows that the framework-protocol approach may serve as a quite effective tool (Hanschel 2003a: p260ff.; Hanschel 2007: 229ff.). Hence, it may reserve questions that cannot be tackled in the beginning for later rounds of negotiations without blocking the initial process, while driving ahead the further negotiation- and rule-making process (Beyerlin and Marauhn 1997: 154; Susskind and Ozawa 1992: 146). The framework-protocol approach may help to generate scientific consensus, to deal with complex equity problems (e.g. the "North–South" divide), and increase awareness of an existing environmental problem (Tolba 1996: 13; Dolzer 1995: 957ff.; Handl 1993: 61). It can make the rule-making process more dynamic, stable, and flexible, and give it a clearer direction. Finally, it can contribute to the dissemination of ideas and possibly even influence the preferences of some actors in the long run. In order to be more effective, the framework convention should at least contain rather concrete provisions as the aims and time-scope of the further regime development, thus pre-structuring the further rule-making process (Beyerlin and Marauhn 1997: 154).[87] But after all, the question whether the framework-protocol approach should be applied and how it should be designed in detail, is to be answered for each regime individually, depending on the concrete constellation of the problem (Munton *et al.* 1999: 238f.).[88]

The framework-protocol approach as applied in the climate regime was preceded by a similar approach in the ozone regime and in the regime on long-range transboundary air pollution (Hanschel 2003a: 86ff., 120ff., 147ff.). However, the institutional design varies considerably: The regime on long-range transboundary air pollution is composed of a framework convention, which was followed by eight protocols, most of them dealing with individual substances.[89] The ozone regime, by contrast, contains a framework convention to be followed by only one protocol, which was in turn developed further by amendments and adjustments many times.[90] The climate regime contains a stronger convention than the ozone regime, but a weaker protocol: Under the UNFCCC, national measures for the stabilization of greenhouse gases have to be taken. The Kyoto Protocol intends to achieve an overall mitigation of six harmful substances by 5 per cent below 1990 levels in the commitment period between 2008 and 2012. Thoms criticizes the fact that the climate regime integrated the developing countries already within the framework convention, while the ozone regime "played the 'sequential game' " correctly, by addressing the more immediate problem of controlling industrialized nations' emissions before offering major incentives for developing nations to join the treaty" (Thoms 2003: 837). However, in the climate regime, an all-encompassing nation involvement including the developing countries was necessary from the start, because countries such as China and India were developing rapidly, causing severe implications for climate change, while others were strongly affected by adverse effects of climate change.

After the Kyoto Protocol entered into force, the climate regime has entered a critical stage:[91] Its rules only cover the first commitment period (2008–2012), requiring states to start renegotiating seven years before the end of that period. Pursuant to the Bali Roadmap agreed upon at COP13, the AWG-KP and the AWG-LCA are still acting in parallel, while their success is limited. COP15 in Copenhagen clearly did not fulfill the high expectations. The ongoing debate hence displays a number of different post-2012 options for the climate regime, both institutionally and in terms of its material contents.[92] Some states criticize the Kyoto Protocol for not getting the developing countries on board, for damaging the economy and for not being based on sound technology.[93] Gardiner maintains that the global cap on emissions should be tightened and the compliance mechanism strengthened, all countries should be integrated and the costs for adaptation measures should also be taken into the equation (Gardiner 2004: 38f.). There are claims that the Kyoto Protocol is seriously flawed (Gardiner 2004: 39), that it should expire in 2012 and that states should redirect the negotiations. Accordingly, one option would be a further development on the basis of the UNFCCC as a loose framework within which states can take voluntary measures that go beyond what is required.[94] It might also be feasible to base further commitments on COP decisions, although this would leave certain doubts as to their legal status (Morgenstern 2009: 241ff.). Another option would be to maintain the framework-protocol approach, but to create several protocols strengthening the obligations by dealing with individual substances (in the similar way as the regime on long-range transboundary air pollution does), or by dealing with different commitment periods. The most prominent idea at COP15 was to create a Copenhagen Protocol, either in addition to or as a replacement of the Kyoto Protocol (Morgenstern 2009). Alternatively, the approach that was followed under ozone regime could be applied, which means that the Kyoto Protocol should be advanced on the basis of its existing provisions with a view towards including further substances and rules for the second commitment period. This last approach is envisaged by the Kyoto Protocol itself, in particular by Art. 3 (9) and Art. 9 KP.

Out of all these options (and there are several others), the last one, which was purported by the negotiation mandate given at COP/MOP1 in Montreal and further pursued at COP/MOP2 in Nairobi, still appears to be the most promising, if combined with a vigorous promotion and implementation of the UNFCCC.[95] The UNFCCC contains some basic obligations, is universally accepted and has been further developed and refined by subsequent COP decisions for a long period now. According to the rationale of the framework-protocol approach, the Kyoto Protocol was adopted which is "by no means in a safe haven, but … certainly in from the cold" (Wittneben *et al.* 2006: 99). While the conferences in Nairobi, Bali and Copenhagen cannot be described as major steps forward, they kept the Kyoto process alive and still leave enough scope for a timely adjustment of the existing targets.[96] The Kyoto Protocol displays a complex and differentiated compromise, resulting in mitigation targets and a number of important institutional devices which should not be re-invented in new agreements.[97] Moreover, it indicates the right direction, since it is geared towards its own constant revision (Kreuter-Kirchhof 2005: 545; DeSombre 2004: 46). Its first commitment period is operational so that the protocol governs current affairs for almost all UNFCCC parties. Member states are required to renegotiate the mitigation targets for subsequent commitment periods and should amend the KP accordingly. Ideally, the weaknesses of this treaty should be corrected step by step within the existing framework, whereas renegotiating the whole package might be a very tedious exercise. Further protocols might become useful in order to deal with specific aspects of climate change in the future,[98] but negotiators should focus mainly on an enhancement and expansion of mitigation targets for the second commitment period, as projected by the Kyoto Protocol.[99]

Still, in order to further develop the Convention as a framework and to integrate the few countries (in particular the United States) that are still only members to the UNFCCC but not to the Kyoto Protocol, the provisions of the UNFCCC should continue to be constantly further developed and implemented, as well. This is the route that was taken by COP11 President Stéphane Dion when pursuing the "dual track" approach (Wittneben *et al.* 2006: 91ff.).[100] This approach, which was endorsed by the Bali Road Map, allows each party to progress at its own speed, while exerting a soft pressure towards progressing more quickly. The only legal caveat is that there must be no conflicting obligations in the two instruments, which parties cannot fulfill simultaneously. In particular, states that are only members to the UNFCCC must not abuse this instrument in order to bypass or circumvent the rules of the Kyoto Protocol. The overall legal conception of the framework-protocol approach which has been applied to the climate regime has proven to be quite a good remedy against such conflicts in other negotiation processes, as the various treaties within environmental regimes and their institutional structure are neatly intertwined. The further development of the UNFCCC should first be exercised by COP decisions, in spite of their potentially ambiguous status. This would apply to issues such as adaptation, transfer of technology and financial support which have been addressed by the Bali Action Plan. Equally, pledges made under the Copenhagen Accord might be fortified by subsequent COP decisions which would allow including countries that have not made any (substantial) commitments under the Kyoto Protocol. When consensus allows for it, some of the results might be integrated into the future Kyoto Protocol, by adding new members and by strengthening existing obligations for the second commitment period. This would be in line with the earlier finding that regime development should mainly be focused on the protocol(s), not on the convention. Other decisions which are rather related to the overarching regime framework may lead to a formal amendment of the UNFCCC. This would avoid side-stepping what has already been achieved in the regime, both institutionally and in terms of commitments. The transition from the non-binding Copenhagen Accord to COP decisions of ambiguous legal validity leading to formal legal amendments of existing treaties would allow for a step-by-step strengthening of commitments. One may contend that this will not get the United States on board as it still refuses to ratify the Kyoto Protocol. However, a new protocol has not been achieved, either, and will be difficult to achieve in the post-Copenhagen negations. COP15 has frustrated hopes that adding a new protocol would convince the United States as well as China and other economies in transition to accept binding targets.[101] While in the regime on long-range transboundary air pollution, it was wise to devise several protocols dealing with individual substances, it seems unwise to try and negotiate a new protocol for every commitment period in the climate regime. Instead, it should be more rewarding to integrate future results into the existing institutional framework, hence striving for a long-term widening and deepening to the Kyoto Protocol. In order to make Kyoto palatable to the United States, its target in the first commitment period (which never became binding) might be deleted or softened and a new target be determined for the subsequent period, potentially in line with its pledge under the Copenhagen Accord. A US commitment might make it easier to convince China and other economies in transition which are already parties to the Kyoto Protocol to accept binding targets for themselves. Obviously, these are rather tentative strategic suggestions which may turn out to be unfeasible in future practice. But even if the only workable long-term strategy to strike a truly global deal including countries that currently decline the Kyoto Protocol should be an entirely new agreement, such an agreement should still be embedded, as far as possible, in the existing framework-protocol structure.[102]

Nothing of this should frustrate regional or even sub-national efforts, such as mitigation targets and emission trading schemes in the European Union, in and between several of the US States and in other places of the world.[103] The most successful current model on flexible mechanisms is probably the EU Emissions Trading Scheme (Hanschel 2008). Such bottom-up approaches may complement to the international top-down approach followed by the climate regime. Horizontal and vertical linkages are important in order to achieve mutual compatibility and hence to expand and fortify the system.[104] In addition, in the light of the dramatic figures of the fourth IPCC report and the failure of COP15 in Copenhagen, *ad hoc* deals of the main contributors to climate change may become even more relevant, such as in the G20, the Major Economies Forum on Energy and Climate or coalitions of the like-minded. But any deal achieved should be connected to the climate regime, in order to safeguard and expand the existing global institutional framework.[105] Contrary to a current popular view, it does not seem feasible to replace the global mode of problem-solving within the UN context by restricted international circles or regional initiatives. Complementary negotiations outside the framework of the climate regime may reduce complexity and opportunities for blockades, which the Copenhagen conference suffered from immensely. However, certain caveats are in order: First, these forums usually operate on consensus, as well, making sure that no state may be voted down on matters gravely affecting its economy. Furthermore, such processes are sometimes not very inclusive, hence suffering from a potential lack of equitableness. Legitimacy may also be a problem since participation is only allowed on invitation, although this might not pertain to the same extent to a coalition of the like-minded. A lack of equity and legitimacy may negatively affect credibility of any deal achieved and hence increase implementation problems. Furthermore, carbon intensive industries might be transferred hence creating a leakage or free-rider problem. Finally, sustainability may suffer from gradually rising emissions in developing countries which may be outside the deal. Nevertheless, such informal forums can help facilitating the climate negotiations if efforts are firmly integrated within or at least related to the UN approach.[106]

Conclusions

The institutional design of the climate regime is strongly inspired by the ozone regime and, to some extent, by the regime on long-range transboundary air pollution. The negotiation-oriented forms of institutionalization have shown their potential to influence the bargaining process, fulfilling functions of dynamics, stabilization, guidance, fairness and acceptance, as well as a model function. They serve to enhance the knowledge base, create an institutional framework, pre-structure the further development and may, in the long run, even be able to influence the preferences of the actors. Hence, they may also help to remove the stumbling blocks of low consensus and high complexity of negotiations to a certain extent. The striking feature of these forms of institutionalization is that they do not require a substantial consensus by the parties as to the solution of the regulatory problem. Their cooperative working mode focusing on implementation, improvement and innovation promotes a constant process of revision which allows them to "learn" and to turn newly gained bargaining results into useful regime amendments. By spurring post-agreement negotiations, they may contribute considerably to the enforcement and the further development of the substantive obligations and thus to the solution of the regulatory problem.[107]

While these effects can be clearly demonstrated in the mentioned predecessor regimes, it is still too early to more than tentatively assert them in the climate regime. The adoption and entry into force of the Kyoto Protocol is (in spite of its weaknesses) evidence of progress which can partially be related back to the institutional set-up requiring parties to continue negotiations within the parameters of the dynamic UNFCCC framework. However, post-Kyoto negotiations have not led to a strengthening of the Protocol's commitments or to a sufficient expansion of its membership so far, nor have they allowed for the adoption of a new protocol. COP15 shows that this has been mainly due to clear differences of preferences and, to some extent, due to the complexity of negotiations. In the aftermath of COP15, firm action is more urgent than ever while time is beginning to run out. The mentioned recent failures may make the institutional design appear dysfunctional. But, in the light of sovereignty concerns, the inherent characteristics of the problem as well as issues of equity and legitimacy, the current approach is indispensable. In spite of recent doubts over whether the UN is still capable to solve the problems of our time, Copenhagen has not shown the end of global negotiations within the UNFCCC context. It may be useful to add further negotiations within smaller forums, but in order to achieve globally valid and hence effective, sustainable, equitable and legitimate results, the current climate regime remains without an alternative.

No institution can completely remedy lack of consensus or high complexity if a complex problem requires a global solution that does not come at low short-term cost. But carefully-designed institutions can facilitate the quest for such a solution, and the framework protocol approach and its negotiation-oriented procedures provide a useful tool to this end which have proven to be effective in other regimes before. Even though climate change has turned out to be a particularly difficult problem to address, lessons on effective regime management can be learned and applied to it. As a consequence, several institutional stumbling blocks can be identified which still hamper the effectiveness of the regime. The most important ones are the lack of truly effective regime development mechanisms in the Kyoto Protocol, problems in the relationship between the COP and the subsidiary bodies including the contact groups, the rules of procedure on how to negotiate within some of the institutions and the lack of equal representation in the contact groups. Other aspects of the institutional design are promising, but still require further time for judgment, such as the CC and the flexibility mechanisms under the Kyoto Protocol. In still other cases, only fine-tuning measures appear necessary: Thus, the procedures for cooperation and renegotiation already operate as important facilitators in the regime, allowing for the necessary forum-shopping and producing important scientific information. Finally, there are ambiguous cases: The role of the scientific bodies, in particular the IPCC, has been crucial to the regime, although non-governmental, scientific experts need to be further integrated into the decision-making bodies. Measures of compliance assistance and capacity building beyond the scope of compliance assistance have developed constantly, and help to promote a global solution to the problem without delaying the North–South conflict involved. However, most of these measures are still not sufficient and need to be developed further.

Chances to remove these stumbling blocks by means of institutional facilitation vary greatly. Generally speaking, the institutional set-up in the climate regime is geared towards negotiation instead of confrontation, so that agreeing upon it (as opposed to agreeing on substantial mitigation or adaptation measures) only requires a rather low substantial consensus. However, the institutional tasks that lie ahead will be much more difficult to tackle. Parties know (and have increasingly learnt through other examples) that changing the institutions may have an important effect on the substantive character of the regime, as well. A stronger and more substantial consensus will probably be needed for the institutional design of the second commitment period. Following an earlier statement of President of COP6, Mr. Pronk, that "the Kyoto Protocol is the only game in town", parties should mainly build on the institutional set-up of the Kyoto Protocol and develop it along the lines of the Montreal Protocol, including a dynamic provision allowing for majority decisions binding all parties as a last resort. They should rely on the continuation of the framework-protocol approach including a stepwise tightening of the existing regulatory framework (if necessary also by concluding additional protocols in the long run). The interaction between the institutions and their respective rules of procedure might be easier to improve, in particular because such organizational matters can partially be influenced by the secretariat and do not always fall within the scope of political bargaining. The flexibility mechanisms are already based on a rather broad consensus, while conflicts on how to make them more effective persist. Finally, the compliance procedure might be interpreted in a dynamic way and developed further by the CC, if this institution manages to achieve a sufficient degree of autonomy.

Although institutions in the climate regime matter, this analysis has shown that the bargaining situation – composed of factors such as the interests and preferences, the problem-structure and the negotiation skills of the actors – is crucial for success. COP15 in Copenhagen has illustrated this very clearly, and it shows that developing countries, in particular emerging economies but also low and least developed countries, have become more self-confident in representing their interests. In principle, this is a positive development, because it ensures equal representation. However, if parties (be it the developing or the developed countries) are simply not willing to make progress in the climate negotiations, no institution set up by them will be able to force them into it. Thus, the effects of the institutional design must not be overestimated. There is no automatism that the right institutions will always help to spur the process and to guide the actors in the best direction. In spite of this caveat, the institutional set-up of the climate regime is already able to provide a mostly positive impact on the negotiations. If refined and changed according to the suggestions made in this chapter, it could facilitate negotiations even further and remove at least some of the existing stumbling blocks. However, some of these changes may take a long time and – sadly enough – could even require the occurrence of further extreme weather events, rising global temperatures and other immediate effects of man-made climate change, which is why adaptation measures may become more and more important, as well.[108] But in the light of recent IPCC findings, public pressure and hopes on the horizon for a stronger engagement of the United States and some other (still hesitant) industrialized countries as well as economies in transition, it appears wise to use the already established institutional structure as a foundation for future negotiations on more effective mitigation targets in order to at least avoid the worst consequences of climate change.

Notes

1 This chapter was essentially completed in 2007. In line with the research format of this book, updates were made as far as considered indispensable to reflect the process leading up to the Copenhagen Conference in 2009, while more recent developments are merely hinted at in the footnotes. During the extended review process regarding this book, the author has published segments of this chapter in Hanschel 2012, which furthermore includes an analysis of the post-Copenhagen process.

2 On this and the subsequent assessment of the framework-protocol approach, see Hanschel (2012); on the importance of post-agreement negotiations, see Spector and Zartman (2003: 271ff.).

3 For a comparative analysis of these regimes, see also Lehmann (2005:73ff.); Wettestad (2005: 219ff.).

4 See in particular the debate between the competing theories of "rational choice" and "reflexive institutionalism", Hanschel (2003: 43ff.).

5 For a comparative analysis of these two regimes, see Thoms (2003: 795ff.).

6 Generally on the Kyoto Protocol, see Oberthür and Ott (1999); for the situation after COP10 in Buenos Aires in 2004, see Ott *et al.* (2005: 84 ff.); for a recent account, see Hanschel 2012. After the latest conference in Durban, Canada has unfortunately decided to leave the KP, see http://www.bbc.co.uk/news/world-us-canada-16151310 (last accessed 28 September 2012). Even worse, all three countries, Japan, Russia, and Canada, have stated that they will not accept new commitments in the KP context, see http://www.guardian.co.uk/environment/2011/dec/16/russia-canada-kyoto-protocol (last accessed 28 September 2012).

7 See http://unfccc.int/meetings/cop_11/items/3394.php (last accessed 28 September 2012); see also Wittneben *et al.* (2006: 90ff.).

8 See COP/MOP Decision 1/CMP.1 (Consideration of commitments for subsequent periods for Parties included in Annex I to the Convention under Article 3, paragraph 9, of the Kyoto Protocol); for the earlier draft decisions proposed by Japan, the EU and G77 plus China and by the President of COP11, see FCCC/KP/CMP/2005/CRP.1, 2, 3 and L.8; for the rather modest results of the AWG session at COP/MOP2, see Sterk *et al.* (2007: 3ff.), see also FCCC/KP/AWG/2006/L.4 para 10.

9 See http://unfccc.int/meetings/cop_11/items/3394.php; see also the Buenos Aires Programme of Work on Adaptation and Response Measures, adopted at COP10, Decision 1/CP.10, Wittneben *et al.* (2006:96 f.); see furthermore Brouns and Langrock (2006: 12); on more recent developments concerning the AF, see Sterk *et al.* (2007: 9ff.).

10 See Decision 1/CP.13 ("Bali Action Plan"), FCCC/CP/2007/6/Add.1.

11 For an account of the scientific assessment of climate change, see Browne (2004: 20ff.). and McKinstry (2004: 3ff.).

12 See furthermore ENB Report COP/MOP 2 Final, Vol. 12, No. 318.

13 See http://unfccc.int/files/meetings/cop_15/application/pdf/cop15_cph_auv.pdf.

For the list of government pledges, see http://unfccc.int/resource/docs/2011/sb/eng/inf01r01.pdf (last accessed 26 September 2012); on the latest Conferences in Cancun (2010) and Durban (2011) which produced the Cancun Agreements, the Durban platform (with the prospect of a new agreement) as well as a commitment to create a second commitment period in the KP in Quatar (2012), see http://cancun.unfccc.int (last accessed 26 September 2012), as well as http://www.guardian.co.uk/environment/2011/dec/12/durban-climate-change-conference-2011-southafrica (last accessed 26 September 2012).

14 See Climate Change 2007: The Physical Science Basis, Summary for Policymakers, Contribution of Working Group I to the Fourth Assessment Report of the Intergovernmental Panel on Climate Change, p. 5, at: http://www.ipcc.ch/publications_and_data/ar4/syr/en/contents.html (last accessed 31 September 2012).

15 For details of these negotiations, see Chasek (2003:187ff.); see also Breitmeier (1996); generally on international negotiation processes regarding environmental issues, see Sjoestedt (1993).

16 See Hanschel (2012). For an assessment of the costs, see the report by Nicholas Stern, Stern Review: The Economics of Climate Change – Summary of Conclusions, at: http://www.hm-treasury.gov.uk/d/CLOSED_SHORT_executive_summary.pdf (last accessed 31 September 2012): "If a wider range of risks and impacts is taken into account, the estimates of damage could rise to 20 % of GDP or more".

17 See The White House, Office of the Press Secretary, 11 June 2001: President Bush Discusses Global Climate Change, http://georgewbush-whitehouse.archives.gov/news/releases/2001/06/20010611-2.html (last accessed 27 September 2012).

18 As to the resulting isolation of the US delegation at COP/MOP1 in Montreal, see Wittneben *et al.* (2006: 100).

19 See http://unfccc.int/files/meetings/cop_15/copenhagen_accord/application/pdf/unitedstatescphaccord_app.1.pdf (last accessed 31 September 2012).

20 See http://unfccc.int/files/meetings/cop_15/copenhagen_accord/application/pdf/chinacphaccord_app2.pdf (last accessed 31 September 2012).

21 See http://unfccc.int/resource/docs/2009/cop15/eng/l07.pdf.

22 In the Cancun Agreements, the parties list and take note of the pledges made since Copenhagen; on the pledges by Annex-I-countries, see FCCC/CP/2010/7/Add.1, FCCC/SB/2011/INF.1/Rev.1.

23 See   http://www.ipcc.ch/publications_and_data/ar4/syr/en/contents.html   (last accessed 31 September 2012).

24 See http://unfccc.int/files/meetings/cop_15/copenhagen_accord/application/pdf/europeanunioncphaccord_app1.pdf (last accessed 31 September 2012).

25 On the scientific evidence, see Pierrehumbert (2006); Barrett, 2006; see furthermore the latest IPCC report Climate Change 2007: The Physical Science Basis, Summary for Policymakers, Contribution of Working·Group I to the Fourth Assessment Report of the Intergovernmental Panel on Climate Change, p. 5: "... leading to a very high confidence that the globally averages net effect of human activities since 1750 has been one of warming ...", at: http://www.ipcc.ch/publications_and_data/ar4/syr/en/contents.html (last accessed 31 September 2012).

26 See the report by Nicholas Stern, Stern Review: The Economics of Climate Change – Summary of Conclusions, at: http://www.hm-treasury.gov.uk/d/Summary_of_Conclusions.pdf (last accessed 31 September 2012): "... the Review estimates that if we don´t act, the overall costs and risks of climate change will be equivalent to losing at least 5 % of global GDP each year, now and forever. If a wider range of risks and impacts is taken into account, the estimates of damage could rise to 20 % of GDP or more."

27 For a detailed account of the changing US position at that time, see Thackeray (2004: 871ff.).

28 Art. 3 (9) merely states that the "commitments for subsequent periods for Parties

included in Annex I shall be established in amendments to Annex B to this Protocol", and that the COP/MOP "shall initiate the consideration of such commitments at least seven years before the end of the first commitment period ...". Art. 9 requires periodical reviews to the Protocol; on the debate concerning these Articles, see Wittneben *et al.* (2006: 91f.); for the results of negotiations on these issues, see Sterk *et al.* (2007: 3ff.).

29 For detailed information on the sessions of contact groups in the climate regime, see the reporting by the Earth Negotiations Bulletin (ENB), at: www.iisd.ca/process/climate_atm.htm (last accessed 31 September 2012).

30 See Decision 27/CMP.1 (Procedures and mechanisms relating to compliance under the Kyoto Protocol).

31 See, however, the more recent account of Mackenzie (2011: 395ff.).

32 Wettestad (1999: 229) classifies this phenomenon as "institutional density".

33 See the homepage of the Climate Secretariat at http://unfccc.int/secretariat/items/1629.php.

34 For the details concerning the CC, see below IV 2 a).

35 It is important to note that the IPCC was not established by the climate regime, but by the World Meteorological Organization (WMO) and the United Nations Environment Programme (UNEP) in 1988 (see http://www.ipcc.ch/organization/organization.htm [last accessed 31 September 2012]), but it is closely linked with the climate regime through various references in the treaties. For recent additions to the institutional setup by the Cancun Agreements, e.g. with regard to the Green Climate Fund, see http://cancun.unfccc.int (last accessed 27 September 2012).

36 For an outline of future NGO strategies in the climate regime see Morgan (2006:1ff.).

37 See Principles 2 and 3 of the Principles Governing IPCC Work, Approved at the Fourteenth Session (Vienna 13 October 1998) on 1 October 1998 and Amended at the 21st Session (Vienna 3 and 6–7 November 2003); for reactions to the IPCC report see http://news.bbc.co.uk/2/hi/science/nature/6323653.stm (last accessed 31 September 2012).

38 See, e.g. http://www.nature.com/n ews/2010/100202/full/463596a.html (last accessed 27 September 2012); on the IPCC reactions see http://www.ipcc.ch/organization/organization_review.shtml#.UDsZYqPMCM8 (last accessed 27 September 2012).

39 See the US Climate Change Science Program at http://www.globalchange.gov (last accessed 31 September 2012).

40 For an earlier reaction of the White House to the latest climate report see http://news.bbc.co.uk/2/hi/science/nature/6323653.stm. For the current US position, see http://unfccc.int/files/meetings/cop_15/copenhagen_accord/application/pdf/unitedstatescphaccord_app.1.pdf (last accessed 31 September 2012).

41 See http://unfccc.int/resource/docs/2009/cop15/eng/l07.pdf (last accessed 31 September 2012).

42 See Principle 2 of the Principles Governing IPCC Work, http://www.ipcc.ch/pdf/ipcc-principles/ipcc-principles.pdf (last accessed 27 September 2012).

43 See Appendix A to the Principles Governing IPCC Work, Procedures for the Preparation, Review, Acceptance, Adoption, Approval and Publication of IPCC Reports, Adopted at the Fifteenth Session (San José, 15–18 April 1999), amended at the Twentieth Session (Paris, 19–21 February 2003) and Twenty-first Session (Vienna, 3 and 6–7 November 2003).

44 For an account of the influence of governments on the IPCC report see spiegel-online, at: http://www.spiegel.de/wissenschaft/natur/0,1518,464048,00.html (last accessed 31 September 2012); for the scientific coverage of climate change within the mandates of the United Nations Environment Programme (UNEP) and the Commission for Sustainable Development (CSD), see http://www.unep.org/climatechange (last accessed 31 September 2012) and http://www.un.org/esa/sustdev/sdissues/climate_change/climate_change.htm (last accessed 31 September 2012); on the role of epistemic communities see generally Haas (1992).

45 Rule 42 contains two alternatives (A and B) concerning majority decisions and possible exceptions to them, which parties never managed to agree upon, see http://unfccc.int/resource/docs/cop2/02.pdf (last accessed 31 September 2012); the specific options for majority decisions concerning treaty amendments will be dealt with below under VI 3 c).

46 See, for example, the report on decision-making with regard to the Cancun Agreements, http://ictsd.org/i/news/bridges/98834 (last accessed 27 September 2012).

47 For an analysis of the country coalitions in the climate negotiations, see Gillespie (2005: 258ff.).

48 See, for example, the suggestions formulated by Agenda 21, A/CONF.151/26/Rev. 1 (Vol. I), p. 9ff.

49 See, for example, FCCC/CP/1995/7/Add.1, Decision 3/CP.1.

50 Decision 22/CMP.1, FCCC/KP/CMP/2005/8/Add.3.

51 Paras. 4, 20ff. of Part I, Annex to Decision 22/CMP.1, FCCC/KP/CMP/2005/ 8/Add.3.

52 See, however, the rather critical remarks by Bafundo (2006: 461ff.); on the role of the implementation committee in the regime on long-range transboundary air pollution, see Hanschel (2003a: 76ff.).

53 See Decision 1/CMP.1 (Consideration of commitments for subsequent periods for Parties included in Annex I to the Convention under Article 3, paragraph 9, of the Kyoto Protocol); for the current members see http://unfccc.int/kyoto_protocol/sta-tus_of_ratification/items/2613.php (last accessed 31 September 2012).

54 FCCC/KP/CMP/2006/L.2, see furthermore ENB Report COP/MOP 2 Final, Vol. 12, No. 318, p. 13; United Nations Press Release: "Spirit of Nairobi prevails as United Nations Climate Change Conference successfully concludes with decisions to support developing countries"; on the current rules of procedure as amended by COP/MOP4, see http://unfccc.int/files/kyoto_protocol/compliance/background/application/pdf/ rules_of_procedure_of_the_compliance_committee_of_the_kp.pdf (last accessed 31 September 2012).

55 For a good overview on the various options for implementation of the compliance mechanism, see MacFaul (2005: 4, 6ff.), who, however, considers a mere COP/MOP decision to be the best option. See furthermore Wittneben *et al.* (2006: 94f.), who suggest that various routes may be taken here. The COP/MOP decided "to commence consideration of the issue of an amendment to the Kyoto Protocol in respect of procedures and mechanisms relating to compliance in terms of Article 18, with a view to making a decision by the third session of the Conference of the Parties serving as the meeting of the Parties to the Kyoto Protocol".

56 See below IV 2 b.

57 Decision 2/CP.1, FCCC/CP/1995/7/Add.1.

58 For an account of the first cases dealt with by the facilitative branch of the CC, see CC-2006-11-1/FB, FCCC/KP/CMP/2006/6.

59 See http://unfccc.int/resource/docs/2009/cop15/eng/l07.pdf (last accessed 31 September 2012), in particular No. 4 and 12.

60 In order to determine adequacy the COP has developed a detailed procedure relying on criterion of "environmental reasonableness" and on the principle of "common, but differentiated responsibilities", see Wolfrum (1999: 128); Cameron (2000: 10).

61 See FCCC/CP/2001/13/Add.1.

62 Para. 4, Section 4, Decision 27/CMP.1, FCCC/KP/CMP/2005/8/Add.3.

63 See http://www.guardian.co.uk/news/datablog/2011/jan/31/world-carbon-diox-ide-emissions-country-data-co2 (last accessed 28 September 2012).

64 For a discussion of equity principles, see Buchner and Lehmann (2005: 45ff.).

65 See the Revised 1996 Guidelines for National Greenhouse Gas Inventories: Reporting Instructions, at: http://www.ipcc-nggip.iges.or.jp/public/gl/invs4.htm (last accessed 31 September 2012).

66 See Copenhagen Accord, at http://unfccc.int/resource/docs/2009/cop15/eng/l07. pdf.

67 For the latest decisions with regard to the Green Climate Fund at the Cancun and Durban conferences, see http://unfccc.int/cooperation_and_support/financial_ mechanism/green_climate_fund/items/5869.php (last accessed 27 September 2012).

68 Decision 5/CP.12, Development and transfer of technologies, FCCC/CP/2006/5/ Add. 1, 26 January 2007 (Advance Version).

69 FCCC/CP/2007/6/Add.1, at: http://unfccc.int/resource/docs/2007/cop13/eng/ 06a01.pdf#page=15.

70 FCCC/CP/2009/L.7, at: http://unfccc.int/resource/docs/2009/cop15/eng/l07.pdf.

71 On the latest decisions with regard to the Technology Mechanism, see http://unfccc. int/cooperation_and_support/technology/items/1126.php (last accessed 27 September 2012).

72 On CDM in relation to developing countries, see Pohlmann (2004).

73 In addition, removal units (RMUS) can be traded in the carbon market, which are gained by land use, land-use change and forestry activities (LULUCF) such as reforestation; for an analysis of emission trading within the European Union, see Pflüglmayer (2004: 102ff.); Hanschel (2008).

74 For open questions concerning the flexibility mechanisms after COP/MOP2, see Sterk *et al.* (2007: 11ff.).

75 See http://unfccc.int/meetings/cop_11/items/3394.php (last accessed 31 September 2012).

76 Decision 2/CMP.2, Implementation of Article 6 of the Kyoto Protocol, FCCC/KP/ CMP7/2006/10/Add.1, 26 January 2007.

77 For an analysis of the EU-ETS, see Anderson and Skinner (2005: 92ff.); Spieth and Hamer (2005:112ff.); Hanschel (2008); on linkages between emissions trading systems, see the publication series of ECOLOGIC at http://ecologic.eu/3080 (last accessed 31 September 2012).

78 Para. 1, Section VI, Annex to Decision 27/CMP.1, FCCC/KP/CMP/2005/8/ Add.3.

79 Decision 1/CMP.1, Consideration of commitments for subsequent periods for Parties included in Annex I to the Convention under Article 3, paragraph 9, of the Kyoto Protocol, FCCC/KP/CMP/2005/8/Add.1.

80 Decision 7/CMP.2, Review of the Kyoto Protocol pursuant to its Article 9, FCCC/ KP/CMP7/2006/10/Add.1, 26 January 2007 (Advance Version); see furthermore Sterk *et al.* (2007: 5f.).

81 As in the COP/MOP1 decision, the AWG merely decided to schedule further sessions "with a view to completing the Work ... as early as possible and in time to ensure that there is no gap between the first and the second commitment periods under the Kyoto Protocol", FCCC/KP/AWG/2006/L.4, paragraph 10.

82 In this regard, the latest conferences in Cancun and Durban provide a ray of hope by setting up a dual process regarding (a) a second commitment period for the KP until 2017 or 2020, and (b) a new universal climate agreement adopted no later than 2015 and intended to enter into force in 2020, see Decision 1/CMP.7, FCCC/KP/ CMP/2011/10/Add.1, as well as Decision 1/CP.17 and Decision 2/CP.17, FCCC/ CP/2011//9/Add.1.. Whether this will be achieved, remains to be seen.

83 Draft Decision 1/CP.11, Dialogue on long-term co-operative action to address climate change by enhancing implementation of the Convention, FCCC/CP/2005/ 5/Add.1.

84 See Decision 1/CP.13, FCCC/CP/2007/6/Add.1.

85 See the Aminoil Litigation, Kuwait v. American Independent Oil Co. (Aminoil), 21 I.L.M. 976 (1982), at p. 1014; see furthermore Ipsen (1999: 98).

86 See Hanschel (2012); see furthermore Hanschel (2007).

87 For the role of ideas in post-agreement negotiations, see Sjoestedt (2003: 89ff.).

88 See Hanschel (2012).

89 See the homepage of the regime at http://www.unece.org/env/lrtap/ (last accessed 31 September 2012).
90 For an in depth analysis of the post-agreement negotiations under the ozone regime, see Chasek, 2003, p. 187ff.
91 See Hanschel (2012).
92 For a current assessment of these options, see Hanschel (2012); see furthermore Bodansky (2004); Aldy *et al.* (2003); Baumert *et al.* (2002); Brouns and Langrock (2006); Ott *et al.* (2005: 85f.). For a recent assessment, see Morgenstern (2009).
93 In a formal letter to Congress President Bush, 2001, stated: "As you know, I oppose the Kyoto Protocol because it exempts 80 per cent of the world, including major population centers such as China and India, from compliance, and would cause serious harm to the US economy. The Senate's vote, 95–0, shows that there is a clear consensus that the Kyoto Protocol is an unfair and ineffective means of addressing global climate change concerns.", see http://climateprediction.net/schools/docs/Bush_Kyoto_letter.txt (last accessed 31 September 2012).
94 For the Russian approach on voluntary commitments see Conclusions on the report of the President on consultations concerning the proposal of the Russian Federation. Proposal by the President, FCCC/KP/CMP/2006/L.6, 17 November 2006.
95 See Hanschel (2012).
96 Sterk *et al.* (2007: 3ff., 14ff.) assert that negotiations should be finished by 2009 in order to allow for a timely amendment of the Kyoto Protocol, including the necessary ratifications, before the end of the first commitment period.
97 See the analysis of Kreuter-Kirchhof (2005: 544f.), stressing inter alia the precautionary principle, the principle of sustainable development, the careful removal of state sovereignty, the principle of common responsibility and new forms of cooperation following a market-oriented approach as crucial elements of the Kyoto Protocol.
98 See for example the proposal by Papua New Guinea on an optional protocol concerning deforestation, Wittneben *et al.* (2006: 93); for the more encompassing new climate agreement envisaged by the Durban Platform for Enhances Action, see Decision 1/CP.17 and Decision 2/CP.17, FCCC/CP/2011//9/Add.1.
99 See Hanschel (2012).
100 See Hanschel (2012); on the COP/MOP 2 negotiations concerning both tracks, see Sterk *et al.* (2007: 3ff., 6ff.).
101 The Durban Platform might justify a revised assessment in the future, depending on its further progress and implementation, see Decision 1/CP.17 and Decision 2/CP.17, FCCC/CP/2011//9/Add.1.
102 To what extent this will happen, depends on how the Durban Platform will interpret its rather open mandate: "to launch a process to develop a protocol, another legal instrument or an agreed outcome with legal force under the Convention applicable to all Parties …", Decision 1/CP.17, para. 2, FCCC/CP/2011//9/Add.1.
103 Generally, see Hanschel (2012); on the EU-ETS, see Hanschel (2008); on emissions trading within the US see, for example, Executive Order S – 3 – 05 by the Governor of the State of California, establishing the following targets: "by 2010, reduce GHG emissions to 2000 levels; by 2020, reduce GHG emissions to 1990 levels; by 2050, reduce GHG emissions to 80% below 1990 levels"; see also Brouns and Langrock (2006: 6); furthermore see the Regional Greenhouse Gas Initiative (RGGI) carried out by several US States and the Renewables 2004 Conference in Germany, Brouns and Langrock (2006: 9f.).
104 On linkages see the publication series of ECOLOGIC at http://ecologic.eu/3080.
105 See Hanschel (2012) with further references.
106 See Hanschel, 2012.
107 On these findings see also Hanschel (2003a: 283ff.).
108 See Financial Times Deutschland, at http://www.ftd.de/politik/international/: klimaschaeden-sind-nicht-mehr-aufzuhalten/167889.html (last accessed 31 September 2012).

References

Aldy, J., Ashton, J., Baron, R., Bodansky, D., Charnovitz, S., Diringer, E., Heller, T.S., Pershing, J., Shukla, P.R., Tubiana, L., Tudela, F. and Wang, X. (2003) *Beyond Kyoto: Advancing the International Effort against Climate Change*. Arlington: Pew Center on Global Climate Change.

Aman, A. (1993) *Administrative Law*. St. Paul, Minnesota: West Publishing.

Aman, A. C. (1995) A Global Perspective on Current Regulatory Reforms: Rejection, Relocation, or Reinvention? *Indiana Journal of Global Legal Studies*. 2 (2). 429–464.

Andresen, S. and Gulbrandsen, L. (2005) The Role of Green NGOs in Promoting Climate Compliance, In Stokke, O., Hovi, J. and Ulfstein, G. (eds.) *Implementing the Climate Regime – International Compliance*. London: Earthscan. 169–186.

Ayres, I. and Gertner, R. (1992) Strategic Contractual Inefficiency and the Optimal Choice of Legal Rules. *The Yale Law Journal*. 101 (6). 729–773.

Bafundo, N. (2006) Compliance with the Ozone Treaty: Weak States and the Principle of Common But Differentiated Responsibility. *American University International Law Review*. 21 (3). 461–495.

Barrett, S. (2006) The Problem of Averting Global Catastrophe. *Chicago Journal of International Law*. 6 (2). 1–26.

Baumert, K., Blanchard, O., Llosa, S. and Perkaus, J. (eds.) (2002) *Building on the Kyoto Protocol – Options for Protecting the Climate*. Washington, DC: World Resources Institute.

Benedick, R. E. (1993) Perspectives of a Negotiation Practitioner, In Sjoestedt, G. (ed.). *International Environmental Negotiation*. London: Newbury Park: Sage. 219–243.

Bernauer, T. (1995) The Effect of International Environmental Institutions: How We Might Learn More. *International Organization*. 49 (2). 351–377.

Beyerlin, U. (1976) Pactum de contrahendo und pactum de negotiando im Völkerrecht. *Zeitschrift für Ausländisches Öffentliches Recht und Völkerrecht*. 36. 407–441.

Beyerlin, U. (2000) *Umweltvölkerrecht*. München: Beck.

Beyerlin, U. and Marauhn, T. (1997) *Rechtsetzung und Rechtsdurchsetzung im Umweltvölkerrecht nach der Rio-Konferenz 1992*, Forschungsbericht 101 06 072, Berlin: Bundesumweltamt.

Birnie, P. W. and Boyle, A. E. (1992) *International Law and the Environment*. Oxford: Oxford University Press.

Bodansky, D. (1992) Managing Climate Change, In *Yearbook of International Environmental Law*. 3. 60.

Bodansky, D. (1993) The United Nations Framework Convention on Climate Change: A Commentary. *The Yale Journal of International Law*. 18 (2). 451–558.

Bothe, M. (1996) The Evaluation of Enforcement Mechanisms in International Environmental Law – An Overview, In Wolfrum, R. (ed.) *Enforcing Environmental Standards: Economic Mechanisms as a Viable Means?* Heidelberg: Springer. 13–38.

Bothe, M. and Rehbinder, E. (2005) *Climate Change Policy*. Utrecht: Eleven International Publishing.

Brown Weiss, E. (1993) International Environmental Law: Contemporary Issues and the Emergence of a New World Order. *Georgetown Law Journal*. 81 (3). 675–710.

Brown Weiss, E. (1996) New Directions in International Environmental Law, In United Nations, *International Law as a Language for International Relations*. The Hague, London and Boston: Kluwer Law International. 271–283.

Brown Weiss, E. (1997) Strengthening National Compliance with International Environmental Agreements. *Environmental Policy and Law*. 27 (4).297–306.

Brown Weiss, E. (1999) Understanding Compliance with International Environmental

Agreements: The Baker's Dozen Myths. *University of Richmond Law Review*. 32 (5). 1555–1585.

Browne, J. (2004) Beyond Kyoto. *Foreign Affairs*. 83 (4). 20–32.

Brouns, B. and Langrock, T. (2006) Kyoto Plus: Start in eine neue Phase internationaler Klimapolitik. *Working Paper for the Conference "KyotoPlus – Escaping the Climate Trap* on 28/29 September 2006 in Berlin.

Bryce, J. (1999) Controlling the Temperature: An Analysis of the Kyoto Protocol. *Saskatchewan Law Review*. 62 (2). 379–414.

Cameron, P. (2000) From Principles to Practice: the Kyoto Protocol, 18:1. *Journal of Energy & Natural Resources Law*. 18 (1). 1–18.

Chasek, P. (2003) The Ozone Depletion Regime, In Spector, B. I. and Zartman, I. W. (eds.) *Getting it Done – Postagreement Negotiation and International Regimes*. Washington, DC: United States Institute of Peace. 187–227.

Chayes, A. and Handler Chayes A. (1993) On Compliance. *International Organization*. 47 (2). 175–205.

Chayes, A. and Handler Chayes, A. (1995) *The New Sovereignty: Compliance with International Regulatory Agreements*. Cambridge, MA: Harvard University Press.

Churchill, R. R. (1991) Controlling Emissions of Greenhouse Gases, In Churchill, R. R. and Freestone, D. (eds.) *International Law and Global Climate Change*. London: Graham & Trotman. 147–163.

Cooter, R. and Ulen, T. (1997) *Law and Economics*, 2nd edn. Reading, MA: Addison-Wesley.

Cullet, P. (1999) Equity and Flexibility Mechanisms in the Climate Change Regime: Conceptual and Practical Issues. *Review of European Community and International Environmental Law*. 8 (2). 168–179.

Davies, P. G. (1998) Global Warming and the Kyoto Protocol. *International and Comparative Law Quarterly*. 47 (2). 446–461.

DeSombre, E. R. (2004) Global Warming: More Common Than Tragic. *Ethics & International Affairs*. 18 (1). 41–46.

Dolzer, R. (1995) Die internationale Konvention zum Schutz des Klimas und das allgemeine Völkerrecht, In Beyerlin, U. (ed.) *Recht zwischen Umbruch und Bewahrung: Völkerrecht, Europarecht, Staatsrecht; Festschrift für Rudolf Bernhardt*. Berlin: Springer. 957–976.

Dudeck, D. and Palmisano, J. (1988) Emission Trading: Why is this Thoroughbred Hobbled? *Columbia Journal of Environmental Law*. 13 (2). 217–256.

Ehrmann, M. (2000) *Erfüllungskontrolle im Umweltvölkerrecht – Verfahren der Erfüllungskontrolle in der umweltvölkerrechtlichen Praxis*, Baden-Baden: Nomos-Verlag-Gesellschaft.

Evans, B. (2004) Principles of Kyoto and Emissions Trading Systems: A Primer for Energy Lawyers. *Alberta Law Review*. 42 (1). 167–178.

Freestone, D. (2005) Introduction: The UN Framework Convention on Climate Change, the Kyoto Protocol and the Kyoto Mechanisms, In Freestone, D. and Streck, C. (eds.) *Legal Aspects of Implementing the Kyoto Protocol Mechanisms: Making Kyoto Work*. Oxford: Oxford University Press. 3–24.

French, D. (1998) 1997 Kyoto Protocol to the 1992 UN Framework Convention on Climate Change. *Journal of Environmental Law*. 10. 227–239.

Frischmann, B. (2001) Using the Multi-Layered Nature of International Emissions Trading and of International-Domestic Legal Systems to Escape a Multi-State Compliance Dilemma. *The Georgetown International Law Review*. 13 (463). 475–480.

Gardiner, S. M. (2004) The Global Warming Tragedy and the Dangerous Illusion of the Kyoto Protocol. *Ethics & International Affairs*. 18 (1). 23–39.

Gillespie, A. (2005) *Climate Change, Ozone Depletion and Air Pollution. Legal Commentaries with Policy and Science Considerations.* Leiden/Boston: Martinus Nijhoff Publishers.

Giorgetti, C. (1998) The Role of Nongovernmental Organizations in the Climate Change Negotiations. *Colorado Journal of International Environmental Law and Policy.* 9 (1).115–132.

Gollnisch, L. P. (1995) *Entwicklungstendenzen im internationalen Umweltrecht.* Aachen: Shaker.

Gündling, L. (1996) Compliance Assistance in International Environmental Law: Capacity-Building Through Financial and Technology Transfer. *Zeitschrift für ausländisches öffentliches Recht und Völkerrecht.* 56. 796–809.

Haas, P. M., Keohane, R. O. and Levy, M. A. (1993) *Institutions for the Earth. Sources of Effective International Environmental Protection.* Cambridge, MA/London: MIT Press.

Handl, G. (1993) Environmental Security and Global Change: the Challenge to International Law, In Lang, W. (ed.) *Environmental Protection and International Law.* London: Graham & Trotman, London. 60–83.

Hanschel, D. (2003a) *Verhandlungslösungen im Umweltvölkerrecht – Eine Untersuchung verhandlungsorientierter Institutionalisierungsformen anhand der Regime über weiträumige grenzüberschreitende Luftverschmutzung, zum Schutz der Ozonschicht und des Klimas.* Stuttgart: Boorberg.

Hanschel, D. (2003b) Environment and Human Rights – Cooperative Means of Regime Implementation. *Yearbook of Human Rights & Environment.* 3. 189–261.

Hanschel, D. (2005) Assessing Institutional Effectiveness – Lessons Drawn from the Regimes on Ozone Depletion and Climate Change, In Riedel, E. and Hanschel, D. (eds.) *Institutionalization of International Negotiation Systems – Theoretical Concepts and Practical Insights.* Mannheim: Mannheimer Zentrum für Europäische Sozialforschung. 11–21.

Hanschel, D. (2007) Negotiating within Legal Frameworks – The Framework-Protocol-Approach as a Model for Effective Post-Agreement Negotiations, In Dupont, C. (ed.) *Négociation et Transformations du Monde – Deuxième Biennale Internationale de la Négociation.* Paris: Publibook. 229–246.

Hanschel, D. (2008) A Legal Analysis of the EU Emissions Trading Scheme – Problems and Prospects, In Rodi, M. (ed.) *Emissions Trading in Europe – Review and Preview – Third International Summer Academy Energy and the Environment,* Universität Greifswald. Berlin: Lexxion Verlag. 69–88.

Hanschel, D. (2012, forthcoming) Institutional Options for the International Climate Negotiations, In Rodi, M. (ed.) *Opportunities and Drivers on the Way to a Low-Carbon-Society.* The Summer Academy Proceedings. Berlin: Lexxion Verlag.

Hasenclever, A., Mayer, P. and Rittberger, V. (1997) *Theories of International Regimes.* Cambridge, UK: Cambridge University Press.

Heintz, R., Kuik, O., Peters, P., Schrijver, N. and Vellinga, P. (1994) Summary and Conclusions on Joint Implementation, In Kuik, O., Peters, P. and Schrijver, N. (eds.) *Joint Implementation to Curb Climate Change – Legal and Economic Aspects.* Dordrecht/Boston/London: Kluwer. 161–204.

Heusel, W. (1991) „*Weiches" Völkerrecht: eine vergleichende Untersuchung typischer Erscheinungsformen des sogenannten "Soft Law" mit Ausnahme der Resolutionen von Organisationen internationaler Organisationen.* Baden-Baden: Nomos-Verlags-Gesellschaft.

Hovi, J., Stokke, O. and Ulfstein, G. (2005) Introduction and Main Findings, In Stokke, O., Hovi, J. and Ulfstein, G. (eds.) *Implementing the Climate Regime – International Compliance.* London: Earthscan. 1–14.

Ipsen, K. (1999) *Völkerrecht,* 4th edn. München: C.H.Beck.

Kelly, M. J. (1997) Overcoming Obstacles to the Effective Implementation of International Environmental Agreements. *The Georgetown International Environmental Law Review.* 9 (2). 447–488.

Knox, J. H. (2004) The International Legal Framework for Addressing Climate Change. *Penn State Environmental Law Review,*. 12 (1). 135–144.

Kreuter-Kirchhof, C. (2005) *Neue Kooperationsformen im Umweltvölkerrecht – Die Kyoto Mechanismen*. Berlin: Duncker & Humblot.

Lang, W. (1995) Compliance with International Standards: Environmental Case Studies – Compliance-Control in Respect of the Montreal Protocol. *The American Society of International Law – Proceedings of the 89th Annual Meeting, 5–8 April 1995*. New York. 206–210.

Lang, W. (1996a) Treaty-Making, Science and Compliance-Control, In Lang, W. (ed.) *The Ozone Treaties and their Influence on the Building of International Environmental Regimes*. Vienna: Oesterreichische aussenpolitische Dokumentation. 93–98.

Lang, W. (1996b) Compliance Control in International Environmental Law: Institutional Necessities. *Zeitschrift für ausländisches öffentliches Recht und Völkerrecht*. 56. 683–695.

Lehmann, J. (2005) A Comparative Analysis of the Long-Range Transboundary Air Pollution, Ozone Layer Protection and Climate Change Regime, In Bothe, M. and Rebinder E. (eds.) *Climate Change Policy*. Utrecht: Eleven International Publishing. 73–102.

Loske, R. and Oberthür, S. (1994) Joint Implementation under the Climate Change Convention. *International Environmental Affairs*. 6 (1). 45–58.

MacFaul, L. (2005) *Adoption of Procedures and Mechanisms relating to Compliance under the Kyoto Protocol: a Guide*. Vertic Brief, 6 November 2005.

Mackenzie, R. (2011) The Role of Dispute Settlement in the Climate Regime, In Brunnée, J., Doelle, M. and Rajamani, L. (eds.).*Promoting Compliance in an Evolving Climate Regime*. Cambridge: Cambridge University Press. 395–417.

McKinstry, R, B. (2004) Laboratories for Local Solutions for Global Problems: State, Local, and Private Leadership in Developing Strategies to Mitigate the Causes and Effects of Climate Change. *Penn State Environmental Law Review*. 12 (1). 15–82.

Malanczuk, P. (1995) Die Konferenz der Vereinten Nationen über Umwelt und Entwicklung (UNCED) und das internationale Umweltrecht, In Frowein, J. A., Steinberger, H., Wolfrum, R. (eds.) *Recht zwischen Umbruch und Bewahrung: Völkerrecht, Europarecht, Staatsrecht; Festschrift für Rudolf Bernhardt*. Berlin: Springer-Verlag. 985–1002.

Marauhn, T. (1996) Towards a Procedural Law of Compliance Control in International Environmental Relations. *Zeitschrift für ausländisches öffentliches Recht und Völkerrecht*. 56. 696–731.

Mason, R. (1995) Joint Implementation and the Second Sulphur Protocol. *Review of European Community & International Environmental Law*. 4 (4). 296–303.

Mitchell, R. B. (1996) Compliance Theory: An Overview, In Cameron, J., Werksman, J. and Roderick, P. (eds.) *Improving Compliance with International Environmental Law*. London: Earthscan. 3–28.

Morgan, J. (2006) NGO Strategies for the Post-2012 Process, *Working Paper for the Conference "KyotoPlus – Escaping the Climate Trap"*, 28/29 September 2006 in Berlin.

Morgenstern, L. (2009) One, Two or One and a Half Protocols? An Assessment of Suggested Options for the Legal Form of the Post-2012 Climate Regime, *Carbon & Climate Law Review*. 3. 235–247.

Munton, D., Soros, Nikitina, E. and Levy, M. A. (1999) Acid Rain in Europe and North America, In Young, O. R. (ed.) *The Effectiveness of International Environmental Regimes – Causal Connections and Behavorial Mechanisms*. Cambridge, MA/London: MIT Press. 155–248.

Oberthür, S. (1997) *Umweltschutz durch internationale Regime – Interessen, Verhandlungsprozesse, Wirkungen*. Opladen: Leske + Budrich.

Oberthür, S. and Ott, S. (1999) *The Kyoto Protocol: International Climate Policy for the 21st Century*. Berlin: Springer.

Ott, H. (1996) Elements of a Supervisory Procedure for the Climate Regime. *Zeitschrift für Ausländisches Öffentliches Recht und Völkerrecht.* 56. 732–749.

Ott, H. (1998) *Umweltregime im Völkerrecht – Eine Untersuchung zu neueren Formen internationaler institutionalisierter Kooperation am Beispiel der Verträge zum Schutz der Ozonschicht und der Kontrolle grenzüberschreitender Abfallverbringungen.* Baden-Baden: Nomos-Verlags-Gesellschaft.

Ott, H. (2001) Global Climate. *Yearbook of International Environmental Law.* 12. 211–221.

Ott, H., Brouns, B., Sterk, W. and Wittneben, B. (2005) It Takes Two to Tango – Climate Policy at COP 10 in Buenos Aires and Beyond. *Journal for European Environmental & Planning Law.* 2 (2). 84–91.

Palmer, G. (1992) New Ways to Make International Environmental Law. *American Journal of International Law.* 86. 259–278.

Parson, E, A. (1993) Protecting the Ozone Layer, In Haas, P., Keohane, R. O. and Levy, M. A. (eds.) *Institutions for the Earth. Sources of Effective International Environmental Protection.* Cambridge, MA: MIT Press. 27–75.

Pierrehumbert, R. (2006) Climate Change: A Catastrophe in Slow Motion. *Chicago Journal of International Law.* 6 (2). 573–596.

Pohlmann, M. (2004) *Kyoto Protokoll: Erwerb von Emissionsrechten durch Projekte in Entwicklungsländern.* Berlin: Duncker & Humblot.

Poitras, J. (1999) Reforming the Convention Amendment Process to Facilitate the Strengthening of Commitments, In Susskind, L. E. and Moomaw, W. (eds.) *New Directions in International Environmental Negotiation.* Cambridge, MA: PON books. 104ff.

Raustiala, K. and Victor, D. G. (1998) Conclusions, In Victor, D. G., Raustiala, K. and Skolnikoff, E. B. (eds.) *The Implementation and Effectiveness of International Environmental Commitments: Theory and Practice.* Cambridge, MA: MIT Press. 659–707.

Richter, R. and Furubotn, E. (1996) *Neue Institutionenökonomik – Eine Einführung und kritische Würdigung.* Tübingen: Mohr.

Riedel, E. (1986) *Theorie der Menschenrechtsstandards: Funktion, Wirkungsweise und Begründung wirtschaftlicher und sozialer Menschenrechte mit exemplarischer Darstellung der Rechte auf Eigentum und Arbeit in verschiedenen Rechtsordnungen.* Berlin: Duncker & Humblot.

Riedel, E. (1996) *The Examination of State Reports – The Monitoring System of Human Rights Treaty Obligations.* Berlin: Berlin-Verlag Spitz.

Riedel, E. (1998) Change of Paradigm in International Environmental Law. *Law and State.* 57. 22–48.

Rubin, J. Z. (1993) Third-Party Roles: Mediation in International Environmental Disputes, in Sjoestedt, G. (ed.) *International Environmental Negotiation.* Newbury Park: Sage. 275–290.

Sand, P. H. (1990) *Lessons Learned in Global Environmental Governance.* Washington, DC: World Resources Institute.

Sands, P. (1995) *Frameworks, Standards and Implementation.* Manchester: Manchester University Press.

Sands, P. (1996) Compliance with International Environmental Obligations: Existing International Legal Arrangements, In Cameron, J., Werksman, J. and Roderick, P. (eds.) *Improving Compliance with International Environmental Law.* London: Earthscan. 48–82.

Sands, P. (2003) *Principles of International Environmental Law.* Cambridge: Cambridge University Press.

Seidl-Hohenveldern, I. (2000) *Völkerrecht.* 17th edn. Cologne, Berlin/Bonn/Munich: Carl Heymanns Verlag.

Sjoestedt, G. (ed.) (1993) *International Environmental Negotiation.* Newbury Park, CA/London/New Delhi: Sage.

Sjoestedt, G. (2003) Norms and Principles as Support to Postnegotiation and Rule Implementation, In Spector, B. I. and Zartman, I. W. (eds.) *Getting it Done – Postagreement Negotiation and International Regimes*. Washington DC: United States Institute of Peace. 89–111.

Spector, B. I. and Zartman, I. W. (2003) Regimes in Motion: Analyses and Lessons Learned, In Spector, B. I. and Zartman, I. W. (eds.) *Getting it Done – Postagreement Negotiation and International Regimes*. Washington DC: United States Institute of Peace. 271–292.

Sterk, W., Ott, H., Watanabe, R. and Wittneben, B. (2007) The Nairobi Climate Change Summit (COP12 – MOP2): Taking a Deep Breath before Negotiating Post-2012 Targets?, *Journal for European Environmental & Planning Law*. 4 (2). 139–148.

Susskind, L. E. (1994). *Environmental Diplomacy: Negotiating More Effective Global Agreements*. New York: Oxford University Press.

Susskind, L. and Ozawa, C. (1992) Negotiating More Effective Environmental Agreements, In Hurrell, A. and Kingsbury B. (eds.) *The International Politics of the Environment – Actors, Interests and Institutions*. Oxford: Clarendon Press. 142–165.

Thackeray, Richard W. (2004) Struggling for Air: The Kyoto Protocol, Citizens' Suits under the Clean Air Act, and the United States' Options for Addressing Global Climate Change. *Indiana International & Comparative Law Review*. 14 (3). 855–903.

Thoms, L. (2003) A Comparative Analysis of International Regimes on Ozone and Climate Change with Implications for Regime Design. *Columbia Journal of Transnational Law*. 41. 795–812.

Tolba, M. K. (1996) The Story of The Ozone Layer, In Lang, W. (ed.) *The Ozone Treaties and their Influence on the Building of International Environmental Regimes*. Vienna: Ministry of Foreign Affairs. 9ff.

Victor, D. G. (1996) The Montreal Protocol`s Non-Compliance Procedure: Lessons for Making Other International Environmental Regimes More Effective, In Lang, W. (ed.) *The Ozone Treaties and their Influence on the Building of International Environmental Regimes*. Vienna: Ministry of Foreign Affairs.

Victor, D. G. (1998) The Operation and Effectiveness of the Montreal Protocol's Non-Compliance Procedure, In Victor, D. G., Raustiala, K. and Skolnikoff, E. B. (eds.). *The Implementation and Effectiveness of International Environmental Commitments: Theory and Practice*. Cambridge, MA: MIT Press.

Werksman, J. (1996a) Designing a Compliance System for the UN Framework Convention on Climate Change, In Cameron, J., Werksman, J. and Roderick, P. (eds.) *Improving Compliance with International Environmental Law*. London: Earthscan. 85–112.

Werksman, J. (1996b) The Conference of Parties to Environmental Treaties, In Werksman, J. (ed.) *Greening International Institutions*. London: Earthscan. 55–68.

Werksman, J. (1998) The Clean Development Mechanism: Unwrapping the ´Kyoto Surprise. *Review of European Community and International Environmental Law*. 7 (2). 147–158.

Wettestad, J. (1999) *Designing Effective Environmental Regimes: The Key Conditions*. Cheltenham: Edward Elgar.

Wettestad, J. (2005) Enhancing Climate Compliance – What are the Lessons to Learn from Environmental Regimes and the EU? In Stokke, O., Hovi, J. and Ulfstein, G. (eds.) *Implementing the Climate Regime – International Compliance*. London: Earthscan. 209–231.

Wittneben, B., Sterk, W., Ott, H. and Brouns, B. (2006) The Montreal Climate Summit: Starting the Kyoto Business and Preparing for post-2012. *Journal for European Environmental & Planning Law. 3 (2)*. 90–100.

Wolfrum, R. (1999) *Means of Ensuring Compliance with and Enforcement of International Environmental Law*. The Hague: Nijhoff.

Yamin, F. (1998) The Kyoto Protocol: Origins, Assessment and Future Challenges. *Review of the European Community and International Environmental Law.* 7 (2). 113–127.

Young, O. R. (1989) *International Cooperation. Building Regimes for Natural Resources and the Environment.* Ithaca: Cornell University Press.

Young, O. R. (1993) Perspectives on International Organizations, In Sjoestedt, G. (ed.) *International Environmental Negotiation.* Newbury Park: Sage. 244–261.

Young, O. R. and Levy, M. A. (1999) The Effectiveness of International Environmental Regimes, In Young, O. R. (ed.) *The Effectiveness of International Environmental Regimes – Causal Connections and Behavorial Mechanisms.* Cambridge, MA/London: MIT Press. 1–33.

15 Verification as a Precondition for Binding Commitments

Facilitation through Trust

Larry MacFaul

Introduction

Global environmental problems constitute a riddle for the international community to solve. Deeply entwined in the process of trying to understand the nature of the problem and finding an effective response strategy is the need to coordinate action between states. Establishing international environmental treaties is a way of coordinating and promoting action. Ideally, these agreements solve the problem of cooperative action in transboundary environmental problems and find a way to manage shared resources effectively. Verification[1] (including all monitoring, reporting and review processes) and compliance procedures (which are intrinsically linked to the verification system) perform certain crucial functions in the formation and implementation of international environmental treaties:

- enhancing cooperation between states, by demonstrating each state's level of effort and by deterring free-riding;[2]
- measuring and promoting overall and individual state's progress towards a treaty's goals.

These functions are interrelated. The emphasis placed on each depends on the particular treaty's aims and provisions. An effective verification and compliance system should prevent free-riding and leave no room for doubt about parties' levels of compliance: parties must not only comply but be clearly seen to comply. If states are confident that the verification system can identify and expose non-compliance, and that instances of non-compliance can be prevented or rectified, they are more likely to agree and adhere to a strong treaty. Moreover, the verification system plays a large role in inspiring confidence among parties in the development and maintenance of a treaty. However, while the verification system must be able to deal effectively with compliance and participation issues, states are unlikely to support a treaty if the system appears disproportionately powerful or intrusive. The ability to measure clearly each state's level of action, in addition to helping to enhance cooperation, shows what progress is being made in reducing the environmental problem being addressed and indicates the adequacy both of parties' efforts and of the provisions in the treaty itself.

The particular characteristics of the environmental problem being faced, the goals agreed on to tackle the problem and the type of provisions, commitments and mechanisms laid down in a treaty will determine what kind of verification system is required. Ideally, such a system includes monitoring and reporting processes, a review process and compliance procedures. The monitoring and reporting processes should show comparably, transparently and accurately how states are progressing towards treaty goals and allow parties to share experience. The review processes should ensure this information is correct and provide helpful feedback to parties. The compliance procedures should use this information to assess parties' adherence to treaty requirements and progress towards its goals, and can contain several ways of promoting and enforcing compliance.

Designing an effective verification and compliance system involves balancing the desire for the highest levels of environmental effectiveness with the monitoring capacity of the parties or other relevant entities. It must be strong enough to ensure environmental integrity but not so tight as to make it impossible to implement, and it should contain procedures for improving monitoring capacity. Compliance provisions can range from management approaches (soft) to enforcement approaches (hard) (Corfee Morlot 1998). Treaties containing only non-binding generalized commitments tend to have less developed and non-demanding compliance systems.

This chapter examines the role of verification and compliance in the climate change regime as seen in a historical perspective. The chapter begins by discussing the verification and compliance systems under the 1992 United Nations Framework Convention on Climate Change (UNFCCC) and the 1997 Kyoto Protocol. It examines progress so far and pathways towards improvement of these systems. It then discusses their role in the future of the climate change regime. There are currently many different proposals for what shape future climate change action should take, ranging from maintaining the present format to using an entirely different structure. A strong verification and compliance system, covering the overall architecture and individual parts, will serve as a vital component in the future of the climate change regime, both as a means of facilitating cooperative action and of assessing the effectiveness of measures taken by parties to tackle climate change. The need to provide and maintain a strong verification and compliance system should therefore play a major part, alongside factors such as environmental effectiveness, economic efficiency and equity considerations, in determining what shape the future climate change regime should be.

Verification systems under the UNFCCC

The UNFCCC

The ultimate goal of the UNFCCC is the stabilization of greenhouse gas[3] concentrations in the atmosphere at a level that would prevent dangerous anthropogenic interference with the climate system. The convention differentiates between parties in terms of the type and strength of their commitments.[4] It includes an aspirational non-binding emissions reduction aim for Annex I parties (developed countries) of jointly or individually returning to their 1990 emissions levels by 2020. The convention contains provisions relating to adaptation to impacts of climate change and also to technology transfer and financial assistance from developed countries to developing countries. It also contains provisions relating to research and systematic observation, and public involvement. The Annex I group includes both the most developed countries (Annex II parties) which are required to provide financial resources for developing countries to undertake emissions reduction activities and implement adaptation measures to the adverse effects of climate change, and countries with economies in transition (EIT parties) which are allowed some flexibility in meeting their commitments and are not required to provide financial support to developing countries. Non-Annex I parties are mainly developing countries. This group also contains a sub-group of least developed countries (LDCs) who are given special consideration under the convention. Several institutions contribute to the functioning of the convention including the Conference of Parties (COP), which is the primary decision-making body; the Subsidiary Body for Scientific and Technological Advice (SBSTA); and the Subsidiary Body for Implementation (SBI).

The monitoring and reporting processes[5]

The convention contains provisions both for monitoring the overall progress towards the treaty's goals and individual parties' progress. Parties can also share their experiences in tackling climate change through these reporting systems. The capacity to monitor emissions varies considerably between parties. In recognition both of this and the need to establish verification systems that are appropriate for the different types of commitments undertaken by parties, the convention differentiates between groups of parties in their monitoring and reporting requirements. Under the UNFCCC, Annex I parties must submit 'national communications' to the UNFCCC Secretariat every three to four years. Non-Annex I national communications are due within four years of the initial disbursement of financial resources from the Global Environment Facility (GEF) for their preparation, with the possibility of a one year extension.[6] National communications are reports detailing what action a party is taking to implement the convention.[7] Parties must follow UNFCCC guidelines, designed to assist them in meeting their commitments under the convention, when drawing up these reports. The guidelines aim at promoting the provision of consistent, transparent, comparable, accurate and complete information in order to enable a thorough review and assessment by the COP of the implementation of the convention and to monitor parties' progress towards the convention's goals. The guidelines should also assist the COP in reviewing the adequacy of the commitments. Separate guidelines are provided for national communications for non-Annex I parties. In addition to assisting these parties in reporting under the convention and encouraging the presentation of information in a consistent, transparent and flexible manner, non-Annex I party guidelines are intended to facilitate the presentation of information on support required for the preparation and improvement of national communications. These guidelines should also ensure that the COP has sufficient information to carry out its responsibility for assessing the implementation of the convention. The secretariat has produced a user manual to assist non-Annex I parties in their usage of the guidelines. As noted above, funding is made available to certain countries to assist in their national communications preparation.

National emissions inventories are the basis of the UNFCCC's verification system. Inventories are essential for assessing the total and individual efforts made to address climate change and progress towards meeting the ultimate goal of the convention as well as compliance under the Kyoto Protocol. They are also needed for evaluating mitigation options, assessing the effectiveness of policies and measures, making long term emissions projections and providing the foundation for emissions trading. Annex I parties must annually submit to the secretariat national inventories of their greenhouse gas emissions and removals for a period covering the base year, normally 1990,[8] up to the last year but one prior to submission. Non-Annex I parties are currently required to submit inventories with their national communications to the extent that their capacities permit.

Annex I inventory submissions are comprised of two parts: the Common Reporting Format (CRF), which is a standardized electronic database, and the National Inventory Report (NIR) which contains information on how the inventory was compiled. Inventories should include enough documentation and data to enable understanding of the underlying assumptions and calculations of all emissions estimates (UNFCCC 2003a). The UNFCCC provides reporting guidelines for inventory preparation to help parties meet their commitments under the convention and the protocol. These guidelines are designed to assist the parties in meeting their reporting commitments as well as to facilitate the process of considering inventories and the preparation of the technical analysis and synthesis documentation and the verification, technical assessment and expert review of the inventory information (see following section). The inventories should be transparent, consistent, comparable, complete and accurate.[9]

Under the UNFCCC guidelines parties must use the Intergovernmental Panel on Climate Change (IPCC) guidelines[10] when compiling their inventories. The IPCC guidelines contain instructions and methodologies for estimating greenhouse gas emissions and removals. Parties may use national methodologies that they consider better able to reflect national circumstances, as long as these methodologies are compatible with IPCC guidelines and guidance on good practice and are well documented. Furthermore, parties are encouraged to develop their own methodologies, including emission factors[11] and activity data, rather than to use IPCC default methodologies which may not always be appropriate for their national contexts, as long as these methodologies are developed in a manner consistent with IPCC good practice guidance.

The inventories provide both national totals and a breakdown of emissions by sector. The sectors included are energy, industrial processes, solvent and other product use, agriculture, land-use change and forestry, and waste. Each of these sectors is sub-divided into several categories. Most emissions estimation is derived from multiplying activity data by emission factors. Direct measurement of individual emission sources is also permitted under the guidelines, but is comparatively rare in this system.[12]

The IPCC provides instructions for quality assurance and quality control (QA/QC) procedures and uncertainty management. Quality assurance is a system of routine technical activities to measure and control the quality of the inventory as it is being developed; for example, accuracy checks on data acquisition. Quality control activities include a planned system of review procedures conducted by personnel not directly involved in the inventory compilation and development process. The IPCC guidelines note that uncertainties[13] in emissions or removals estimates vary widely between different sources and sinks and between the different greenhouse gases, as well as between states reporting the same gases and sources (depending on the approach and data used and the level of detail).

The IPCC provides information on certain specific verification procedures that parties can use to check the reliability of their emissions inventories (see Box 15.1). This information will be of use to the inventory compilation agencies, any independent inventory review arranged by parties in order to check and improve its inventory and for the expert review teams (see following section). Some parties have arranged bilateral inventory reviews in order to enhance checks on, and improvement of, their inventories.

The IPCC verification procedures should provide inputs to improve the inventory process, build confidence in emissions estimates and trends and help improve scientific understanding related to inventories. The tools available for inventory verification at the national level include comparison with other national emissions data, direct source testing and comparison with national scientific publications. At the international level, inter-country comparisons for the same year can be made involving comparison of activity levels and aggregated emission factors, as well as comparison of emissions trends or trends in input data. Estimated uncertainties and intensity indicators can also be compared between countries. Top down and bottom up estimates can be compared. An inventory can also be compared with other independently compiled international datasets.[14] However, the data sources used by the international datasets are often not completely independent of one another or from the dataset used to calculate a national inventory. The IPCC notes that some of the comparison processes do not always provide verification of the actual data but rather the reliability of the data. However, inventory reviewers could use these techniques to identify inconsistencies or areas needing more detailed verification. The IPCC also notes that the amount of time that inventory agencies can spend on these independent verification activities will depend on resources available to them, and an evaluation of the value of these activities compared with others ways of improving their inventory.

Other verification techniques suggested by the IPCC include comparison with atmospheric measurements at the local, regional or global scales. This may involve local and regional atmospheric sampling, continental plumes monitoring (which can provide an indication of emissions on a broad scale), satellite observations (which can allow quasi-continuous concentration profiles for part or all of the globe), and global dynamic approaches which measure trends in time in atmospheric concentrations of particular compounds indicating changes in the global balance between sources and sinks. Comparison of inventory estimates could also be made with international scientific publications, for the purpose of checking the inventory quality. Comparisons with global or regional budgets could be used for updating the global budget or providing feedback to inventory developers, or both. Finally, global or regional totals could be compared against atmospheric concentrations, changes in concentrations or against isotopic signature analysis. The use of remote sensing systems and atmospheric concentration measurement in relation to emissions inventories will be discussed in greater detail in Section 4.1.

Box 15.1 IPCC guidelines for verification of national inventories

A. Checks:

- Check for discontinuities in emission trends from base year (usually 1990) to end year.

B. Comparisons of emissions and other such features:

- Compare the Reference Approach for CO_2 emissions from fuel combustion with other approaches.
- Compare inventory emissions estimates by source category and gas against independently compiled national estimates from international databases.
- Compare activity data against independently compiled estimates and perhaps activity data from countries with similar source categories and sectors.
- Compare (implied) emission factors for source categories and gases with independent estimates and estimates from countries with similar source categories and sectors.
- Compare sector intensity estimates of selected source categories with estimates from other countries with similar source categories and sectors. If necessary, calculate emission intensity estimates based on international statistical compendia.

C. Comparisons of uncertainties:

- Compare uncertainty estimates with those from reports of other countries and the IPCC default values.

D. On-site measurements:

- Perform direct source testing on *key source categories*, if possible.

Source: IPCC Good Practice Guidance and Uncertainty Management, 2000.

The greater the level of detail and sectoral breakdown within an inventory the clearer the indication will be of what activities or entities are producing emissions and in what quantity. Greater detail also allows greater precision in formulating methods for reducing emissions. However, the difficulty of drawing up an inventory rises according to the level of detail required. Consequently the IPCC system, being global in scope, uses a tiered structure of increasing levels of detail and allows for a more simple approach than some other inventory systems.[15] Drawing up a national inventory is a complex task that requires significant financial resources as well as strong technical and institutional capacity. While other inventory systems were designed only for domestic purposes or to suit a group of countries that is relatively homogenous in terms of development, the IPCC had to cater for a diverse range of countries with different institutional, political, technical, geographical and economic circumstances. Nevertheless, the system is effective, extensive and also has the capacity to improve and evolve over time. This multilateral approach to emissions monitoring provided for under the UNFCCC means that states' emissions estimates are comparable, since they must all use IPCC or IPCC-compatible methodologies. The use of comparable methodologies (coupled with the formal review processes discussed in the following section) reduces the chance of intentional or unintentional under- or over- estimation and allows confidence to grow in the regime. Moreover, the sectoral tracking of man-made emissions included in the system can be used to assess more closely the effectiveness of policies.

The review process

Both national communications and inventories are subject to a formal review process. This review process is a key factor in making the UNFCCC and Kyoto Protocol verification system effective and robust. The review process should both check reports for accuracy and adherence to UNFCCC guidelines and also improve their quality and comparability.

National communications are subject to an in-depth review by an international team of experts chosen from a roster and coordinated by the secretariat. The review, usually conducted by desk-based study and a country visit, results in an in-depth review report which facilitates assessment by the COP of a party's level of implementation of commitments. These reports should also make comparison of information between parties easier. In addition, the secretariat prepares a 'compilation and synthesis report' that summarizes the most important information in parties' national communications. Non-Annex I parties' national communications are not subject to in-depth review but information from them is drawn up into a compilation and synthesis report.

The national inventories of Annex I parties are subject to a review process that aims to provide a technical assessment of the inventory and to check its adherence to reporting guidelines. The review should ensure that the COP has adequate and reliable information on inventories and emissions trends and should provide the COP with an objective, consistent, transparent, thorough and comprehensive technical assessment of the quantitative and qualitative inventory information.

The review process is also meant to assist Annex I parties in improving the quality of their inventories. The review process includes an initial check for completeness and correct format, the synthesis and assessment of basic inventory information across parties and a preliminary assessment of individual parties' inventories, including identification of potential problem areas. Finally, there is an individual review of each inventory by international expert review teams (ERTs). As of 2003, annual review of national inventories became mandatory. The individual review can be carried out in three ways: by in-country review, by centralized review (which takes place at the UNFCCC Secretariat) or by desk review (for which the experts work from their home countries). ERTs are meant to contain sufficient expertise to be able to cover all economic sectors, and membership of the teams is balanced between Annex I and non-Annex I parties. The nomination of experts by parties coupled with geographically balanced representation helps to reassure parties that the review is objective. A code of practice was adopted at COP9, (2003, Milan), for the treatment of confidential information in the review process. The code covers reviewers' access to, and handling by the secretariat of, confidential information. In addition, reviewers must sign an agreement concerning, among other issues, appropriate conduct during the review. The code includes reference to conduct related to the protection of confidential information. An inventory review training scheme was introduced by the secretariat in 2004 to promote broader participation from parties and increase the number of experts. New experts who have been nominated must complete training and pass an examination before they can be invited to participate in a review. Inventories, national communications, review documents and guidelines are publicly available on the UNFCCC website.

The COP itself is mandated to review the implementation of the convention and also the adequacy of convention commitments for developed countries.[16] Since the convention only contains non-binding generalized commitments, the verification system, while expansive, does not include a rigorous compliance system.

Verification systems under the Kyoto Protocol

The Kyoto Protocol

The Kyoto Protocol was adopted in order to boost action on climate change. The protocol negotiations were fraught and this treaty has remained contentious for some parties as well as some non-parties. Since the problem of climate change is highly complex, and anthropogenic contributions to climate change are made throughout a state's economy, the political and economic stakes concerning what type of, and how much action to take, are high. Disagreement on how to tackle the problem has pervaded efforts to coordinate action, especially regarding stronger forms of commitments. Nevertheless, negotiations ended successfully with an agreement on a clear new set of measures to tackle climate change. The protocol came into force in 2005 but has not been ratified by all parties to the UNFCCC.

The Kyoto Protocol established absolute or 'flat-rate' emissions reductions commitments for Annex I parties defined in terms of change from an agreed base year. Each Annex I party[17] was given a quantified emissions limitation or reduction target, with a view to reducing their overall emissions by at least 5 per cent below 1990 levels by the end of the protocol's first commitment period, 2008–2012. These targets were arrived at by negotiation between parties with no specific logical basis used to differentiate commitments (Grubb 2004).

Under the protocol, parties must take domestic action to meet these targets but can also use the so-called flexible mechanisms set out under the treaty which were established in order to lower the overall costs of meeting the emissions targets (see Box 15.2).

Box 15.2 The Kyoto Protocol's flexible mechanisms

- Clean development mechanism (CDM), under which Annex I parties can implement sustainable development projects that lead to emissions reductions in non-Annex I parties and thereby earn certified emissions reduction (CERs).
- Joint Implementation, under which Annex I parties can implement emissions reduction projects in other Annex I parties and earn emissions reduction units (ERUs).
- Emissions trading, which permits Annex I parties to trade emissions units – assigned amount units (AAUs), removal units (RMUs), CERs and ERUs – with other Annex I parties.

The protocol also contains provisions relating to other general commitments (including promoting environmentally friendly technology transfer, research and public awareness) and minimizing impacts on developing countries. It also sets out a list of policies and measures for tackling climate change relating to actions such as enhancing energy efficiency and promoting renewable energy. Phrasing related to commitments on policies and measures is largely non-binding, reflecting parties' unwillingness to take on obligations that would impinge on their choice of methods in how to tackle climate change (Grubb 2004).

The protocol demanded a verification and compliance system that built on and developed the convention's system in order to cope with its wider range of measures and more specific targets. The result is a system which is both complex and thorough: complex, since it must monitor compliance with emissions reduction targets and parties' participation in the 'flexible mechanisms', and thorough, since it not only needs to guarantee environmental integrity but also accurately and transparently regulate a new emissions trading system. Indeed, confidence in the quality of monitoring, reporting, review processes and compliance procedures, including in their ability to identify and deal with problems, is critical to the validity of trading for environmental management and the financial viability of the commodity being to be bought and sold. Furthermore, comparability of data is crucial not only for comparing parties' levels of effort and compiling consistent data, but also because trading will mix states' inventories (Corfee Morlot 1998).

The monitoring and reporting processes

The Kyoto Protocol specifies that Annex I parties must have in place a 'national system' for estimating greenhouse gas emissions and removals at least one year before the start of the first commitment period. The national system must conform to strict criteria. National systems comprise all institutional, legal and procedural arrangements for estimating greenhouse gas emissions and removals. Parties must also establish a national registry (an electronic database) to account for their emissions trading units (ERUs, CERs, AAUs and RMUs).[18] The secretariat will run an independent transaction log to ensure the integrity of transactions by checking them against the trading conditions set out under the protocol. Parties must include additional information on these and other matters in their national communications and their annual greenhouse gas inventories. Parties also have to calculate their 'assigned amount', which is the total amount of greenhouse gases a party can emit during a commitment period. The assigned amount is calculated using the party's quantified emission limitation or reduction commitments. Parties must fulfil monitoring and reporting requirements before they are eligible to participate in the flexible mechanisms.[19] There are also extensive provisions governing the use of each of the flexible mechanisms, in particular how they should be monitored.[20] Annex I parties should also submit a report by 2006 on demonstrable progress made towards meeting their commitments up till 2005.

The compliance system

In order to promote and assess parties' adherence to their emissions reduction commitments and the reporting requirements, an intricate compliance system was developed. This system augments the monitoring, reporting and review processes by formally determining adherence or non-adherence to the monitoring and reporting standards and emissions reduction obligations set out under the treaty. Compared with most international environmental treaties the system is particularly elaborate. However, it is as yet untested, its legal nature is unresolved and ultimately its power may be found wanting, although there are ways to strengthen it.

At the core of this system is the Compliance Committee which consists of a Facilitative Branch and an Enforcement Branch. Each branch has ten members elected to it by the Conference of Parties serving as the Meeting of Parties to the Kyoto Protocol (COP/MOP). Membership of the Compliance Committee is balanced between Annex I and non-Annex I parties. The Facilitative Branch provides advice and assistance with the aim of promoting compliance and also 'early-warning' of cases where a party may be in danger of not complying with its emission targets. The Enforcement Branch determines cases of non-compliance with emission targets, reporting requirements and eligibility to participate in the flexible mechanisms. It can apply certain 'consequences' in cases of non-compliance. The Committee's mechanisms are triggered when it receives 'questions of implementation',[21] either in reports by the ERTs, or from a party with respect to itself or from any party with respect to another party as long as it has corroborating information.[22] Each branch bases its work on information from several sources: ERT reports; the party itself; a party which has submitted a question of implementation regarding another party; reports of the COP, the COP/MOP and the COP subsidiary bodies; and the other branch. Intergovernmental organizations and non-governmental organizations (NGOs) may also submit factual and technical information. Each branch may seek expert advice. In addition, subject to rules relating to confidentiality, information considered by the branch should be made available to the public. However, the branch may decide of its own accord or at the request of the party concerned, not to release information until its decision has become final. Any party about which a question of implementation is raised has the opportunity to represent itself through writing and at hearings.

The scenarios in which the Compliance Committee can apply consequences and the nature of these consequences are listed below:

- When a party fails to meet the monitoring and reporting requirements, it must develop a compliance action plan.
- When a party fails to meet one or more of the eligibility requirements for the flexible mechanisms, it will have its eligibility suspended in accordance with the relevant provisions.[23]
- When a party exceeds its assigned amount:

  1. A number of tonnes equal to 1.3 times the amount in tonnes of excess emissions can be deducted from the party's assigned amount for the second commitment period.
  2. A compliance action plan must be developed by the party.
  3. A party's eligibility to transfer quotas can be suspended.

In addition, ERTs can apply adjustments to inventory data (including base year data) if data is unavailable or if the inventory has not been prepared in accordance with the IPCC guidelines. If the party disagrees with the ERT decision, the issue will be forwarded to the Compliance Committee to resolve.

Finally, parties can appeal to the COP/MOP on a decision of the Enforcement Branch but only if the party believes it has been denied due process. The COP/MOP can only overturn the decision of the Enforcement Branch with a 2/3 majority.

Compliance assessment takes place before, during and after the commitment period. As mentioned in the previous section, parties must have their national system and registry in place before the commitment period begins, and both must be checked for suitability. They must also supply data needed to calculate their assigned amount and must have submitted their most recently required inventory. Furthermore, parties must submit a pre-commitment period report to the secretariat by 2007 to demonstrate adherence to the protocol's preconditions. During the commitment period, ERTs will check inventories against the UNFCCC criteria. A party may lose eligibility with respect to the flexible mechanisms during the commitment period. After the commitment period, there is an additional period of 100 days during which parties can make final transactions to bring themselves into compliance. Parties must submit a report on the additional period. Compliance assessment can only take place once the ERTs have access to all the commitment period inventories. Since there is a time-lag of two years in inventory preparation, compliance assessment for the first commitment period will not take place until 2015. The consequences can then be applied in cases of non-compliance.

Evolution of the compliance system

The concentration of effort on compliance issues has tended to be on the Protocol rather than on the implementation of the convention. Although the convention demands emissions reductions and also includes reporting requirements, the lack of specificity of its obligations makes assessment of its implementation difficult. In 1998, it was suggested that a multilateral consultative committee (MCC) should be formed to give advice to parties concerning their implementation of the convention. However, the COP failed to agree on the MCC's composition and size so it never came into being (Wang and Wiser 2002). Thus while the COP is mandated to review the implementation of the convention, a systematic compliance system was not established. However, the obligations established under the Kyoto Protocol necessitated a clear compliance system. The protocol text only included the outline of the treaty, the details being left to later meetings to elaborate. Negotiation of the system was often turbulent. It was designed by the Joint Working Group (JWG) on Compliance (established in 1998) which received views from parties on what the composition of the compliance system should be, and also heard suggestions from representatives from the secretariats of other international institutions and conventions, as well as from two NGOs.[24] Negotiations at COP6 part II (mid-2001, Bonn) had resulted in agreement on several elements of the Kyoto Protocol, but divisions arose over the nature and strength of the compliance regime and several technical issues which were not settled by the end of the meeting. When parties resumed talks at COP7 (late-2001, Marrakesh), the divisions existed chiefly between the European Union (EU) and the G77/China, on the one hand, and the 'umbrella group'[25] on the other. On the important issue of compliance, the US had sided with the EU and G77/China in wanting a strong compliance system rather than with its fellow members of the umbrella group. However, when the US pulled out of the protocol in early 2001, not only did its departure reduce support for a strong compliance system but it gave the remaining members of the umbrella group a far stronger bargaining hand than they previously had because the protocol's entry into force had become dependent on their ratification (ENB 2001a and 2001b; Wang and Wiser 2002). Members of the umbrella group's objections concerned, among other issues, eligibility on use of the flexible mechanisms, reporting of sink data and use of sink credits. The ability of one party to raise a question of implementation regarding another party was also objected to by members of the umbrella group. Following the fierce negotiations over these issues, a deal was reached at the end of the meeting which allowed the 'Marrakesh Accords' to be adopted. These accords provide the detailed mechanisms for how the Kyoto Protocol should work in practice.

There is a range of options available to promote compliance with international environmental treaties (Chayes and Chayes 1995; Corfee Morlot 1998; Raustiala and Victor 1998). The optimal means of ensuring compliance is a judicious balance between management (soft) and enforcement (hard) approaches. Some treaties will require more emphasis on one approach than the other. Management approaches which can identify and address compliance problems include national reporting and review processes, consultation and negotiation and mediation and conciliation. Beyond these approaches are those involving a 'carrots and sticks' approach including financial and/or technical support and issuing warnings or cautions. Facilitation, encouragement and compliance assistance are key elements in many environmental treaties. Enforcement approaches include making funding conditional on compliance, the suspension of rights or privileges, trade measures, financial penalties or economic sanctions.

The Kyoto Protocol contains several of these options. The division of the Compliance Committee into the two branches – facilitation and enforcement – reflects the dual function a treaty should provide. The Facilitative Branch provides advice and assistance. On the enforcement side, the system provides consequences designed to stimulate both restitution and deterrence. Restitution is achieved by requiring a party to make up for its insufficient emissions reductions in the following period. Deterrence is provided by multiplying this amount by 1.3 and through the suspension of eligibility to transfer quotas (Hagem *et al.* 2003).[26]

Other international environmental regimes focus on facilitative measures in relation to compliance. For instance, the ozone layer protection regime (including the 1985 Vienna Convention for Protection of the Ozone Layer and its 1987 Montreal Protocol on Substances that Deplete the Ozone Layer) provides for technical, financial and reporting assistance, issuing cautions, and suspension of treaty rights. It is the first two of these – softer measures – that have been generally preferred. However, this facilitative approach has been supported by firmer measures. Within the 1979 Convention on Long-Range Transboundary Air Pollution (CLRTAP), an assistance based approach to non-compliance has been used. Its Implementation Committee examines progress in all protocols, publishes reporting scores, and gradually increases pressure on non-compliant parties through a 'name and shame' policy. The explicitness of this naming procedure can in part be seen as the result of parties allowing more severe language to be used because they have reached a certain level of trust with each other in the regime. Indeed, there appears to be a trend in environmental treaties towards tougher review procedures in the form of more explicit naming and shaming, and tougher language being used by environmental convention bodies (Wettestad 2005). It may also be characteristic of environmental treaties that parties may invoke the compliance procedures themselves, in the expectation of receiving assistance and perhaps a more lenient response if enforcement procedures are applied.

The climate change regime has higher economic and political stakes than any other environmental treaty and more tools available for review processes, compliance assessment and enforcement than many other treaties. Enforcement, whether related to failure by a party to meet the monitoring and reporting requirements or to a party exceeding its emissions reduction target, may therefore play a larger role in its compliance promotion than in other treaties. However, current enforcement measures provided for in the treaty may not be sufficient to deter free-riding.

The more facilitation and assistance that is provided, the better the treaty will function in terms of progress towards its emissions reduction goals and in fostering a more cooperative atmosphere between parties. The emphasis should therefore be on management and facilitation, backed up by credible use of the penalties. A particularly challenging situation will arise concerning the deduction of 1.3 times the excess emissions in the case of non-compliance by a party that is severely struggling to meet its target due to unavoidable factors (in other words, unintentionally) and when facilitation measures have proved to be insufficient. Finally, it is important to note that difficulties in assessing compliance are most likely to occur over the quality and appropriateness of methodologies used by parties in emissions estimates.

Compliance system problems

Although the compliance system provided for in the Marrakesh Accords is regarded as particularly innovative and elaborate, it may be weakened by disagreement over its legal nature; problems inherent in its structure; parties' monitoring capacity; uncertainty over the future of the protocol regime; and the lack of effective enforcement provisions.

Legal nature

Parties could not reach agreement on the legal nature of the climate regime. This decision has been deferred until the first COP/MOP. Article 18 of the Kyoto Protocol states that compliance procedures and mechanisms 'entailing binding consequences shall be adopted by means of an amendment to this Protocol'. The amendment would only bind parties that ratified it. Efforts were made to get round this problem and streamline procedures relating to compliance decisions. However, the withdrawal of the US led to the disintegration of these efforts (Wang and Wiser 2002). Instead of an amendment, a COP decision could be taken on the issue, but this would not necessarily make the compliance regime consequences legally binding.

While the emissions targets themselves can be considered legally binding, the binding nature of the compliance regime is as yet unresolved. It is encouraging that parties agreed to the details of the protocol's compliance mechanisms, as this indicates a serious commitment to the principle of adhering to extensive verification and compliance procedures; consequently, it would seem that parties are less likely to consider free-riding. However, it is a matter of concern that the legally binding nature of the system has provoked such controversy. A speedy resolution of the matter would set the commitment period and the run up to it on a clearer and more solid foundation.

Credibility

Within the current structure of the system lies a problem relating to the likelihood that the stipulated consequences allowed for under the compliance regime will be applied when a party has exceeded its target. The credibility of the threat to apply the consequences may be weakened when these penalties would also have a negative economic effect on the states from which the members of the Enforcement Branch are drawn. The imposition of the first and third consequences (deduction of 1.3 times excess emissions and suspension of eligibility to make transfers, respectively) on a given party will have different effects on the economies of complying parties in terms of the quota (trading units) price and on price differences between different fossil fuels and emissions intensive products. This may lead to the possibility of strategic considerations on the part of members of the Enforcement Branch in deciding whether to apply consequences or not. Furthermore, there is the possibility that a party may withdraw from the treaty when threatened, which could have a variety of economic effects on complying parties. Suggestions have been made as to how to improve the compliance mechanisms to eliminate this possibility. For instance, as an alternative to the third consequence, if a party is found to have exceeded its allowance, then part of this allowance could be transferred to other parties. This alternative serves to overcome a situation where the Enforcement Branch is reluctant to punish non-compliance. However, since this alternative gives all compliant parties more allowances, it may make the Enforcement Branch too eager to punish, thus the possibility of strategic behaviour is not eliminated. Another option is to define a penalty per unit of exceeded emissions independent of quota trading. However, while the 1.3 penalty remains, there is still a possibility of strategic voting, though it may be greatly reduced (Hagem *et al.* 2003; Hagem and Westskog 2005). Of course, the problem of strategic behaviour will not arise if Compliance Committee members serve in their individual capacity as they are meant to, and do not take national interests into account. The selection of the candidates for the Enforcement Branch should therefore be scrutinized carefully to avoid 'political' appointments.

Problems with uncertainties

It has been suggested that uncertainties in emissions estimation could present difficulties for making accurate judgements over whether parties have met their emissions reduction targets or whether the overall Kyoto Protocol goal has been reached (Berntsen *et al.* 2005). The presence of uncertainties means that parties' reported emissions could be different from their actual emissions, so while a party appears to demonstrate fulfilment of commitments, its actual emissions exceed its target. The level of uncertainties in emissions trends can be about equal to (or sometimes even more than) parties' typical emissions reduction obligations. When a party's obligations are met with a margin less than the level of uncertainty, judgement of whether a party has achieved its goal would appear to be problematic. However, the compliance procedures take this problem into account: if an inventory has been approved by an ERT, uncertainties no longer feature in compliance assessment. Uncertainties are also unlikely to be a divisive issue concerning the overall emissions reduction goal. As mentioned above, the quality and appropriateness of methodologies used in emissions estimates is more likely to cause difficulties in compliance assessment than uncertainties. That is not to say that the presence of uncertainties is not important, rather that there are procedures in place for dealing with them so that the reporting and compliance systems can work efficiently and effectively. Furthermore, uncertainties will be reduced as parties improve their inventories and will be made easier to assess when uncertainty management systems are fully implemented. It will be important to target methodological problems when improving inventories from year to year, in order to minimize disputes over non-compliance.

External Factors

Factors external to the structure of, and capacity in, the verification system also take their toll on its efficacy. The current uncertainty over the future of the post-2012 climate regime reduces the potency of the compliance regime. For instance, the threat of imposing a 1.3 emissions deduction penalty in the second commitment period appears somewhat empty if it seems likely that there is not going to be a second commitment period or if what takes the place of the second commitment period does not resemble the first in any way. In addition, an atmosphere of resignation may pervade compliance activities if it is widely believed that there will be no available process to carry out compliance assessment after 2012. Furthermore, uncertainty over the future of the climate change regime also affects confidence in the carbon trading market.

There are, however, positive external factors. Actors such as NGOs, stakeholders and the public can, and sometimes do, play a role in reviewing and promoting compliance in international environmental agreements. Within the UNFCCC and Kyoto Protocol processes, NGOs have been particularly active both in reviewing action by states and in attempting to influence the development of the compliance regime. As discussed earlier, there are opportunities for NGO involvement in compliance assessment within the formal compliance system. Furthermore, NGOs may try to promote compliance with the protocol through strategies external to the formal compliance procedures. Other external actors, such as international organizations, also promote action and compliance by providing assistance.

Deeper Problems

Until the uncertainty over the future of the climate change regime is gone, compliance with the treaty will rely on the goodwill of parties. However, a closer analysis of the treaty reveals that goodwill may be a key factor in compliance issues regardless of the fate of the current regime: it turns out that the compliance system appears to have far blunter teeth than its elaborate consequences would suggest. Listed below are inherent weaknesses in the compliance system regarding enforcement:

- Punishment is forever delayed: if the Enforcement Branch applies the 1.3 deduction penalty for the next commitment period, a state could conceivably put off the punishment repeatedly to each subsequent commitment period.
- If a party believes that it will fall into non-compliance in the first period, it may attempt to negotiate a weak emissions limit in the next period to make up for any penalty that is applied.
- A party can avoid compliance with the compliance penalty itself.
- A party could simply withdraw from the treaty, or not ratify an amendment concerning compliance procedures, or make its participation conditional on being allowed a high emissions ceiling (Barrett 2003; Wang and Wiser 2002; Wiser and Goldburg 2000).

Ultimately, the compliance system in the current climate change regime lacks any way of making determined non-compliance, withdrawal and non-participation more costly for a state than complying or participating. With its current provisions, the regime is reliant on states wishing to comply or participate, so that the integrity of the regime is preserved. Continued dedication of this sort is far from certain, especially when the first commitment period and the reality of the emissions reduction commitments begin to bite. Moreover, there are prominent examples demonstrating the option states have for non-participation in the treaty. However, there may be ways of deterring non-compliance and non-participation among the enforcement approaches already noted, namely trade measures. The next section discusses the applicability of these options to the climate change regime.

Trade measures

Although trade measures would appear to be a powerful way of readjusting incentives both in order to encourage compliance by parties and to discourage non-participation, they are not included in the tool kit of the climate change regime compliance system. Nevertheless, application of trade measures to compel a state to participate or to improve compliance levels could take place outside of the current climate change regime, either unilaterally or by a coalition of states.

Possible trade measures include trade bans, tariffs or border tax adjustments. Trade measures are used within some international environmental treaties to promote participation and compliance. Two examples are the Montreal Protocol and 1973 Convention on International Trade in Endangered Species of Wild Fauna and Flora (CITES). The threat of trade measures under the Montreal Protocol appears to have induced several non-parties to participate and played a limited role in increasing compliance (Barrett 2003 and OECD 2000). In addition, trade measures have discouraged non-participation and played a limited role as sanctions for non-compliance in CITES. In these cases, the use of trade measures is limited to trade in the same areas that these treaties cover. Lessons for the climate regime can be learned from the experience of other treaties which use trade measures. However, the differences between regimes must be taken fully into account. Consideration of the use of trade measures must be set against the backdrop of the World Trade Organization (WTO). The WTO lays down certain principles relating to free trade that trade measures may impinge on. However, it also contains an article on exceptions (Article XX) which permits trade measures on environmental grounds in certain cases.

Multilaterally coordinated trade measures, which are mandated or authorized by international agreements, are less vulnerable to challenge under international trade rules than unilateral action, particularly if the target state is a party to the agreement in question (Stokke 2005). Indeed, trade measures in multilateral environmental agreements are unlikely to be challenged internationally. One reason for this is the undesirability of calling into question a treaty signed by many states (OECD 2000). It should also be noted that trade measures are asymmetrically available because their effectiveness depends on the size and diversity of the enforcer's and target state's economy, and therefore they favour states that are more powerful (Stokke 2005). In order for the threat to use trade measures to be credible, it must either originate from a powerful state or a coalition of states with approximately the same incentives. Furthermore, when a powerful state believes that unilateral use of trade measures is in its interest, the threat to use such measures may be more credible than action involving a coalition, since it does not depend on the consent of other countries. In a situation where trade measures are applied to increase participation or improve compliance in the Kyoto Protocol, the way in which the state applying the trade measures and its target are treated under international law would depend on whether the enforcer and target state are parties to either the WTO or the Kyoto Protocol, or both, or neither. Non-members of the WTO will receive the least legal protection, while a non-complier with the Kyoto Protocol would be more shielded than a non-party to the Kyoto Protocol (Stokke 2005). Hovi (2005) concludes that measures applied without the mandate of an international regime can enhance the deterrent effect of a regime but it may also operate to undermine the regime's legitimacy. Greater cooperation and coordination between international regimes, and in particular, between the climate change regime and the, could mitigate potential conflict between them (Stokke 2005).

There is likely to be growing debate in the international community over the use of external or internal trade measures to promote participation or compliance. There are already murmurs at the international level of a move towards the use, or at least threat to use, trade measures against non-parties to the Kyoto Protocol.[27] Such murmurs may grow, since there is no other equally powerful recourse currently provided for under the treaty. External measures for this purpose or to discourage non-compliance could lead to unintended consequences, such as a harder anti-treaty stance, being adopted. In particular, such measures could weaken domestic action on climate change in non-participating countries, which would in turn reduce the likelihood of the state participating in international action.

Nevertheless, measures kept as a distant threat may serve to demonstrate the seriousness with which participating states take the issue.

While parties were previously reluctant to introduce stronger compliance measures into the regime, that does not mean that they will remain so. Non-participation in the Kyoto Protocol is currently the primary area of concern for the international community. However, once the first commitment period is underway, compliance problems may become the chief problem. This could in turn lead to graver non-participation issues. It is possible that some states may consider it necessary, either due to their perception of the gravity of the climate change problem or for economic reasons, or both, for the regime to have more stringent penalties than it is currently equipped with to ensure that withdrawal or persistence in non-compliance is not in a party's interest. If introduced, such measures should be a last resort. Overuse of penalties could be damaging to the maintenance of a cooperative atmosphere within the regime by creating an antagonistic stance between states. Incorporation of trade measures into the regime may not be easy. The high economic and political stakes in the climate change regime and the complexity of both the problem and breadth of action that must be taken to solve it, coupled with the large number of types of product that could be covered under trade measures, would require agreement on a particularly sophisticated and powerful sanction system to resolve the compliance and participation problem.[28]

It should be noted that financial penalties and a compliance fund were considered as a means of preventing or finding a remedy to non-compliance (Wang and Wiser 2002). However, without any stronger means of enforcement, these procedures would meet the same difficulties in terms of potential compliance avoidance by parties, as exist in the present compliance system.

Progress so far and improvement pathways

The current climate change verification system as described earlier is the result of more than a decade of intense negotiation, technical development and institutional capacity building within the UNFCCC, Kyoto Protocol and IPCC processes. This system, constantly evolving, forms the backbone of both the convention and the protocol.

Monitoring, reporting and reviewing the greenhouse gases and sectors under the Kyoto Protocol are formidable tasks. The level of difficulty in carrying out these tasks varies considerably between gases and sectors. Certain sectors, such as Land-Use, Land-Use Change and Forestry (LULUCF), are particularly difficult to monitor accurately. However, difficulties in collecting activity data and use of appropriate emission factors affects the whole inventory compilation process. Nonetheless, Annex I parties have steadily improved both their annual inventories and also national communications with respect to the quality and timeliness of submissions. Timeliness of submission is a key factor in the verification system, as delays can hinder the assessment of implementation and compliance.

Many Annex I parties' recent inventories are of good quality. They are more complete than former ones were, both in terms of the explanations supporting submissions and the number of years for which estimates are available (UNFCCC 2004a). Furthermore, the number of parties making late submissions has decreased substantially. Parties have gained experience in running their national systems, though several still have some way to go in their development. Setting up and running national systems for effective national inventories is a complex process requiring a considerable amount of expertise and resources. The composition of, and arrangements for, each party's national system differ widely, based on factors such as the nature of its economy, the structure and size of its bureaucracy and the way in which political and bureaucratic authority is devolved within the state. Types of entities involved in preparing an inventory can include government departments and agencies and, in some cases, research facilities and private entities as well. The primary factors determining the difference in the levels of development between parties is the degree of experience in emissions monitoring a party has had, how well resourced and coordinated the relevant institutions are and the level of technical capacity and experience that personnel have. Some parties had a head start in this field as a result of previous domestic or regional monitoring activities.

The national systems and inventories of EIT parties tend to be less developed than those of other Annex I parties but have also improved greatly in recent years. However, they often require institutional improvements and sometimes lack a complete time series of data. Problems with completeness, transparency and the need to establish QA/QC procedures are also present. Furthermore, base year data[29] for some of these parties are currently deficient. In order to improve further their inventories, EIT parties require capacity building assistance. In particular, they need greater technical expertise in personnel, institutional strengthening and greater financial resources. While Annex II parties' national systems are more developed, they still need improvement. Again, completeness and transparency are principal problem areas, in particular the need to explain choice of methodologies when the methodologies differ from those of the IPCC and the use of emissions factors. Often there is a need for a higher tier of detail to be used for key sources. Implementation of QA/QC procedures and quantitative uncertainty analysis needs to be improved in varying degrees by all Annex I parties. With regard to inventory improvement, the degree to which Annex I parties specify clear improvement plans and implement their improvement plans or the ERT recommendations varies widely (UNFCCC 2004b).

Inventory preparation can be an arduous task for non-Annex I parties. Inventory compilation problems lie in the unavailability, inaccessibility and/or quality of data in energy, agriculture, and Land-Use Change and Forestry sectors, as well as the lack of disaggregated data required to apply the IPCC methodology. Data on land use and forest cover are often out of date. Problems with the activity data can result from inadequate data collection and/or management systems. These parties need stronger institutional capacity and coordination for collection, archiving and management of data for preparing the inventory and for the systematization/standardization of activity data. Many parties expressed the need to improve and update their inventories but would require financial and technical assistance to carry this out. Almost all these parties received external support in preparing their inventories (UNFCCC 2002, 2003b). Herold (2003) identifies several key problems for non-Annex I inventory preparation: the lack of a continuous inventory system resulting from inventory teams only working temporarily at a project basis; non-availability of activity data collected on a continuous basis in many sectors; lack of information on methods and data sources used (posing problems for time-series consistency if teams change and if the national communication itself does not provide adequate guiding information); and the lack of individual review for these parties which means there is no targeted feedback for them to use to improve their inventories. However, Herold (2003) also notes that non-Annex I participation in the ERTs and the ERT training programme will increase capacity for inventory preparation in those experts' home countries.

Indeed, the review processes provide a powerful means of improving the quality of emissions monitoring for all parties through the feedback from, and experience gained by, experts. However, their effectiveness is hampered by the large quantities of data and methodologies they have to examine in a limited time-span (Rypdal *et al.* 2003). The more capacity that can be built into the roster of experts (through the training scheme) and the greater the number of measures implemented to strengthen the review process (for instance, by allowing reviewers more time) will go far to improve the quality of monitoring and reporting by parties as well as more accurate review reports.

Monitoring and reporting standards should increase over time as experience and capacity grows. Experience from other air pollution conventions shows that an institutional warm-up period is needed. Reporting under the Convention on Long-range Transboundary Air Pollution (CLRTAP) was initially lacklustre but improved over time, partly as a result of the establishment of an Implementation Committee which focused on this problem. So too with the ozone regime which also suffered from teething troubles in its reporting system but later experienced steady improvement. In this case the ozone secretariat has assisted considerably in improving the standard of reporting (Wettestad 2005). It is clear that assistance and facilitation provided or coordinated by convention bodies or other actors catalyses the process of improvement. In addition to the UNFCCC review process, workshops are a source of assistance to Annex I parties. EIT parties, in particular, receive support from various international organizations. When the Compliance Committee begins its work, the Facilitative Branch can start to assist parties in their monitoring and reporting requirements leading up to the first commitment period. The UNFCCC Secretariat facilitates the provision of assistance to non-Annex I parties. These parties should also receive financial assistance towards meeting their reporting commitments. Other organizations, sometimes in cooperation with the secretariat, also provide some assistance, including workshops. In addition, a consultative group of experts (CGE), established by the COP, gives advice to non-Annex I parties on ways to improve their national communications. These efforts need to be maintained and expanded if a truly comprehensive and effective global monitoring system is to be established in the short to medium term.

Annex I national communications have improved with each round of submissions. In particular, reporting on policies and measures has expanded, though problems relating to transparency, terminology and categorization of information linger. Timing of national communications submission has also improved. It is difficult to assess progress in non-Annex I national communications since very few parties have submitted their second report. However, the fact that a large number have submitted their first is encouraging.

National registries are presently under development. Parties are again at different levels of development. However, consultations held before UNFCCC meetings have shown that cooperation and the exchange of information and experience can help both states which are advanced in registry development and those that are in the early stages. The secretariat plays a primary role in facilitating registry development.

Remote sensing systems

A state's monitoring system can be improved in many ways, several of which have been outlined earlier. Remote sensing systems can offer an additional way to improve emissions monitoring.

Much of the activity in remote sensing related to climate change is directed towards earth observations involving the monitoring of many different indicators, for instance, sea level rise and ice cover. However, remote sensing can also provide much needed information for the LULUCF sector. These systems can be used to identify land-use and land cover in 1990 to assist in base-line calculations, and can be used to monitor afforestation, reforestation and deforestation (ARD) activities in the first commitment period and thereafter (Aschbacher 2002). Rosenqvist *et al.* (1999) identify as areas where remote sensing can be applied: the provision of systematic observations of relevant land cover; support to the establishment of carbon stock baselines; detection and spatial quantification of change in land cover; quantification of above ground vegetation biomass stocks and associated changes; and mapping and monitoring of sources of anthropogenic methane. Remote sensing data already provides some input to national inventory compilation for this sector and is permitted by the IPCC. Although the capacity to provide data for 1990 was limited, there has since been a significant increase in capacity and this trend is likely to continue. Recently, there has been a surge in the number of earth observation activities geared towards specific end-uses, in particular (but not exclusively), in the number of satellites. Major efforts to coordinate earth observation monitoring systems have recently been launched which will further improve capacity in this field. In particular, the Group on Earth Observation (GEO) has drawn up a ten-year implementation plan for a Global Earth Observation System of Systems (GEOSS) which will coordinate land-, sea-, air and space-based instruments.[30] Satellite data has the advantage of being independent, repeatable and comparable. However there is a need for states to agree on appropriate methodologies and standardization of practices to convert data into the required parameters (Aschbacher 2002). This process has begun under the auspices of GEOSS.

While states can use earth observation systems to build up national activity data and to verify their own carbon stock estimate, formal independent verification of states' inventories using satellite data would be difficult to establish. Some states may feel this is too great an infringement on state sovereignty. There would also be several legal issues concerning use of remote sensing that would need to be examined. Remote sensing of this kind should be regarded as a support mechanism for the UNFCCC and Kyoto Protocol, at least for the moment (Aschbacher 2002; Rosenqvist *et al.* 1999).

Remote sensing can also be used for measuring atmospheric concentrations of emissions. There must be sufficient knowledge about the atmospheric transport of gases, and about sink and loss processes, in order to run models to estimate the locations and amounts of emissions. Given the difficulties associated with drawing up emissions inventories and reviewing them, it is worthwhile considering to what extent remote sensing, which can provide independent and objective measurements, can augment or could even replace the inventory compilation and compliance system.

Measuring atmospheric emissions concentrations suffers from its own set of uncertainties which include complications caused by natural emissions and loss processes. Emissions of carbon dioxide, methane and nitrous oxide have a high level of uncertainty, especially regarding carbon dioxide from LULUCF activities and nitrous oxide from soils. On the other hand, uncertainties are lower for halogenated gases that do not have natural sources and have long lifespans in the atmosphere.

At the national level, the use of inverse modelling to make estimates based on measurements of atmospheric concentrations and numerical transport modelling suffers from several obstacles including insufficient measurement networks, inaccurate measurement techniques, lack of knowledge of natural sources and sinks, insufficient quality of meteorological data, and coarse representation of key transport processes in transport models. These obstacles give rise to uncertainties and a level of geographical resolution in the calculated distribution of emissions that prevent inverse modelling from being used either to replace inventories or for compliance assessment on its own (Rypdal *et al.* 2003; Berntsen *et al.* 2005). Apart from these technical problems, there would be considerable difficulties – political, logistical, methodological and organizational – in agreeing on and establishing an atmospheric monitoring system which could, on its own, monitor emissions and be used to assess parties' emissions compliance. The same is true for a regional emissions trading scheme. However, even though remote sensing systems cannot replace inventories or be used on their own for compliance assessment, they can perform verification of emissions in certain regions on a coarse spatial scale. Unrecognized emissions sources could also be identified using these systems. In the future, verification of emissions could take place on an increasingly finer resolution. However, the obstacles referred to above need to be overcome. In particular, more ground-based stations are needed as well as better network techniques and improved modelling. These stations can be used in combination with operations involving aircraft equipped with measuring instruments and satellite data (Rypdal *et al.* 2003; Berntsen *et al.* 2005).

The measurement of atmospheric concentrations at the global level is simpler than at the regional or national level, since gas transport uncertainties are not an influencing factor. Remote sensing systems could therefore be used for assessing the overall effectiveness of the Kyoto Protocol regime, as well as overall levels of implementation. These systems could consequently serve as a tool to support decision-making in negotiations over the regime's adequacy. If a specific atmospheric concentration level is set in future climate negotiations, such systems could be used to help determine both whether parties as a whole were on course to meeting this objective and, at the end of a specified time-period, whether the objective had been met. Remote sensing data of this type could also be used for exposing deficiencies in the reporting procedures (Rypdal *et al.* 2003; Berntsen *et al.* 2005).

It is clear that the verification system of the UNFCCC and Kyoto Protocol has both strengths and weaknesses and that there are many possible ways to rectify these weaknesses. However, the international community stands at a crossroads over coordinated action on climate change. There is a lively wide-ranging debate over the future of the climate change regime among governments, research institutes, NGOs and the business sector. The chief questions are: to what extent, if at all, should the Kyoto Protocol format be kept, and what would be the consequences in each different possible regime scenario? There are many different proposals on what form the future regime should take. The following sections examine some of the more prominent proposals, focusing on verification issues.

Verification and the future of the climate change regime

The Kyoto Protocol is sometimes criticized as having little environmental effect and for not involving developing countries in emissions reduction commitments. However, it should be noted that, first, the Kyoto Protocol was designed to be an iterative process: the 2008–2012 commitment period is meant to be only the first of many, during which commitments could be deepened (made more stringent), and second, the UNFCCC requires that developed countries take the lead in tackling climate change. The Kyoto Protocol reflects this requirement. This feature does not prevent other countries from taking on commitments of their own when they are ready at some point in the future. There are many options facing the international community for the post-2012 period, ranging from keeping all the elements of the Kyoto Protocol to introducing an entirely different structure. Many different proposals have been drawn up which attempt to create a more economically efficient, environmentally effective and a more equitable regime. Currently, there is little consensus on what the future climate change regime should be. Some states lean towards a structure similar to the Kyoto Protocol while others advocate significantly different formats (see Introduction chapter).

Negotiators have so far followed the principle that mobilizing international efforts to tackle climate change requires absolute emissions reduction commitments regulated by a sound verification system. A different type of international agreement with different goals will clearly require its own type of verification system. Assessment is needed both of the role verification and compliance would play in the overall structure of a particular regime and how each specific instrument under it could be successfully monitored.

Innovation and developments in the business world will be key drivers of emissions reductions. However, companies too have divergent opinions of what form the future of the climate change regime should take. One unifying theme binds all those companies that are prepared to engage in action on climate change, namely a desire for certainty in the future structure of the climate change regime and homogeneity in regulatory and market structures, so that they can make appropriate long-term investment plans.[31]

The following sections do not give a definitive answer to the question of what the future regime or verification system should look like. Instead, they seek to shed light on what the verification issues involved in each proposal might be and make suggestions as to what may be the preferable format. Other costs and benefits of the proposals will only be outlined briefly here.

Future climate change regime proposals

Proposals for the climate change regime usually consider one or both of the following elements depending on how broad their scope is: types of commitments, or participation and differentiation criteria. The key aim is to broaden (in terms of numbers of participating countries) and deepen the regime.

A long-term target?

The UNFCCC aims for stabilization of greenhouse gas concentrations at a level that would prevent dangerous climate change occurring. However, it does not specify what this concentration level should be. Nor does the Kyoto Protocol consider a long-term target. Indeed, there is currently no clear long-term target agreed on by the international community.

On the one hand, setting such a target could have many benefits. It might stimulate international negotiations. It would guide efforts and facilitate measurement and assessment of progress and facilitate evaluation of how much action is required. By setting a target, the risks of climate change may be addressed more effectively. By providing a clear signal to markets, businesses would be given confidence and investment in clean technology would be stimulated. Finally, it could promote public awareness and participation. On the other hand, achieving international consensus on a long-term target could be difficult, since opinions on many aspects of the climate change problem and perceptions of risk differ widely, as do states' national priorities. Attempts to negotiate such a target could conceivably stall negotiations and lead to inaction.

There are several different indicators that could be used to establish a long-term target including emissions, concentrations, radiative forcing, temperature levels or impacts. Each indicator has its own set of advantages and disadvantages. Alternatives to internationally negotiated targets include hedging strategies or an informal target. Some analysts as well as several governments have proposed various targets using various indicators. In the absence of an internationally negotiated target, it is possible that a target proposed by influential states could gradually become a guide for action (Torvanger *et al.* 2004).[32]

Types of target[33]

If states decide to continue with a target-based approach to tackling climate change, there are several alternatives to the fixed flat-rate emissions reduction targets established under the Kyoto Protocol. Each different type of target has its own benefits and costs. Fixed targets, such as those found in the Kyoto Protocol, are comparatively easier to monitor and to assess compliance with. Their fixed nature provides greater certainty on emissions levels but also greater cost uncertainty. Another option is for all states to use dynamic targets, that is, targets indexed to a variable. Indexed targets tend to provide more cost certainty, but with less certainty on emissions levels. Intensity targets, defined as a ratio of greenhouse gas emissions to gross domestic product (GDP), are a form of indexed target.[34] Targets of this type may present greater monitoring and compliance problems than fixed targets, since two variables have to be monitored rather than just one. Furthermore, measurement of GDP can be controversial and difficult in developing countries (Grubb 2004; Philibert *et al.* 2003). Finally, while dynamic targets appear to be compatible with emissions trading, they are unlikely to prove as simple as trading with fixed targets. Alternatively, developed countries could use a price cap on a quantitative target – where supplementary permits are available in unlimited quantity at a fixed price. With this type of target, high cost certainty is achieved but at the expense of certainty on emission levels. Emissions trading may also be difficult to implement with this system. For developing countries, non-binding targets, or no-lose targets, are possible, as these states are unlikely to accept binding or fixed targets. A no-lose target means that a developing country is able to sell on to an emissions trading scheme[35] if its emissions are below a given target level, but would not incur any penalties if its emissions go above that target.

Alternatively, economic sectoral targets could be established and could include any of the quantitative instruments outlined earlier. While these kinds of targets can ease some competitiveness concerns, they may raise others in their place. States also lose the flexibility to choose where to make emissions reductions. A sectoral approach raises several questions regarding monitoring, reporting, review and compliance procedures. If the sectoral targets were mandatory, national governments would have to be involved in order to ensure that targets were complied with in their own country and that data could be collected and assessed for veracity. CDM-style sectoral target proposals have received considerable attention; however, they may present accounting and compliance assessment difficulties if emissions reductions are specifically linked to policy action.

Alternatives to targets

While the target-based approach is the focus of current efforts to tackle climate change, it is not the only option. One alternative is a 'policies and measures' format, in which states could agree to implement certain policies in a 'pledge and review' system. However, even with a review system in place, it may be difficult to compare states' efforts in this type of scheme. Technology agreements are another option. Such agreements may appear attractive at first glance but face a number of potential problems related to technology choice and capture, international coordination problems linked with competitiveness concerns, negotiating complexity due to the vast number of different eligible technologies, and finally, inconsistency with market economics (Grubb 2004). Indeed, technology agreements may be more costly than an emissions trading based system (Philibert *et al.* 2003) and, on their own, may fail to stimulate the global market sufficiently (Philibert 2004; see also Barrett 2003). A target-based approach is instead needed to galvanize innovation, development, and dispersion. Technology agreements could, however, sit within a wider climate change regime framework. The complexity of monitoring a technology agreement would depend on the number and type of industries or products under consideration and the number and type of standards agreed. Other options include research and development commitments and internationally harmonized carbon taxes.

Emissions reduction targets and their alternatives vary in terms of their economic and environmental predictability and outcome. Different parties will find certain targets or alternatives more appealing than others depending on their perception of the climate change problem, national priorities and the nature of their economy. However, the degree to which these targets and the alternatives can be monitored and accounted for effectively, and their suitability for compliance assessment, also varies. Therefore it will be crucial to incorporate both evaluation of the implementation issues related to these approaches and assessment of their operationality into the debate on the future of the climate regime as early as possible.

Differentiation and allocation criteria

If the climate change regime is to secure broad participation, it must be seen to be fair. There are several different concepts of equity in relationship to the climate change regime. These include responsibility for the problem; equal entitlements; capacity to act; basic needs fulfilment; comparability of effort; and consideration of future generations (Aldy *et al.* 2003). Differentiation of action between countries could be based on any one or several of these equity principles, and use a variety of indicators (for instance, per capita GDP and per capita emissions) to order them into groups. Commitments or benefits which are appropriate to each group's characteristics can then be allocated, and could range from various forms of binding emissions reduction targets and the provision of financial assistance, to policies and measures obligations, non-binding targets or the receipt of financial assistance. Countries could move from one group to another by meeting certain economic and developmental graduation criteria which would apply when a country has met a criterion, such as a specific level of per capita GDP or per capita emissions.

The UNFCCC and Kyoto Protocol differentiate between Annex II, Annex I and non-Annex I parties. Each group has its own particular set of obligations. However, as suggested above, it is possible to devise more elaborate criteria which would differentiate more subtly between parties. The 'multi-stage' proposal has recently gained prominence and provides a pathway for developing countries to participate in the regime as well as an overall regime structure. It offers an approach using four stages of progressively stronger types of commitments for groups of parties. The differentiation criteria are based on per capita emissions, and an emissions allocation system is provided which may include targets that reflect national circumstances (Bodanksy 2004). Countries graduate from one stage to the next based on emissions per capita. In the 'South-North Dialogue' Ott *et al.* (2004) suggest that states could be classed as newly industrialized countries, rapidly industrializing countries, other developing countries, least developed countries, and Annex I and Annex II. Each different group would have its own responsibilities, commitments and/or benefits. Differentiation between countries and for determining commitments would be based on a country's potential to mitigate, responsibility to mitigate and capacity to mitigate (using cumulative per capita emissions, the Human Development Index and an indicator of potential derived from carbon dioxide/GDP and greenhouse gas per capita). Countries graduate when they pass over the threshold of each particular stage's criteria. An alternative approach is the 'triptych' proposal which is sector and technology oriented and calculates quantified emission limitation objectives. It establishes different burden-sharing rules for different economic sectors: convergence of per capita emissions in the domestic sector, efficiency for energy intensive industry and carbon intensity targets for the power generation sector. Objectives can be adjusted to suit desired levels of carbon dioxide concentrations. National targets are calculated by adding together the sectoral targets. Compliance assessment is based on the national rather than sectoral targets. This approach is intended to reflect the concerns of 'equity', needs and circumstances of developing countries, cost effectiveness and sustainable development (Groenenberg *et al.* 2004).[36] Another proposal that has received considerable attention is 'contraction and convergence' (Global Commons Institute 2005). This proposal establishes a global trajectory towards a specific concentration level of carbon dioxide. Under this proposal, all countries agree an annually reviewable target and then work out the rate at which emissions must contract in order to reach it. Allocations of carbon dioxide converge by a specific date from current emissions to allowances that are proportional to national populations (equal per capita emissions). The proposal is based on the principle of equal per capita emissions but does not take national circumstances into account. Emissions would be divided among regions and nations by negotiation.

The proposals for novel systems of differentiation and allocation vary in terms of complexity and operationality. Some of the proposed indicators that would be used in these systems could be contentious and raise difficult questions for negotiators. While a greater level of complexity could broaden participation, make action more suitable for each state and consequently lead to deeper commitments, the problem in establishing a transparent and robust international system for such systems will increase with the number of indicators used and the degree to which they are measurable and contestable. Implementation of new types of target would have to factor in new verification and compliance needs for all states using them. Proposals with new graduation, differentiation and allocation schemes, including different types of target, would have to include specific criteria governing standards of emissions monitoring and reporting for each stage or group in order to make compliance assessment and emissions trading possible. The development and implementation requirements for some of these proposals could therefore be daunting. The capacity to successfully implement and manage these systems would have to be rapidly built up in all states, especially in developing countries, with financial and technical assistance from developed countries. New international monitoring, reporting, review and compliance structures would need to be agreed on and implemented to successfully regulate these systems.

Each proposal now in front of the international community will be evaluated according to criteria such as environmental effectiveness, economic efficiency and concepts of equity. Given that verification and compliance systems build trust, measure progress and promote cooperation and action, their level of applicability to each proposal and the ease with which they can be implemented and maintained must also be included in the main criteria used to assess the alternative proposals.

Conclusion

The challenge facing the international community is to cultivate a climate change regime that is environmentally effective, economically efficient, equitable, and in which participation is broadened and commitments deepened. There are many options facing the international community for the post-2012 period ranging from keeping all the elements of the Kyoto Protocol to introducing an entirely different structure. The 2009 Climate Conference addressed verification and non-compliance, but no major achievements were made.

Effective verification and compliance systems are crucial for facilitating cooperative action between states and measuring and promoting progress towards treaty goals. The UNFCCC is a proven forum for coordinating action on climate change. Its verification systems should be maintained and continuously improved so that individual and total progress on tackling climate change can be measured with increasing accuracy. Initiatives for improving these systems should be enhanced to catalyse this process. The Kyoto Protocol expanded this system into a highly sophisticated verification and compliance system to reflect the complexity both of the treaty itself and the problem it addressed. The post-2012 structure could build on the positive components of the UNFCCC and Kyoto Protocol, in particular the verification system.

Since the current verification and compliance system of the Kyoto Protocol has both strengths and weaknesses, negotiators will have to assess the pros and cons of the potential remedies to the weaknesses in the compliance system as well as the options for improving the verification system. Experience in using these systems will help this process. However, negotiators will have to set the discussion of any improvements and remedies to the verification and compliance system firmly in the context of whichever proposals for the future regime gain prominence during negotiations.

A target-based approach which builds on the UNFCCC and Kyoto Protocol seems most likely to facilitate comparability of effort, measurement of progress and compliance assessment. It could be designed to broaden and deepen action through an equitable and workable system of differentiation and graduation. The framework could include several different types of targets. Such a system should promote cooperative action between states. Technology, research and development, and policies and measures agreements, could all have a role to play in the regime, as part of the wider framework. Whatever approach is adopted, it will be important to avoid a structure that, either by its particular characteristics or by its complexity, prevents effective monitoring and compliance assessment.

The development and implementation of the current climate change regime verification and compliance system has already taken several years, and the process is ongoing. Discussions and decisions over appropriate future structures, including new types of target and differentiation and allocation procedures, will have to take into account the time it would take to develop and implement new verification systems, which could be considerable. Early consideration is needed of states' capacity for verification systems. The institutional, legal, technical and resource base of all states, and especially EITs and developing countries, will need to be carefully assessed when weighing up which approaches are feasible and what needs to be done in order to successfully and swiftly implement effective verification systems.

Acknowledgements

I would like to thank Cedric Philibert and Joseph Aschbacher for their valuable comments on this chapter and all those who helped with my research. The views expressed herein are, of course, the author's, with whom responsibility for any errors or shortcomings rests.

Notes

1 In the climate change field, the term 'verification' can also refer more specifically to the checking of greenhouse gas emissions reports for reliability.
2 Free-riding can include non-compliance with, or non-participation in, a treaty.
3 The UNFCCC covers all greenhouse gases not covered by the 1987 Montreal Protocol to the United Nations Convention on Protection of the Ozone Layer. The focus of the monitoring and reporting processes and the Kyoto Protocol emissions reduction targets is on carbon dioxide, methane, nitrous oxide, hydrofluorcarbons, perfluorocarbons and sulphur hexafluoride. In order to promote effective policy formulation on climate change, the Intergovernmental Panel on Climate Change (IPCC) provides a measurement system for comparing the potential of each greenhouse gas to contribute to global warming. The global warming potential (GWP) of each gas shows how much it contributes to global warming in relation to the reference gas, carbon dioxide, which has a GWP of 1.
4 'The Parties should protect the climate system for the benefit of present and future generations of humankind, on the basis of equity and in accordance with their common but differentiated responsibilities and respective capabilities. Accordingly, developed country parties should take the lead in combating climate change and the adverse effects thereof' (Article 3 of the UNFCCC).

5 The UNFCCC also promotes climate observation and monitoring. A primary component of this is parties' contribution to the Global Climate Observing System (GCOS) of the World Meteorological Society which was set up to provide comprehensive observations required for monitoring the climate system; for detecting and attributing climate change; for assessing the impacts of climate variability and change; and for supporting research towards improved understanding, modelling and prediction of the climate system.

6 The issue of the timing and frequency of the developing countries' national communications has proved difficult to reconcile and was only resolved in May 2005.

7 National communications are expected to include: information on national circumstances pertaining to greenhouse gas emissions and removals; greenhouse gas inventory information; policies and measures; projections and overall effect of policies and measures; vulnerability and impact assessment and adaptation measures; account of financial resources or assistance and transfer of technology; report on research and systematic observation; information on education, training and public awareness.

8 Certain exceptions can apply whereby parties use a different base year.

9 All UNFCCC guidelines are available on the UNFCCC website http://www.unfcccc.int.

10 All IPCC guidelines are available on the IPCC website httl://www.ipcc.ch.

11 The IPCC guidelines define emissions factors as providing a representative rate of emission for a particular activity level under a particular set of operating conditions.

12 This is because these systems may be inappropriate for certain sources, for instance mobile sources such as cars or wide area sources in agriculture. In addition, for those sources for which direct measurement is suitable (for instance, industry and some energy sources) direct measurement systems can be costly to install (though less costly if there is a pre-existing system which only has to be modified to measure another gas). Also, the systems should be checked against certain standard criteria. Furthermore, direct measurement is not always more accurate than emissions calculation.

13 IPCC guidelines note that uncertainties may be caused by, among other factors, differing interpretations of source or sink categories or by the use of simplified representations with 'averaged' values, especially emissions factors and related assumptions to represent characteristics of a given population. Uncertainty in socio-economic activity data and in the understanding of the basic processes leading to emissions and removal may also affect estimates.

14 There are several other bodies that compile international emissions datasets including the International Energy Agency (IEA); the Carbon Dioxide Information Analysis Centre (CDIAC); the Global Emissions Inventory Activity (GEIA) Centre; and the Emission Database for Global Atmospheric Research (EDGAR).

15 For instance, CORINAIR (Core Inventory of Air Emissions), the European emissions inventory programme.

16 The convention also contains provisions related to the settlement of disputes concerning the interpretation or application of the convention. Appeals can be made to the international Court of Justice (ICJ), but, according to Wang and Wiser (2002), since proceedings can take a long time and are confrontational, states are reluctant to pursue this route in multilateral environmental agreements.

17 The EU, as a regional economic integration organization, was also given an overall target (as well as each member state having its own specific target).

18 Parties must keep a minimum level of units in the registry (known as the 'commitment period reserve') to prevent them from 'over-selling'. A different registry is used for issuance and distribution of CERs.

19 The state is eligible if: it is a party to the Kyoto Protocol; its assigned amount has been calculated and recorded; it has in place its national system; it has in place its national registry; it has submitted annually the most recent required inventory; it submits certain supplementary information on its assigned amount.

20 CDM projects must be approved by 'designated national authorities'. Participants must prepare a project design document that includes a description of the proposed monitoring methodology and baseline. The project must then be validated by an operational entity and registered by the relevant convention body. It must be monitored and a monitoring report must be submitted to the operational entity for verification that will then certify the emission reductions. Parties involved in JI projects must inform the UNFCCC secretariat of their national guidelines and procedures for approving these projects as well as monitoring and verification procedures. In order to fast track the process, JI projects can take place when a host party meets only some of the eligibility requirements. In this case project participants must submit a plan to an independent body (accredited by the relevant convention body) for evaluation and have established an appropriate baseline and a monitoring plan.

21 A question of implementation is a problem pertaining to language of mandatory nature in the relevant UNFCCC guidelines influencing the fulfilment of commitments.

22 In the climate change regime, the UNFCCC Secretariat cannot raise questions of implementation.

23 An expedited procedure exists for parties to have their eligibility restored.

24 The organizations were the World Trade Organization and International Labour Organization, while the treaties were the Montreal Protocol and the Convention on Long-Range Transboundary Air Pollution. The two NGOs were the Centre for International Environmental Law (CIEL) and the World Wide Fund for Nature (WWF) (Wang and Wiser 2002).

25 Australia, Canada, Japan, New Zealand, Norway, Russia, Ukraine and the US.

26 The Kyoto Protocol uses the same dispute resolution mechanism as the UNFCCC, see Endnote 16.

27 A Member of the European Parliament (MEP) recently argued that, by refusing to ratify Kyoto, the US is shielding its exporters from the costs of averting climate change, effectively handing them a state subsidy. The MEP called on the European Commission (EC) to raise the issue at the World Trade Organization (WTO) at the earliest opportunity and to demand authorization to impose countervailing duties or border taxes on US imports into the European Union (EU). However, the EC said it currently had no intention of imposing trade measures or raising trade distorting effects of Kyoto Protocol compliance at the WTO. On the other hand, recent activities in the European Parliament Environment Committee may indicate a change in attitude on this matter. The Environment Committee has adopted a resolution which called on participants in the climate change talks held in May 2005 to explore the scope for imposition of trade sanctions on states that do not ratify the Kyoto Protocol. Following this, the European Parliament adopted a resolution calling on the EC to consider the possibility of using border adjustment measures on trade (News Release from MEP Caroline Lucas, 9 March 2004; European Parliament Committee on the Environment, Public Health and Food Safety, 2005; European Parliament, 2005).

28 Indeed, Barrett (2003) argues against using the Montreal Protocol as a precedent for the climate regime and that trade measures are unsuitable for enforcing the Kyoto Protocol. According to Barrett, the Montreal Protocol is much easier for parties to implement and comply with, and the adverse impacts from ozone depletion are more clearly perceived, and the benefit-cost ratio of action in the Montreal Protocol is much higher than that of the climate change regime; consequently parties are more likely to participate and comply. Barrett further argues that sufficiently severe trade measures under the Kyoto Protocol are not credible, border tax adjustments impractical or ineffective and they may expose a tension between liberalized trade and environmental protection. However, a recent study by the Swedish National Board of Trade found there is scope for trade related measures pursuant to the Kyoto Protocol to be in accordance with WTO rules, although they do note some areas of concern (Swedish National Board of Trade 2004).

29 Problems with base year data makes calculation of assigned amounts difficult.
30 At COP9 (December 2005, Buenos Aires), a decision was taken calling for GCOS to collaborate with the *ad hoc* Group on Earth Observations to implement a plan for global climate observations. GCOS involves the use of many different types of land, sea, air and space based monitoring instruments.
31 It should be noted that certain initiatives have recently been developed in response to the need for corporate greenhouse gas emissions reporting. These include the Greenhouse Gas Protocol of the World Business Council for Sustainable Development (WBCSD) and the World Resources Institute (WRI) and a new International Organization for Standardization (ISO) standard on greenhouse gas accounting and verification. In addition, domestic and regional emissions trading schemes require detailed emissions accounting and reporting by companies and verification of the reports.
32 The EU has in fact suggested a target of limiting global average temperature increase 2°C above pre-industrial levels (EC Communication 2005; European Council 2005) and outlined the likelihood of meeting that target at various concentration levels.
33 This section is based on Philibert *et al.* (2003); Philibert and Pershing (2002); Aldy *et al.* (2003).
34 However, Dudek and Golub (2003) argue that intensity targets do not necessarily provide greater certainty about costs. They also argue intensity targets may raise the costs associated with a given emission level by making it harder to implement emissions trading, and, finally, that the introduction of variables such as gross domestic product (GDP) into the negotiations may well stall them.
35 The emissions trading scheme would also have to involve states (developed countries) with firm targets in order to ensure there are buyers.
36 The EU used a triptych approach to work out its burden sharing agreement of greenhouse gases.

References

Aldy, J., Ashton, J., Baron, R., Bodansky, D., Charnovich, S., Diringer, E., Heller, T., Pershing, J., Shukla, P., Tubiana, L., Tudela, F. and Wang, X. (2003) *Beyond Kyoto – Advancing the international effort against climate change.* Arlington: Pew Centre on Global Climate Change.
Aschbacher, Josef. (2002) Monitoring Environmental Treaties Using Earth Observation, In Verification Yearbook 2002. London: VERTIC. 171–186.
Barrett, S. (2003) *Environment and Statecraft.* Oxford/New York: Oxford University Press.
Berntsen, T., Fuglestvedt, J. and Stordal, F. (2005) Reporting and Verification of Emissions and Removals of Greenhouse Gases, In Stokke, O., Hovi, J and Ulfstein, G. (eds.) *Implementing the climate change regime.* London: Earthscan. 85–105.
Bodansky, D. (2004) *International Climate – Efforts beyond 2012: A Survey of Approaches.* Arlington: Pew Centre on Global Climate Change.
Chayes, A. and Chayes, A. (1995) *The New Sovereignty: Compliance with international regulatory agreements.* Cambridge and London: Harvard University Press.
Corfee Morlot, J. (1998) Ensuring compliance with a global climate change agreement. *OECD Information Paper,* ENV/EPOC(98)5/REV1.
Dudek, D. and Golub, A. (2003) Intensity targets: pathway or roadblock to preventing climate change while enhancing economic growth? *Climate Policy.* 3 (S2). S1–S156.
Earth Negotiations Bulletin (2001a), Chasek, P. (ed.) Summary of the Resumed Sixth Session of the Conference of Parties to the UNFCCC, 16–27 July 2001. International Institute of Sustainable Development. 12 (176).
Earth Negotiations Bulletin (2001b), Chasek, P. (ed.) Summary of the Seventh Conference

of Parties to the UNFCCC, 29 October–10 November 2001, International Institute of Sustainable Development. 12 (189).

EC Communication (2005) Communication from the Commission to the Council, the European Parliament, the European Economic and Social Committee and the Committee of the Regions, Winning the Battle against Climate Change, {SEC(2005) 180}. Brussels, 9 February 2005 COM (2005) 35 Final.

European Council (2005) Presidency Conclusions, Brussels, 23 March 2005, 7619/1/05 REV 1.

European Parliament (2005) Resolution on the Seminar of Governmental Experts on Climate Change, P6_TA-PROV(2005)0177, PE 357.354.

European Parliament Committee on the Environment, Public Health and Food Safety (2005) Amendments 1–20, Draft Motion for a Resolution, The Seminar of Governmental Experts on Climate Change, PE 375.672v01-00, (PE 357.580v01-00).

Global Commons Institute website, 2005, www.gci.org.uk.

Groenenberg, H., Blok, K. and van der Sluijs, J. (2004) Global Triptych: a bottom-up approach for the differentiation of commitments under the Climate Convention. *Climate Policy*. 4 (2). 153–175.

Grubb, M. (2004) Kyoto and the Future of International Climate Change Responses: From Here to Where? In Nishioka, S. (ed.) *International Review for Environmental Strategies*. 5 (1). IGES, Japan. 15–38.

Hagem, C., Kallbekken, S., Maestad, O. and Westskog, H. (2003) Tough Justice for Small Nations, *CICERO Working Paper*, 2003:01.

Hagem, C. and Westskog H. (2005) Effective Enforcement and Double-edged Deterrents: How the Impacts of Sanctions also Affect Complying Parties, In Stokke, O., Hovi, J, and Ulfstein, G. (eds.) *Implementing the Climate Change Regime*. London: Earthscan. 107–120.

Herold, A. (2003) Current Status of National Inventory Preparation in Annex I Parties and non-Annex I Parties, OECD/IEA, COM/ENV/EPOC/IEA/SLT(2003)7.

Hovi, J. (2005) The Pros and Cons of External Enforcement, In Stokke, O., Hovi, J. and Ulfstein, G. (eds.) *Implementing the Climate Change Regime*. London: Earthscan.129–145.

News Release from MEP Caroline Lucas, 9 March 2004.

OECD (2000) Trade Measures in Multilateral Environmental Agreements. OECD Industry, Services & Trade, 2000. 1999 (26).

Ott, H., Winkler, H., Brouns, B., Kartha, S., Mace, M. J., Huq, S., Kameyama, Y., Sari, A., Pan, J., Sokona, Y., Bhandari, P., Kassenburg, A., Lèbre La Rovere, E. and Rahman, A. Atiq (2004) *South-North Dialogue on Equity in the Greenhouse*, Eschborn: Deutsche Gessellschaft für Technische Zusammenararbeit (GTZ) GmbH.

Philibert, C. (2004) Lessons from the Kyoto Protocol: Implications for the Future, In Nishioka, S. (ed.) *International Review for Environmental Strategies*. 5 (1). IGES, Japan. 311–322.

Philibert, C. and Pershing, J. (2002) Beyond Kyoto – Energy Dynamics and Climate Stabilisation. OECD/IEA 2002, IEA Publications, Paris, France.

Philibert, C., Pershing, J., Morlot, J.C., and Willems, S. (2003) Evolution of Mitigation Commitments: Some Key Issues, OECD and IEA Information Paper.

Raustalia, K. and Victor, D.G. (1998) Conclusions, In Victor, D.G., Raustiala, K. and Skolnikoff, E.B. (eds.) *The Implementation and Effectiveness of International Environmental Commitments: Theory and Practice*. MIT Press, Cambridge. 659–707.

Rosenqvist, A., Imhoff, M., Milne, A. and Dobson, C. (eds.) (1999) Remote Sensing and the Kyoto Protocol: A Review of Available and Future Technology for Monitoring Treaty Compliance. Report of a workshop, October 20–22. Ann Arbor, MI: Ann Arbor.

Rypdal, K., Stordal, F., Fuglestvedt, Jan S. and Berntsen, T. (2003) Assessing compliance with the Kyoto Protocol: Expert reviews, inverse modelling, or both? *CICERO Working Paper*, 2003:07.

Stokke, O. (2005) Trade Measures, WTO and Climate Compliance, In Stokke, O., Hovi, J and Ulfstein, G. (eds.) (2005): *Implementing the climate change regime*. London: Earthscan. 147–165.

Swedish National Board of Trade (2004): Climate and Trade Rules – harmony or conflict? http://www.kommers.se/binaries/attachments/3430_Climate%20and%20Trade%20Rules.pdf.

Torvanger, A., Twena, M. and Vevatne, J. (2004) Climate policy beyond 2012, CICERO Report 2004:02.

UNFCCC (2002) Fourth compilation and synthesis of initial national communications from Parties not included in Annex I to the Convention, available through http://unfccc.int/cop8/latest/3_sbi.pdf, last accessed 31 August 2012.

UNFCCC (2003a) Counting emissions and removals, greenhouse gas inventories under the UNFCCC, available through http://unfccc.int/resource/docs/publications/counting.pdf, last accessed 31 August 2012.

UNFCCC (2003b) Fifth compilation and synthesis of national communications from Parties not included in Annex I to the Convention, available through http://unfccc.int/resource/docs/2003/sbi/13.pdf, last accessed 31 August 2012.

UNFCCC (2004a) Issues relating to greenhouse gas inventories, available through http://unfccc.int/ghg_data/ghg_data_unfccc/items/4146.php, last accessed 31 August 2012.

UNFCCC (2004b) Reports of the individual review of greenhouse gas inventories 2004, available through http://unfccc.int/national_reports/annex_i_ghg_inventories/review_process/items/2762.php, last accessed 31 August 2012.

Wang, X. and Wiser, G. (2002) The Implementation and Compliance Regimes under the Climate Change Convention and its Kyoto Protocol. RECIEL 11 (2) 2002. Blackwell Publishers Ltd, Oxford, UK and Malden, US.

Wiser, G. and Goldburg, D. (2000) Restoring the balance. Report prepared for the World Wildlife Fund.

Wettestad, J. (2005) Enhancing Climate Compliance – What are the Lessons to Learn from Environmental Regimes and the EU? In: Stokke, O., Hovi, J. and Ulfstein, G. (eds.) *Implementing the climate change regime*. London: Earthscan. 209–231.

16 Difficulties of Benefit-Cost Analysis in Climate Negotiations

Stumbling Blocks for Reaching an Agreement

Charles Pearson

Introduction

Neither the 1992 United Nations Framework Convention on Climate Change (FCCC) nor the 1997 Kyoto Protocol rest on a benefit cost (BC) foundation. The FCCC objective is to stabilize greenhouse gas concentrations "at a level that would prevent dangerous anthropogenic interference with the climate system." Neither costs nor benefits are explicit in setting the objective, although members are admonished to take policies that are cost effective (i.e. least cost). The Kyoto Protocol established mandatory greenhouse gas emission targets for Annex I countries for the first commitment period (2008–2012). But there was no attempt to defend either the aggregate Annex I reduction target (averaging 5 per cent below 1990 levels), or the allocation of emission reductions among countries, on a benefit cost basis.[1]

The absence of BC input into policy cannot be explained by the absence of analysis. A number of first generation models had been published.[2] Nordhaus (1993), for example, estimated that to stabilize emissions at 1990 levels, the voluntary target associated with the FCCC would have had a net present value *loss* of US$ 7 trillion as compared with the optimal policy. In a simple model, Maddison (1995) calculated that a business as usual policy (i.e. no global warming policy) would beat a policy of stabilizing emissions at 1990 levels by US$ 3 trillion. Not all studies were adverse to aggressive action. Cline (1992), for example, reports a benefit-cost ratio from substantial near term abatement measures to increase from 0.86 to 1.52 when the social rate of time discount was changed from 3 to 1.5 per cent. In any event, the strengths, weaknesses, and results of BC analysis were thoroughly thrashed out in the 1996 Intergovernmental Panel on Climate Change Second Assessment Report and were, together with scientific information, neatly assembled and available before Kyoto.[3]

Perhaps this neglect of BC analysis in policy is as it should be. Perhaps the global warming *problematique* is not amenable to a calculus of costs and benefits. Perhaps insistence that policy pass a BC test would be a stumbling block to an effective response, and other approaches would yield better results.

Perhaps. But it would be remarkable if BC analysis played a truly negligible role in climate policy. Indeed, at first glance, it appears to be an almost ideal tool. It has come to be the single most powerful and widely used economic technique to evaluate public expenditure and public policy. BC has many attractive features: It is grounded in mainstream economic theory (welfare economics); it champions efficiency; it is adaptable to decisions involving time and scale; and, through shadow pricing, can accommodate distorted market prices and non-market effects. At the risk of glossing over numerous assumptions, BC can provide a single number for decision makers, the net present value (NPV) of the project or policy.[4] Finally, BC can be joined to climate, environmental, and ecological analysis in Integrated Assessment Models.

And indeed a closer look suggests that benefit cost analysis has not been totally neglected. While the Kyoto Protocol is not grounded in benefit cost, it has been subjected to numerous such assessments, many of which are critical. Tol (1999) and Nordhaus and Boyer (2000) are examples. At the political level, President Bush, in explaining his opposition to the Kyoto Protocol, explicitly cited prospective damage to the US economy as well as an unfair burden on the US:" I oppose the Kyoto Protocol because it exempts 80 per cent of the world, including major population centers such as China and India, from compliance, and would cause serious harm to the US economy."[5] On the other side, for Russia the prospective opportunity to benefit from sales of excess carbon permits as permitted by Kyoto, thus improving its national level BC calculations, was a factor in its eventual ratification.[6] Finally, benefit cost continues to be employed to evaluate modifications and alternatives to Kyoto. For example, Buchner and Carraro (2003) analyze whether expansion to include China would alter US benefit cost calculations so as to induce US participation in Kyoto.

This paper identifies the strengths and weaknesses of BC in formulating climate policy. Section II lays the foundation by presenting the basic assumptions of BC analysis. Section III considers the validity of these assumptions in light of the distinctive characteristics of global warming. Section IV introduces some alternatives to BC, and Section V concludes. A full understanding of the role of BC in the global warming debate is especially desirable in light of recent studies reporting lower marginal damages than earlier studies showed (Tol 2005; Mendelsohn 2005; Pearce 2005). At first glance these results suggest less urgency in confronting global warming. But because major issues concerning the treatment of catastrophe, the incorporation of option values, equity weighting, measuring existence values, and the need for time varying discount rates, have yet to be fully sorted out, the underlying assumptions and techniques of BC analysis need elaboration. A systematic inclusion of these complexities could easily modify the tentative consensus that now exists, and could possibly support early aggressive action.

Benefit Cost and its assumptions

BC is a decision tool designated to improve efficiency and social welfare in government policies and projects. It compares social welfare with and without the policy (project). If B_i is the benefit of a policy to an individual and C_i is the cost, the decision criterion is net benefits summed over all affected individuals, N, or

$$NB = \sum_{i=1}^{N} (B_i - C_i)$$

If NB > 0, the policy passes the BC test. If the effects are experienced over time, both costs and benefits must be compared at the same time, generally present values. Thus the criterion is

$$NPV = \sum_{i=1}^{N} \sum_{t=1}^{T} \frac{(B_i - C_i)}{1 + r}$$

Where **t** is time, **T** some future time when the effects of the policy cease, and **r** is the discount rate. If net present value is positive (NPV > 0), the policy (project) passes the basic BC test.[7]

If the scale or intensity of a policy (project) is a choice variable, BC may seek out the optimal scale at which NPV is maximized. Given smooth, differentiable functions, this is where marginal benefits equal to marginal costs,

MC = MB

Thus for optimality, two tests must be passed,

MC = MB
NPV > 0

Finally, in dynamic optimization models applied to global warming, the marginal cost of abating the last (marginal) ton of the greenhouse gas is set equal to the discounted value of the stream of damages avoided by not emitting that ton (the marginal benefit), and this is repeated at each point in time to build an optimal time path of abatement. At that point, the analyst would check that the aggregate discounted benefits of the program exceed the discounted costs. Even when optimality is not sought, BC analysis of various policy proposals conveys useful information: a finding that NPV > 0 indicates that the policy results are an improvement in the efficiency with which resources are allocated.

BC analysis requires that both benefits and costs are in the same metric – monetary units. If it is not possible to identify and monetize benefits, the fallback technique is least cost (cost-effectiveness) analysis. In this approach a quantifiable target is selected, presumably in a political process (for example reduction in the incidence of malaria, limiting discharge of oil into the marine environment, or setting a limit on arsenic in drinking water), alternative measures to achieve the objective are identified, and the least cost method is selected. Least cost analysis improves efficiency in designing and implementing policy and projects, but cannot guarantee improvement in social welfare.[8]

Benefit cost rests on a number of assumptions. It is useful to expound them before turning to their validity in light of the unique features of global warming.

Assumption 1: It is possible to finesse the problem of interpersonal welfare comparison using the Kaldor-Hicks hypothetical compensation test.

Here is the problem. Modern welfare economics does not allow us, as economists, to make interpersonal welfare comparisons (Persky 2001). If, as a result of policy, individual **i** gains US$ 10 and **j** loses US$ 5, we cannot conclude, based on objective economics, that social welfare has improved. But virtually every public policy and public investment will make some people better off and others worse off. It is impractical, in almost all cases, to fully compensate those who are harmed.

If left there, public policy would be virtually paralyzed. The resolution to this dilemma, which was suggested many decades ago and which undergirds all BC today, is the Kaldor-Hicks hypothetical compensation test: Does the policy in question generate sufficient benefits so that the winners *could* compensate the losers and still have something left? If so, the action is justified even if compensation if not paid. The practical effect of the K-H test is huge. Economists could concentrate on their principal concern, efficiency, and finesse equity (distributional) questions. The justification for the K-H test is that society has a political system for achieving distributional objectives. Equity/distributional aspects of policies could be stripped away and consigned to ethics and the political process, freeing economics to analyze efficiency. This "sleight of hand" was assisted by the second welfare theorem, which states that, in a competitive equilibrium, redistribution to achieve distributional objectives need not conflict with efficiency.[9]

Assumption 2: The existing distribution of income is acceptable.

Benefit cost monetizes benefits and costs based on estimates of what individuals are willing to pay (WTP) for benefits, and willing to accept (WTA) in compensation for costs (or loss of benefits).[10] But WTP depends on ability to pay, or income. In similar fashion, WTA also depends on income. Two different distributions of income will produce two different estimates of WTP, two different patterns of demand, two different measures of consumer surplus (or prices), two different calculations of aggregate benefits and aggregate costs, and hence two different estimates of NPV. Indeed, depending on income distribution, NPV may be estimated to be positive or negative.

This is important. BC analysis, which purports to be about efficiency, is contingent on income distribution, an equity concern. To accept BC results as legitimate, one must have confidence that income distributional objectives can and are being attended to through other means. If they are not, BC bears the additional burden of contributing to society's distributional objectives. This could be accomplished by weighing the distributional effects (high weights for benefits to poor, low for rich) but the weights themselves would be based on ethical, not economic foundations.[11]

Assumption 3: The effects of policies (projects) can be reasonably anticipated and monetized.

This is a double assumption. BC handles uncertainty through a variety of techniques: sensitivity analysis (testing to determine which of many assumptions have significant impact on NPV); switching values (the percentage change in an input value that would switch NPV between positive and negative); calculation of expected value based on objective or subjective probabilities, and Monte Carlo style analysis.[12] These techniques are satisfactory in many cases but less so when the extent of uncertainty increases, and may not be appropriate in situations involving a low (but unknown) probability of very high catastrophic damages.

Full BC analysis (not least cost analysis) requires monetization of the main benefits and costs. The starting point is often market prices, which can be adjusted for distortions through various shadow pricing techniques. BC originated from a partial equilibrium tradition, but major policies and projects require more complex general equilibrium modeling to capture interactions. Market prices are not available for many significant effects, however, especially in the areas of human health, life, and the environment. The two main approaches for determining non-market values are "surrogate market techniques" in which value is inferred from a substitute market and contingent valuation, which is a survey-based approach.[13]

Assumption 4: Agreement can be reached on a social rate of time discount (SRTD).

A discount rate links values over time. Efficient resource allocation requires that the costs and benefits of a project (policy) must be compared after conversion to the same point in time, which in practice is present value. Thus a discount rate is needed. Moreover, there is general agreement that, for efficiency, the same discount rate should be used to evaluate projects across sectors (e.g. transportation, health, environment). In this way, the marginal present value benefit of a dollar of government spending is equalized across sectors.

Finding the "correct" social rate of time discount has been contentious for almost a century, and involves theoretical, empirical, and ethical elements. The longer the time horizon is, the greater the impact of the discount rate, and the more critical the choice. A popular framework for considering the discount rate is:

$$SRTD = \rho + \mathbf{nc}$$

where ρ is an "impatience" parameter describing how urgently we prefer present versus future consumption (some times known as the utility discount rate), $\mathbf{n}$ is an estimate of elasticity of marginal utility with respect to marginal consumption, (a measure of how rapidly incremental utility decreases as consumption increases, or how much better the first sip of wine tastes than the last), and $\mathbf{c}$ is expected percentage increase in per capita consumption. Unfortunately, the first term, ρ, cannot be directly observed. Moreover, some observers accept that individuals may exhibit impatience, but dispute that *society* is or should be impatient. Roy Harrod (1948), a seminal figure in the revival of growth economics, claimed that any positive pure time preference (ρ) was "a polite expression for rapacity and the conquest of reason by passion" (cited by Manne 1995). The second parameter, $\mathbf{n}$, expresses the strength of diminishing marginal utility and is also not directly observable ($\mathbf{n}$ can also be thought of as a measure of aversion to inequality). Moreover when $\mathbf{n}$ is inferred from surrogate evidence, for example a progressive tax structure, at best it would shed light on intra-generational, not intergenerational, utility, and it is its role in estimates of intergenerational welfare that is of direct interest. Finally, **c**, the expected increase in per capita consumption, is generally considered positive. But declining per capita consumption is not unknown, and if **c** is sufficiently negative, the discount rate itself could be negative. The implication then would be a societal preference for future over present consumption.

These factors – ρ, **n**, and **c** – are thought to form society's willingness to exchange present consumption for future consumption, as reflected in the discount rate. The *ability* to transform present into future consumption via real investment is measured at a point in time by the marginal return to capital. In a perfect, distortion free world (no taxes, perfect competition, etc.), a single discount rate would reflect both preferences for present versus future consumption and the marginal return to capital.[14]

There are three additional complications. First, while the discount rate contributes to efficiency in a discounted utilitarian framework, it also has strong intergenerational equity implications. As we will see, discounting the damages to future generations of global warming may adversely affect intergenerational equity. Second, "high" discount rates may compromise the sustainability of real income or environmental resources, or both. Discounting and the benchmark of sustainability may be in conflict. Finally, although virtually all analyses assume a constant discount rate throughout the life of a policy or project, there is no strong economic justification for this assumption.

In addition to these four fundamental assumptions of BC analysis, one should keep in mind some of the other basic postulates of modern welfare economics: that utility functions are independent; that preferences are exogenous and stable; and, most fundamentally, that maximizing utility is *the* objective.

The global warming challenge to Benefit Cost analysis

Global warming has three unique features that challenge conventional BC analysis: the exceptionally long time horizon; the high degree of scientific and economic uncertainty (not unrelated to the time scale); and the international character of the problem. For orderly examination, we call these Features 1, 2, and 3. Table 16.1 summarizes in matrix form the four assumptions, the three features, and the principal points of conflict.

Assumption 1, Feature 1

In the absence of aggressive greenhouse gas abatement, carbon emissions might rise from the current annual flow level of 6.9 gigatons (gtC, billion tons carbon) to 30.3 gtC by 2100. However, with a strong shift to non-fossil fuel technology, emissions might peak about mid-century at about 12 gtC before declining to about 5 gtC in 2100 (IPCC 2001). In either event, because global warming reflects accumulated atmospheric stock, the positive flow over the next century guarantees that

Table 16.1 Summary of assumptions, features, and principal points of conflict

Unique Features of GW / Basic Assumption of BC	1. Exceptionally Long Time Horizon	2. High Degree of Scientific & Economic Uncertainty	3. International Character of the Problem
1. Interpersonal Welfare Comparisons can be finessed with K-H Compensations test	X		
2. The Existing Distribution of Income is Acceptable			X
3. The Effects of Policies Can be Anticipated and Monetized		X	
4. A Social Rate of Time Discount (SRTD) Can be Agreed Upon	X	X	X

temperatures will increase during this period and for many years beyond. Indeed, global temperature and rising sea level from thermal expansion of the oceans will continue for hundreds of years after atmospheric stabilization. Thus the relevant time frame for analyzing action or inaction *today* is several centuries.

This time scale poses a direct challenge to the first BC assumption, that the K-H hypothetical compensation test allows us to relegate equity concerns to the political process and concentrate on efficiency. The time scale is clearly intergenerational. There are no secure political institutions or mechanisms through which the present generation can compensate generations in the far future for the consequences of global warming. All attempts at setting up a sinking fund for compensating future victims are subject to plunder in intervening decades.[15] The intergenerational welfare transfer problem is symmetrical. There is no easy way in which future generations (at least some of whom are apt to be wealthier than we are) can compensate us if we sacrifice with costly greenhouse gas abatement measures today.[16]

Thus the potential for compensating losers, which is at the heart of traditional BC, and which allows focus on efficient resource allocation, is compromised. In Mishan's terms, we confront a *potential* potential Pareto improvement (Mishan 1988: 300), and the prospects for actual compensation became a matter of hope and faith. The net result of this compromise is not to toss out BC analysis, but to acknowledge that intergenerational equity concerns must be explicitly addressed in formulating global warming policy. BC needs to be tempered by equity (distributional) analysis. Efficiency, based on discounted utilitarianism, is not sufficient.

Assumption 2, Feature 3

Assumption 2 is that the existing distribution of income is acceptable (i.e. "just" or "fair"). This may be justified when analyzing policies (projects) at the national and sub-national level in countries with democratic institutions. But the essence of global warming is its international character. There is no world government through which democratically determined distributional objectives are attended to. This is a problem that goes to the core of BC analysis. We need monetary values for benefits and costs, but monetary values depend on willingness *and* ability to pay, and ability to pay reflects actual the international (and internal) distribution of income, which is *not* the result of global democracy in action. Hence, estimates of benefits and costs based on actual international of distribution of income are tainted.[17]

The problem can be illustrated in a very specific context. It is widely agreed that global warming will cause additional deaths, and these deaths will occur disproportionably in poor countries. The standard way to monetized increased mortality in to estimate the value of a statistical life (VOSL). Estimates of VOSL in the United States, derived mainly from hedonic wage models, tend to cluster around US$ 6 million (Viscusi and Aldy 2003). If this value is scaled by the per capita income difference between the United States and India, an estimate of willingness to pay per life in India would be less than US$ 500,000.[18] If this value were then used for projected Indian deaths, we would be left in the ethically uncomfortable position of low-balling global warming damages, hence weakening abatement policy and increasing deaths in poor countries, all because damages in poor countries are given low monetary weight. Indeed, the larger the fraction of damages borne by the poor countries, the lower the monetized damages, the weaker the abatement policy, and the more damages they will suffer.

The problem of course is not restricted to loss of life, but extends to all global warming damages. It follows that the weight given to damages to low-income persons and countries may need to be adjusted upward. A reasonable response is to use *equity weighting* to correct for an unjust (unfair) distribution of income.[19] In general, equity weights will be highest for global warming damages borne by the poorest. Such equity weights can significantly increase the aggregate benefits of greenhouse gas abatement and support a more aggressive climate policy (Frankhauser *et al*. 1997; Azar and Sterner 1996; Azar 1998; Tol 2005). Equity weighting is not without problems, however. The weights are derived from ethical systems mediated by politics, and not from economic analysis. And to weigh willingness to pay in poor countries in global warming analysis, without using equity weights for other projects and policies, can itself create distortions. Note that if a society has a strong egalitarian preference (inequality aversion), it will weight damages to the poorest very heavily. But to the extent that the aversion to inequality merely reflects the notion of diminishing marginal utility of consumption, the same society would choose a large **n** in its social rate of time discount. *Ceterius paribus* this supports a "high" discount rate and therefore a relatively heavy discount for future damages. The two tend to offset each other and the net effect could be a wash.[20]

Assumption 3, Feature 2

Benefit cost presumes that the with-without effects of policy can be identified, quantified, and monetized. This is problematic in the case of global warming due to pervasive uncertainties. A simple listing of the main analytical steps underscores the multiple sources of uncertainty.

1. Baseline emissions of greenhouse gases derived from estimates of future population increase, economic growth, and energy use are needed.
2. The relations between emissions (a flow) and atmospheric concentrations (a stock) must be specified, including estimates of carbon sinks and atmospheric decay rates for the various gases.
3. The effects of ambient concentrations on climate (temperature, weather systems, etc.) need to be estimated.
4. Changes in ecosystems arising from climate change need to be estimated. Regional details are needed. Effects on crops, forestry, fisheries, etc. are modeled.
5. The time profile of social impacts on market sectors (e.g. agriculture) and non-market sectors (e.g. water, disease and health) are identified. At this point optimal adaptation measures should be identified and costed (i.e. seawalls, increased air conditioning, crop substitutions, etc.)
6. All impacts – market sectors, non-market sectors, ecosystem effects, use value, non use "existence" value, etc. – need valuation (monetization), and the valuation should be relevant to the time and location of the impacts.
7. The time path of the mitigation costs is needed (which requires estimates of technology change).

This is a formidable task considering the time scale and the possibilities for technical change, and intervening changes in the structure of the production and consumption.

Uncertainty raises several other interesting questions. For example, does uncertainty surrounding potentially irreversible effects (extinction of species, altered ecosystems, sea level rise) support more aggressive abatement actions now? On the one hand, the prospect of irreversibility increase the option value of early aggressive abatement. If, after a period of time, our information improves (and uncertainty diminishes) and the costs of global warming are found to be large, the "stable" climate option is to be maintained. But the option argument for precautionary policy is not an open and shut case. Presumably, if aggressive costly action is delayed, our information base expands and, should global warming costs prove to be modest, costly (irreversible) sunk costs in abatement capital would have been avoided. In short, the irreversibility/option argument cuts two ways.

Another uncertainty issue in the global warming literature is whether the standard BC approach to uncertainty of maximizing expected utility is technically possible. Tol (2003) notes that, for expected value calculations, the variances of net present marginal costs and benefits must be finite. In his analysis, he finds a possibility that some regional growth rates might be negative due to high sensitivity to climate change. In turn, negative growth implies a negative discount rate, and with negative discount rates, present value damages of global warming might be unbounded (nonfinite). In that event, in a technical sense no NPV can be calculated. This complication need not disqualify BC analysis, however, but it does point up a need for supplemental policies to assist severally impacted regions (Yohe 2003).

The final and more serious issue in the global warming–BC–uncertainty nexus is the catastrophe question. Two such scenarios are the melting of the West Antarctic Ice Sheet and the possible disruption and shut down of the thermocline circulation in the Atlantic Ocean (the loss of the Gulf Stream). Is BC an appropriate tool for formulating a response to a small, but unknown, probability of catastrophic global warming damages?

There is no easy answer to the catastrophe question. One can argue that in situations of potentially catastrophic consequences, the precautionary principle, rather than BC, should rule. Standard practice suggests that BC analysis should be risk-neutral in most situations, under the assumption that in a large portfolio of relatively small projects, risks even out. But when the effects are potentially large, as they are for global warming, risk aversion rather than risk neutrality is more appropriate. But resort to the precautionary principle is not totally convincing. How much should be spent to reduce the (unknown) risk? How is risk aversion to be measured and how do we incorporate it in expected utility calculation? The issue is unsettled. We provide a sampling. Jeroen C.J.M. van den Bergh (2004) castigates virtually all BC analysts for staying within the framework of net present value even in the face of extreme uncertainty and catastrophic dangers. He advocates a "qualitative" BC approach supporting the precautionary principle. Howarth (2003) considers climate stabilization as a form of insurance (risk management). He finds that expected monetary benefits of climate stabilization should be adjusted upwards by as much as 554 per cent to account for risk aversion. Howarth also advocates discounting at a risk free rate of return of 0.4 per cent per year, stating that the higher discount rates that are often used reflect a risk premium that is inappropriate in global warming analysis. Pizer (2003) is concerned with a somewhat different question – the appropriate control instrument, a price (tax) measure or a quantity (cap and trade or permit) measure, when considering uncertain catastrophic damages. His analysis draws on the seminal paper by Weitzman (1974), which showed that, with uncertainty, tax measures are more efficient than quantity measures if marginal damages are relatively constant when compared to marginal costs.[21] The possibility of catastrophic damages challenges the assumption of constant marginal damages from greenhouse gas emissions. A catastrophe suggests a sharp kink in the marginal damages of greenhouse gas emissions (i.e. a sharp kink in the abatement benefit function). In that case, theory suggests that quantitative measures should trump price (tax) measures. Pizer's analysis does not support this, finding instead that price and quantity controls have roughly equal efficiency. His analyses do conclude, however, that with the possibility of catastrophic damages: "Catastrophe avoidance is much more important than efficient catastrophe avoidance."

In concluding this section, few studies that have attempted to formally incorporate the possibility of catastrophic damages in the calculations are to be noted. Nordhaus and Boyer (2000) is an exception. Their conclusion when catastrophe is included point toward a more aggressive abatement effort than otherwise would be the case. Nordhaus and Boyer assume the probability of catastrophe rising with temperature, the damage (income loss) from catastrophe to vary by region, and a relatively high-risk aversion factor. From these assumptions they calculate "willingness to pay" estimate to avoid catastrophe. These vary from 0.45 per cent of GDP for the US to 1.9 per cent of GDP for OECD Europe and India for a 2.5°C temperature increase. If, however, temperature were to increase by 6°C, the estimate rises to 2.53 per cent of GDP for the US and 10.79 per cent for GDP for OECD Europe and India.[22] While a 6°C increase is at the outer limits of current estimates, the calculations do point up the nonlinearity of potential damages. At a more general level Ingham and Ulph (2005) summarize the catastrophe–benefit cost conundrum as follows "the balance of results [from catastrophe modeling exercises] would suggest that the possibility of catastrophic effects should lead to reduction in current emissions, possibly a very drastic reduction" (p. 70). Finally, a recent study by Hallegatte (2005) was noted, which concludes that, because of their structure, traditional modeling of extreme events in climate models greatly underestimates their economic and welfare costs. More specifically, she concludes that a more appropriate modeling structure could more than double the estimated cost of large weather extreme events, again suggesting more aggressive abatement efforts.

Assumption 4, Features 1, 2, 3

All three features of global warming – the long time horizon, the unprecedented level of uncertainty, and the international character of the problem – complicate the question of the appropriate discount rate. And, as one might imagine, the discount rate determines whether an immediate, aggressive abatement strategy is warranted or not. To give some perspective, one million dollars in damages one hundred years from now has a present value of US\$ 7,569 when discounted at 5 per cent and US\$ 369,664 when discounted at 1 per cent.

Consider first the international character of the global warming problem. Whose discount rate should be used in calculating the optimal time path of emissions reductions? In traditional BC analysis, conducted *within* a country, the discount rate emerges from a constellation of country specific parameters – rate of return on marginal investment, preference for present versus future consumption, expectations of future per capita income growth, and so on. The presumption is that the appropriate discount rate for "poor" countries is higher than for "rich" countries for two reasons. First, at poverty (subsistence) levels, survival implies a strong preference for present versus future consumption. Second, the relative scarcity of capital for investment suggests that its rate of return is relatively high. Thus there is a well-founded proposition that social rates of time discount differs among countries, and specifically are lower in rich countries than poor countries. The obvious implication is that rich country discount rates support a much more aggressive greenhouse gas abatement policy than do poor country discount rates. The consensus on relative damages between rich and poor countries, however, is that the poor have more to lose than the rich.

Thus the dilemma: if poor country discount rates are used in optimization models, we wind up with a weaker abatement regime and higher relative dangers for poor countries. If rich country discount rates are used, the abatement regime is stronger and the global warming damages borne by future generations in poor countries are less. But an even more efficient allocation of resources would appear to be to invest in accelerating development now, to lessen the vulnerability and increase the adaptation capacity of currently poor countries. Stated somewhat more broadly, the choice of the discount rate is not independent of the existing and prospective distribution of income, and again undermines the view that efficiency can be separated from equity.

In addition to the international character of global warming, uncertainty itself complicates the discount rate issue. Presumably the discount rate reflects both the marginal rate of return on capital and individuals' preferences for current versus future consumption. In principle, it is future returns to capital and future preferences that are needed. The uncertainty surrounding these determinants increases exponentially as one looks forward to 2100 and beyond. In an important contribution, Weitzman (1998) has shown, however, that it is the expected value of the discount factor, not that discount rate, that properly measures the expected present value of future benefits. (The discount factor is $1/(1+r)^n$ where r is the discount rate and n the number of years.) Applying this insight to global warming, Newell and Pizer (2003) find that a correct treatment of discount rate uncertainty almost doubles the expected present value of benefits from abating greenhouse gas emissions.

Moreover, there is increasing evidence that a lower discount rate should be used for the far future, as compared to the near future. Such a time variant discount rate, known as hyperbolic discounting, has the effect of increasing the net present value of preventing global warming, and thus justifying an early, aggressive abatement policy.

Perhaps the most profound challenge to discounting as practiced in conventional BC arises directly from the time scale of global warming. Schelling (1995) points out that abatement expenditures today will benefit future generations, unborn and unknown. Thus the ρ term in the discount rate, measuring our impatience for our own future consumption, is simply irrelevant. How can this generation be impatient for a generation in the twenty-second century to enjoy its consumption? Instead, our investment in abatement today should be considered a transfer of welfare to generations distant in time, not a measure of our desire for our current versus future consumption. Furthermore, Shelling argues that BC optimization models, in calculating and incorporating $(n)(c)$ – the adjustment for projected increases in per capita income and the estimate of the elasticity of utility with respect to consumption – fail to disaggregate as to *whose* consumption is increasing. His argument is that the rich countries will pay for abatement over the next several decades, and the beneficiaries will mainly be generational descendents of today's poor countries. Despite per capita consumption growth in these countries, they may well still have lower per capita incomes than rich countries in the distant future. If so, abatement efforts now represent an income transfer from today's rich to tomorrow's less rich. With the assumption of diminishing marginal utility of consumption, which is the key reason for including $(\mathbf{n})(\mathbf{c})$ in the discount rate equation, a transfer from rich to less rich suggests that a discount rate based on marginal utility comparison would be negative. In short, when one considers policies spanning many generations, the conventional determinants of the discount rate, which are the impatience parameter and the adjustment for assumed rise in per capita consumption, no longer provide trustworthy guides in BC optimization models.

Alternatives

There are basically two alternatives to a full-blown dynamic optimization benefit cost analysis that still make essential use of BC. The first is policy analysis, or the policy evaluation approach. A particular policy, however formulated, is evaluated for its benefits and costs. There is no attempt to determine if the policy is optimal. For example, the emissions targets established in Kyoto Protocol constitute a policy. The cost of meeting these targets can be estimated. The temperature and climate effects of meeting the targets, as compared to a baseline scenario, are estimated, the socio-economic effects are identified, and then translated into monetized values. These are the benefits. Benefits and costs can then be compared. Alternatively, instead of setting emissions targets, a policy of limiting atmospheric CO_2 concentrations to, say, two times pre-industrial levels can be proposed and evaluated. In that case and working backwards, the atmospheric concentration target determines emission limits and ultimately costs; working forward concentration also determines temperature and climate impacts, and hence benefits.

These BC based policy evaluation studies are useful but have three drawbacks. First they do not maximize net benefits, as optimization models purport to do. Second, they encounter many of the same difficulties in estimating benefits that optimization models do (i.e. appropriate discount rate, equity weighting, etc.). Third, they provide little information on the correct timing of policy intervention.

The second major alternative to full-blown dynamic BC analysis is to set one or more constraints and proceed to solve a constrained optimization problem. For example, at a high level of abstraction, there can be conflict between the objective of dynamic efficiency (maximizing net present value) and sustainability, where sustainability is defined as non-declining per capita welfare (Pezzey 1992). In that event, society may seek a global warming strategy that maximizes net present value subject to a sustainability constraint expressed either in terms of maintaining the flow of real income or the stock of physical, human, and natural capital that produces the flow of income. The ethical basis for the constraint would be the notion that future generations have "rights" to resources, including natural capital and climate. This is tantamount to establishing an inter-temporal endowment of resources based on each generation's rights, and only then seeking a dynamically efficient solution (Woodward and Bishop 1995). Operationalizing this approach, however, runs into serious difficulties, especially in determining the degree of substitutability/complementarity between natural capital (i.e. climate) and physical social and human capital in both production (e.g. agriculture), and in direct consumption (individual utility functions). Still, full optimization BC models confront much the same difficulty.

The constraint(s) need not be as grand and abstract as sustainability. The tolerable windows approach (TWA, a.k.a. guardrail approach) is often defined in physical terms, and was developed in response to the scientific and socio-economic uncertainties that characterize global warming. In one formulation, the "tolerable window" was set in terms of maximum temperature increase and rate of change – 2°C and < 0.2°C per decade – and a maximum emission reduction rate of 4 per cent per year, to avoid severely disruptive economic consequences (Bruckner and Schellnhuber 1999). These constraints can then be converted to admissible and inadmissible emission reduction time paths, thus restricting, but not eliminating policy options. At this point, remaining options can be evaluated for cost-effectiveness. Broadly speaking, the FCCC itself reflects the spirit of the TWA when it takes as its objective stabilizing greenhouse gas concentrations at levels that would prevent dangerous interference with the climate system. If the concept of "stewardship" can be defined in quantitative terms, it, too, would fit within the tolerable windows approach. Finally, proponents of TWA stress that the identification of the constraints that frame the tolerable window "are ultimately not the task of science, but rather have to be specified within a negotiation process between politicians, scientists, economists, etc." (Petschel-Held *et al.* 1999: 324).

Notes

1   Kyoto did provide for three flexibility mechanisms – emissions trading, joint implementation, and the clean development mechanism – which contribute to a "least cost" response.
2   One of the first analyses of climate change was d'Arge (1975) cited by Spash (2002: 158). The conjecture of carbon emissions and global warming traces to Svante Arrhenius, a Swedish scientist who, in 1895, estimated that a doubling of CO_2 in the atmosphere would increase global temperature by 4 to 6°C. (McKibben and Wilcoxen 2002: 110).
3   Tol (2005) summarizes 103 estimates of marginal damage costs contained in 28 published studies.
4   Alternatively the internal rate of return (IRR) or benefit cost ratio, B/C.
5   Letter from President Bush to Senators Hagel, Helms, Craig, and Roberts, 13 March 2001.
6   One study showed net benefits to Russia if at least 4 billion tons of carbon could be "sold" through trades (Buchner and Dall'Olio 2005: 9). Other considerations, including prospective admission to the World Trade Organization, also played a role.
7   An alternative criterion is that the Internal Rate of Return, IRR – the discount rate such that NPV = 0 – exceeds the opportunity cost of capital.
8   Obviously good BC analysis also seeks out least cost alternatives as part of the analysis.
9   The standard terminology goes like this. An allocation of resources is *Pareto optimal* if it is not possible to reallocate resources and make one individual better off without making another worse off. (Note: there are many Pareto optimal allocations and it is not possible to rank them based on economic science.) An action is a strict *Pareto improvement* if at least one individual is indeed better off and no one is harmed. An action is a *potential Pareto improvement* if winners *could* compensate losers and still be better off, the K-H test.

10 More precisely BC may attempt to measure WTP for benefits, WTA compensation for being denied benefits, WTP to avoid costs, WTA compensation for bearing cost or combinations of the above. There are serious theoretical and empirical issues concerning discrepancies between WTP/WTA estimates but they do not concern us here.

11 For an early attempt to formalize weighting, see Squire and van der Tak (1975).

12 Monte Carlo methods are statistical techniques used to investigate outcomes of interacting stochastic variables.

13 Surrogate market techniques include hedonic wage models valuing risks to health and life; housing price models valuing environmental amenities; and travel cost models to infer value to recreational activities.

14 Due to taxes, etc., the marginal return to capital generally exceeds the rate of return to savers (individuals' time preference). In that event, it is desirable to weight the sources of funding for a project or policy by the fraction that reflects reduced private consumption, and the fraction that reflects reduced private investment.

15 A "sinking fund" consisting of technical advances is least likely to be plundered.

16 Lind (1995) does describe a mechanism through which a future generation could in theory compensate the present generation for productive investments undertaken now to benefit the future. But the mechanism applies to private goods subject to secure property rights. Global climate investment would not qualify due to its public goods character.

17 Azar (1999) states this forcefully: "A situation where the richest billion people live in abundance, and the poorest billion people suffer from chronic hunger, can by no reasonable standards be considered a global welfare maximum." Foreign aid reflects a desire by some to redistribute income. But the amounts are not the result of a democratic international political process.

18 Viscuci and Aldy cite an Indian study based on Indian data, not scaling, showing VOSL at US$ 1.2 to US$ 1.5 million.

19 If income distribution is considered fair, no weighting is needed even if income is not equal among all individuals.

20 But see Shelling's argument below.

21 See Pearson (2000) for explanation.

22 According to Nordhaus and Boyer, OECD Europe's vulnerability arises from possible shifts in ocean currents and agricultural impacts, and India's vulnerability from the role of monsoons in agriculture and from potential health impacts.

References

Azar, C. and Sterner, T. (1996) Discounting and Distributional Considerations in the Context of Global Warming. *Ecological Economics*. 19. 169–184.

Azar, C. (1998) Are Optimal Emissions Really Optimal—Four Critical Issues for Economists in the Greenhouse. *Environmental and Resource Economics*. 11. 301–315.

Azar C. (1999) The timing of CO2 emissions reductions: the debate revisited. *International Journal of Environment and Pollution*. 10 (3/4): 508–521.

Barrett, S. (2005) Kyoto Plus, In Helm, D. (ed.) *Climate Change Policy*. Oxford: Oxford University Press. 282–303.

Bruckner, T. and Schellnhuber, H. J. (1999) Climate Change Protection: The Tolerable Windows Approach, available through http://www.environmental-expert.com/articles/article69/article69.htm, last accessed May 2006.

Buchner, B. and Carraro, C. (2003) China and the Evolution of the Present Climate Regime, *Climate Change Modeling and Policy*, Fondazione Eni Enrico Mattei. Nota Di Lavoro 103.2003.

Buchner, B. and Dall'Olio, S. (2005) Russia and the Kyoto Protocol: The Long Road to Ratification. *Transition Studies Review*. 12 (2). 349–82.

Cline, W. R. (1992) *The Economics of Global Warming.* Washington, DC: Institute for International Economics.

Fankhauser, S., Tol R. and Pearce, D. (1997) The Aggregation of Climate Change Damages: A Welfare Theoretic Approach. *Environmental and Resource Economics.* 10. 249–266.

Hallegatte, S. (2005) Accounting for Extreme Events in the Economic Assessment of Climate Change, *Climate Change Modeling and Policy*, Fondazione Eni Enrico Mattei. Nota Di Lavoro 1.2005.

Howarth, R. B. (2003) Catastrophic Outcomes in the Economics of Climate Change. *Climatic Change.* 56. 257–263.

Ingham, A. and Ulph, A. (2005) Uncertainty and Climate Change Policy, In Helm, D. (ed.) *Climate Change Policy.* Oxford: Oxford University Press. 43–72.

IPCC (2001) Climate Change 2001: Mitigation. Third Assessment Report of the Working Group III: Mitigation. Geneva: IPCC.

Lind, R. C. (1995) Intergenerational Equity, Discounting, and the Role of Cost-Benefit Analysis in Evaluating Global Climate Policy. *Energy Policy.* 23 (4/5). 379–389.

Maddison, D. (1995) A Cost-Benefit Analysis of Slowing Climate Change. *Energy Policy.* 23 (4/5). 337–346.

Manne, A. S. (1995) The Rate of Time Preference: Implications for the Greenhouse Debate, *Energy Policy.* 23 (4/5). 391–394.

McKibbin, W. J. and Wilcoxen, P. J. (2002) The Role of Economics in Climate Change Policy, *Journal of Economic Perspectives.* 16 (2). 107–129.

Mendelsohn, R. (2005) The Social Cost of Greenhouse Gases: Their Value and Policy Implications, In Helm, D. (ed.) *Climate Change Policy.* Oxford: Oxford University Press. 134–152.

Mishan, E. J. (1988) *Cost Benefit Analysis: An Informal Introduction*, 4th edn. London: Unwin Hyman.

Newell, R. G. and Pizer, W. A. (2003) Discounting the Distant Future: How Much Do Uncertain Rates Increase Valuations? *Journal of Environmental Economics and Management.* 46. 52–71.

Nordhaus, W. D. (1993) Reflections on the Economics of Climate Change. *Journal of Economic Perspectives.* 7 (4). 11–25.

Nordhaus, W. D. and Boyer, J. (2000) *Warming the World: Economic Models of Global Warming.* Cambridge, MA: MIT Press.

Pearce, D. (2005) The Social Cost of Carbon, In Helm, D. (ed.) *Climate Change Policy.* Oxford: Oxford University Press. 99–133.

Pearson, C. S. (2000) *Economics and the Global Environment.* Cambridge: Cambridge University Press.

Persky, J. (2001) Retrospectives: Cost-Benefit Analysis and the Classical Creed. *Journal of Economic Perspectives.* 15 (4). 199–208.

Petschel-Held, G., H.-J. Schellnhuber, T. Bruckner, F. L. Tóth and K. Hasselmann (1999) The Tolerable Windows Approach: Theoretical and Methodological Foundations. *Climatic Change.* 41 (3–4). 303–331.

Pezzey, J. (1992) Sustainable Development Concepts An Economic Analysis, *World Bank*, Environmental Paper Number 2.

Pizer, W. A. (2003) Climate Change Catastrophes, *Resources for the Future*, Discussion Paper 03–31 May.

Schelling, T. C. (1995) Intergenerational Discounting. *Energy Policy.* 23 (4/5). 395–401.

Spash, C. L. (2002) *Greenhouse Economics: Value and Ethics.* London: Routledge.

Squire, L. and van der Tak, H. (1975) *Economic Analysis of Projects.* Washington, DC: World Bank.

Tol, R. S . J. (1999) The Marginal Costs of Greenhouse Gas Emissions. *Energy Journal.* 20 (1). 61–82.

Tol, R. S. J. (2003) Is the Uncertainty About Climate Change Too Large for Expected Cost-Benefit Analysis? Climatic Change. 56. 265–289.

Tol, R. S. J. (2005) The Marginal Damage Costs of Carbon Dioxide Emissions: an Assessment of the Uncertainties. *Energy Policy.* 33. 2065–2074.

van den Bergh, J. (2004) Optimal Climate Policy is a Utopia: from Quantitative to Qualitative Cost-Benefit Analysis. *Ecological Economics.* 48. 385–393.

Viscusi, Kip W. and Aldy, J. E. (2003) The Value of a Statistical Life: A Critical Review of Market Estimates Throughout the World, *National Bureau of Economic Research*, Working Paper 9487 (February).

Weitzman, M. L. (1974) Prices versus Quantities. *Review of Economic Studies.* 41 (4). 477–491.

Weitzman, M. L. (1998) Why the Far-Distant Future Should be Discounted at Its Lowest Possible Rate. *Journal of Environmental Economics and Management.* 36. 201–208.

Woodward, R. T. and Bishop, R. C. (1995) Efficiency, Sustainability, and Global Warming. *Ecological Economics.* 14. 101–111.

Yohe, G. W. (2003) More Trouble for Cost-Benefit Analysis. *Climatic Change.* 56. 235–244.

17 Proposal for Insurance for Facilitation of Adaptation

Joanne Linnerooth-Bayer, M. J. Mace and Reinhard Mechler

Introduction

Article 4.8 of the United Nations Framework Convention on Climate Change (UNFCCC) and Article 3.14 of the Kyoto Protocol call upon Parties to consider actions, including insurance, to meet the specific needs and concerns of developing countries arising from the adverse effects of climate change and/or the adverse impacts of the implementation of response measures. To date, there is little understanding or agreement within the climate community on the role that insurance and other risk-transfer mechanisms can play in implementing these Articles. This paper addresses this role by discussing opportunities and obstacles for the climate negotiation regime to support insurance mechanisms for weather-related extremes in disaster-prone countries.

Precedents already exist for donor-assisted risk-transfer programs in highly exposed developing countries. The World Food Programme, for example, has purchased insurance based on a rainfall index to help fund its aid response and protect the livelihoods of farmers subject to severe droughts in Ethiopia (*Herald Tribune* 2006). The World Bank provides low-interest capital back-up to the Turkish Catastrophe Insurance Pool to make insurance more affordable to property owners (Gurenko 2004). It also provides a contingent credit facility for hedging disaster risks absorbed by the Colombian government (World Bank 2005). Oxfam GB, a British development, relief, and campaigning organization, is subsidizing the premiums of a microinsurance program in India that provides financial protection to vulnerable families against a variety of natural hazards (Krishna 2005).

These programs aim at providing incentives – some stronger than others – for participants to reduce their risks. Assisting risk-transfer programs in developing countries not only can encourage risk prevention, but can leverage limited development aid budgets and free recipient countries from dependence on the vagaries of post-disaster assistance.

Despite tangible benefits from supporting adaptation to the impacts of climate change through insurance-related instruments, vulnerable developing countries face many challenges in negotiating financial support for this purpose. A major stumbling block has been a call on the part of member countries of the Organisation of Petroleum Exporting Countries (OPEC) for parallel treatment with regard to the use of insurance to address the impacts of response measures on oil revenues. Other negotiation stumbling blocks include concerns regarding the absence of a clear quantitative linkage between climate change and extreme event risks, concerns about setting precedents for legal liability, and the current lack of political will on the part of some developed countries to absorb a portion of developing country losses from weather extremes. The importance of overcoming these impediments and putting concrete proposals on the negotiating agenda has been emphasized by the first Executive Director of the UNFCCC, who considers implementation of Article 4.8 as "one of the most critical aspects" of the climate-change negotiations (Capdevila 2000, quoted in Barnett and Dessai 2002).

We begin in section 2 with a discussion of the state of knowledge regarding the role of climate change on current and future weather-related disaster losses, and how the international community responds to these losses. In the third section, we present a number of promising options for implementing Article 4.8's insurance provision, and we support these options with concrete examples of successful donor-supported insurance projects. We follow with a discussion of obstacles or "stumbling blocks" in supporting these actions within the UNFCCC framework. We conclude by arguing that these obstacles are surmountable and that, indeed, slow progress is being made.

Climate change, extreme weather impacts and disaster assistance

Climate specialists warn that shifts in *average* weather conditions may be less disruptive than increased weather *variability*, which can result in more frequent and/or intense droughts, windstorms, floods, landslides, avalanches and other hazards. The Intergovernmental Panel on Climate Change (IPCC) concludes with a high degree of confidence that the risk of extreme weather events will increase as the climate changes (IPCC 2001, Chapter 8). Although it is difficult to separate the role of a changing climate from the many other factors influencing losses from weather events (Mileti 1999), evidence suggests that the climate influence or signal may already be present, for instance, increasing the risk of extreme precipitation at mid- and high-latitudes (Schönwiese *et al.* 2003), extreme floods and droughts in temperate and tropical Asia, severe dry events in the Sahel and southern Africa (IPCC 2001), and tropical cyclone activity in the Atlantic and the Pacific region (Emanuel 2005). Yet the problem of uncertainty remains: climate scientists are currently unable to quantify the extent that climate change has worsened the impacts from disasters or led to disasters.

Irrespective of climate change, shifting land-use practices, population growth and concentration of capital in high-risk areas have dramatically increased the economic damages from disasters of all types. Losses are almost seven-fold greater from the 1960s to the 1990s, and insured losses about 25-fold (Munich Re 2006). The poor and vulnerable bear the largest relative burden of disaster losses. In a sample of large natural disasters over the period 1980–2011, fatalities per event were higher in low income countries (see Figure 17.1) (Munich Re 2012).

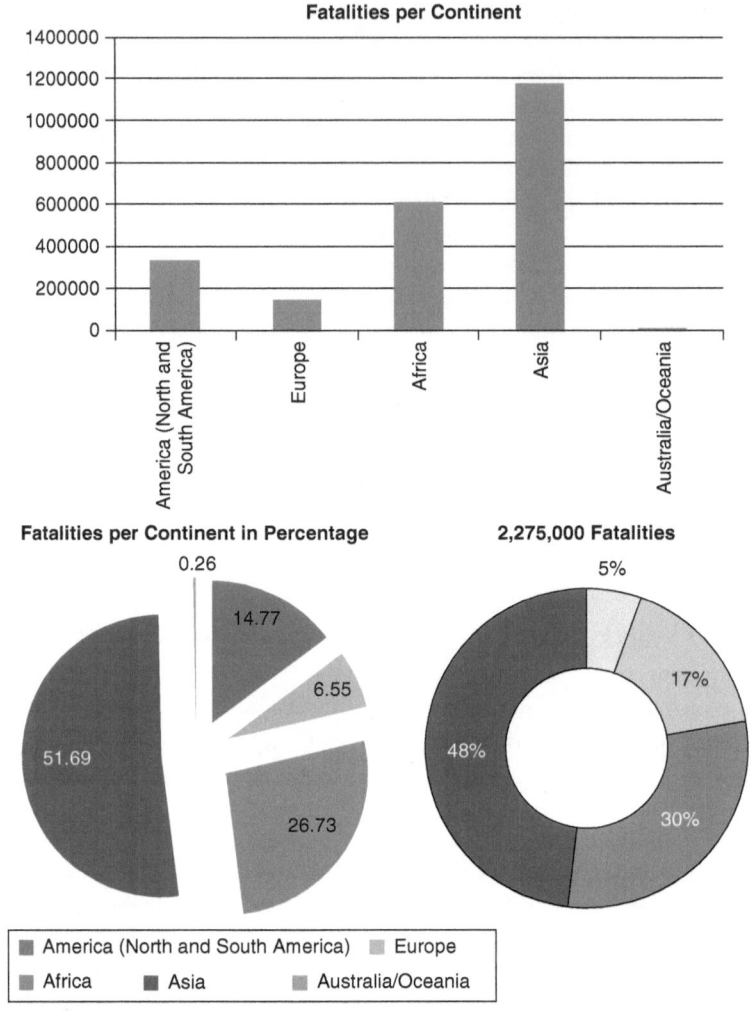

Figure 17.1 Fatalities per continent and income, 1980–2011

Moreover, more than 78 per cent of all fatalities are recorded in the low middle income and low income economies (Munich Re 2012).

For reasons of equity and efficiency, the climate regime should consider supporting insurance-related mechanisms to enhance the adaptive capacity of developing countries in addressing the impacts of extreme weather events (Linnerooth-Bayer *et al.* 2005). Although scientists cannot establish the extent to which climate change has contributed to any specific disaster, as discussed earlier, there is strong evidence that climate change is increasing risks and losses from climate-related extremes. Since many developing countries that are highly exposed to the impacts of extreme weather events have contributed little to the greenhouse gas emissions that contribute to climate change, equity arguments call for shifting some responsibility to the developed world. Switching to pre-disaster assistance, even at extra cost, is also an efficient long-term strategy because of its potential to reduce vulnerability to the impacts of climate change and thus reduce the longer term need for humanitarian aid.

Humanitarian needs following disasters are often not fully met. While well-publicized disasters attract large donations (e.g. over US\$ 10 billion pledged after the 2004 Asian tsunami), humanitarian aid reported by the OECD Development Aid Committee is generally only a small percentage (usually under 10 per cent) of disaster losses in recipient countries (Linnerooth-Bayer and Amendola 2000). Moreover, promises often fall short of actual outlays. Two years after the 2001 earthquake in Gujarat, India, assistance from the central reserve fund and international sources had reached only 20 per cent of original commitments (World Bank 2003). More worrying, post-disaster assistance discourages governments and individuals from taking advantage of high returns on preventive actions (Mechler 2005). With insurance or other risk-transfer instruments, developing country governments will rely less on debt financing after a disaster, and find it easier to attract foreign investment if assured funds exist for repairing critical infrastructure post-disaster. Pre-disaster donor support can also provide poor households, farmers and businesses with access to affordable means to spread risks, which will in turn secure their livelihoods, improve their creditworthiness and contribute to poverty reduction. For many, an insurance contract is a more dignified means of coping with disasters than relying on (or begging for) the generosity of donors after a disaster strikes.

Already a number of innovative projects are in place that demonstrate the viability of donor-assisted insurance programs, and which can provide examples to the climate change community in implementing Article 4.8's insurance provisions (Linnerooth-Bayer *et al.* 2005). These projects are discussed in the next section.

Opportunities for addressing insurance under Article 4

During the negotiation of the UNFCCC, a proposal for an international insurance pool to compensate victims of sea-level rise was put forward by the Alliance of Small Island States (AOSIS) (Intergovernmental Negotiating Committee 1991). The proposed pool was to draw its revenue from mandatory contributions from industrialized countries, using a formula based on GNP and CO_2 emissions. Although the proposal was not adopted, it influenced the ultimate wording of Article 4.8, which calls upon all Parties to:

> give full consideration to what actions are necessary under the Convention, including actions related to funding, insurance and the transfer of technology, to meet the specific needs and concerns of developing country Parties arising from the adverse effects of climate change and/or the impact of the implementation of response measures, especially on: (a) Small island

countries; (b) Countries with low-lying coastal areas; (c) Countries with arid and semi-arid areas, forested areas and areas liable to forest decay; (d) Countries with areas prone to natural disasters [...] (h) Countries whose economies are highly dependent on income generated from the production, processing and export, and/or on consumption of fossil fuels and associated energy-intensive products.

(UNFCCC Art. 4.8)

For insurance mechanisms to be successful in addressing the needs of highly exposed countries and marginal communities within these countries, they must be affordable and accessible to households, farmers and governments in highly exposed regions and countries. Affordability in many cases will only be possible with international support. Four examples of promising approaches are set out below; each has been raised within the climate change negotiations by developing countries (UNFCCC 2004).

Public-private insurance systems

A number of developed countries, including the US, France, Norway, Japan and New Zealand, have legislated public-private insurance systems for natural disasters. These systems differ in form, but many combine private and public responsibility for disaster losses and incorporate incentives and responsibilities for loss prevention. For example, the French system ensures affordability to highly exposed, low-income communities with cross subsidies within the system and government backup capital that reduces the costs of commercial reinsurance. In the case of the US flood insurance program, the federal government fully underwrites the risks but offers contracts only to communities that have taken prescribed risk-reduction measures. With donor assistance, such systems could be made affordable and viable for middle-income developing countries.

A recent innovation exists in Turkey where, for the first time, the international community is providing support for a public-private national earthquake insurance program. The recently launched Turkish Catastrophe Insurance Pool (TCIP) is based on a government mandate that makes earthquake insurance policies obligatory for all property owners, who pay a fee to a privately administered public fund. This fee is based on their risk zone and the type of construction of their property. The TCIP exempts poor households in rural areas, but for all others it will fully replace post-disaster government assistance (and might be criticized for this lack of social solidarity). The system is viable because a multilateral financial institution, the World Bank, provides back-up capital to the TCIP in the form of a contingent loan with highly favourable conditions. If the TCIP were instead to reinsure its catastrophe risk purely through the commercial market, insurance premiums would probably be significantly higher and out of reach for many apartment owners.

The TCIP can serve as an example for the international climate regime. UNFCCC Parties could assure back-up capital to national schemes to render these public-private systems financially viable in middle-income developing countries.

Improving on the TCIP, this support might be conditioned or enhanced upon the adoption of preventive measures, such as reducing the vulnerability of homes, schools and other public buildings.

Regional insurance pools

The Turkish insurance system would not be appropriate for small countries or island states that have too few residents to spread risks sufficiently. These countries might consider forming a regional pool, and several options have been proposed by Pollner (2000) and others for the hurricane-exposed small island states of the East Caribbean. The proposed options envisage a larger role for private insurers and the inclusion of cover for infrastructure and other public assets. Although a Caribbean pool has not been implemented due mainly to institutional incompatibilities among the proposed countries, the severe hurricane seasons of 2004 and 2005 has rekindled interest. As in Turkey, however, these opportunities can only function with governments and multilateral institutions supporting their development. The climate change regime could play an important role in providing this support under Article 4.8.

Microinsurance

Public-private pooling schemes of the type described for Turkey and the East Caribbean are not an option for countries at the lower end of the development spectrum that lack private insurance infrastructure. Microinsurance schemes, which avoid the high transaction costs of commercial insurance, are becoming an option for transferring risks from the private sector. Microinsurance is oriented to small-scale cash flows and often combines subsidies (or, as in India, cross-subsidies from other lines of insurance) with voluntary contributions from groups within a community (Mechler *et al.* 2006).

Disaster risks have rarely been considered explicitly as a niche for microinsurance because of the high uncertainties and large potential losses. One pilot case has been implemented in the coastal Andhra Pradesh region, where microinsurance cover for multiple hazards is provided as a part of the Disaster Preparedness Program. Coverage under this scheme is extended currently to more than a thousand vulnerable families. The British NGO Oxfam GB has been instrumental in setting up the scheme and is paying 50 per cent of the premium (Krishna 2005).

Disaster insurance is only viable with large capital reserves, reinsurance or extensive diversification, all of which add to its cost. The problem for small-scale insurers, besides achieving sufficient scale, is thus offering low-cost policies (Brown and Churchill 2000). Other traditional risk-pooling schemes, such as asset pooling and kinship networks, also break down for the same reason in the context of geographically-concentrated disasters that affect all households and family members at the same time (POVCC 2003).

The dilemma to developing-country entrepreneurs offering low cost weather hedges and microinsurance is, therefore, how to pay the price of capital or reinsurance for backing these schemes while at the same time keeping them afford-able to their low-income clients. By providing all or part of the necessary back-up capital, the climate community has another opportunity to leverage adaptation funding in addressing needs under Article 4.8.

Government risk transfer

Governments, in addition to households and farmers, can benefit from low cost risk- transfer instruments. Developing country governments frequently cannot raise sufficient funds after a disaster to repair critical infrastructure and provide relief to the private sector (Otero and Marti 1995). A study for the Inter Ameri-can Development Bank showed that El Salvador, Colombia and the Dominican Republic can meet only 60 per cent, 62 per cent and 82 per cent, respectively, of their anticipated post-disaster expenses in the event of a 100-year (.01 probability) disaster (Freeman *et al.* 2003).

Vulnerable country governments have the option of purchasing commercial insurance, but premiums can be prohibitively high. Auffret (2003) points out that in the Caribbean region, catastrophe insurance premiums were about 1.5 per cent of GDP during the period 1970–1999 while average losses per annum (insured and uninsured) accounted for only about 0.5 per cent of GDP. As an alternative to commercial insurance, Mexican authorities have reinsured their national catastro-phe insurance fund (FONDEN) with a mix of financial instruments involving tra-ditional reinsurance and a catastrophe bond (Cardenas 2006). A "cat bond" is an instrument whereby the investor receives an above-market return when a specific catastrophe does not occur in a specified time (e.g. an earthquake of magnitude 7.0 or greater on the Richter scale in the vicinity of Mexico City over a one-year period) but sacrifices interest or part of the principal following the event. The gov-ernment's disaster risk is thus transferred to international financial markets that have many times the capacity of the reinsurance market.

There is large scope for international risk transfer for governments that cannot form a viable insurance pool of taxpayers within their borders; however, there are also substantial costs inherent in establishing risk-transfer mechanisms at the governmental level. Although Mexico, a middle-income developing country, will finance its catastrophe bond with its own means (technical assistance has been pro-vided by the World Bank over the last years for devising a risk transfer strategy), a similar but donor-assisted bond is an option for poorer countries. Still another possibility is a contingent credit arrangement. For a small upfront payment, Colombia – as a case in point – has the right to access a World Bank loan up to a fixed amount should a disaster occur (World Bank 2005). Subsidizing catastrophe bonds and other forms of public risk transfer, and linking this assistance to risk pre-vention, is yet another possible way the climate community can assist vulnerable developing countries in minimizing the adverse effects of climate change under Article 4.8.

Stumbling blocks for implementing Article 4.8

While many possibilities exist for assisting developing countries through insurance and addressing their needs and concerns under Article 4.8, there are stumbling blocks to the creation of necessary insurance funding mechanisms. These include:

- the complicating role of OPEC member countries' demands for parallel support to address the potential financial impacts of climate change mitigation efforts on fossil-fuel producing economies;
- difficulties in estimating disaster risk as a basis for designing risk-transfer instruments, and scientific hurdles in assigning the contribution of climate change to disaster risks;
- concerns about legal liability on the part of developing countries for climate-related disaster losses in the developing world;
- lack of broad political support to provide sufficient financial assistance to developing countries to increase their adaptive capacity to climate change impacts.

The role of OPEC member countries

Article 4.8 of the UNFCCC calls for consideration of actions related to insurance "to meet the specific needs and concerns of developing countries arising from the *adverse effects of climate change and/or the impact of the implementation of response measures …* ." (emphasis added). This Article's textual linkage between the adverse effects of climate change (for example, the impacts of more intense weather-related events) and the impacts of response measures (for example, the financial impact on oil-producing countries of measures taken by oil consuming countries to reduce their greenhouse gas emissions) has been problematic. Just as AOSIS seeks financial assistance, insurance and the transfer of technologies within the climate negotiations, to help small island states adapt to a changing climate, OPEC countries seek compensation for lost revenues from reduced use of fossil fuel.

Although seemingly unrelated, negotiations on these two issues have long been intertwined, which has deadlocked progress on both issues (Barnett 2001). The linkage continues despite views that these two categories of impacts are different in kind, scope and temporal scale, and different in the nature of the communities impacted (Barnett and Dessai 2002). Vulnerable communities exposed to sea level rise, threats to their water supplies or an increased intensity of hazards have played little role in creating these physical threats. In sharp contrast, the implementation of response measures can be expected to affect economies that have played a direct role in contributing to climate change (and that have benefited from this role) through fossil fuel production or fossil fuel consumption. While the poorest countries have little financial and technological capacity to cope with the impacts of climate change, according to Barnett and Dessai (2002), OPEC countries can cope with the impacts of response measures by acting as a cartel and by self-supporting economic diversification.

Moreover, while climate change is an urgent issue for many vulnerable countries, it is increasingly difficult to argue that OPEC countries will suffer any time soon from the economic impacts of measures taken to reduce greenhouse gas (GHG) emissions from fossil fuel consumption. Reportedly, OPEC countries are presently pumping to capacity, and crude oil prices now exceed US$ 60 a barrel. The International Energy Agency (IEA) forecasted that demand would grow by 1.8 per cent in 2006, and noted that consumption rose 1.2 per cent in 2005 and 3.8 per cent in 2004, more than twice the average growth of the last decade (International Energy Agency 2006).

The credit crisis of 2008 and 2009 led to a slower energy use growth to 2030 than projected in 2007 (IEA 2008: 92).A comparison with the situation today suggests sluggish economic growth could restrict annual oil demand growth to 0.9 mb/d in 2012 and 0.8 mb/d in 2013, with demand averaging 89.6 mb/d and 90.5 mb/d, respectively (International Energy Agency, 2012). However, the IAE (2008: 78) sees the overall trends broadly unchanged: the dominance of fossil fuels and a rising share of emerging economies in global energy consumption. In a reference scenario which assumed no new government policies beyond those already adopted by mid-2008, world primary energy demand expands by 45 per cent between 2006 and 2030 – on average rate of growth of 1.6 per cent per year. Fossil fuels account for 80 per cent of the world's primary energy mix in 2030 – down only slightly on today (IAE 2008: 78).

Some developing countries have an interest in both issues, for example, drought-prone countries with economies dependent on revenues from coal exports or energy-intensive goods, or countries whose poor lack access to cleaner energies and will suffer from increasing fossil fuel costs (ECO 2004a). Nevertheless, it is a widely-held view that by advocating "equal progress" in negotiations on both aspects of Article 4.8, OPEC countries have held developing countries that are vulnerable to the impacts of climate change hostage in their attempts to obtain financial and political gain from the UNFCCC process (Dessai 2004; ECO 2004a). A still more sinister view is that OPEC's strategy is an effort to disrupt and derail the climate process as a whole (ECO 2004b).

Demands made by OPEC members of the G77/China coalition of developing countries have created tension within the group. OPEC countries have consistently sought parallel treatment of the issues of adverse effects of climate change and the impact of response measures in coalition positions, including those that seek action on adaptation. In the context of insurance, this is reflected in OPEC calls for insurance mechanisms to address the impacts of response measures, which are linked to AOSIS calls for insurance mechanisms to address the adverse effects of climate change (ECO 2004a). When G77/China positions are brought forward from the group for further negotiation with developed country Parties, this linkage has the unfortunate effect of obstructing progress on adaptation to the direct impacts of climate change – which is of concern to all developing countries. Developed countries, witnessing increasing demand for fossil fuel resources and experiencing increasing prices for these resources, understandably have little interest in discussing further compensation for OPEC countries as part of the package presented by developing countries. Yet, when developing countries meet resistance in their calls for attention to basic adaptation needs, this dynamic ultimately impacts the willingness of some developing countries to consider commitments to abate GHG emissions and therefore disrupts progress within the climate regime as a whole.

As Barnett and Dessai (2002) argue, a solution to the challenge of negotiations under Article 4.8 ultimately rests with G77/China, which must find a way forward within its heterogeneous group. These authors suggest that G77/China, in light of its long-term goals, de-link the two issues by prioritizing the issue of "adverse effects" over that of "impact of response measures", and thus rethink its past strategy of carrying both issues forward in parallel.

There are some grounds for cautious optimism concerning this problematic linkage. As will be discussed below, the Marrakesh Accords established a Special Climate Change Fund that partially de-links assistance for adaptation measures and response measures. The fund provides support for certain adaptation measures, including capacity building and institutional capacity for preventative measures, planning, preparedness and management of disasters related to climate change (UNFCCC 2001a). Notably, the fund makes no provision for compensating for lost oil revenues, but includes support for economic diversification for Parties included in Article 4.8(h) – "countries whose economies are highly dependent on income generated from the production, processing and export, and/or on consumption of fossil fuels or associated energy-intensive products."

Risk assessment

Developed country negotiators have expressed concern about two aspects of insurance-related strategies for reducing developing country impacts from climate change: the difficulty in assessing risks from infrequent weather-related events and the problem of determining the contribution of climate change to these risks. In the face of large uncertainties in assessing disaster risks and the uncertain effects of climate change on these risks, can a framework for the implementation of insurance-related activities be developed?

Uncertainty is an inherent and essential condition for an insurance contract, but it is important to distinguish between two types of uncertainty. The first is temporal and spatial: when and where (or to whom) will an event occur? Without this uncertainty, insurance is not possible. For example, slowly developing catastrophes, such as sea-level rise, are predictable and thus uninsurable in the traditional sense; however, a fund can be created to compensate victims (Tol 1998). The second type of uncertainty concerns the confidence one has in the risk estimates. The more uncertain or ambiguous these estimate are, the more cautious insurers will be in offering policies.

In the absence of large sets of historical data, advanced risk modelling simulation techniques have increased the confidence insurers place in risk estimates and greatly enhanced the insurability of catastrophic risks (Kozlowski and Mathewson 1997; Bier *et al.* 1999; Clark 2002; Boyle 2002). Although risk assessments can be very resource intensive, by drawing attention to risk and prevention measures they can be useful beyond the pricing of insurance contracts. In developing countries, however, there may be little data on infrastructure lifelines, property values and sensitivity of structures. Moreover, it is difficult to project these risks into the future by taking account of land-use changes, population movements, climate change and many other factors. While many uncertainties will remain, catastrophe modelling can improve the insurability of risks and, therefore, can aid in the development of a framework for the implementation of insurance-related activities.

Liability and compensation

We noted in Section 2 that available scientific evidence indicates a strong likelihood that global warming will result in more frequent and severe extreme weather events, but scientific evidence regarding the influence of increased greenhouse gases in the atmosphere on *current* floods, droughts and other extremes is only just emerging. Some industrialized countries, among them the US, appear concerned that any consent to support for insurance mechanisms might be interpreted as a *de facto* recognition that climate change is, indeed, *currently* contributing to increased disaster losses. An acknowledgment of responsibility for a share of impacts from extreme events, it is further feared, might open the door to liability suits brought by impacted countries, who might then seek compensation for damages far beyond anticipated contributions to adaptation funding. Some observers, for example, Verheyen (2005), argue that international law already provides a basis for liability for climate change impacts, which lies in widely accepted customary law such as the no-harm rule, and that many international declarations and agreements support this view.

The implementation of insurance-related actions need not be viewed as an acknowledgment of responsibility for climate impacts. Instead, these actions can be viewed as a pragmatic response to the articulated "specific needs and concerns" of vulnerable developing countries under Article 4.8. A strong argument can be made for the usefulness of insurance-related mechanisms in enhancing adaptive capacity, even in the absence of a direct causal link established between increasing levels of greenhouse gases in the atmosphere and particular extreme weather events.

The United States, sometimes with the support of other developed countries, has downplayed or dismissed linkages between climate change and extreme weather events in the climate negotiations. This has been seen in efforts by United States negotiators to resist the inclusion of the term 'insurance' in decisions of the UNFCCC Parties, reject any decision text linking climate change and extreme weather events, and oppose consideration of the outcomes of related international meetings held in other fora and their relevance to discussions within the UNFCCC negotiations (e.g. 2005 World Conference on Disaster Reduction, International Meeting for the Ten-Year Review of the Implementation of the Barbados Programme of Action for the Sustainable Development of Small Island States) (see ECO 2004b; Earth Negotiations Bulletin 2004a; Earth Negotiations Bulletin 2004b; Earth Negotiations Bulletin 2005). This defensive strategy presents yet another challenge to an open discussion of the ways in which insurance mechanisms can assist in reducing the vulnerability of developing countries to the impacts of climate change, and the ways in which the UNFCCC can assist in this process.

Establishing financial resources

At the Seventh Conference of the Parties (COP7) to the UNFCCC in 2001, the Marrakesh Accords established three new funds: a Least Developed Country (LDC) Fund, a Special Climate Change Fund (SCCF) and an Adaptation Fund (the latter under the Convention's Kyoto Protocol) (see Mace 2005; Barnett and Dessai 2002; Dessai and Schipper 2003). Each of these funds addresses aspects of developed country commitments on adaptation. For developing countries, the creation of these funds was important for their acceptance of a much weakened Kyoto Protocol (Dessai 2003). It was agreed in Marrakesh that predictable and adequate levels of funding shall be made available to developing country Parties to meet the commitments of the Convention through these funds, the Global Environment Facility, and bilateral and multilateral sources (UNFCCC 2001a).

Developed country contributions to the LDC Fund and SCCF are voluntary, rather than binding. The Adaptation Fund is financed from a share of the proceeds of credits earned from Clean Development Mechanism (CDM) mitigation projects, as well as other voluntary contributions. CDM projects reflect investments in developing countries that generate certified emission reduction credits that can be used by developed countries toward their Kyoto targets. Hence contributions are automatic, but vary with demand for CDM credits and may result from private investments rather than governmental contributions. Contributions to the LDC Fund, SCCF and Adaptation Fund have been made since Marrakesh (Verheyen 2005; Mace 2005), but substantial funding has yet to be committed.

The creation of the SCCF was important in signalling a degree of political will to implement Article 4.8 and its related Kyoto Protocol provisions for the broad group of developing countries. The SCCF will finance activities in four areas: adaptation; technology transfer; sector-specific mitigation activities; and activities to assist countries with economies highly dependent on income from fossil fuel production or consumption in diversifying their economies. The fund is thus associated with both the issue of adverse effects of climate change and the impacts of response measures (cf. Barnett and Dessai 2002). Significantly, among many other adaptation activities, it will support capacity building for preventive measures, planning, preparedness and management of climate-related disasters.

The establishment of the SCCF forced the G77/China to set priorities for this new funding channel. At COP9, in 2003, adaptation to the adverse impacts of climate change was recognized as 'top' priority, followed by technology transfer, mitigation and economic diversification. The discourse on the prioritization of the SCCF was a useful step forward in implementing Article 4.8. By separating issues of adverse effects of climate change and adverse effects of response measures, further discussions on how to operationalize the fund for these different purposes was able to proceed separately.

Negotiations on the Adaptation Fund under the Kyoto Protocol (UNFCCC 2001b) began in December 2005, with the first meeting of the Kyoto Protocol Parties. It remains to be seen how these negotiations will be influenced by OPEC countries, or countries concerned with direct linkages between climate change and extreme weather events (see ENB 2005).

Substantially more capital will be needed than presently exists in the SCCF to support the kind of insurance-related activities discussed in this paper. The Adaptation Fund, once it becomes operational, is likely to have far greater resources to address the needs of particularly vulnerable developing countries. This fund certainly could address preparedness for and management of climate-related disasters, though negotiations have only just begun on how the fund will evolve. However, because not all major developed countries contribute to the SCCF or to the Adaptation Fund, identification of an appropriate, adequate and equitable funding channel presents one of the most challenging aspects of implementing insurance-related mechanisms in response to Article 4.8.

Conclusions

UNFCCC Parties have viable options for supporting insurance instruments in developing countries in response to Article 4.8. Important precedents for donor-supported risk-financing schemes already exist: the public-private partnership in Turkey, microinsurance for disasters in India and public asset insurance for Colombia. This does not mean that a fund for proactive disaster support could be immediately operational. To the contrary, there are many obstacles. Yet, these obstacles, we have argued, are surmountable. Already negotiators are making progress on de-linking support for adaptation to climate impacts from OPEC demands for financial assistance for lost oil revenues. Moreover, a shift in emphasis from compensation for climate change impacts to insurance to enhance adaptive capacity may offer an important step for reassuring those concerned about legal liability. Climate insurance is becoming a possibility with recent developments in catastrophe modelling and innovative pilot programs implemented with support from the international donors and financial institutions. Finally, and most significantly, climate change funds now exist that can address support for preparedness and management of climate-related disasters.

References

Auffret, P. (2003) Catastrophe Insurance Market in the Caribbean Region – Market failures and Recommendation for Public Sector Interventions. *World Bank Policy Research Working* Paper No. 2963.

Barnett, J. (2001) *The Meaning of Environmental Security: Ecological Politics and Policy in the New Security Era.* New York, NY: Zed Books.

Barnett, J. and Dessai, S. (2002) Articles 4.8 and 4.9 of the UNFCCC: adverse effects and the impacts of response measures. *Climate Policy.* 2. 231–239.

Bier, V., Yacov, M., Haimes, Y., Lambert, J. H., Matalas, N.C. and Zimmerman, R. (1999) A Survey of Approaches for Assessing and Managing the Risk of Extremes. *Risk Analysis.* 19 (1). 83–94.

Boyle, C. (2002) Catastrophe Modeling: Feeding the Risk Transfer Food Chain, *Insurance Journal*. 25 February, available through http://www.insurancejournal.com/magazines/west/features/2002/02/25/18828.htm, last accessed 29 August 2012.

Brown, W. and Churchill, C.F. (2000) Insurance Provision in Low-Income Communities, Part II: Initial Lessons from Microinsurance Experiments for the Poor, http://www.mip.org.

Cardenas, V. (2006) *Personal communication*. 26 January 2006.

Clark, K. M. (2002) The Use of Computer Modeling in Estimating and Managing Future Catastrophe Losses. *The Geneva Papers on Risk and Insurance*. 27. 181–195.

Dessai, S. (2003) The Special Climate Change fund: origins and prioritisation assessment. *Climate Policy*. 3. 295–302.

Dessai, S. (2004) An Analysis of the Role of OPEC as a G77 Member at the UNFCCC, Report for WWF, December 2004.

Dessai, S. and Schipper, E. L (2003) The Marrakech Accords to the Kyoto Protocol: Analysis and Future Prospects. *Global Environmental Change*. 13.149–153.

Earth Negotiations Bulletin (2004a) Earth Negotiations Bulletin, Vol. 12, No. 258, 16 December, 2004, available through http://www.iisd.ca/download/pdf/enb12258e.pdf, last accessed 28 August 2012.

Earth Negotiations Bulletin (2004b) Earth Negotiations Bulletin. 12 (260) 20 December, 2004, available through http://www.iisd.ca/download/pdf/enb12260e.pdf, last accessed 28 August 2012.

Earth Negotiations Bulletin (2005) Earth Negotiations Bulletin. 12 (281), 29. November, 2005, available through http://www.iisd.ca/download/pdf/enb12282e.pdf, last accessed 28 August 2012.

ECO (2004a) OPEC Puts Self Interest Ahead of G77 Solidarity, *The Climate Action Network UN Climate Conference Newsletter*. CX (4), 9 December 2004.

ECO, 2004b Saudi Arabia – A Poor Developing Country? *The Climate Action Network UN Climate Conference Newsletter*. CX (7), 13 December 2004.

Emanuel, K. (2005) Increasing destructiveness of tropical cyclones over the past 30 years. *Nature*. 436 (4), August. 686–688.

Freeman, P. K., Martin, L. A., Linnerooth-Bayer, J., Mechler, R., Warner, K. and Pflug, G. (2003) *Disaster Risk Management: National Systems for the Comprehensive Management of Disaster Risk and Financial Strategies for Natural Disaster Reconstruction*. Washington, DC: Inter-American Development Bank.

Gurenko, E. (2004) Introduction, In Gurenko, E. (ed.) *Catastrophe Risk and Reinsurance: A Country Risk Management Perspective*. Haymarket: Risk Books. xxi–2.

Herald Tribune (2006) In a first, UN agency buys insurance against droughts, *Herald Tribune*. 9 March 2006.

International Energy Agency (2006) Oil market report. 14 March 14 2006. Paris: IEA.

International Energy Agency (IEA) (2008) World Energy Outlook 2008 Edition, Paris: IEA.

IPCC: Houghton, J. T., Jenkins, G. J. and Ephraums, J. J. (eds.). (2001) Climate Change 2001, The Scientific Basis, Contribution of Working Group I to the IPCC Third Assessment Report. Cambridge: Cambridge University Press.

Kozlowski, R.T. and Mathewson, S. (1997) A primer on catastrophe modelling. *Journal of Insurance Regulation*. Spring. 322–341.

Intergovernmental Negotiating Committee (INC) (1991) A/AC.237/Misc.1/Add.3. UNFCCC, Second session, 19–28 June 1991. Geneva, Switzerland.

Linnerooth-Bayer, J., and Amendola, A. (2000) Global change, natural disasters and loss sharing: Issues of efficiency and equity. The Geneva Papers on Risk and Insurance. 25. 203–219.

Linnerooth-Bayer, J., Mechler, R. and Pflug, G. (2005) Refocusing Disaster Aid. *Science.* 309 (5737). 1044–1046.

Krishna, H. (2005) Insurance for Vulnerability Reduction. Based on Oxfam's experience of using Insurance as a strategy for Disaster Risk Reduction in Coastal Andhra Pradesh. Background paper for World Bank Institute on-line training course "Financial Strategies for Managing the Economic Impacts of Natural Disasters in India."

Mace, M. J. (2005) Funding for Adaptation to Climate Change: UNFCCC and GEF Developments since COP7. *Review of European Community and International Environmental Law.* 14 (3). 225–246.

Mechler, R. (2005) Cost-benefit analysis of natural disaster risk management in developing countries. Working Paper, Eschborn: Deutsche Gesellschaft für Technische Zusammenarbeit.

Mechler, R., Linnerooth-Bayer, J. and Peppiat, D. (2006) *Microinsurance for Natural Disasters in Developing Countries: Benefits, Limitations and Viability.* Geneva: ProVention Consortium.

Mileti, D. (1999) Disasters by Design. Washington, DC: Joseph Henry Press.

Munich Re (2006) Topics Geo Annual review: Natural catastrophes 2005. Munich: Munich Re.

Munich Re (2012) Natural Catastrophes in Economies at Different Stages of Development. January. Munich: Munich Re.

Otero, R. C. and Marti, R.Z. (1995) The impacts of natural disasters on developing economies: implications for the international development and disaster community, in Munasinghe, M. and Clarke, C. (eds.) *Disaster Prevention for Sustainable Development: Economic and Policy Issues.* Washington, DC: World Bank. 11–40.

Pollner, J. (2000) Managing catastrophic risks using alternative risk financing & insurance pooling mechanisms. Washington, DC: World Bank.

POVCC (2003) Climate Change and Poverty, Joint Agency Paper, World Bank/BMZ/DFID etc. (<http://www.worldbank.org> – post-consultation draft June 2003).

Schönwiese, C. D., Grieser, J. and Tromel, S. (2003) Secular change of extreme monthly precipitation in Europe. *Theoretical and Applied Climatology.* 75. 245–250.

Tol, R. S. J. (1998) Climate change and insurance: a critical appraisal. *Energy Policy.* 26. 257–262.

UNFCCC (2001a) FCCC/CP/2001/13/Add.1, p. 43, decision 7/CP.7 (Funding under the Convention).

UNFCCC (2001b) FCCC/CP/2001/13/Add.1, p. 52, decision 10/CP.7 (Funding under the Kyoto Protocol).

UNFCCC (2004) Proposed Draft Text of the Co-Chairs of the Contact Group Implementation of Article 4, Paragraphs 8 and 9, of the Convention, Progress on the implementation of activities under decision 5/CP.7, Draft Conclusions at 3, (10 June 10 2003); Views on activities, programmes and measures in the areas listed in paragraph 2(c) and (d) of decision 7/CP.7, FCCC/SBI/2004/Misc.6, Submissions from Parties, at 7 (Submission by Saudi Arabia), 4 October 2004.

Verheyen, R. (2005) *Climate Change Damage and International Law: Prevention Duties and State Responsibility.* Leiden: Martinus Nijhoff Publishers.

World Bank (2003) Financing rapid onset natural disaster losses in India: a risk management approach. Washington, DC: World Bank Group.

World Bank (2005) *Colombia – Second Disaster Vulnerability Reduction Project.* Report No.AB908. Washington, DC: The World Bank.

Part IV

Conclusion

Conclusion

Strategic Facilitation of Climate Talks

Gunnar Sjöstedt and Ariel Macaspac Penetrante

The ultimate purpose of this project is to propose and assess significant approaches to strategic facilitation of the UN negotiation on climate change. The research strategy has been to identify concrete obstacles to the climate talks themselves that may be downgraded, eliminated, or circumvented with the help of the strategic facilitation measures proposed. The use of this approach means that not all circumstances obstructing the negotiations and making it difficult for negotiating parties to reach constructive agreements are considered to be relevant targets for strategic facilitation, as the term is understood in this project.

First of all, one must bear in mind that the special purpose of this project is to discuss and assess strategic facilitation measures to address long-living quasi-structural impediments in the climate negotiation. Obstacles that are entirely tied to a particular situation are not considered in the analysis. Consequently, measures suitable for tactical facilitation, such as quick fixes or mediation, are similarly not discussed. Such measures, though important, pertain only to tactical facilitation. However, a broad facilitation approach may combine both tactical and strategic facilitation measures.

There are many important impediments to the environmental, economic, and political backdrop of the climate negotiation, for instance, the difficulty of coping with the growing concentrations of greenhouse gases in the atmosphere – from the technical point of view of the real catastrophic effects of climate warming, the actual economic costs of introducing more renewable energy into an economy, and the existence of configurations of stakeholders with extensive and differing interests in national energy policies. Technological development is yet another important background factor constraining the climate talks and their successful outcome. For example, further technological breakthroughs regarding solar energy or hydrogen gas as an energy source would dramatically change the external setting for the climate change negotiations.

In this project, however, it should be recalled that the key concept of *stumbling block* exclusively represents impediments that are intrinsic to the climate negotiation as such. This is a combination of *negotiated issues, actors/strategies, process, structure* and *outcome*. Naturally, these negotiation elements may be closely associated with parts of the external setting of the climate negotiations but are, in principle, separate from it and usually have different qualities. For example, the issue of climate change on the agenda of the climate talks is obviously closely associated with the physical reality of climate change and its consequences for humans and ecosystems.

However, climate change as a negotiated issue in the pre-Kyoto or post-Kyoto process is not a direct or a perfect representation of the physical phenomenon of climate warming. The issue of climate change is a construction by the negotiating parties (governments) and other actors such as, for instance, scientists of the IPCC, political parties, NGOs, and other opinion builders. In fact, an important part of the climate negotiation is concerned with how climate change should be constructed and framed for the UN negotiation. Thus, in this project strategic facilitation targets climate change only in its guise of a social/political construction placed on the negotiation table. Similarly, according to the analytical logic used in this project, only the direct manifestations, or reflections, of backdrop circumstances in the climate talks are potential stumbling blocks. For example, the difficulty scientists have of predicting the cost–effectiveness of new mitigation technologies may appear as an uncertainty problem in the climate talks, impeding bargaining about emission reductions.

To be useful for practical purposes, counsel regarding strategic facilitation must be very clear with regard to its objective and meaning and must also be precise with regard to its target(s) in the climate negotiation. The analyses of the various cases studied in this book demonstrate that important stumbling blocks may be highly specific within the context of the issue of climate change. For example, parties have faced several difficulties in following the 1996 IPCC guidelines for national GHG inventories, limiting the transparency and comparability of the reporting (see also Watson *et al.* 1996; Noble *et al.* 2000). These problems have not only obstructed the negotiation for instance on land use change and forestry (LULUCF), particularly the issue of "sinks," in a clearly discernible way, but have also had an effect on the climate talks in general, although this has been more diffuse and generalized.

A number of distinct stumbling blocks that are related to different elements of the climate negotiation have been identified in the case studies of the project. These negotiation obstacles are assessed in the first part of the Conclusion. An ensuing section proposes and assesses strategic facilitation approaches that are meant to reduce negative effects of specific stumbling blocks as far and as cost-effectively as possible.

Structure-related stumbling blocks

The structure of the climate talks pertains to both their external and their internal environment.

External structural elements

Various factors pertaining to the external environment of the climate talks support or constrain them. Here belongs, for example the state of the world economy: boom or recession; the risk for international conflict and destabilization in important regions of the world; the power structure of the international system with its diffuse conditioning of the climate talks.

However, in one sense, the international distribution of power and similar systemic factors are largely irrelevant in a discussion about facilitation measures in the climate negotiations. Like other structural properties of the international system, the power structure can be modified only by rather long-term social, economic, and political processes whose interplay is difficult to understand.

Coping with power relations and power structure in negotiation performance may still be a meaningful target for strategic facilitation. For example, both exhibiting and responding to political leadership in the climate negotiation requires a profound understanding of how the prevailing power structure conditions this negotiation. Negotiation parties also need to be aware of how existing power relations constrain political choices and proposed agreement texts. A strategic outlook considering prospects for the future must consider likely changes in the international power structure.

In some cases, changes in structural power conditions may simply co-vary with alterations in the power position of individual leading actors such as the United States, the EU, China, India, Brazil, or the G77. Some of these powers grow relatively stronger and others relatively weaker.

Other changes occurring in the power structure become stumbling blocks because they produce political instability internationally or because they generate uncertainties in policymaking. The role and performance of developing countries at the Climate Conference in Copenhagen is an illustration of this. In 2009, it was clear to many observers that developing countries had become more assertive during the last decade. This development had manifested itself in both the climate talks and the parallel Doha round in the WTO. Still, it was difficult to specify to what degree and in what regard the influence of developing countries had actually increased. This uncertainty was probably one of the explanations of the diplomatic mistakes made in the Copenhagen Climate Conference, which provoked the walkout of developing countries from the plenary meeting.

The notion of international power structure implies a systemic perspective on international power relations between states and non-state actors.[1] Although the strength of individual actors needs to be considered, this factor should be put into a systemic context. The incapacity of a negotiation party to make such power assessments accurately is in itself a stumbling block in the climate talks, because it impedes a government or other key decision maker to develop an appropriate long-term policy in the climate talks. Inaccurate assessments of the relevant power structure underpinning the climate talks may lead to serious mistakes in the conduct of the negotiation or the planning of a COP. Seemingly, erroneous assessments in this regard contributed to the outcome of the 2009 Copenhagen Climate Conference, which is regarded by many people as a failure.

A critical erroneous estimation before Copenhagen was the true power position of the EU in the circle of leading countries at COP15, which was clearly overestimated in some quarters, and particularly in Western Europe. The EU governments had actually planned a leadership role at the 2009 Climate Conference.

Surprisingly, another common miscalculation concerned the rise of China as a leading power in the climate talks, which seems to have been underestimated in Copenhagen. This assessment failure was astonishing because a clear sign of China's growing power in world politics had been visible and commented on for years (Buzan and Foot 2004).

The organizational leadership of the Copenhagen Conference made a similar mistake when it presented developing countries with a take-it-or-leave-it proposal concerning the proposed Copenhagen Accord. These tactics had been used before in multilateral talks, for example, in the World Trade Organization (WTO) and its predecessor the General Agreement on Tariffs and Trade (GATT). The GATT/WTO negotiation has been highly illustrative of how developing countries have been treated in multilateral talks. In several rounds, a large part of the text of a final agreement, including binding rules, was developed in the Organization for Economic Cooperation and Development (OECD) and then transferred to the trade organization. It should be recalled that, currently, the OECD has a membership of 30 industrialized countries, whereas 139 states were parties to WTO in February 2010.[2] Thus, only a little more than 20 per cent of the parties to the WTO are members of OECD. This gives an indication of the traditional hierarchical relationship between developed and developing countries in the international trade negotiations, which in turn has been akin to how the power structure of the climate talks was understood before the 2009 climate conference. This error had strong negative repercussions on the High Segment – ministerial – talks in Copenhagen.[3]

Internal structural elements

To a large extent, organizational bodies, and the knowledge and the norms and procedures that they contain, epitomize significant structural elements of the climate talks. This organizational structure gives indispensible support. It carries out tasks that are necessary for the conduct of the complex, recursive, and long-drawn-out climate talks.

Part of the internal structure offers an organizational setting for conferences, workshops, and other types of meetings. Informal and formal rules and procedures are linked to these organizational bodies and bring order and practical direction to the negotiation. For example, there are rules and procedures prescribing how collective decisions are to be taken in the negotiation. There is also a shared knowledge, indeed a sort of tradition, advising chairs and other officers of the negotiation how to perform their respective function.

The internal structure includes bodies with advanced secretarial duties, which prepare meetings in the negotiation process, make reports from meetings and are responsible for keeping and updating the collective memory of the climate negotiation organization.

IPCC, the Inter-Governmental Panel on Climate Change, is a key component of advanced organizational machinery whose task is to analyze and assess the climate issue: what it is, what problems it causes, and how the warming of the atmosphere can be abated. This internal analytical and strategic intelligence organization has, to date, had a critical role in the climate negotiation. It has organized the buildup of a consensual knowledge that has represented something much more important than background information. In the climate negotiation, the continuously updated consensual knowledge has in reality provided a sort of leadership as it has included, or implied, policy recommendations drawn from competent and therefore authoritative scientific analysis.

Generally speaking, the internal organizational structure seems to have been fairly well adapted to its purpose of supporting the UN climate negotiation. Nevertheless, it has also exhibited a number of deficiencies, some of which can be regarded as stumbling blocks that will confront future negotiating parties in the climate talks. Expressed in general language, there are two categories of stumbling blocks related to the internal structure of the climate talks that need to be highlighted. One kind concerns the negotiation effectiveness of particular organizational bodies that are part of the overall internal structure of the climate talks, such as the Conference of the Parties to UNFCCC (COP) or the Meeting of the Parties to the Kyoto Protocol (MOP). Another category of stumbling blocks pertains to the negotiation effectiveness of the entire organizational structure constructed for the purpose of supporting the negotiation on climate warming. This particular ineffectiveness problem has much to do with the difficulty of coping with organizational complexity in the climate talks.

Lack of negotiation effectiveness of particular organizational bodies

The negotiation institutions of the climate talks are a manifestation of the UN model of negotiation organization. This means that the climate negotiation organization is less adapted to the special needs of negotiating binding and costly commitments in a multilateral setting than the WTO, which represents an alternative model for a negotiation organization.

A comparison between the WTO model and the UN model indicates that the latter suffers from relative negotiation ineffectiveness. At the 2009 Copenhagen Conference, this negotiation ineffectiveness was particularly highlighted by the work of the Plenary of the COP meeting and the High Level Segment of COP/ MOP. These top institutions of the Climate Conference certainly produced significant results in the climate-regime-building process. Notably, they also increased the attention to and awareness of the climate problem by bringing 118 heads of state or government to the negotiation table in Copenhagen. With the Copenhagen Accord they also established clear guidelines and objectives for the post-Copenhagen negotiation. However, the Plenary/High-level Segment part of the negotiation organization did not manage to produce what they were intended to bring into being: an agreement including binding commitments to reduce greenhouse gas emissions.

This failure has a number of explanations, many of which pertain to the power game involving the leading nations in the climate talks. Agreement on binding emission regulations could not be reached in Copenhagen simply because the great powers had diverging interests and key parties such as China and the United States did not want, and did not have, to make any significant concessions in this regard.

However, the negotiation organization constructed along the lines of the UN model may also be blamed for the perceived failure in Copenhagen. The results achieved in this round of the climate talks correspond to what is usually attained, and also expected, at high-level meetings in the UN system, namely joint political declarations. That is what the negotiation organization is designed to primarily produce, not binding, precise and costly commitments. For example, the draft texts discussed in Copenhagen were not developed enough, and there was too little time available for decisive multilateral bargaining on detail, which in turn contributed to impairing the quality of the formal leadership of the Plenary High-level Segment of COP/MOP in Copenhagen. One example is that the chair had to risk asking for a decision on the proposed text of what became the Copenhagen Accord, although there was very little time left for reflection and assessment within delegations and consultations between them. Many governments of developing countries were not consulted at all.

The IPCC exhibits another kind of stumbling block related to a specific negotiation body, whose importance seems to have increased in recent years: of the tendency for authority and trustworthiness to decrease.

According to its Charter the tasks of IPCC are:

> to assess on a comprehensive, objective, open and transparent basis the scientific, technical and socio-economic information relevant to understanding the scientific basis of risk of human-induced climate change, its potential impacts and options for adaptation and mitigation. IPCC reports should be neutral with respect to policy.

During the pre-Kyoto negotiation, the IPCC performed this mission satisfactorily and had a key role in establishing and developing the consensual knowledge needed in the climate talks. The knowledge and information transferred from IPCCC into the negotiation process had sufficient usability and authority derived from the perceived reliability of the data, analyses, and assessments provided by the IPCC.

This situation has changed in the last decades (Skodvin, Chapter 12). The IPCC has become increasingly criticized for quality failures in some of its reports. A number of incorrect propositions and assessments in IPCC documents have come to light in the climate debate and in the media. One example is the assertion in the 2007 IPCC Climate Report that the glaciers of Himalaya will have disappeared by the year 2035, whereas the mainstream world scientific community believes that this will not happen before the year 2350. A number of other inaccurate statements concerning, for example, the rainforests in South America and decreasing precipitation in Africa have appeared in IPCC reports for the same reason. The relevant parts of the reports have not been peer reviewed and have in some cases been taken from campaign material from an environmental NGO.

Inaccurate data and conclusions in IPCC publications are obviously a problem in their own right, but reports in the media that IPCC is using unprofessional and politicized working methods represent a greater quandary. There are signs that the good reputation and trustworthiness of the IPCC are being undermined. This would be a problem for any international organization but is especially harmful for the IPCC, which is a high-profile, single-focus organization whose existence depends on its own reports. The seriousness of this development is revealed by the response by the IPCC itself. By February 2010, the IPCC had decided to set up an independent commission with the task of investigating how its own capacity to respond to future challenges can be improved (von Hall 2010).

One of these challenges is to develop better methods of handling social scientific knowledge, which needs to be drawn into the climate negotiation more extensively because of the stronger focus on adaptation to climate warming and the special problems of poor developing countries. As seen in a negotiation process perspective, a greater role for social science in the climate talks represents a growing problem. Working methods and institutions of the IPCC were originally designed to cope with natural science, which in important respects is different from social science. Therefore, there needs to be an investigation of whether the IPCC needs to transform itself to be able to cope with new social scientific knowledge and information required at the negotiation table.

Negotiation effectiveness of the entire organizational structure

The high complexity of the internal organizational structure of the climate negotiation represents a significant stumbling block. The complexity problem has tended to increase over time, with a proliferation of new committees and bodies as a response to a widened agenda in the development of the post-Kyoto negotiation following the ratification of the Kyoto Protocol, the establishment of the Bali Action Plan, and the preparation for the 2009 Copenhagen Climate Conference. There are considerably more organizational bodies in the climate regime at the present time than in 1992 when the UN Framework Convention on Climate Change (UNFCCC) was signed.

The climate negotiation unfolds in the UN system of rules, norms, and procedures. The relationship between the special climate institutions and the overall UN system is complicated and evolving. The overall UN structure only represents a general setting for the climate talks. The UNFCCC represents the fundamental pillar of the evolving climate regime. It gives general guidelines and offers a basic frame of reference for the climate negotiations. The formal role of UNFCCC in the UN system is somewhat unclear as it is not a formal subsidiary of the UN. Within the UNFCCC frame, special institutions have been created to support three functional tracks representing:

1. Scientific, technical and socio-economic assessment related to climate warming and its consequences, primarily undertaken in the Intergovernmental Panel on Climate Change (IPCC 1988);
2. Development of hard and soft law on the basis of UNFCCC at COP/MOP meetings;
3. A financial mechanism for a number of environmental conventions, the Global Environmental Facility (GEF1991).

Much of the discussions on the consolidation or reinforcement of the climate regime has taken place in UNFCCC, which has functioned as a major setting for the UN negotiation on climate change in the whole period from 1992 until the present time. Negotiation to develop the Framework Convention on Climate Change takes place at yearly Conferences of the Parties (COP meetings) of which the latest was the 2009 Conference in Copenhagen, COP15. COP is serviced by an international secretariat with a staff of more than two hundred international civil servants and almost as many scientific and technical consultants temporarily engaged in the climate talks.

In the post-Kyoto talks, a partly separate institutional framework was established to support the development of the UN climate regime along the guidelines of the Kyoto Protocol. Discussions and bargaining on the further development of the Kyoto Protocol have been made in Meetings of the Parties (MOP) to this binding agreement on the reduction of emissions of greenhouse gases. Since the climate conference in Montreal in 2005, MOP sessions and COP meetings have taken place simultaneously at the yearly climate conferences. A number of other bodies are also part of the institutional structure in which the climate negotiation unfolds. The Subsidiary Body for Scientific and Technological Advice (SBSTA) and The Subsidiary Body for Implementation (SBI) are of particular importance.

The increase in institutions set up to support the climate negotiation is not a problem in itself. It may rather be seen as a realistic functional response to the growing scope of the agenda of the climate talks. However, "institutional proliferation" adds to other problems that may be thought of as stumbling blocks, notably cumbersome coordination across different negotiation bodies. Essentially, the building of an international climate regime by means of multilateral negotiation in the UN context before the Copenhagen Conference unfolded on two different pathways without a satisfactory formula to bring these two tracks wholly together.

One path pertained directly to the UNFCCC representing a broad regime approach but essentially excluding binding and costly commitments to reduce greenhouse gas emissions. The second track looked for a cap solution to reduce greenhouse gas emissions, in line with the norms and principles of the UNFCCC regime but by developing the essentially separate climate regime based on the Kyoto Protocol and its binding governmental commitments regarding emission cuts. In one sense, the organizational setting for the climate talks is fully integrated. Formally, the Kyoto Protocol is an extension of the UNFCCC and therefore part of it. However, UNFCCC and the Kyoto Protocol are to some extent conflicting evolving regimes because of the political realities prevailing in the climate talks. The US government developed a strong negative attitude and position against the Kyoto Protocol in spite of having signed it. One reason is quasi-ideological: the Kyoto Protocol represents a policy of relying on international rules to curb emissions of the greenhouse gases causing climate warming. For the government of the United States as well as of other countries, binding regulations are entirely unacceptable. Governments recommending such a market approach accept future negotiation on the climate regime in the context of UNFCCC but prefer to circumvent the Kyoto Protocol.

In contrast, before the Copenhagen Conference, the EU and other like-minded participants in the climate talks considered the Kyoto Protocol to be a critical element of an international climate regime and emphasized the need to strengthen and further develop its binding rules on emission reductions. Their preferred regime-building strategy was a continued planned framework/protocol approach.

The split between negotiation approaches, and a consequential divide in the organization supporting the climate talks, indicates a significant and potential structure-related stumbling block in the climate talks. This rift was clearly manifested in the period following the establishment of the 1997 Kyoto Protocol and particularly during the second term of the Bush Administration. At this time there were clear signs that the US government was developing and pursuing a long-term strategy to develop an international climate regime with roots in the UNFCCC but with critical differences from the Kyoto regime with its emphasis on formal regulation of emission cuts.[4]

As well as its straightforward confrontational position on binding regulations, the regime model supported by Washington had other special features distinguishing it from the Kyoto approach. Notably, the US/UNFCCC approach was less multilateral than the basic Kyoto approach. The energetic leadership that Washington exercised within the US/UNFCCC context differed strongly from the US performance in the Kyoto segment of the overall climate negotiation. The Kyoto process was genuinely multilateral in character, which was not the case with the US/UNFCCC approach. The US strategy was rather to build up the US/UNFCCC regime as a network of bilateral relations around the United States.

The differences between the Kyoto and the US/UNFCCC parts of the overall climate regime negotiated in the United Nations concerned two absolutely fundamental issues: first, whether formal rules should be used to halt climate warming and, second, the extent to which the regime should have a bilateral or a multilateral profile. This internal inconsistency was so significant that it became a formidable stumbling, the impact of which grew as the post-Kyoto stage of the climate talks unfolded. The 2009 Copenhagen Climate Conference revealed its full significance.

Another structure-related stumbling block pertains to the place of the climate talks in the overall UN system and organizational culture. The background of the UN regime is conducive to many tasks that have to be performed in the climate talks, for example, the debate about principles or positioning on particular issues or the taking of nonbinding resolutions at plenary meetings. However, a comparison with the well trimmed negotiation machinery of the WTO, which is outside the UN, indicates that the UN institutional setting does not give optimal support to bargaining about binding commitments regarding reductions of greenhouse gas emissions. This critical part of the climate negotiation has not been addressed at COP meetings, either in its pre-Kyoto or in its post-Kyoto stage, but in separate negotiation organized outside the UN institutions.

Issue-related stumbling blocks

It may seem trivial to note that parties to the UN negotiation on climate change have consistently had considerable difficulties in coping with this issue since it was put on their agenda in the mid-1980s. It is clearly less trivial to pin down what intrinsic qualities of the climate issue systematically cause these problems and how. Nor is it always clear how intrinsic issue properties can be distinguished from other factors that pertain to other negotiation elements, rather than the climate issue itself. For example, a common proposition is that climate change has proven difficult to negotiate because this issue is so politically controversial. It is difficult to refute this suggestion which, however, has more to do with what leading negotiation parties achieve than with any inherent properties of the climate issue.

According to the theoretical logic accepted in this project, political controversy regarding the climate problems cannot be regarded as an intrinsic quality of the climate issue. It is a function of stark collisions of interest between among governments, as well as among other types of actor (e.g. NGOs) engaged in the climate talks. The genuine intrinsic qualities of the climate problems follow from how they have been framed and constructed. A problem is that intrinsic issue qualities are harder to discern than highlighted clashes of interest regarding the political handling of climate warming.

The negotiated issue of climate change represents the *raison d'être* for the climate talks, and obviously its point of departure. However, issues cannot be regarded as given and permanent conditions for the climate negotiation. Issues are constructions by the parties and are accordingly neither given by nature nor fixed once they have been established on the agenda. In fact, to bring issues or issue details into the negotiation and to frame or reframe them is part of the continuous strategic power game of the climate talks. Governments that are part of the climate negotiation want to control the formation and development of negotiated issues as far as possible. Competition regarding this control is hard, although many nations taking part in the climate talks have little leverage in this part of the strategic power game driving the climate talks.

Nevertheless, negotiating parties in the climate talks share a number of common and quasi-structural problems associated with how the climate issue has been collectively defined in the climate talks. These issue-related stumbling blocks are particularly related to:

1. The pronounced transboundary character of the climate problem;
2. Exceptional uncertainty;

3. The extreme values at stake;
4. Extreme issue complexity;
5. Immeasurability problems;
6. External horizontal linkages between climate change and other negotiated issues.

The pronounced transboundary character of the climate problem

The global impact of climate warming entails far-reaching international cooperation and policy coordination. The pronounced transboundary nature of the climate issue is therefore a strong motive for successful negotiation and a powerful driver in the process. The many governments who "share" the negative consequences of climate warming have a strong incentive to cooperate to diminish these threats, although this motivation may become partly offset by other circumstances directly related to the climate issue.

A major problem with the transboundary dimension of the constructed climate issue is the great variation of negative or disastrous effects of climate warming, in combination with their asymmetrical distribution across the nations of the world (see Schwarts and Randall 2003). Government perceptions of the climate issue vary considerably across nations because of the different catastrophic conditions, both existing and potential, around the world driven by climate warming. For example, some regions and countries fear a rising sea level, others are alarmed about sinking groundwater levels, and still others anticipate more frequent and more forceful hurricanes and tornados that will cause unacceptable losses of lives and property. This asymmetry of the actual consequences of climate warming across nations, and thus divergent perceptions from region to region, is likely to take on the features of a stumbling block as it impedes coordinated policy action at the international level.

Differences of *threat perceptions* are, in turn, enhanced by various other factors such as, for example, a varying capability across nations to understand the climate problem properly for the purpose of developing an instrumental climate policy in the climate talks.

Diverging *risk perceptions* regarding climate warming and its consequences across negotiating parties is one of the causes of another major problem – stumbling block – in the climate talks. The lack of a sufficiently precise long-term goal can be expected to have growing negative impacts in the post-Kyoto talks as the stakes increase, compared with the UNFCCC and pre-Kyoto stages of the climate talks.

The extreme transboundary character of the climate issues also amplifies the cumbersome negotiation problem of attributing responsibility for climate warming to individual nations and ultimately also to particular companies or other economic actors. Attribution of responsibility for reducing climate warming is driven by various political factors such as perceived national interests or ethical norms (e.g. equity and other aspects of fairness and justice). However, ultimately this critical negotiation task depends on an assessment of how large emissions of greenhouse gases a certain country, a certain industry, or a certain company have generated during a given historical period of time. The stark transboundary character of the climate obstructs such assessments. Greenhouse gases causing climate warming are generated by a multitude of sources situated in all countries of the world, and are being generated in manifestly different volumes. To build up precise and trustworthy emission inventories is a complicated and technically burdensome task whose complexity represents an important stumbling block. For example, the associated uncertainty problems can be regarded as a significant stumbling block in its own right.

Exceptional uncertainty

Most negotiated environmental issues are typically characterized by a high degree of uncertainty. In the case of the climate issue, uncertainty is not only extreme but also unusually complex for any multilateral negotiation. Parties are confronted with high uncertainty in all significant dimensions of the climate problem, its causes, its manifestations, its disastrous effects, as well as instrumental measures to cope with it.

Scientists are not certain exactly what warming effect will be produced by a given concentration of greenhouse gases (GHGs) in the atmosphere. It is not yet possible to map the degree of certainty that is acceptable to politicians regarding how greatly the different emission sources have contributed to increasing atmospheric greenhouse concentrations or how carbon is removed from the atmosphere each year.

MacFaul (Chapter 15) notes that, in the inventories of greenhouse gas emissions, signatory states to the UNFCCC and the Kyoto Protocol have undertaken to communicate in the context of the negotiation on climate change. The inventories made in some countries remain of such a low quality that they do not really contribute to reduce uncertainty about greenhouse gas emissions and their distribution around the globe. There is now an emerging consensual knowledge about what kinds of disastrous effects can be expected to be produced by climate warming, such as a rising sea level around low islands or coastlands, inundations, desertification, or more frequent and stronger hurricanes. However, no certain predictions can be made about the occurrence of climate-driven events in a particular country or geographical region. For example, when policymakers ask whether disastrous hurricanes will hit their country in the next five or ten years, they will not get a meaningful answer.

The general uncertainty problem is amplified by the long time frame in which the climate problem, as well as mitigation measures, has to be viewed. These circumstances condition a predicament which has been referred to as "negative perceptions of the immediate outcome" (Faure and Rubin 1993: 23). Short-term precise and certain mitigation and adaptation costs are compared with long-term, relatively diffuse and highly uncertain benefits of stabilizing greenhouse concentrations in the atmosphere.

The extreme and complex uncertainty characterizing the climate issue has obstructed the negotiation on this topic in various ways. Negotiating parties with a low capability to analyze and understand the climate issue have had difficulties in determining what their interests are with regard to various sub-issues derived from climate warming as a negotiated issue such as commitments to reduce greenhouse gas emissions, "sinks," or the verification mechanisms of the climate regime. A vague perception of interests due to uncertainty problems obstructs the determination of party positions in the climate talks, particularly in negotiation groups where problem solving and construction of texts for a binding agreement takes place in the realm of "editing diplomacy" in which precise, final formulations of a text are determined. The divergence of risk perceptions on the part of negotiation parties has been sustained or even reinforced as the climate talks have unfolded. In turn, this has complicated the bargaining process in the climate talks. Hence, an issue like "sinks" has been difficult to address in a constructive way. The capacity to monitor, measure, and estimate the carbon removed by "sinks" is an unrealized condition for calculating how much credit a given nation should be given for LULUCF activities, for example, extension of forest land. In a more general sense, the most powerful impeding effect of high and complex uncertainty is that it contributes to increasing the reluctance of responsible policymakers to make costly, binding commitments, such as a new undertaking to cut emissions of greenhouse gases by at least 20 per cent by the year 2020.

Extreme values at stake

There is a broadly shared international understanding that unabated climate warming will lead to losses of extreme values, for example, due to changed patterns of precipitation, devastation of ground water resources, inundations, or hurricanes (Parry *et al.* 2007). In some countries, large areas of land will be lost due to a rise of the sea level. Some island nations may be doomed to disappear altogether (*ScienceDaily* 2007).

Effective measures of mitigation or adaptation to climate warming will also be extremely costly (see Metz *et al.* 2007). A far-reaching reorganization of many critical societal activities is required in, for example, agriculture, industrial production processes, the generation of electricity, or the transport of goods and people. Expected mitigation costs will probably increase considerably in the post-Kyoto talks as compared to the negotiation leading to the Kyoto Protocol. It is likely that targeted emission reductions of around 20 per cent or more in the post-Kyoto talks and the negotiation beyond the Copenhagen meeting will become considerably more costly – more hurting – than the 5–8 per cent reductions negotiated in the pre-Kyoto talks.[5]

Growing stake values have already affected the climate negotiation negatively in various ways in the talks producing the Kyoto Protocol. Hence, conflicts of interest between negotiating parties tend to become amplified or intensified as the stakes of the climate talks increase. Parties also tend to become more reluctant to commit themselves to binding agreements. For these and other reasons, the extreme values at stake in the climate talks will probably represent a formidable stumbling block in the post-Kyoto period.

Extreme issue complexity

The high complexity, which has been typical for many environmental issues addressed in multilateral negotiation, is particularly pronounced in the case of the climate talks. This complexity is a reflection and corollary of the complicated natural processes that produce climate warming, and also ultimately causes their devastating effects. For example, referring to the difficulty of understanding the "sink"[6] issue (LULUCF), Bolin and Sukumar (2000) argue that the limited understanding of the dynamics of the carbon cycle within ecosystems is a big problem. Forests are dynamic systems where cycling of nutrients and carbon takes place. The cycles depend on, and are regulated by, a complex arrangement of physical, chemical, and biological factors that determine the flows, quantities and distribution of matter and energy within the ecosystem (Bolin and Sukumar (2000).

Conceived of as a constructed and negotiated issue, climate change rests on a multi-disciplinary foundation that cuts across the social, ecological, economic, and political spheres. The IPCC and many analysts of climate change look for a multi-dimensional approach which in principle may have ramifications for all sectors of society (Churie Kallhauge and van Well, Chapter 10).

The negotiated issue of climate change consists of a multitude of different components representing causes, manifestations, and effects of climate warming, as well as abatement approaches including, for example, the application of high technology, international regulations, and variety of economic instruments. Policymakers and negotiators depend on incessant continuous supply of scientific and technological knowledge/information from numerous academic disciplines and research areas to be able to recurrently reframe, analyze, assess, and tackle the problem of climate warming.

Issue complexity has no doubt been a stumbling block both in the talks leading to UNFCCC and the Kyoto Protocol, as well as in the current post-Kyoto talks. Many poor developing countries have almost entirely lacked the capacity to cope with high issue complexity. For example, these nations have not been able to mobilize the necessary scientific knowledge and expertise to back up offensive climate diplomacy (Najam *et al.* 2003). Most developing countries did not have enough personnel resources to be active in all committees and working groups addressing specific issues on the agenda of the climate talks. Numerous poor nations were unable to attend two parallel negotiation sessions, regardless of what topics were aired. The strongly diverging negotiation capability of the Parties was in itself a complicating factor in the pre-Kyoto stage of the climate talks (Gupta 1997). The importance of this handicap is now increasing in the post-Kyoto talks due to the need to integrate developing countries further into the emerging climate regime. Unless developing countries are granted a greater say in the climate talks, they are unlikely to accept a binding commitment to reduce greenhouse gas emissions.

The disparity in the capacity of negotiating parties to cope with issue complexity has also manifested itself as an important problem – a stumbling block – in the climate talks. Methods to cope with issue complexity that are viable and useful for high-capacity nations (notably, the member states of the OECD and a few other nations) are often less useful, or even not relevant, for low-capacity nations such as numerous developing countries. This situation represents important constraints for the effective implementation of new negotiation methods and support of the negotiation process. For example, facilitation in the form of process construction may become counter-productive if such reform methodically favors one category of nations (e.g. OECD Member States).

To cope with issue complexity, nations with a capacity to retain a useful negotiation capability have needed to send large composite national delegations to meetings including different kinds of experts at the side of the traditional diplomats, such as scientists, international lawyers, and experts on national security. The need to coordinate preparations and conduct of the negotiations on various technically complex subtopics of the climate issue has been an obstacle for many individual delegations. Such capability problems for individual countries easily develop into process obstacles.

Immeasurability problems

Some elements of the climate problem can in principle be measured with great exactness in quantitative terms such as emissions of CO_2 from a given source, say, a power plant. This situation is, however, rather exceptional.

Other more complex constituents of the climate issue are more difficult, or even impossible, to assess with exactitude and a reasonable degree of certainty, such as the actual value of what is destroyed by the disastrous effects of climate warming. In this regard, only fairly general and relatively uncertain risk assessments are attainable. The benefits of mitigation measures are even more cumbersome to assess, as they will ensue only in the longer term (several decades), are hard to specify, and are highly uncertain. These circumstances give rise to a critical immeasurability problem in the climate talks; the difficulty – or complete unfeasibility – of making a cost–benefit assessment of how cost-effective mitigation and adaptation measures are in terms of stabilized greenhouse concentrations in the atmosphere or reduced climate warming; two of the principal objectives of the climate talks. It is argued that only the difficulty of discounting costs and benefits into the same time dimension represents an obstacle in the negotiation on climate change, as it impedes a comparison of costs and benefits as seen in a long term perspective (Pearson, Chapter 16).

The immeasurability problem represents a major stumbling block. It contributes to making many governments hesitant, or outright reluctant, to accept costly commitments to strengthen and sharpen the climate regime, even if they have a general interest in doing so.

External horizontal linkages

A special problem of the climate issue is its numerous linkages to other established issue areas on the chessboard of international politics. The issue of climate change represents an interface between many different policy sectors such as environmental policy, energy policy, transport policy, or agricultural policy. These inter-issue linkages need to be considered in the climate talks. Hence, the political requirement of addressing climate change in the context of sustainable development puts pressure on negotiating parties to manage the horizontal issues to which climate is linked or that it is part of. This is a complicating factor in the post-Kyoto talks, a noteworthy stumbling block.

As a rule, environmental regime building has unfolded in a comparatively narrow issue area with respect to delimited topics like, say, ozone depletion or desertification. Generally speaking this strategy has been successful. It has produced a large number of international agreements, each of which is covering a clearly delineated issue. In common with other environmental issues, climate change has been addressed as a delimited sector separated from other environmental problems. The purpose of this general strategy, which has generated a large number of international environmental treaties, has been to reduce complexity and to attain a strong focus in each particular negotiation. It has, however, become increasingly difficult, unrealistic and ultimately also counter-productive to keep the climate issue area secluded. Over time, the number of external horizontal connections requiring consideration by negotiating parties has increased. These linkages are of different types and therefore call for different treatment in the climate talks. There are physical linkages in the sense that problems and opportunities appearing in the climate talks are wholly or partly caused by real world phenomena pertaining to other international regimes than that of climate change (Sokolov, Chapter 7). Accordingly, there is a need to harmonize negotiation on climate change with regime building in some other environmental areas such as desertification or stratospheric ozone depletion, as well as with certain political processes and regimes in the socio-economic area, notably the World Trade Organization. The significance of external linkages into areas outside the environmental sector has increased because of the requirement to develop the climate regime in line with the aims of sustainable development. Another factor is obviously the call for massive financial resources that are needed both for effective mitigation measures and satisfactory adaptation to climate warming, particularly in poor developing countries.

Strong institutional factors hamper an effective handling of horizontal linkages that are connected to the climate talks and the international climate regime. The need to deal with such horizontal issues represents a major, and growing, stumbling block in the climate talks. The separation of issue areas is strongly embedded in state-controlled institutions at all levels of policymaking. Coping with issue linkages is for the most part just as demanding at the national governmental level as it is in international negotiation.

Issue-related stumbling blocks: General tendencies of change

Parties need to constantly look for ways to make the climate issue more easily negotiable. To become effective, facilitation should systematically target issue properties that are of special or increasing importance, particularly the problems of complexity, uncertainty, and immeasurability.

The upgrading of sinks (land use and forestry) and adaptation as negotiated sub-issues in the post-Kyoto climate talks have tended to increase both the uncertainty and the complexity dimension of the climate issue. Knowledge diplomacy, which is of such a critical importance in the climate talks, has become more complex and demanding.

Negotiated climate themes associated with the mitigation of climate warming have been constructed and reconstructed with the help of a more or less continuous critical input of natural scientific knowledge/information. This intra-process communication in the climate talks that largely unfolded in the IPPC functioned very well in the pre-Kyoto negotiation (Skodvin, Chapter 12). The IPCC served as an effective instrument for the interaction and exchanges of scientists and policy-makers/negotiators. The common understanding of the climate issue that resulted from these continuous exchanges facilitated negotiation and even functioned as a driver in the process. However, this favorable situation has been somewhat altered as the post-Kyoto talks of negotiating parties has shifted somewhat from mitigation to adaptation and the talks on "sinks", a development that tends to have a significant negative impact on the negotiation process. When the use of social science in the climate negotiation grows more wide-ranging, it will also become more difficult to uphold the clear borderline between interpretation of science and policymaking that was so important in the construction of consensual knowledge at the pre-Kyoto stage. There is a risk that the new consensual knowledge that needs to be constructed about sinks and adaptation, as well as other sub-issues, will not attain sufficient authority and legitimacy to support and drive the climate talks in the same way, and to the same degree, as acknowledged natural scientific knowledge was able to do in the pre-Kyoto talks.

Actor-related stumbling blocks

Nations participating in the climate talks have sometimes deliberately created stumbling blocks because too much progress has been contrary to their true interests. Thus, since the ratification of the Kyoto Protocol the US has systematically performed the role of "brakeman" by delaying advancement in the negotiation on binding regulations regarding emission reductions. For the US this has been a better strategy than leaving these talks. By choosing a brakeman role rather than an exit strategy, Washington has been able to categorically refuse to accept compulsory regulations and at the same time remain in the negotiation on how such rules should be designed.

The 2009 climate conference in Copenhagen demonstrated that other nations other than the US are now becoming sufficiently strong to have a significant impact on the climate negotiation by performing as a brakeman. China is one such country and three others are Australia, Brazil, and India.

A stumbling block which emerges when countries like China or the US act like this in the climate talks does not reflect a failing performance capacity or unforeseen consequences of negotiation performance. It is rather the successful outcome of strong diplomacy at the unit level. In such a situation there may possibly, but not likely, be room for a traditional third-party mediator.

Proposals for facilitation measures proposed by third parties should be more useful concerning other types of actor-related stumbling blocks that are *not* sought by the negotiation parties involved. Parties to the climate talks have a varying capacity to negotiate effectively in the climate talks. Some negotiation parties lack an adequate negotiation capability altogether. Under some conditions, a relatively insufficient negotiating capacity characterizes both weak and strong states as well as nongovernmental organizations, and may hence represent significant stumbling blocks. One set of problems concern the difficulties many poor developing countries have in promoting their true interests in the technically complex climate talks. The scarcity of authoritative political leadership in periods and situations where the United States have chosen to perform a rather passive role.

Failing negotiation capacity of developing countries

Many weak developing countries have great difficulties in defending or even understanding their interests in the climate talks because they lack the necessary resources and capabilities (Gupta 1997; Richards 2001; Heller and Shukla 2003). Developing countries do not have large enough government budgets to permit them to send sufficiently competent delegations to all meetings in COP/MOP or other settings where inter-governmental negotiations on climate warming take place. In fact, at the Conferences of the Parties to UNFCCC, many weak countries were not able to be active in two parallel sessions. Delegations of, for example, African countries have typically been conspicuously smaller than the negotiation teams sent by developed OECD member states, which often include a considerable number of diplomats and experts representing various useful kinds of expertise.

Lack of adequate negotiation capabilities is damaging not only to the individual developing country but to the whole process of climate negotiation, where their full participation is increasingly needed. Because of their inadequate negotiation, capability numerous developing countries have simply not been able to develop sufficiently nuanced and flexible positions to remain continuously truly involved in the unfolding negotiation process.

To some degree the relative powerlessness of developing countries in the climate talks has been compensated for by membership in one or more coalitions of states that have been active in the climate talks. One example is various coalitions of island states in the Pacific Ocean that are seriously threatened by a rising sea level, such as South Pacific Island States and the Alliance of Small Island States (Chasek 2005).

Another example of coalition support is the G77 which buttresses individual developing countries in all kinds of cooperation and negotiation unfolding in the UN system. G77 assistance to individual governments has been far from negligible but has still not enabled most developing countries to take new offensive positions in the climate talks. Generally speaking, G77 has mostly helped developing countries to obtain acceptance for exceptions from binding and costly commitments in the Kyoto Protocol and other negotiated agreements attained in the climate talks. In June 2006, 164 countries had signed the Kyoto Protocol but only 40 governments were included in its Annex B where governments have made binding and costly commitments to reduce their greenhouse gas emissions.

In the pre-Kyoto talks, special and favorable treatment of developing countries in the form of exceptions from binding commitments signified a major ambition of weak developing countries. This is not an optimal approach either for the negotiation as a whole or for these particular nations as it tends to reduce their possibilities of influencing the negotiated buildup of a global climate regime.

Essentially, to seek exceptions with regard to the climate regime means to deliberately retain a peripheral role in the process of climate negotiation. The approach of seeking "across-the-board" exceptions from binding commitments in the climate regime is, at best, a short-sighted tactical way of avoiding imminent binding sacrifices which, however, are gains that will often prove very costly as seen in a longer time frame. The price for short-term (probably superficial) gains in the form of exceptions from binding rules is crippled influence in the regime building process.

Furthermore, a "seeking exceptions strategy" is likely to become increasingly unacceptable for those countries that have actually made costly commitments to reduce emissions of greenhouse gases in the post-Kyoto Protocol or elsewhere. In many cases *exceptions* are seen as representing a *free ride*. In the spirit of this kind of transitional thinking, various industrialized countries are likely to put harder pressure on developing countries to start bringing down their greenhouse gas emissions and to incorporate themselves more solidly into the global climate regime evolving in the UN system.

The United States has clearly expressed a preference for this policy even before signing the Kyoto Protocol in 1997. When the United States recently returned in earnest to the negotiation about the Kyoto Protocol after the election of Barak Obama as new President in 2008, its position on the issue of exceptional treatment of developing countries had actually hardened. US offers regarding emissions cuts have become increasingly strongly linked to requests for reciprocal offers by other individual states, not least by the leading developing countries like Brazil or India.

From the situation sketched above regarding weak states, a number of scenarios can be deduced which are likely stumbling blocks in the post-Kyoto negotiation.

First, if developing countries continue to demand "special and favorable" treatment in the form of continued exceptions to binding and costly commitments, there is a risk that the climate negotiation will develop into a trench war between the G77 member nations and the developed OECD countries that are part of Annex B of the Kyoto protocol. Either the continued climate talks within the UN will become seriously obstructed or the countries refusing to accept commitments to reduce GHGs will be pushed out of the building of the UN climate regime for all practical purposes, although most of them may probably be able to preserve a formal affiliation.

Second, if weak developing countries accept to negotiate emission reductions for binding commitments and refrain from "special and favorable" treatment, the climate negotiation will be disturbed by their failing capability to negotiate effectively in the same context, and under the same conditions, as the more influential industrialized countries. Therefore, the weak developing countries, as well as their partners in the OECD group, have a common interest in helping them become more capable parties in the climate negotiation.

The leadership problem

An international political process like the climate talks is of extreme complexity. In one sense the talks represent a gigantic management problem. To function well, or even at all, an effective *administrative leadership* is needed for guidance and order. This kind of leadership is provided by the relatively large number of people who serve as conference officers performing roles such as Chair, Secretary/Secretariat or *Rapporteur* from, for example, major negotiation sessions or working groups addressing complicated technical issues.

No allegations concerning critical deficiencies in the administrative leadership of the climate negotiation have been put forward either by governments or by academic analysts. Some shortcomings have been observed which are typical of the UN system (see Hanschel, Chapter 14). Skillful "chairmanship" is a scarce and special capability, particularly at negotiation sessions where binding decisions are taken by consensus. However, in the climate talks, as in the UN institutions, the expected high chair capability has to be combined with geographical representativeness when administrative leadership is elected. This procedural rule means inevitably that some *conference officers* have little experience from chairing in the UN system and low skill to do so. Another problem is that the chairs of some committees running on parallel tracks addressing interdependent issues may have a different capacity to guide and support the ongoing negotiation work. If the pace of the work process and/or the quality of its outcome varies too much, coordination between the committees concerned will be hampered.

Only strong governments can provide political leadership to guide, manage, and drive the whole process of the climate talks, either acting individually or in a fairly integrated coalition of states like the EU.

A political leader in the climate negotiations is distinguished by his/her its ability to carry out some, or all, of the following tasks:

1. Lead the work of framing and defining the issues to be addressed in the negotiation;
2. Present authoritative proposals in the context of agenda setting, issue clarification;
3. Put forward guiding ideas pertaining to problem solving regarding issues as well as the advance of the negotiation process;
4. Lead final negotiation sessions during a COP/MOP meeting characterized by bargaining and the exchange of concessions into a collective decision as to whether a proposed text for an agreement should be adopted or not;
5. Table grand proposals for a negotiated solution to the overall problem addressed in the negotiation, draft texts to final issue-specific accords, or a

final comprehensive agreement incorporating all issues addressed at a particular session of the climate talks;

6. Manage the negotiation process as it unfolds;
7. Drive the negotiation process by means of interaction with other negotiating parties.

A comprehensive leadership includes a capacity to carry out all these functions effectively, as well as a faculty to switch from one task to another when the situation so requires. The historical record demonstrates that only a few negotiation parties can aspire to a complete leadership role. However, a number of actors in the climate process are able to fulfill one or more than one of the seven leadership tasks. For example, several small North European states are fully competent to "present proposals in agenda setting and issue clarification"; "to put forward guiding ideas pertaining to problem solving" or to "table grand proposals for a negotiated solution".[7]

One complication is that even if such interventions and submissions by small states are technically feasible and innovative with regard to contents and form, they are usually not sufficiently authoritative to have an impact on the negotiation. Interventions by small states are simply not backed up by enough political power. At least two leadership tasks, *to lead final negotiation sessions* and *to drive the negotiation process*, are usually completely reserved for the great powers. A major stumbling block related to the function of political leadership in the climate talks is the continued dependence on an American leadership role to reach a binding agreement.

The UNFCCC talks and much of the pre-Kyoto negotiation were to a large extent driven by a political leadership that was provided by the United States and exercised by means of both carrots and sticks. The climate talks developed in patterns that were similar to those that emerged in many multilateral talks after World War II and had also been dominated by the United States. The Clinton Administration played a major role in the negotiation on the Kyoto Protocol which was supported by the US until Congress refused to ratify it. This development also led to Washington withdrawing completely from its *leadership* role and more consistently starting to perform as a *brakeman*. As a result, the climate negotiation entered into a leadership crisis, as no other negotiating party had the capacity to fully take over the role of a driving *leader* (Andresen, Chapter 8). The European Union tried to take over the responsibilities of leadership but was only moderately successful in realizing this ambition, although it was a crucial player in negotiation on the Kyoto Protocol. When in 2001 the United States refused to ratify the Kyoto Protocol and therefore partly withdrew from the Kyoto regime, a power vacuum was created which the EU only managed to fill, but only to a limited degree.

Assuming that the US will remain outside the Kyoto regime, the need for a leader role to be performed by the EU will be reinforced. There is simply no other country or coalition of states able to challenge the EU leader role when the US performs as a brakeman. The G77 + China may be able to exhibit a strong veto power when decisions are taken in the negotiation, but they are not able to lead in problem solving or to drive the process (Edwards 2002).

However, the external constraints for an effective EU leadership role in the climate talks are more likely to increase rather than to decrease in the years to come. First, the dramatic growth in the number of EU member states in the last decades obstructs the actor capability of the EU.

Second, the demands on a leader in the climate talks can be expected to grow as the post-Kyoto talks evolve. Negotiation on upgraded compulsory emission reductions covering an expanding set of greenhouse gases will make the talks on mitigation increasingly politically sensitive because of the enormous economic values at stake. A stronger focus on adaptation to climate warming will make the agenda of the climate talks more complex and, for this reason, more demanding for all participating countries, and particularly for an actor or coalition aspiring for leadership.

Third, the United States is not likely to remain passive outside the Kyoto process but is developing a competing regime-building strategy that, if fully successful, will make the Kyoto process insignificant or perhaps bring it to a standstill (Kanie 2005).

The guidance of the climate talks is not only performed by exceptionally strong and competent negotiating parties (like the United States) but also individuals elected to serve as chairpersons, or other functionaries, serving the negotiation. To pinpoint exactly what such negotiation officers are able to do to move the negotiation process forward, the leadership needs to be broken down into functional subcategories. This can be done in various ways and with reference to different theoretical perspectives. One practical approach is simply to distinguish what kind of actions a leader can, and needs, to perform to live up to the requirements of this role.

The participation of NGOs in the climate talks

The increasing participation on the part of NGOs in multilateral processes and institutions was a general development before COP15 in 2009. In some cases this change has been sweeping, as illustrated by the current international trade regime of the WTO. Before the GATT was transformed into WTO in 1994, no single NGO had access to the proceedings of the international trade talks. When the Sixth Ministerial Conference of WTO took place in Hong Kong in 2005, around 650 NGOs attended (Accreditation Centre 2005).

The climate talks have also been characterized by a mounting number of participating NGOs This involvement of more NGOs has had a number of positive implications. Individual NGOs, or groups of NGOs, have in the past carried out important tasks in these talks and other environmental negotiations (Betsill and Corell 2007). For example, NGOs have provided useful expertise and intelligence. They have supported verification systems and represented weak groups such as women in poor countries. More NGOs in the climate talks is likely to increase the volume of services that they make available. However, the NGO contributions

Table C.1 Number of NGOs attending climate talks 1995–2011

COP1 1995	COP3 1997	COP4 1998	COP6 2000	COP6bis 2001	COP7 2001
165	236	148	275	219	294
COP8 2002	COP9 2003	COP10 2004	COP11 2005	COP12 2006	COP13 2007
168	267	226	362	246	335
COP14 2008	COP15 2009	COP16 2010	COP17 2011		
464	794	594	665		

have a price, primarily because NGO participation contributes to the complexity of the negotiation process but also holds it back. These negative consequences of NGO participation can be regarded as a stumbling block that needs to be eliminated, at least to such a degree that they are balanced out by positive effects.

The Copenhagen Climate Conference demonstrated that if NGO participation is badly organized this may provoke negative reactions that become so persistent that taken on the character of stumbling blocks.

Process-related stumbling blocks

The process of the climate talks represents the entire pattern of interaction during the whole life of the climate negotiation. Evidently, stumbling blocks pertaining to other dimensions of the climate negotiation – issues, actors, and structure – are likely to have a negative impact on the process. For example, if a large number of governments lack sufficient competence to negotiate critical aspects of the climate issue in a meaningful way, the process of negotiation will become protracted and also ineffective in terms of producing adequate results.

However, the process perspective is important because it offers a holistic systemic view on party interaction that is particularly important in strategic thinking about climate talks facilitation. The climate talks do have a number of basic features that are problematic and that need be addressed at a systemic–process–level. These stumbling blocks pertain to the complexity of the negotiation, and its long duration, sometimes ineffective decision-making, and asymmetrical patterns of culture and power.

Strategic facilitation approaches

The focus of this study is on *facilitation with a strategic purpose* which aims to create conditions that are "conducive to reaching agreement" (Hopmann 1996: 231) in climate talks that may continue for many years. The project is hence concerned with the long-term development of the UN talks on climate change. Many successful facilitation measures being carried out at present will possibly not generate any positive effect on the climate talks until COP/MOP meetings that will occur 5–10 years into the future.

The qualitative difference between short-term and long-term facilitation needs to be emphasized. In contrast to strategic facilitation, tactical facilitation measures are intended to have a positive impact on the situation in which they are carried out. For example, measures to immediately involve developing countries more in decision making at the Copenhagen Climate Conference could probably have prevented their spectacular daylong walkout.

There are sometimes opportunities to establish fruitful links between tactical and strategic facilitation, which may generate important synergy effects. *Mediation* offers an illustration. This kind of third party intervention is a process of conflict management related to, but distinct from, the parties' own negotiations, where those in conflict seek, or accept an offer of, assistance from an outsider (whether an individual, an organization, a group, or a state) to change their perceptions or behavior without resorting to physical force or invoking the authority of law (Bercovitch and Rubin 1992). For the parties, mediation is a voluntary form of conflict management and the adversaries retain their control over the outcome of their conflict.

Mediation essentially has a tactical purpose and may have different forms, depending on the circumstances. A basic approach of a mediator is to act as a go-between, simply facilitating communication between mutually hostile parties. Another mediation method is to use superior power and the threat of sanctions to make parties move closer to an agreement (Rubin 1981). A third approach is to use superior technical knowledge and information to help parties discover joint interests and feasible solutions to negotiation problems (Stenelo 1972).[8] In the context of the climate talks, this form of mediation seems to have the greatest potential for constructive couplings with strategic facilitation. Two scenarios come to mind as illustrations of this contingency.

In one scenario, tactical mediation between leading powers in a particular negotiation round (e.g. the Copenhagen Conference) may focus on interests and "hard" policies and find it supports and paves the way for strategic facilitation; this is because striving to identify or specify common great power objectives results in advanced knowledge diplomacy (Bolin 1994; Biermann 2000; Kjellén 2007; Sjöstedt 2009). The essence of this scenario is that mediation efforts in the short term may help to clarify what forward-looking knowledge diplomacy can do to narrow the differences between the countries in question.

In another scenario there is a reverse relationship between tactical and strategic facilitation. In this case the efforts to determine the common interests of important negotiating parties by means of continuous advanced knowledge diplomacy will facilitate tactical mediation between the same states in future negotiation rounds. One approach is the establishment of consensual knowledge, which has been fundamental during the whole process of building the climate process.

Other forms of tactical facilitation than mediation may also become linked to a long-term facilitation strategy. Coupling or decoupling of issues has often been used as a tactical method to avoid, or get out of, a deadlock in a particular negotiation session such as a COP or MOP meeting. Coupling of issues is a way of using trade-offs in a multilateral negotiation to reach a comprehensive agreement. Decoupling of issues is a way of simplifying bargaining on detail and of paving the way for knowledge-based compromises between key adversaries.

Both tactical coupling and decoupling of negotiated issues in the climate talks may have intended or unexpected positive consequences for long-term facilitation strategies. For example, both coupling and decoupling may represent the beginning of a process of fundamental issue reframing.

A basic task for strategic facilitators is to find ways to handle the variety of major issue-related stumbling blocks such as those identified in this study:

1. the pronounced transboundary character of the climate problem;
2. exceptional uncertainty;
3. the extreme values at stake;
4. extreme issue complexity;
5. unmeasurability problems;
6. external horizontal linkages between climate change and other negotiated issues.

To tackle issue-related stumbling blocks, a facilitator can conceivably also address other basic elements of a negotiation: actors/strategies, process, and structure.

To a large extent, facilitation measures targeting issue-related problems would represent the transfer into the negotiation process of the results of targeted research and studies to clarify the above problems and notably those relating to *transboundary issue character, issue complexity*, and *issue linkages* (1, 4 and 6). Research work may also be undertaken to formulate concrete proposals on how to actually cope with at least some of the other issue-related stumbling blocks, such as the *uncertainty problems, issue complexity*, the *immeasurability problems* as well as *issue linkages* (2, 4, 5 and 6). For example, natural scientists need to continue to actually reduce uncertainties regarding specific aspects of the climate issue, such as inventories of greenhouse gas emissions or specific consequences of climate warming (e.g. those of a rising sea level in a particular geographical area, say, the land in the downstream Nile area).

However, for strategic facilitation, such studies of various critical aspects of the climate issue have to be integrated into a clear negotiation perspective. Strategic facilitation measures may differ with respect to immediate target, influencing mechanism, and operational objective. For practical purposes, it is meaningful to distinguish between, on the one hand, strategic facilitation targeting the climate talks at large and, on the other, facilitation measures being directed at specified elements of the evolving long-term climate-regime-building process. This outlook has two dimensions pertaining to a unit and systems level of analysis, respectively. The unit level includes facilitation measures targeting actors of the climate talks. The systems level pertains to process and structure and, particularly, the interface and interaction between these two dimensions of the negotiation on climate change. However, both unit and systemic level assessment of measures of strategic facilitation have to be made against the background of a discussion of strategic facilitation targeting the UN climate negotiation as a whole.

A general approach to the building of a post-Kyoto climate regime

When the post-Kyoto negotiation on climate change was formally launched at COP11 in Montreal (2005), the intention was to establish and use a direct continuity from the Kyoto Protocol to a future new agreement of the same character covering a period after 2012. This plan implied the continuance of the framework/ protocol approach, which was *de facto* established when UNFCCC was accepted at the UNCED Conference in Rio de Janeiro 1992.

After 1997, the essence of the original framework/protocol strategy was to build further on UNFCCC and the Kyoto Protocol, thereby widening the scope of the agenda, extending the circle of countries making binding pledges to reduce greenhouse gas emissions, and raising the average level of agreed emission cuts stated in international regulations.

Conceived of as a strategy, the framework/protocol approach has several features that may be useful to rely upon for strategic facilitation purposes (Hanschel, Chapter 14).

• It makes it possible to shift difficult questions to a later stage of the negotiations without blocking the process.
• It facilitates the task of building up a necessary scientific consensus and helps to raise awareness of the climate agenda around the world and to make the rule-making process more dynamic, stable, and flexible.
• It gives the regime building process a clearer direction, contributes to the dissemination of ideas and possibly even influences the preferences of some actors in the long run. Additional protocols following the Kyoto agreement might be useful to deal with specific aspects of climate change in the future, which have not been coped with effectively in the past (such as specific substances in greenhouse gases, financial questions, etc.).

The problematical political situation due to the refusal of the United States and other leading nations to accept the Kyoto Protocol, highlighted again at the 2009 Climate Conference in Copenhagen, requires the framework/protocol approach to be applied with flexibility and imagination if it can be applied at all. It has therefore been suggested that, in the future, the post-Kyoto climate talks should be organized in such a way that two (or even more) separate protocols can be negotiated and established in parallel processes but still be associated with one another. Thus, in the post-Kyoto/Copenhagen talks, one protocol may continue to be designed as a direct extension of the Kyoto Protocol and another may deal with research and development of new technology (Barrett 2003; Brandt in Chapter 6). Eventually the two different approaches may become integrated to the degree possible. Continued development of UNFCCC represents a third potential avenue toward a more advanced and developed climate regime, even if there is little movement forward on the track leading from the Kyoto Protocol to more demanding binding regulations on emissions reductions.

Brandt (Chapter 6) argues that a separate technology protocol can be constructed to become self-enforcing. One way of doing this would be to introduce a so-called matching mechanism, with a formula being established that makes each country's contribution to costs of the agreement dependent on three factors: the total level of accepted expenditures, its own special circumstances (e.g. share of historic emissions) and, finally, the contribution of other parties. Such an arrangement can become self-enforcing if the abatement cost for all countries decreases sufficiently, because of a positive association between financing, research effort, and technology development. In this way, a tipping point effect can be exploited when a sufficient number of countries join the technology protocol. The remaining negotiating countries will have an incentive to become a member of the agreement, and no signatory state will have a motivation to leave it (Brandt, Chapter 6).

External facilitators may be engaged to design a concrete plan for a bi- or pluri-agreement approach as a contingency plan in case political obstacles created by a continued US refusal to take part in a continued "pure" Kyoto process make more conventional framework/protocol negotiations more or less meaningless. The situation will become even more difficult if other great powers such as China also choose to support the US position in the future. The outcome of the 2009 Climate Conference in Copenhagen seems to confirm that a "pure" long-term framework/protocol approach is becoming increasingly unrealistic.

A major challenge for negotiating parties in the climate talks is to cope effectively with the need to take sustainable development into account in the building of a climate regime. A particularly difficult problem in this regard is that of taking horizontal issues and issue linkages across different regimes into consideration. The problem of sustainable development can be addressed in two different ways, one of which is technically undemanding whereas the other one is highly difficult to accomplish at the climate negotiation table.

The simple approach is to insert declarations of intent or principle into negotiated texts, for example, in the preamble to a declaration. The more demanding method is to develop international regulations, which can give clear direction to concrete policy action. An actual facilitation approach is to draw lessons across sectors from specific topics on the agenda of the climate talks. For example, the negotiated issue of "sinks" (formally the agenda item of Land Use and Land Use Change and Forestry, LULUCF) may be used to demonstrate how a negotiation approach can be developed to satisfy the requirements conditioned by the norm of sustainable development. Guiding examples can be found in LULUCF activities, which are meant to contribute to the conservation of biodiversity and to the sustainable use of natural resources. Another example is the inclusion of social and environmental assessment in the Clean Development Mechanism (CDM). The work on sustainable forest management in other contexts than the climate negotiation, such as the Convention on Biological Diversity (CBD) and the International Tropical

Timber Organisation (ITTO), may be addressed in discussions on LULUCF, hence indicating how cross-regime fertilization can be accomplished and used.

Facilitation at the unit level: Targeting negotiating parties

Because of its high complexity, the climate issue represents a formidable challenge for negotiators from all countries. For a full understanding of the climate issues and their implications at different stages of the negotiation, it is no longer sufficient for a party to draw knowledge and information from the national scientific community in the traditional way. Looking forward, it is clear that international negotiation on climate change and similar issues call for a reformed diplomatic approach (Kjellén, Chapter 2).[9]

To some extent, diplomats-generalists leading national delegations need to be replaced by diplomats-specialists with a more thorough education and training in understanding the climate issues, so as to communicate more effectively with their own national scientific advisors' expertise of international organizations and of other delegations. It is no longer adequate for negotiators in the area of climate change to "learn a script," as is a widely accepted discourse on climate change. A skilful negotiator needs to have a more genuine and deeper knowledge about the climate issue to accurately evaluate assessments, propositions, and arguments made by other actors in the negotiation, and to be able to put forward her/his own creative and "smart" proposals.

Other pertinent changes in the traditional approach to negotiation on climate warming relate to the effective use of modern information technology in the communication processes in which the national climate delegation is involved. This, in turn, has important consequences for how delegations at the table can report to the capital and how instructions can be given from the capitals to the delegations at any point in time.

With a new diplomatic approach, the dialog between a government and its delegation in place at the table can be much more flexible and adaptive to changing circumstances in an evolving negotiation than is possible with traditional multilateral diplomacy. Accordingly, a further shift from traditional to new diplomacy would probably produce a significant facilitation effect, particularly if this change were coordinated, or at least if it more or less followed the same track in all participant countries. Unfortunately, such a development is unlikely, at least in the short term, because of the great disparities between strong and weak parties in the climate negotiation regarding accessible relevant capabilities and resources. In many weak countries, the transition from traditional to new climate diplomacy would require substantial financial and other forms of assistance from the international community.

Capacity building in weak negotiating parties

Numerous weak developing countries are, in reality, unable to defend their interests in the complex climate negotiation, effectively or at all. They may not even be able to discern what their true interests are in specific negotiation situations where they need to take a position. At the same time, the Copenhagen Climate Conference showed once more that weak developing countries are becoming more assertive in the climate talks, as well as in other global negotiations.

Clearly, facilitation measures targeting the weak disenfranchised countries in the southern hemisphere are justified by the global norms of justice and fairness. However, more influential parties also have an interest in enhancing the negotiation power of weak states. Empowerment of developing countries represents an important approach to integrating them more strongly into the international climate regime. States that have been marginalized in the climate talks are likely to become more willing to make commitments to undertake costly measures to cope with climate warming if they begin to feel that they have a growing chance of influencing the regime-building process.

Capacity building represents a major instrument for long-term facilitation of the negotiation performance of weak countries taking part in the climate talks. This is not a new issue in the climate talks. The need for capacity building in disenfranchised states was acknowledged in the 1992 Framework Convention on Climate Change (UNFCCC, Article 9) and reinforced in the 1997 Kyoto Protocol (Article 10e). Churie Kallhauge and van Well (Chapter 10) recall that, since the 1999 meeting of the Conference of the Parties in Bonn (COP5), capacity building was part of the recurrent – or rather continuous – agenda of the yearly COP meetings at a ministerial level.

Programs of capacity building have in the past typically "focused on technical assistance and human resources development." It has been "an inherent part of many development assistance policies and is regarded as an essential ingredient for the assurance of sustainable development" (Churie Kallhauge and van Well, Chapter 10). Hitherto, capacity building in weak countries in the context of the climate talks has had three major functions:

1. To ease the transfer of critical knowledge pertaining to the climate issue to individual weak countries
2. To enhance the ability of the authorities in these states to understand the impact of climate warming on nature and human conditions in their respective countries
3. To improve the capabilities and effectiveness of institutions in weak countries that are responsible for analysis, decision-making and policy implementation concerning the climate problem.

These common forms of capacity building will probably continue in the forthcoming negotiation on climate warming and also presumably remain on the formal agenda of future sessions of COP/MOP after the 2009 Copenhagen Conference. However, this project considers that capacity building in the climate talks also needs to be given some new directions (Churie Kallhauge and van Well, Chapter 10).

Capacity building projects should give more emphasis to multilateral negotiation techniques. They should have a clear focus on particularly demanding negotiation circumstances in which weak parties have especially great difficulties with respect to maneuvering. One example is the handling of complex scientific knowledge/information in *problem solving* in the search for a settlement or in situations where consensual knowledge is constructed or revised.

Training sessions organized for facilitation purposes may review the prerequisites for successful negotiation, analyze power asymmetries at the negotiating table, and suggest how to deal with incomplete information. Such exercises should cover different levels and characteristics of complexity, coalition building, and group dynamics. Through a series of interactive training sessions, participants will apply skills and techniques in role-play exercises, simulations, and in video and working group analyses broadcast on the Internet (Churie Kallhauge and van Well, Chapter 10).

Capacity building should aid targeted governments to understand and cope with general and recurrent features of the climate talks. However, it should also focus on specific complex issues of critical importance not only for the country concerned but also for the progress of the negotiation as a whole. For example, MacFaul (Chapter 15) points at the need to increase the capacity of many countries to handle national inventories of greenhouse gas emissions more effectively. Weak countries have a particular need for more expertise, better equipment, and ultimately more financial resources. The elimination of unsatisfactory national inventories can be expected to generate important positive spin-off facilitation effects. Verification procedures can be made more effective which, in turn, is expected to improve compliance with the Kyoto Protocol, or other binding commitments to reduce the GHG emissions. Ultimately, more effective verification and more reliable compliance are likely to facilitate the establishment of future binding and costly commitments, as effective and reliable verification procedures decrease the risk that other parties will not honor an agreement (the classic *free-rider problem* of international cooperation).

Compliance assistance (both by means of transfer of technology and financial assistance) can likewise make weak developing countries more willing to accept costly commitments in the climate talks. There are numerous states that are not able to live up to the provisions of the climate regime without external support from regime institutions or other nations. According to UNFCCC, Art. 4 (3), the industrialized countries have to bear the agreed full costs for the provision of relevant data and to pay for the agreed full incremental costs for the mitigation and adaptation measures, as required. UNFCCC Art. 4 (4) grants further support to the particularly vulnerable states. UNFCCC Art. 4 (5) provides a transfer of technology. UNFCCC Art. 11 sets up a financial mechanism, which is further borne out in the Kyoto Protocol.

Thus, Article 11 of the Kyoto Protocol requires the industrialized states to provide new and additional resources for the developing countries. Compliance assistance may help countries fulfill their reporting obligations, which requires a rather complex process of data collection. The guidelines issued by IPCC help to structure and thus facilitate the reporting, but there are also grave technological and financial difficulties in getting the necessary data. Compliance assistance under the climate regime is quite substantial, but also remains insufficient after the 2009 Climate Conference. For example the mechanism of financial assistance around the Global Environment Facility needs further fine-tuning.

Capacity building in the climate talks is generally meant to target an individual party, usually an individual developing country. However, most individual poor countries will not attain any real influence in the climate talks, even if their negotiation capacity becomes significantly improved. Weak countries can only become influential as a member of a coalition of states. Accordingly, an important objective of capacity building should be to enhance the negotiation strength of weak-country coalitions in the climate talks, including the G77 that incorporates 132 developing countries (The Group of 77 at the UN 2012).

The G77 represents a major problem because this coalition has significant political strength in a general sense, at the same time as having conspicuous weaknesses when it performs as a party to the climate talks. Capacity building programs should particularly strive to phase out deficiencies pertaining to three basic negotiation functions: joint knowledge building, joint problem solving, and exchange of concessions regarding non-quantified values pertaining to complex issues. When it is performing any of these functions, the G77 derives little or no advantage of its biggest asset in general UN diplomacy, which is voting power, as decisions in the climate talks are taken by consensus. The G77 may have a certain veto power in some situations because of its general political weight, but risks being excluded from inner-circle negotiation if it tries to pursue a blocking strategy under the wrong circumstances. In the past, it has been common to exclude developing countries from parts of an international agreement by permitting exceptions to certain binding commitments accepted by developed countries. The WTO is one example and the Kyoto Protocol is another. It is true that, in general, developing countries have sought exceptions to certain binding commitments in international regimes (e.g. tariff cuts in GATT/WTO). However, this observation should not conceal another reality, namely, that exceptions have been associated with continued marginalization in multilateral negotiations.

Knowledge building will continue to be a critical element of the climate talks in the post-Copenhagen phase. For example, targeted research will be required to reduce critical uncertainties regarding the climate issue, its causes, and consequences. In the pre-Kyoto negotiation, most developing countries were able to contribute very little to this knowledge building because they lacked expertise in this area and were dependent on knowledge dissemination from developed countries. As an organization, the G77 did not have a significant knowledge-building capability and could therefore give little support to individual weak countries.

However, in the early stage of the post-Kyoto/Copenhagen talks, a number of changes are discernible which are likely to enhance the capacity of the G77 and other coalitions of developing countries to play a more significant role in the construction or reconstruction of issues to be included in the negotiation agenda.

First, some of the larger and more advanced developing countries have now acquired a considerably stronger scientific competence regarding the climate issue than they had when the UN negotiation on climate change formally started, some

20 years ago. This may increase the power of the G77 in the post-Kyoto talks. The greater assertiveness of developing countries at the 2009 Climate Conference was one indication of this development.

Second, in the post-Kyoto talks, it had become rather less difficult for national policymakers and negotiators to access knowledge and information on climate warming and its harmful effects. When the formal climate negotiation first started in the 1980s, solid knowledge about climate change was still quite limited in the world scientific community and even scarcer among policymakers and negotiators in most nations. Weak countries performing a peripheral role in knowledge-building activities in climate talks had less access to available knowledge/information than active and influential participants. In the ongoing post-Copenhagen talks the situation will be somewhat different. There is now a large pool of knowledge/ information in the literature and in organizational memories (archives and computerized databases) that all actors can draw on under more or less the same conditions. Much of this is accessible through the institutions serving and supporting the climate talks, such as IPCC and the secretariat assisting the recurrent COP/ MOP sessions.

The expected stronger focus on the adaptation to negative consequences of climate warming will also affect the capacity of the G77, and individual developing countries, to handle knowledge/information effectively in future negotiations. Data on adaptation issues in individual countries (such as poverty, freshwater resources, or human health) are potentially more accessible and manageable in countries with a weak analytical capacity than knowledge about GHGs. Adaptation issues also have a higher priority in developing countries than topics related to mitigation. This is a favorable condition for greater funding of data collection and analysis in this area.

There are thus a number of measures able to enhance the capacity of the G77 to deal with the knowledge/information needed in the climate talks:

(i)   More resources and technical assistance with regard to knowledge management to secretariats and other institutions servicing and supporting the G77
(ii)   Development of more effective lines of communication between the secretariats and individual member states of the G77
(iii)   More regional forums and more extensive forum activities for the exchange of knowledge/information and the coordination of data gathering and analysis.

Facilitation measures similar to those supporting knowledge building are likely to have a positive effect where the G77 lacks the necessary capability to negotiate effectively, notably problem solving and skilful exchange of concessions in negotiation on certain complex issues. Successful facilitation measures designed to aid knowledge building may generate positive spill-over effects on problem solving and skilful exchange of concessions.

Thus, *problem solving* depends on necessary expertise and sufficient access to critical knowledge/information. Effective coordinated joint action for effective problem solving also requires that delegations of different countries be able to communicate what they consider to be their main concerns in a commonly understood language code (ethnic and professional). In the past, the mechanisms for inter-party exchange of knowledge/information were not sufficient to secure a satisfactory performance in problem solving on the part of G77 member states. Usually, coordination of G77 performance has only been possible in general political/ ideological confrontations such as debates in main UN bodies where specificity on all elements of a complex argument was unnecessary.

Exchange of concessions in the climate talks has been difficult for many weak countries which do not fully understand the issues as they have been framed for the negotiation. Therefore, their policymakers and negotiators have been reluctant to take a position in favor of precise, costly, and binding commitments. This has tended to strengthen the tendency of the G77 to take positions in the climate talks that are based on a very low common denominator of member country interests and objectives and thus are diffuse and general. Improved procedures for knowledge building, in combination with enhanced mechanisms for exchange of knowledge/ information *within* the G77, also opens the way for a more instrumental and effective participation in various sessions of concession trading in the climate talks, where weak countries have had little influence and a very limited ability to defend their own interests.

Facilitation of leadership in the climate talks

Paradoxically, some of the strongest actors (individual nations and coalitions of states) in the climate talks are sometimes too weak to take on a leadership role. External facilitators cannot address this major stumbling block in the same way as they can design empowerment approaches to support weak developing countries. Strategic facilitation may nonetheless support failing leadership in the negotiation by focusing on specific leadership functions, and notably on *process management*. This approach cannot produce miracles but may still help to boost process dynamics.

Process management may be enhanced by measures aiming at building the capacity and increasing the responsibilities of elected negotiation officers (especially Chairs). Such a scheme would particularly target candidates for a chair at the plenary meetings of COP/MOP sessions or of other main negotiation groups. A practical measure would be to set up rosters of chair candidates from all participating countries which are revised, say, before each or every second COP/MOP meeting. Listed persons would be invited to take part in special training exercises focusing on process management.

Changes in the organizational framework supporting the organization of the negotiation process could be made to facilitate the taking on of a leadership role. The effects of such measures are likely to be moderate or nil, because an effective leadership role is always dependent on superior resources and capabilities. However, increased effectiveness in process and structure is likely to decrease the risk of disturbances in the negotiation due to poor process management which can increase the burden of leadership.

Integrating nongovernmental organizations

ECOSOC specifies that "any international organization which is not established by intergovernmental agreement shall be considered as a nongovernmental organization (Yamin and Depledge 2004). Most of what these organisations do in the negotiations can be summarized as six types of activity:

(i.)   Lobbying government delegates
(ii.) Circulating information and position papers
(iii.) Working with the media
(iv.) Hosting side events
(v.)   Making interventions during debates
(vi.) Monitoring developments in the negotiations.

A significant approach to facilitating the climate talks will be to attain more intensive and far-reaching participation of NGOs. The essence would be to particularly enhance four roles that NGOs may perform in the climate negotiation; that of *controller, advisor, legitimator,* and *activist* (Kanie, Chapter 9).[10] Another objective would be to reduce the incentives for NGOs to participate in demonstrations near where a Climate Conference is taking place, as such demonstrations have tended to generate violence and police confrontations.

A basic step in a facilitation approach addressing NGOs would be to develop guidelines to deepen and extend their participation in the climate talks, with consideration given to the functions they may carry out in the process and the role(s) they may perform in the interaction with other negotiation parties. Concretely, this may mean that NGOs are given different privileges of access to meetings and the right to make statements and intervene in discussions, depending on their category of organization.

Kanie (Chapter 9) points out that a NGO performing effectively as a *controller* helps to increase the accountability of governmental delegations to their domestic constituencies. "By observing the negotiations, NGOs can obtain the information they need to follow the negotiation process, monitor the positions of governments, develop their own stance and report back to their members (Depledge 2005:217). For some delegations the presence of controller NGOs in a meeting may be a constraint, particularly if the NGO belongs to their national political scene. In one scenario, a government is reluctant to make concessions in the negotiation because a powerful domestic pressure group will be alerted to the fact by NGOs who will criticize this.

However, the presence of a NGO *controller* in a negotiating group may also have an opposite effect, by making it more difficult for a government to oppose a proposal for an agreement which is supported by domestic public opinion informed by domestic NGOs. Furthermore, NGO *controllers* may contribute to facilitating the climate negotiation through contributing to more effective verification of the implementation of any accords made. NGO *controllers* do not require extended rights of intervention at sessions of the climate talks. Improved access is sufficient (Kanie, Chapter 9).

The technical competence of many NGOs, combined with their access to international or national scientific communities, allows this type of organization to perform an *adviser role* in the climate talks. An experienced student of NGOs claims that NGOs "usually have better expertise than many governments and, thus, can assist and clarify the issues in the issue definition stage. During the negotiations themselves, governments can benefit from updated scientific, technical and human-focused information prepared by the non-governmental community. As some governments have found, a non-governmental perspective may shine new light on a contentious issue" (Chasek 2001: 231). Furthermore, some NGOs, or coalitions of NGOs, have a better overview of the whole process of climate talks, particularly in developing countries, For example, the *Earth Negotiations Bulletin* (ENB) accurately presents itself as "a balanced, timely and independent reporting service that provides daily information in print and electronic formats from multilateral negotiations on environment and development." The ENB is published by the International Institute for Sustainable Development (IISD), a non-profit organization based in Winnipeg, Canada, but with an office in New York City, two blocks from the United Nations.

A reform which gives NGO advisors more access to contact groups and other closed meetings in the climate negotiation may have important facilitation effects, particularly if some are given a right of intervention. Competent scientific NGOs will be able to contribute to knowledge building and issue construction as well as to problem solving in the negotiation.

Legitimator and activist

NGOs performing as either legitimator or activist may either disturb or support the climate talks depending on the situation. A disturbance may occur if NGO activities are not well received by any government participating in the climate talks. In contrast, a facilitation effect may be produced if, for example, an NGO legitimator gives normative backing to a proposed agreement text or if an activist NGO adds new arguments in its support.

Facilitation at the systems level: Targeting process and structure

Pre-structuring of systemic elements represents a critical part of a facilitation strategy for the climate talks and may have far-reaching consequences. Institutional design, or redesign, is an instrument to achieve pre-structuring of the climate negotiation.[11] For example, meaningful organizational changes of the IPCC may affect the building and distribution of knowledge about the climate problem that, in turn, may engender more favorable conditions for problem solving and agreement in future talks.

However, the concept of institutional design needs to be joined with another concept that is less familiar in the literature on international negotiation, namely, *process design*. This approach assumes that a complex negotiation process, like that of the climate talks, can to some extent be planned and deliberately constructed in advance to create specific process characteristics, for instance, openness and transparency. Seen in a strategic facilitation perspective, institutional design is in principle subordinated to process design. Institutional effectiveness should be regarded as a means of making the process of climate regime building–negotiation–effective.

It is meaningful to think in terms of process design in the context of strategic facilitation because complex multilateral negotiation like the climate talks tend to progress in certain fairly predictable patterns. The logic of negotiation conditions a recursive process development. Hence, all multilateral negotiations move along a similar sequence of process stages from pre- to post-negotiation. Certain functions are repeated in all negotiations such as construction of consensual knowledge, problem solving or (re-)distribution of disputed values (e.g. territorial, economic, human, or cultural values). Expected recurrent patterns and features represent an entry point to process design as an instrument of strategic facilitation.

For practical purposes, strategic facilitation looking at the process and structure of the climate talks will be addressed in the same subsection of this chapter. However, first, a special part of the problem of process design will be focused on, addressing: how to change the goal structure guiding the climate talks in order to ease the process.

Specifying the goal structure

The 2009 Copenhagen Accord set a 2°C target for maximum permitted average temperature increase in the atmosphere. Many supporters of the Kyoto Protocol complained that this non binding goal formulation represented a step backward in the process of building a global climate regime. It should, however, also be noted that the specification of the 2°C target in Copenhagen was a new accomplishment for a COP/MOP meeting and therefore also represented a small step forward, even if the 2009 Climate Conference mostly highlighted the goal structure as a major problem area in the climate talks.

Establishing precise goals for the post-Kyoto/Copenhagen talks represents a crucial strategic stage of the negotiation on climate change. Participating nations wish to control this part of climate talks as far as possible. Goal setting is at the heart of the power game of nations and it both drives and constrains the regime-building process. Nevertheless, external facilitators may perform an important role in this connection. For example, one facilitator task is to point out the requirements that an overall goal structure for the whole negotiation needs to meet in order to function as an effective driver of the process. It is not obvious what these circumstances are. To obtain clarification, special studies may be required, the results of which may be communicated into the climate talks as part of a strategic facilitation operation.

External facilitators may also help to distinguish the consequences for the climate negotiation of the establishment of a more specific long term objective for the climate talks. Article 2 of UNFCCC spells out such an overriding goal. It stipulates that the aim of the climate negotiation is "stabilization of greenhouse gas concentrations in the atmosphere at a level that would prevent dangerous anthropogenic interference with the climate system" (Brandt, Chapter 6; Pershing and Tudela 2004). One way of sharpening this objective is to actually define numerically what threshold concentrations of greenhouse gas concentrations in the atmosphere are that cannot be surpassed. This has long been the subject of international discussions.

Usually, CO_2 emissions have been highlighted and the focus has been set on three alternative concentration thresholds: 450 ppm, 550 ppm and 650–1000 ppm.[12] Assessing these three levels of aspiration for the year 2100, Pershing and Tudela (2004) consider that 550 ppm is the alternative that is most likely to be accepted as a viable solution. The 650–1,000 ppm option is acceptable for most governments in terms of costs but is still unacceptable to many because of the long-term catastrophic effects that it would produce. In contrast, the 450 ppm alternative is "the most stringent long-term target that might likely be achieved under current circumstances" for scientific/technical reasons, but is unlikely to become accepted because of the large cost required to reach this low concentration level (Pershing and Tudela 2004: 23). This makes the 550 concentration aim a feasible option from a political point of view.

It is not up to external facilitators to determine whether the threshold level for greenhouse gases should be 450 ppm, 550 ppm, or 600–1,000 ppm. However, their mission could be to clarify the crucial relationship between goal specification and process effectiveness. In this connection, it is useful to distinguish between two kinds of result from goal specification, diffuse and specific facilitation effects respectively.

Diffuse facilitation effects may have an impact on climate policy in general at all levels of decision-making including in the international climate negotiation. For example, Pershing and Tudela (2004) make a fairly long list of potential, general positive consequences following from the establishment of a long-term precise target:

1. Provide the international community with a clear statement of the goal to which near- and medium-term efforts must be geared;
2. Increase a broader awareness of long-term consequences of current policy measures;
3. Make it more feasible to measure progress in the climate policy area;
4. Induce technological change. By providing a clear signal to markets, businesses would be given confidence and investment in clean technology would be stimulated;
5. Mobilize societies to undertake further steps to cope with the climate threat;
6. Promote mobilization for policy making in the climate area (MacFaul, Chapter 15; Pershing and Tudela 2004: 13–14).

Diffuse facilitation effect from the establishment of a clear long-term objective can be expected to stimulate international climate cooperation and negotiation in a general sense, but will also help to generate various *specific facilitation effects*. One reason is that agreement on a long-term objective requiring that concentrations of greenhouse gases in the atmosphere can only be permitted to, say, a level of 550 ppm represents a switch from relative to absolute aims in the negotiation on emission reduction, which can be expected to facilitate the negotiation according to lessons learned in the past.[13] The agreement in Copenhagen on a 2°C target has to be assessed in this perspective.

Several other environmental negotiations on emission reductions than the climate talks were initially also driven by relative objectives. A good case for comparison with the climate talks is the negotiation on long-range transboundary air pollution in Europe (so-called acid rain), which led to the 1979 Convention on Long-Range Transboundary Air Pollution. In line with a framework/protocol regime-building strategy this Convention has been extended by means of eight protocols.[14] The use of the IIASA RAINS model in combination with the Critical Load Approach in the protocol negotiations provided a scientifically framed criterion for the eventual determination of an absolute goal for the negotiation, namely, the critical load of a given pollutant that a specified geographical area could sustain.[15] Emissions of the negotiated pollutant had to be limited to a level that permitted the achievement of this objective. The RAINS model described long distance movements of pollutants in the atmosphere and the deposition of pollutants in a particular geographical area. The critical load analysis pointed out how much acid rain had to be reduced to cope with the problem of acid rain and thereby also identified long-term absolute targets for required emission reductions.

There is strong reason to believe that this procedure for determining consensual long-term absolute targets for sulfur and other substances to be included in the regime on Long-range Transboundary Air Pollution in Europe contributed the important results eventually produced by the negotiation (Bäckstrand 2001). The existence of an absolute objective gives clear direction to a negotiation on emission reductions and establishes criterion benchmark for evaluating the offers to reduce emissions that governments place on the negotiation table.

The example of Long-range Transboundary Air Pollution indicates that the establishment of an absolute, precise, and authoritative long-term objective can be expected to have a strong facilitating effect on the post-Kyoto talks. This aim does not imply a negotiation strategy to immediately achieve the long-term aim in the post-Kyoto/Copenhagen talks. A more realistic approach is to have long-term objectives guiding a further development of the international climate regime, in line with the procedures of a framework/protocol approach. Therefore, the long-term objective should be framed and formulated in such a way that sub-goals can easily be determined for each stage of a recursive regime building process such as individual COP/MOP sessions.

A consensual link between short-term targets and a generally accepted long-term objective for the climate talks may become a powerful favorable prerequisite for coping successfully with the extreme uncertainty and complexity characterizing the negotiation on climate change. In this regard especially two facilitation effects need to be considered:

1. A consensual, long term and absolute objective for the climate negotiation will guide efforts and facilitate measurement and assessment of progress in the negotiation;
2. It will also facilitate evaluation at any point of time of how much and how costly actions are required to reach satisfactory concentrations of greenhouse gases in the atmosphere.

An important task of external facilitators is to indicate and evaluate alternative criteria for success in the international regime building process. For example, Pershing and Tudela (2004) summarize the following alternative criteria, which have all been proposed and discussed in the climate talks:

1. Magnitude of *emissions* of greenhouse gas into the atmosphere;
2. Average *temperature* in the atmosphere;
3. *Impacts* of climate warming on societies, economies, and eco-systems;
4. *Human activities* leading to emissions of greenhouse gases causing higher average temperatures in the atmosphere globally and regionally (Pershing and Tudela 2004).

An important task of facilitators engaged in the climate talks would be to prepare concrete approaches and methods to substitute one (or more) of these criteria of ultimate goal attainment for that of greenhouse gas concentrations in the. In this regard an important difference should be noted between, on the one hand, criteria (1) and (2) and, on the other, criteria (3) and (4).

In their capacity of goal variables, *magnitude of emissions* and *average temperature in the atmosphere* are comparable to greenhouse gas concentrations in the atmosphere, in the sense that complete goal achievement has a clear and precise meaning and can be measured quantitatively. This is not the case with the goal variables *impacts of climate warming* and *human activities leading to emissions of greenhouse gases*. In these latter cases, a clear and measurable threshold separating acceptable from non-acceptable *impacts* or *activities* cannot be based only on scientific grounds. However, these two goal variables have other advantages and possibilities in the negotiation. *Impacts* will, under all circumstances, be needed in the part of the climate negotiation that will address the mounting issue of adaptation to harmful effects of climate warming. *Activities* may become used as a goal variable on a proposed separate technology protocol if a bi- or pluri-protocol approach to the climate negotiation is used.

If the spirit of the 1997 Kyoto Protocol survives mounting political strife in the climate talks, signatory states that have accepted binding commitments are likely to adopt a strong political ambition to pull non-signatory states deeper into the Kyoto regime. For this reason, criteria for differentiation of costly obligations need to become included into the goal structure. It is, however, not realistic to require that all developing countries move directly from "no binding commitments" in the Kyoto Protocol to "full binding commitments" in an agreement following Copenhagen. On the basis of an acceptable equity criterion, some mechanism needs to be developed for a gradual or step-wise undertaking of "full binding commitments" which can only be realized during a sequence of yearly COP/MOP meetings.

There are various different concepts of *equity* that may be included in the climate change regime:

1. Responsibility for the problem;
2. Equal entitlements;
3. Capacity to act;
4. Basic needs fulfillment;
5. Comparability of effort;
6. Consideration of future generations.

A careful choice has to be made between these equity options. Differentiation between countries could be based on any one or several of them. A variety of indicators can be used to order them into groups, for instance, *per capita GDP* or *per capita emissions*. Commitments or benefits which are appropriate to each group's characteristics can then be allocated, and could range from various forms of binding emissions reduction targets and the provision of financial assistance, to policies and obligations, non-binding targets or the receipt of financial assistance.

The "multi-stage" proposal provides both a pathway for developing countries to participate in the regime and an overall regime structure. It offers an approach using four stages of progressively stronger types of commitments for groups of parties. The differentiation criteria are based on per capita emissions, and an emissions allocation system is provided, which may include targets that reflect particular national circumstances.

Altering process and structure

A multitude of factors including extraordinary high stakes, extreme issue and process complexity, in combination with serious uncertainty problems, made the pre-Kyoto climate talks complicated, difficult, and protracted. It is important to discover all possible ways and means to improve the dynamics of the negotiation on climate change. External facilitators may have an important role in achieving that objective.

One approach to enhancing process dynamics in the climate talks would be to pre-structure the climate talks in detail as far as this is possible and is not counter-productive.[16] This kind of organizational planning should particularly target major sessions such as COP/MOP meetings. A key activity would be to have external facilitators and other independent experts, led by the secretariat, draft a plan on how the issues could best be subdivided. Such subdivision or fragmentation of problem areas and issues based on substantive analysis may, in turn, pave the way for pre-structuring of the negotiation process in the form of rational sequencing of sub-issues. Preparations can then be made to address sub-issues in a time-tabled consecutive order. This consecutive order of sub-issues can be used to set up a consecutive order of meetings in contact groups and other negotiation bodies with the problem area concerned on its agenda (say, LULUCF). This procedure would contribute to reducing the impact of one important obstacle in the pre-Kyoto negotiation, ineffective participation of developing countries in contact groups because of too many simultaneous meetings.

The handling of scientific knowledge and other complex information represents a critical factor in the climate talks. The system for knowledge management constructed for the negotiation on UNFCCC and the Kyoto Protocol was evidently a precondition for the noteworthy accomplishments of these talks. However, facilitation measures could be instrumental in sustaining, or preferably developing, the procedures and institutions for bringing scientific and other complex knowledge into the negotiation. Authors of chapters in this book have put forward a number of concrete measures that can be undertaken to attain that objective.

One suggestion is the establishment of a formal space for the interaction between scientists and policymakers in the context of the negotiation on climate change. Some space serving this function is currently in place in the negotiation organization as it looked like during the pre-Kyoto talks. One example is national delegates attending the IPCC plenary. Another case is experts who are invited to workshops, side events, and SBSTA sessions. However, new forms need to be developed for the useful and effective involvement of scientists in the negotiation work. Options could include the targeted participation of scientists in different meetings, science briefings with high level officials, workshops, questions and answer sessions, and others.

Facilitators with a special competence in the communication of scientific knowledge can be placed at different spots in the negotiation organization. Such experts can sit in contact groups as advisors to the chairs. These facilitators could contribute to enhance process dynamics in the climate talks in at least three different ways:

1. To ease exchanges in the group which are dependent on the input of scientific knowledge;
2. To provide information concerning the current bargaining situation in other contact groups hence facilitating inter-group coordination;
3. To contribute to strengthen the leadership function in some of the contact groups.

There is a need to introduce some institutional reform into the IPCC machinery to avoid the exchanges of scientific knowledge becoming excessively politicized. There was a risk of this in the pre-Kyoto stage of the climate negotiation, but it is likely to increase in the continued post-Kyoto talks. The ongoing elevation of the issue of adaptation to climate warming on the agenda of the climate talks represents a major driver of such a development (Skodvin, Chapter 12).

Skodvin (Chapter 12) points out that the impact of this adjustment is likely to become increased by a change – both needed and expected – in the approach to addressing adaptation. The top-down approach, relying on the input of natural sciences like biology and geophysics, needs be revised because it has focused too much on predictions of future climate impacts and has neglected to consider social, behavioral, or other obstacles in the adaptation process. The anticipated switch to a stronger focus on social and political factors that determine a country's or a region's current state of vulnerability will highlight problems areas like poverty reduction, diversification of livelihoods, protection of common property resources, and strengthening of collective action (Skodvin, Chapter 12).

To cope effectively with the new approach to negotiation on adaptation, the IPCC needs to develop new instruments (institutions and procedures) to handle increasing inflows of social scientific knowledge/information. Care should be taken to design these mechanisms so that the authority of social scientific knowledge/information is sustained as far as possible, while at the same time avoiding unnecessarily rigid, detailed, time-consuming, and inflexible rules of procedure. It has been suggested that a powerful permanent forum of independent experts that can discuss issues without having to consider national constraints should be set up.

How issues are interrelated on the agenda may have a considerable impact on process dynamics in the forthcoming climate talks. Separation, combination, and sequencing of issues may function as powerful tactical devices in any multilateral negotiation including the climate talks to move out of the blind alley in which it finds itself. Delinking or linking issues in the negotiation may also have a strategic impact. To address issues on the agenda in a certain order is not only a ploy to avoid problems in the negotiation; sequencing can also contribute to facilitating the whole negotiation in the longer term. Thus, in the climate talks, the issue of verification may generate significant facilitation effects if it can be handled at an early stage of the negotiation process.

Verification can be described as compliance control. It is a technique to find out if, and to what extent, signatories of an international agreement have lived up to the commitments they have made in the treaty concerned. MacFaul (Chapter 15) emphasizes that this function of deterring signatory states from free-riding is interlinked with another basic function of a verification system: "measuring and promoting overall and individual state's progress towards a treaty's goals" (Mac-Faul, Chapter 15).

MacFaul concludes that the verification system plays a large role in inspiring confidence among parties in the development and maintenance of a treaty such as the Kyoto Protocol. The IPCC provides information on certain specific verification procedures that parties can use to check the reliability of their emissions inventories. MacFaul explains that the IPCC verification procedures are meant to improve the inventory process, build confidence in emissions estimates and trends, and help improve scientific understanding related to inventories. Estimated uncertainties and intensity indicators can be compared among countries.

The general positive effect of improved verification can spill over from compliance control into strategic facilitation of the entire climate negotiation. The logic behind this assumption is that, if governments believe that the risk of other countries not honoring their obligations under a new climate protocol is decreasing, they will become more likely to adhere to it themselves. An effective verification system in the climate regime can be expected to produce other facilitation effects, in particular, contributing to reducing some of the formidable uncertainty problems.

Looking deep into the compliance system that has been built up within the climate regime, MacFaul notes that, although it has existed for some time, it retains certain shortcomings that need to be removed. For example, the inventories of verification information are not as transparent, consistent, comparable, complete, and accurate as they should be. Strong efforts should be made to improve the verification system as soon as possible; preferably, the time schedule for the talks on these issues should be set in such a way that results can be attained before the negotiation on mitigation gets under way. One possible beneficial effect will be that these talks on emission reductions will become facilitated at least to some extent. An improved verification system contributes to reducing the uncertainty problems or rather helps negotiating parties handle them in problem solving, exchange of concessions, and agreement.

MacFaul suggests a number of measures to improve the crucial enhancement of the verification mechanisms of the climate regime. Thus he underlines that, ideally, a verification system includes satisfactory monitoring and reporting processes, a review process, compliance procedures, and procedures for improving monitoring capacity. The inventories of verification information should be transparent, consistent, comparable, complete, and accurate. There are guidelines for verification performance that may be used as facilitation instruments. These should be developed and communicated more effectively to governments participating in the climate negotiation. Moreover, the IPCC instructions for quality assurance should be sustained and strengthened. The quality assurance is a system of routine technical activities to measure and control the value of the inventory as it is being developed, such as accuracy checks on data acquisition. Quality control activities include a planned system of review procedures conducted by personnel not directly involved in the inventory compilation and development process. Thus, there may be a role for external facilitators in this context.

The notion of "sinks" is derived from the natural ability of trees, other plants. and the soil to soak up carbon dioxide and temporarily store it, in other words. take it out of the carbon cycle. The biomass and the soil function as "sinks" when they absorb more carbon than they emit. "Sinks" emerged as an issue in the pre-Kyoto negotiation. based on the rationale that a nation participating in the climate talks should have the right to deduct its "sink" capacity when its obligation to reduce CO_2 emissions is determined in the negotiation.

It has been argued that the addition of "sinks" to the agenda has contributed to facilitating the climate talks. However, to produce positive facilitation effects, the concept of "sinks" as such needs to be further clarified, as well as the approach to dealing with it in the climate negotiation. A clear approach to sinks could prevent conflicts of interpretation and avoid discussion on the transfer of LULUCF to the technical discussions in which political decisions cannot easily be taken. Two proposals can be made:

First, the post-Kyoto negotiation is to rely on, and highlight, the work of the Subsidiary Body for Scientific and Technological Advice (SBSTA) on LULUCF and thereby also "sinks". Decisions on "sinks" incorporated into the Kyoto protocol could – and should – be further considered in the context of the Convention as a basis for discussions on future climate change regimes. This approach has the potential facilitating effect in that it is consonant with a bi- or pluri-protocol process in the post-Kyoto negotiation: the development of the "sink" part of the international climate regime will not entirely be locked into the continuation of the Kyoto process.

However, flexibility mechanisms and other actions such as negotiation on "sinks" should not contribute to reducing the effort to stabilize GHG concentrations in the atmosphere. A guiding principle should be that climate warming mitigation has to be achieved mainly through reducing greenhouse gas emissions. Any issue that could have impacts on the establishment of targets for the mitigation negotiation, like LULUCF, should be addressed, together with the overall agreement on commitments.

The second proposal is that changed procedural rules, modified consensual negotiation methods, and institutional redesign represent a basic element of a grand facilitation strategy for the post-Kyoto climate talks. One ambition would be to make meetings in the institutions supporting the climate talks more effective.

Time could be saved at COP/MOP Plenary Meetings if stricter timetables were introduced with enforced time limits for each speaker. Still more time could be gain if a more business-like culture was developed in COP/MOP, lifting many unnecessary ceremonial statements out of Ministerial Meetings and other high level sessions in the climate talks. The High Segment meeting at COP11 in Montreal (2005) can serve as a case of illustration.

All Heads of Delegation started their intervention with more or less the same words (in English or French) thanking the governments of Canada and the state Quebec for organizing the COP/MOP meeting, praising the beauty of Montreal and thanking the Mayor of the same city for hosting the conference. If one single delegate (such as the doyen, the oldest Minister) expressed such ceremonial statements at least 30 seconds would be saved in each other interventions which, in turn, would mean that more than one hour and twenty minutes of speaking time could be allocated to other purposes which is a considerable amount of time (even if the High Segment session starts at 10 o'clock in the morning and ends at 10 o'clock in the evening).

More time could be gained if other more or less ceremonial statements were eliminated or shortened. Many statements made at COP/MOP Plenary Meetings are not only general in character, they also repeat statements about a given country's position which have been put forward at earlier meetings during the conference concerned. For example, one part of such a position presentation made by a high level person at the very end of the conference – indeed on the final day – is to explain something that all other delegations know: first, that it accepts the draft final agreement text and, second, why. Such declarations may be significant when they are referred to in the domestic context of the country concerned but, as a rule, do not contribute much substance to the conclusion of a COP/MOP meeting. The position and concerns of individual negotiating parties can be, or has already been, put on the official record, and can be communicated to constituencies outside the negotiation process by other means.

The time saved by the elimination of ceremonial or other unnecessary interventions at COP/MOP Plenary Meetings can be used to increase the value of its contribution to both problem solving and value distribution in the large decision making sessions of climate talks. A partly reformed plenary meeting with regard to process design may produce several kinds of facilitation effects:

1. It may contribute to increased openness and better implementation of decisions taken at COP/MOP meetings by enhancing the role of certain selected NGOs and perhaps other representatives of the civic society including business companies;
2. It may give room for a more continuous input of scientific knowledge/information to the discussions in Plenary Meetings when this is wanted;
3. It will create better conditions for a penetrating discussion at the Plenary level about horizontal questions related to different issues on the agenda for the climate talks, as well as about issue linkages connecting the climate negotiation with political processes going on in other contexts.

Hanschel (Chapter 14) proposes facilitation measures to enhance consultation and negotiation in the contact groups, calling for a procedure on how to negotiate within the contact groups that also contains provisions about equal representation of all coalitions. However, rules of procedure have to leave a substantial margin of discretion so that the necessary flexibility of negotiations is ensured. Financial support for adequate capacity-building measures is necessary to enhance the participation of weak (developing) countries in the contact groups. The organization of the work of contact groups should take the capacity problem of weak countries into consideration. Preferably, issues on the agenda of some contact groups addressing questions of special interest for developing countries should be addressed sequentially, rather than simultaneously, to facilitate active participation of weak nations. The need for more time for the contact groups would represent a useful investment.

A second line of facilitation measures would be to expand and improve horizontal exchange of knowledge/information between contact groups and other negotiation bodies under the Plenary Meetings in the organizational hierarchy of COP/MOP negotiation institutions. In this way, effective coordination across different sub-issues to climate change would be facilitated hence improving the conditions for constructive agreements at COP/MOP meetings (Hanschel, Chapter 14).

An orthodox application of the framework/protocol approach would mean sequentially created separate international agreements, all of which refer to an initial framework agreement.[17] In this way, one greenhouse gas after another can stepwise be brought into the climate regime in separate protocols. However, the various agreements underpinning the climate regime are not entirely unconnected. A new accord may have, or should have, consequences for the interpretation or effectiveness of earlier agreements, and notably the initial Framework Agreement

(UNFCCC). This means that one or more section of UNFCCC needs to be rene-gotiated and eventually amended.

Reinforced and otherwise improved renegotiation and amendment procedures may have important facilitation effects on the future climate talks.[18] Ott (1998: 50) claims that renegotiation procedures contribute to the maintenance of the pro-cess of negotiation, and rule-making procedures for renegotiation have proven to facilitate negotiations by pushing them further ahead, leading them into the right direction, and allowing for the necessary treaty regime amendments (see also Hanschel, Chapter 14).

Facilitated amendment procedures may relate to the treaty itself, to annexes to the treaty, and to new treaties that are connected to the initial one. They facilitate further negotiations in regimes that depend on a dynamic redevelopment of exist-ing rules to react to changing technical and scientific knowledge. Changes that are brought about by qualified majority decisions enter into force six months after the decision was taken. According to Hanschel (Chapter 14), the amendment proce-dures in the climate regime are too weak and need to be reinforced.

Methods of strategic facilitation

The analysis above demonstrates that strategic facilitation in the climate talks can have many different forms and objectives, depending on the stumbling block that is to be targeted regarding issues, actors, negotiation process, or structure. This variation needs to be reflected in composition of the tool box available for strategic facilitation. Thus, measures aiming at strategic facilitation need to be capable of accomplishing quite different operational tasks:

1. To enhance the capacity of negotiating parties to understand and deal satis-factorily with the climate issue in analysis, assessment, bargaining and deci-sion making;
2. To help weak stakeholders in the climate talks (e.g. least developed countries) to voice their concerns and to defend their interests more effectively;
3. To suggest neutral measures that, if effective, will contribute to make the negotiation process smoother, less time consuming or less resource-demanding. A facilitation measure is neutral if it does not significantly favor one or more particular parties to the disadvantage of other parties;[19]
4. To enhance the capacity of negotiation institutions to support the climate talks instrumentally and effectively.

Each of these four tasks implies a set of important guidelines as to how a strate-gic facilitation approach in the climate talks should be developed in theory and implemented in practice. It would be helpful to design facilitation measures that are linked to any one of the four main tasks. However, a full-fledged facilitation strategy for the climate talks needs to integrate the attainment of all four objectives as far as possible. This requires a comprehensive outlook covering all the differ-ent types of stumbling blocks that may arrive on the scene at some point during the climate negotiation. Trade-offs and other types of linkages between different measures need to be considered and integrated into the comprehensive facilitation strategy to the extent that this is practically feasible.

A complete facilitation strategy should have a capacity to generate considerable synergy effects if different measures are put into constructive interaction with one another. For example, positive effects of capacity building in weak developing countries will become amplified if they are coordinated with well matched institutional reform of the negotiation bodies in the UN system (Churie Kallhauge and van Well; Hanschel Chapter 14).

Another critical issue is how strategic facilitation proposed by external parties may have a real impact on the climate negotiation. Any useful plan for strategic facilitation needs to include practical ideas of how it can be somehow implemented at the systemic level pertaining to the overall negotiation on climate change. A classical question is evoked: how should constructive interaction between scientific advisors, facilitators, and policymakers best be organized? The problem is further complicated because, in this case, facilitators are not natural scientists but represent the society of social scientists and are more knowledgeable about the process of the climate negotiation than about the climate issue itself. So far, access to the climate talks by such process experts has been less than that of natural scientists who are experts on climate warming (causes, manifestations, and consequences) and who have become officially and formally engaged in the climate talks with the help of the IPCC and other negotiation bodies focusing on natural scientific questions.

Proposals for useful long-term facilitation of the climate talks address how facilitators could be given a greater role in the negotiation. This can be accomplished in different ways but only with due consideration taken of the special obstacles that will probably be encountered by social scientific facilitators in the climate talks. A number of concrete propositions may be put forward regarding the implementation of the operational version of a strategic facilitation plan.

Implementation of strategic facilitation

Strategic facilitation, as conceptualized and developed in this project, has to date essentially been a factor lacking in the climate talks. There are no institutions or routines in the negotiation system with the task of proposing or actually carrying out a long-term facilitation strategy. The notion of strategic facilitation is not widely acknowledged. Accordingly, not only the design but also the implementation of strategic facilitation requires new ideas, new methods, and new instruments to achieve effective external support of the climate talks.

Recall that strategic facilitation should ultimately serve the common interests of all negotiating parties even when individual parties are targeted. Actors undertaking strategic facilitation in the climate talks would be third parties that have been given formal access to the negotiation to differing degrees, for example, independent experts, scientists or representatives of the secretariats of international organizations. Facilitators formulate advice and communicate it to the parties in the context of the negotiation.

The ways in which facilitators communicate with the climate negotiation are naturally associated with the role that they have been given in or have taken in this context. An important factor distinguishing facilitator roles is how strongly they are directly involved in ongoing climate talks. Some facilitator roles require a very high direct involvement, whereas medium or low involvement is sufficient for others.

High direct involvement in the climate talks

In this case, a basic requirement is that a facilitator is actually "sitting at the negotiation table", although he or she may be doing different things. There are essentially two facilitation roles representing high direct involvement in the climate talks: mediator and coach.

One contingency is, hence, that a facilitator is engaged as a mediator at a particular meeting – or a whole negotiation session – in which a deadlock is likely to occur. Facilitators that have been assisting council meetings in the European Union represent a model that, with some modifications, may also be used in the climate negotiation. Such mediation often has a short-term time frame, typically trying to help negotiation parties get out of a deadlock in the current talks. However, mediation measures may also have a long-term strategic significance by paving the way for continued negotiation in the future. One example is when a mediator finds a way to combine two competing approaches to problem solving in particular issue areas, for example "sinks" or verification. Another case is the framing or reframing of issues on the agenda of the UN climate talks, in order to attain a more consensual understanding of them.

Coaching is a second form of direct involvement of a facilitator in the climate talks. Coaches have a generally acknowledged and accepted mission to give direct support to individual parties in the climate talks in order to facilitate the negotiation process. Three possibilities come to mind.

First, the participants of the climate talks may see that they have a common interest to accept competent coaches to help certain weak developing countries to negotiate more effectively, to prevent them from becoming stumbling blocks in the climate negotiation. Many poor countries can be expected to want more say in the negotiation before they accept binding commitments to reduce greenhouse gas emissions.

Second, the community of negotiating parties may find it to be in their interest to accept to finance coaches to be put into the administrative support structure of important coalitions like G77. One example is that coaches may help the staff of G77, which is very small, to communicate faster and more effectively with regional groups and coalitions outside G77 in the climate negotiation.

Third, negotiating parties may accept commissioning facilitators having a deep knowledge and understanding of multilateral negotiation to give coach support to negotiation functionaries (such as chairs and members of the secretariat of the climate regime) in preparation for, and during, negotiation sessions for example at COP/MOP meetings.

Medium direct involvement in the climate talks

Capacity building is a traditional form of negotiation facilitation in the climate talks that was formally highlighted already in the 1992 Framework Convention (UNFCCC). This type of capacity building typically has the form of an aid flow of essentially financial resources from a facilitator to the target for a facilitation operation. Facilitators are in this case usually international organizations and ultimately national governments that are also participating in the climate talks. However, in principle, in their capacity as donors, organizations and governments are involved in the climate talks, but only at arm's length.

The transfer of resources to weak countries may have the purpose of general empowerment but it may also have much more specific aims, for example, compliance assistance or help to send a larger delegation to the climate talks.

Capacity building may also represent enhancement of a country's capabilities of relevance for negotiating climate warming. For example, training sessions organized for facilitation purposes may review the prerequisites for successful negotiation, analyze power asymmetries at the negotiating table, and suggest how to deal with incomplete information. Through a series of interactive training sessions, participants will apply skills and techniques in role-play exercises, simulations, and in video and working group analyses broadcast on the Internet.

Low direct involvement in the climate talks

Independent analysts and research centers may offer useful, indeed important, strategic facilitation, in spite of the fact that they have a low direct involvement in the climate talks. This kind of facilitator may be thought of when special units of analysis have been attributed a high degree legitimacy by negotiating parties, partly due to their acknowledged competence regarding topics related to the climate talks in combination with a record of being impartial and objective. The special units of analysis need to be established close and functional working relations with institutions and actors representing the climate negotiation process. Through these channels, different kinds of advice and suggestions may be transmitted to pre-selected receivers in the climate talks, such as the secretariat of the climate negotiation or critical chairs during a COP/MOP meeting.

Examples of facilitation proposals from the special units of analysis are 1) a plan to redesign the IPCC in order to enhance its capacity to handle social scientific knowledge and information; 2) a proposed institutional reform to establish more effective mechanisms for the coordination of negotiation work carried out in a continually increasing number of negotiation bodies; 3) development of better procedures for the participation of non-state actors in the climate talks, NGOs, and the business sector.

A final word

This book has *not* provided a clear formula for a coherent strategy for the long-term facilitation of the climate negotiation. It has argued for the importance of making stronger efforts in this direction while at the same time it has demonstrated the difficulty of this task due to the enormous complexity of the climate issue and the negotiation addressing it. One explanation for this complexity is the heterogeneity of the obstacles – the stumbling blocks – confronting parties to the climate talks. This means that a comprehensive facilitation strategy needs to use a variety of approaches and methods. This book is primarily a plea for studies on strategic facilitation of complex multilateral negotiations and a proposed agenda for research.

Notes

1 A variety of organisations are presently becoming increasingly significant non-state actors in international politics, for example, business companies, nongovernmental organisations (NGOs) and large cities, as well as rebel groups and terror networks.

2 Recall that, to become a member of WTO, a state needs to live up to a number of requirements regarding the structure and the performance of the national economy, which explains why the membership of the trade organisation is considerably smaller than the United Nations.

3 Recall that some of the great powers in the climate talks (notably the EU and the US) have indicated a willingness to accept emission reductions in the range of 60%–80% by the year 2050. Of course, it is debatable how credible this vision is.

4 This strategy was spelled out at a seminar organized by the United States in the context of COP11 in Montreal 2005. Speakers and potential speakers at this meeting were the US chief negotiator in the climate talks, other high-level officials from the US administration, as well as a few senators and business leaders with a special interest in the climate issue.

5 Recall that some of the great powers in the climate talks (notably the EU and the US) have indicated a willingness to accept emission reductions in the range of 60%–80% by the year 2050. Of course, it is debatable how credible this vision is.

6 Recall that *sink* refers to the capacity of sea water, land and biomass to absorb CO2 in the atmosphere.

7 The European negotiation on acid rain (long-range air pollution) is an interesting case. In the early stages of these talks, Norway and Sweden performed effective leadership roles in the face of opposition from many other countries. Critical elements of their power base were superior knowledge about acid rain combined with important skill in multilateral negotiation.

8 See Stenelo (1972). One example referred to in this book is Sweden's mediator role in the UN Disarmament negotiation during the 1950s and the 1960s.

9 Climate change is an extreme but not unique case. Also other issue areas (concerning environmental or economic affairs) represent a similar challenge to the traditional diplomatic approach.

10 Note that the perspective here is different from the usual one when NGO roles are discussed. Typically role analysts in this sense address the question how, and to what degree, NGOs can influence policy making. This study is concerned only with NGO roles related to facilitation of the climate negotiation.

11 Regarding institutional design see for example Goodin (1996) and Koremeno (2001).

12 Ppm = parts per million.

13 A typical relative objective is that country commits itself to reduce CO_2 emissions by a certain percentage, say 8, 20, or 30 per cent.

14 These protocols, which extended the number of pollutants covered by international regime based on the Convention were signed in 1984, 1985, 1988, 1991, 1994, 1998 (two protocols) and 1999.

15 The Regional Air Pollution Information and Simulation (RAINS) model is a tool for analyzing alternative strategies to reduce acidification, eutrophication, and ground-level ozone in Europe.

16 Recall the trap of planning: preparing for one line of action tends to eliminate alternative lines of action. The more the plan is developed, the stronger is this tendency of exclusion.

17 Recall the model case concerning air pollution in Europe in which new agreements added new pollutants to the initial framework agreement (the 1979 Convention on Long-range Transboundary Air Pollution) which only covered one category, sulfur producing acid rain. The new protocols added nitrogen oxids (1988), volatile organic compounds (1991), heavy metals (1998), persistent organic pollutants (1998), and ground-level ozone (1999).

18 *Renegotiation procedures* are laid down in a UNFCCC Art. 4 (2) d and 7 (2) and Art. 9 of the Kyoto Protocol. *Amendment procedures* are formulated in UNFCCC, Art. 15 (3), Art. 15 (4); Art. 16 (2), Art. 21 and the Kyoto Protocol Art. 20, Art. 20 (4), 20 (4), Art. 16 (3) and (5).

19 Facilitation measures are often not perfectly neutral but can still be acceptable if their bias is small and they are helpful for the negotiation in a general way.

References

Accreditation Centre (2005) Hong Kong Ministerial Conference 2005, available through http://www.wto.org/english/thewto_e/minist_e/min05_e/list_ngo_hk05_e.pdf, last accessed 29 August 2012.

Bäckstrand, K. (2001) *What can Nature Withstand? Science, Politics and Discourses in Transboundary Air Pollution Diplomacy.* Lund: Department of political science, Lund University.

Biermann, F. (2000) *Science as Power in International Environmental Negotiations: Global Environmental Assessments Between North and South.* ENRP Discussion Paper 2000–17. Cambridge, MA: Belfer Center for Science and International Affairs, Kennedy School of Government, Harvard University.

Bercovitch, J. and Rubin, J. (1992) *Mediation in International Relations: Multiple Approaches to Conflict Management.* New York: St Martin's Press.

Betsill, M. and Corell, E. (2007) (eds.*) NGO Diplomacy. The Influence of Nongovernmental Organizations in International Environmental Negotiations.* Cambridge: MIT Press.

Bolin, B. (1994) Science and Policy Making. *Ambio.* 23 (1). 25–29.

Bolin, B. and Sukumar, R. (2000) Global Perspective, In Watson, R.T., Noble, I.R., Bolin, B., Ravindranath, N.H., Verardo, D.J. and Dokken, D.J. (eds.*) Land Use, Land-Use Change, and Forestry.* A Special Report of the Intergovernmental Panel on Climate Change. Cambridge, UK: Cambridge University Press. 25–51.

Buzan, B. and Foot, R. (eds.) (2004) *Does China Matter? A Reassessment: Essays in Memory of Gerald Segal.* London: Routledge.

Chasek, P. (2001) NGOs and State Capacity in International Environmental Negotiations: The Experience of the Earth Negotiations Bulletin. *Review of European Community and International Environmental Law.* 10 (2). 168–176.

Chasek, P. (2005) *Margins of Power: Coalition Building and Coalition Maintenance of the South Pacific States in the UN System.* Paper presented at the Annual Meeting of the International Studies Association 2005, Hilton Hawaiian Village, available through http://www.allacademic.com/meta/p69563_index.html, last accessed 20 August 2012.

Depledge, J. (2005) *The Organization of Global Negotiations: Constructing the Climate Change Regime.* London and Sterling, VA: Earthscan.

Edwards, J. (2002) *Has the European Union Exercised Leadership in the International Climate Change Regime since the Hague Conference?* Bruges: Department of European Political and Administrative Studies, College of Europe.

Faure, G.O., Rubin, J.Z. (eds.) (1993) Culture and Negotiation. Newbury Park: Sage Publications.

Goodin, R. (1996) *The theory of Institutional Design.* Cambridge: Cambridge University Press.

Gupta, J. (1997) *The Climate Change Convention and Developing Countries. From Conflict to Consensus.* Dordrecht: Kluwer Academic Publishers.

Von Hall, G. (2010) Experter Synar Klimatpanelen (Experts Scrutinize the Climate Panel), *Svenska Dagbladet.* 27 February 2010.

Heller, T. and Shukla, P.R. (2003) *Development and Climate: Engaging Developing Countries.* Arlington VA: Pew Centre on Global Climate Change.

Kanie, N. (2005) *Current Policy Direction and the Beyond 2012 Climate Regime. Climate Regime Implications of the EU and the US Directional Leadership.* Paper presented at the Annual Meeting of the International Studies Association 2005, Hilton Hawaiian Village, available through http://www.valdes.titech.ac.jp/~kanie/old/050117EU%20and%20US%20climate%20leadership%20for%20ISA%20%28kanie%29.pdf, last accessed 30 August 2012.

Kjellén, B. (2007*) A New Diplomacy For Sustainable Development: The Challenge Of Global Change.* New York: Routledge.

Koremeno, B. (2001) The Rational Design of International Institutions, *International Organization.* 55 (4). 761–800.

Metz, B., Davidson, O.R., Bosch, P.R., Dave, R., and Meyer, L.A. (2007) (eds.) *Climate Change 2007, Mitigation of Climate Change.* Contribution of Working Group III to the Fourth Assessment Report of the Intergovernmental Panel on Climate Change. Cambridge, UK/New York: Cambridge University Press.

Moravcsik, M. (1976). *Science Development: The Building of Science in Less Developed Countries.* Bloomington: International Development Research Center, Indiana University.

Najam, A., Huq, S. and Sokona, Y. (2003) Climate negotiations beyond Kyoto: Developing Countries concerns and interests. *Climate Policy.* 3. 221–231.

Noble, I., Apps, M., Houghton, R., Lashof, D. Makundi, W., Murdiyarso, D., Murray, B., Sombroek, W. and Valentini, R. (2000) Implications of Different Definitions and Generic Issues, In Watson, R., Noble, I., Bolin, B., Ravindranath, N.H, Verardo, D., and Dokken, D. (eds.) *Land Use, Land-Use Change and Forestry.* Special Report of the IPCC. Cambridge, UK: Cambridge University Press.

Ott, H. E. (1998) The Kyoto Protocol – Unfinished Business. *Environment.* 40 (6) 11–47.

Parry, M.L., Canziani, O.F., Palutikof, J.P., van der Linden, P.J., and Hanson, C.E. (eds.) (2007) *Climate Change 2007,* Impacts, Adaptation and Vulnerability, Contribution of Working Group II to the Fourth Assessment Report of the Intergovernmental Panel on Climate Change. Cambridge, UK/New York: Cambridge University Press.

Pershing, J. and Tudela, F. (2004) A Long-Term Target: Framing the Climate Effort, In Aldy, J., Ashton, J., Baron, R., Bodansky, D., Charnovitz, S., Diringer, E., Heller, T., Pershing, J., Shukla, P.R., Tubiana, L., Tudela, F. and Wang X. (eds.) *Beyond Kyoto. Advancing the International Effort against Climate Change.* Arlington, VA: Pew Center on Climate Change. 11–36.

Richards, M. (2001) *A Review of the Effectiveness of Development Country Participation in Climate Change Convention Negotiations.* London: Forest Policy and Environment Group, Overseas Development Institute.

Rubin, J. (1981) *The Dynamics of Third Party Intervention. Kissinger in the Middle East.* Westport: Praeger Publishers.

Rubin, J. (1992) Conclusion: International mediation in context. In *Mediation in international relations: Multiple approaches to conflict management*. Bercovitch, J. and Rubin, J. (eds.). London: MacMillan. 252–268.

Schwarts P. and Randall, D. (2003) *An Abrupt Climate Change Scenario and its Implications for the United States National Security*. Washington, DC: Pentagon.

ScienceDaily (2007) *IPCC Report: Climate Proofing Small Islands*, 11 April, available through http://www.sciencedaily.com/releases/2007/04/070410135159.htm, last accessed 29 August 2012.

Sjöstedt, G. (2009) Knowledge Diplomacy: The Things We Need to Know to Understand It Better. *Pin Points*. 33.

Stenelo, L.-G. (1972) *Mediation in International Negotiation*. Lund Political Studies. 14. Lund: Studentlitteratur.

The Group of 77 at the UN (2012) About the Group of 77, available through http://www.g77.org/doc/, last accessed 30 August 2012.

Watson, R., Zinyowera, M. and Moss, R. (eds.) (1996) Technologies, Policies and Measures for Mitigating Climate Change, Technical Paper, Working Group II, Geneva: Intergovernmental Panel on Climate Change, available through http://www.ipcc.ch/pdf/technical-papers/paper-I-en.pdf, last accessed 29 August 2012.

Yamin, F. and Depledge, J. (2004) *The International Climate Change Regime. A Guide to Rules, Institutions and Procedures*. Cambridge, UK: Cambridge University Press.

Annex

The Evolution of COP

Tanja K. Huber

Today's scientific literature on international cooperation and negotiation on climate change has found a new and growing focus of attention. Institutions like UNFCCC, IPCC and COP have pushed their way into numerous publications, political discussions, and the education of today's and tomorrow's generations. These institutions are strongly connected with one another and it is difficult to talk about one of them without mentioning the others. UNFCCC, IPCC and COP, and a number of other institutional bodies, are part of the same evolving climate regime. To obtain a full understanding of how this international regime has developed over the last decades, the whole pattern of interaction between these institutions and how it has unfolded over time must be considered. However, an overview of main occurrences in just one of these institutions, the Conference of Parties (COP), gives a general historical account of how the overall multilateral negotiations on climate change have evolved in the last 25 years.[1]

The UNFCCC Convention

In June 1992, the UN Conference on the Environment and Development was held in Rio de Janeiro, Brazil. One hundred and fifty-six countries signed the proposed United Nations Framework Climate Change Convention (UNFCCC). The objective of the Convention was "the stabilization of greenhouse gas concentrations in the atmosphere at a level that would prevent dangerous anthropogenic [man-made] interference with the climate system".[2]

COP1

Two years after signing, the Convention entered into force. In the following year, the first Conference of Parties (COP1) was held in Berlin, Germany (28 March–7 April 1995). Like all COPs which were to follow, it was organized by the UN-linked Climate Change Secretariat, a full-time service body responsible for the smooth operation of the conference which also remains active between sessions. Every year, the conference is held in a different UN region, presided over by a senior official or minister from the hosting state or region.

The COP acts as the supreme body of the Convention and has the goal of clarifying the functions carried out by the 1995 established Subsidiary Body for Scientific and Technological Advice for the Convention (SBSTA) and its relationship with the Intergovernmental Panel on Climate Change (IPCC). Besides the SBSTA, two other temporary bodies were established at COP1: the *Ad hoc Group on the Berlin Mandate* and the *Ad hoc Group on Article 13 (AG13)*. The first was formed to establish a process to negotiate strengthened commitments for developed countries (which would finally lead to the adoption of the Kyoto Protocol), while the second was formed in consideration of Article 13 of the Convention, namely, to find ways of assisting governments in overcoming difficulties they might encounter while fulfilling their commitments. The aim was to present the report of the second group at COP4.

COP2

In 1990, the United Nations General Assembly (UNGA) requested the IPCC to draft its First Assessment Report, a document consisting of the written findings of three Working Groups (I, II, III), each with an individual focus on:

I) the scientific assessment of topics such as greenhouse gases and observed climate variations and change;

II) the scientific understanding of climate change impacts on water, forests and agriculture;

III) strategies for mitigation and adaptation processes. In 1995 the IPCC concluded its Second Assessment Report (SAR). In comparison with the first report, a new subject area was now being investigated: the socio-economic aspects of climate change. Again, three Working Groups drafted their reports.

The findings of the SAR were presented at COP2, which took place 1996 in Geneva, Switzerland (8–19 July). Ministers and other heads of delegations present at the conference recognized and endorsed the SAR, noting the discernible evidence of human influence on the global climate, the impact of climate change on ecological systems and socio economic sectors, and the need for continued information exchange between the IPCC and bodies of the Convention.

COP3

The third conference was held in Kyoto, Japan in 1997 (1–11 December). That year, the ongoing Berlin Mandate process had finally led to the adoption (by consensus) of the Kyoto Protocol. The conference acknowledged the IPCC's continued work and encouraged the preparation of a Third Assessment Report (TAR) with yet another issue to be addressed: the policy relevance of climate change. As a result, Working Groups II and III were adjusted. However, TAR was not to be completed before 2001, the year of the next COP.

The Kyoto Protocol was drafted both to strengthen the international response to climate change and to support the Convention by encouraging a more advanced implementation of existing commitments and setting new emission targets for developed countries for the post-2000 period (i.e. until 2012). However, the flexible Kyoto Mechanisms (e.g. "Clean Development Mechanism", "Joint Implementation", "Emission Trading", etc.) are known to provide loopholes which allow countries to circumvent the targets. Enforcement of the Protocol was only to be made upon ratification by the majority of the parties. (At COP3, the goal was set to accomplish this by the time of the World Conference on Sustainability in Johannesburg, September 2002).

COP4

The fourth COP meeting took place in Buenos Aires, Argentina (2–13 November 1998) concluding with the adoption of the "Buenos Aires Plan", an action plan with fixed deadlines for efficient progress regarding the mechanisms, compliance and policy issues. Unfortunately, the loopholes of the Protocol could not be closed and the danger of not achieving the set emission targets slowly became visible.

COP5

At the fifth conference, held in 1999 in Bonn, Germany (25 October–5 November), the discussion of the parties focused on enabling decisions as well as the setting of a timetable for completing the outstanding details of the Kyoto Protocol by the time of COP6 in 2000. The flexible mechanisms again posed too big a challenge and, thus, no real progress could be made.

COP6

In the year 2000, the sixth conference was organized in Den Haag, The Netherlands (13–24 November). As it did not bring any success on a consolidated agreement on key issues such as sinks, the Kyoto mechanisms, sanctions for non-compliance and the inclusion of developing countries, the parties concluded with the decision to suspend the meeting and reconvene in 2001.[3]

The second COP6 was once again held in Bonn (16–27 July, 2001). This time, the meeting was successful, producing a historical outcome: despite the US withdrawal from the Protocol and the hesitancy of several other countries, the parties were able to reach an agreement. All necessary preparations were finally rolling for the long-awaited ratification and eventual implementation of the Kyoto Protocol. This was a crucial issue, as the conference was running the risk of suffering a severe loss of image. However, decisions on the mechanisms, sinks and compliance were still not made by the end of the meeting and the discussion had to be postponed until COP7 in Morocco in late October.

COP7

The TAR, begun in 1997, was finally completed when the parties met for the seventh time. The venue of the meeting was Marrakesh, Morocco (29 October–9 November 2001). The Parties of COP7 agreed on a "package deal", which brought three years of work under the earlier established Buenos Aires Plan of Action to a close, with decisions on rules for compliance, the usage for reporting such data, sinks, among others, under the mechanisms. Furthermore, the *Marrakesh Ministerial Declaration* was adopted which aimed to be a decisive input to the World Summit on Sustainable Development in Johannesburg. Additionally, the IPCC was encouraged to continue its work and asked to prepare a Fourth Assessment Report (FAR), to be completed by 2007.

COP8

In late autumn 2002, the eighth session of the COP took place in New Delhi, India (23 October–1 November). It concluded with the recognition of the Delhi Ministerial Declaration on Climate Change and Sustainable Development. The Declaration appealed to all countries to ratify the Kyoto Protocol as soon as possible, referring to the alarming climate prognosis in TAR. The focus of the overall process had now shifted to implementation.

COP9

Milan, Italy, was the venue of the ninth COP (1–12 December 2003). Together, the SBSTA and the Subsidiary Body for Implementation (SBI) held their nineteenth session. The decisions of the meeting centred on institutions and procedures of the Kyoto Protocol and on the implementation of the UNFCCC. The need for prompt operationalization of the Special Climate Change Fund and the Least Developed Countries Fund, with priority on adaptation, was voiced. New emission guidelines were set resulting from reports on changes in carbon concentrations (due to land-use changes and forestry). A major success was the agreement on modalities and scope for carbon absorbing forest management projects in CDM – which now completed the package adopted two years earlier in Marrakesh.

COP10

In 2004, the tenth conference was held once again in Buenos Aires (6–17 December), ten years after the UNFCC had entered into force. The Convention and the recollection of the accomplishments made during the past decade naturally served as a core theme of the meeting. The Buenos Aires Programme of Work on Adaptation and Response Measures was adopted and the Convention Secretariat was given the task of organizing an informal seminar of governmental experts the following May in Bonn (with a focus on adaptation, mitigation, policies and measures). The Kyoto Protocol, ready to enter into force by 16 February 2005, was another core theme of the conference, bringing the US back into the discussions on mitigation (after their withdrawal in 2001), now that the focus had shifted to timelines beyond 2012.[4]

COP11

The eleventh COP was held in Montreal, Canada (28 November–9 December 2005). The conference also served as the first Meeting of the Parties to the Kyoto Protocol (MOP1). Both events attracted more than 7,000 participants from the Convention's 189 parties and nongovernmental organizations (NGOs), intergovernmental organizations and media.[5]. The current status of the ratification process of the Protocol, and efforts to find ways of accelerating implementation actions, were of great interest to everyone attending.

COP12

In 2006, the twelfth COP was held in Nairobi, Kenya (6–17 November). The conference was also host of the second meeting of the Parties to the Kyoto Protocol (MOP2). In addition, the conference served as the twenty-fifth session for both the Subsidiary Body for Scientific and Technological Advise (SBSTA25) and the Subsidiary Body for Implementation (SBI25). It also hosted the second session of the Ad Hoc Working Group on Further Commitments for Annex I Parties under the Kyoto Protocol (AWG2). Representatives of 180 Parties to the UNFCCC attended. Among other topics, the Global Environment Facility, the operating entity of the financial mechanism of the Convention, raised concerns. The parties requested the Facility to simplify its funding procedures and to improve the efficiency of the process through which developing countries can implement their commitments under Art. 4, paragraph 1 of the Convention.

COP13

The thirteenth COP Meeting and third Meeting of the Parties to the Kyoto Protocol took place from 3–15 December 2007 at the Bali International Convention Center, Indonesia. More than 12,000 participants, mostly representatives of NGOs, economists, science bodies and the media witnessed the development of the conference's highpoint, the adoption of the Bali Road Map: a number of forward-looking decisions designed to reach a secure climate future. Part of this is the Bali Action Plan, which directs the course for a new negotiating process. Most importantly, the plan aims to enable the implementation of the Convention through long-term cooperation action which also goes beyond 2012. For that purpose, yet another Ad Hoc Working Group was established,[6] with its first meeting scheduled for spring 2008 and a deadline for completion by 2009.

COP14

Poznan, Poland was the venue for the fourteenth conference, held from 1–12 December 2008. At the same time, the conference served as the meeting of the parties to the Kyoto Protocol in its fourth session. The convening parties agreed that the Adaptation Fund Board was to be given a legal capacity to directly support developing countries. Other important topics discussed included finance, technology, disaster management and the reduction of emissions from deforestation and forest degradation (REDD). To "shape an ambitious and effective international response to climate change",[7] parties agreed to provide a negotiating text to the UNFCCC Bonn Climate Change Talks in June 2009, with the aim of finalizing it at COP15 in Copenhagen. Another preparation tool for the upcoming conference of parties was the ministerial round table on a shared vision on long-term cooperative action on climate change.

COP15

The Climate Conference in Copenhagen that took place from 7–19 December 2009 needs to be especially highlighted for several reasons:

1. It was the first attempt to achieve a binding agreement on climate change to come into effect after the first commitment period of the Kyoto Protocol to the UNFCCC ends in December 2012.
2. Copenhagen was seen as a point of departure for the future negotiation and regime-building process, which is addressed in this project.
3. Its perceived political importance was high, as indicated by the presence in Copenhagen of more than a hundred heads of state or government.
4. It created a broad awareness that the climate negotiations as a whole will need to be of long duration if they are to have a satisfactory braking effect on climate warming.

One principal aim of the Copenhagen Conference was to reach an agreement on a new framework for tackling rising greenhouse gas (GHG) emissions that would enter into effect at the end of 2012 after the expiry of the first commitment period of the Kyoto Protocol to the Climate Convention. Binding commitments regarding cuts of GHG emissions were meant to be linked to the would-be Copenhagen framework.

Before Copenhagen, there were various differing but often high expectations as to what should come out of the COP15 meeting. The 15 November 2009 Leaders' Statement issued after the Asia-Pacific Economic Cooperation (APEC) left the impression that only a "political framework" was possible in Copenhagen. Many delegations anticipated correctly an arduous future negotiation process and sensed that there was a need for a "two-step" process to reach a final climate treaty "at the earliest" in 2010 (Schuenemann 2009). A main reason for the "failure" of the talks in Copenhagen to reach the goals adopted in the Bali Road Map was the inability (or unwillingness) of the major emitters such as Brazil, China, South Africa and the United States to reach a compromise regarding their opposition against commitments to reduce GHG emissions.

In the view of both parties and observers, the organization of the Conference by its Danish hosts was "chaotic" and "under overwhelming pressure". Developing countries were kept outside important informal talks.

The Copenhagen Conference agreed "to take note of" the so-called Copenhagen Accord. This text had been drafted by the heads of state of the United States and the BASIC bloc countries (Brazil, South Africa, India and China). The main issues covered by the Copenhagen Accord can be summarized as follows:

- Reaffirmation of the ultimate objective of the UNFCCC that greenhouse gas concentrations in the atmosphere should be stabilized at a level that would prevent dangerous anthropogenic interference with the climate system.
- Recognition of the scientific view that the increase in global temperature should be maintained below 2° Celsius on the basis of equity and in the context of sustainable development.
- Call for an assessment of the implementation of the Copenhagen Accord to be completed by 2015, including strengthening the long-term goal in relation to limiting temperature rises to 1.5° Celsius.
- Commitment sought from Annex I Parties to mitigate emissions, by implementing individually or jointly "economy-wide emissions targets for 2020" by 31 January 2010.
- Delivery of reductions and finance by developed countries to be measured, reported, and verified (MRV) in accordance with COP guidelines. However, this strategy is constrained by the lack of binding quantitative commitments with respect to emission reductions in the post-Kyoto period.

COP16

In Cancun (2010), 193 countries came together and demonstrated a renewed commitment to the struggle against global warming. The Cancun Agreements are a detailed set of visionary yet pragmatic principles that make important strides to begin implementing the accord reached in Copenhagen the year before. The countries gathered in Cancun made progress on emissions reductions, greater transparency, forest preservation and the creation of the green fund to help mobilize much needed investments throughout the world.[8]

COP17

The Durban conference in 2011 represents still another step forward in the direction towards a continuation of the Kyoto Protocol. A"soft" deadline was set under the Copenhagen Accord for countries to submit emissions reduction targets be seen. Durban produced a document whose character is similar to that of both the Bali Road Map and the Copenhagen Accord but under a new headline, the Durban platform.[9] This non-binding agreement calls for revitalized negotiations for the new agreement on emission reductions which should not be concluded later than 2015, resulting in a new binding agreement that will take effect from 2020.

The Durban Conference led to institutional/organizational developments which may become important in the longer term; the creation of the Adaptation Committee, which will provide advice and ensure coherent action on adaptation and the establishment of a Technology Executive Committee, to facilitate the development of low-carbon technologies.

The meeting in South Africa also tackled the conflict between developed and developing countries in a constructive way. It decided to establish the Green Climate Fund (US$ 100 billion per year or more by 2020) whose principal function would be to support climate policies and activities in developing countries.

COP1–17: An example for a regime-building process

More than ten years have passed since UNFCCC entered into force, years filled with scientific evidence of the ongoing change in global climate. Each COP Meeting had specific goals and outcomes, yet they can all be seen as intensive episodes in a continuous regime-building process, with activities taking place between the formal sessions.[10] One can observe a line of continuity of plans and ideas, often initiated long before the actual meeting. The significance of one COP session is sometimes only understandable when looking at the outcome of the following COP. The Kyoto Protocol and post-Kyoto timelines and decisions are good examples of this phenomenon. Adopted in 1997, the Kyoto Protocol could not be implemented until 2005, when the treaty was finally ratified. However, post-Kyoto negotiations had already been started and related issues and unanswered questions continuously discussed (also between COP meetings). This means that the period after Kyoto Protocol adoption brought two tracks of new negotiations: 1) the post-agreement negotiations of the Kyoto Protocol, and gradually at the same time, 2) the pre-negotiations for the post-Kyoto period (i.e. for post-2012). One can see the dual function of the later COP meetings, starting with COP11 in Montreal, in a similar light. On the one hand, they have functioned as regular COP sessions, and on the other hand, they have also served as hosts for the Meetings of Parties (MOP1, 2, etc.), initiating a whole new series of negotiation.

Climate warming poses one of the greatest challenges of the twenty-first century. Only now do we seem to have understood the seriousness of the situation and the fact that there will be changes that we can no longer prevent from happening and also that – in order to prevent *further* changes – we must act now. The world needs to take action. The overview of seventeen COP meetings warrants two reflections. First, nothing is lost. So far, we are on a slow but steady path to progress in the climate talks. Second, the COP meetings bear witness to the complexity of the climate negotiation and clearly demonstrate how cumbersome they are for negotiating parties. We need to find ways to facilitate the process as much as possible in the future.

Notes

1 Most information is based on material distributed by UNEP, IPCC, http://unfccc.int, www.accc.gv.at, http://bmu.de, http://greenpeace.de, www.iisd.ca.
2 Source: http://unfccc.int.
3 Some conservative US stakeholders even suggested that the failure of the conference was a clear indication that the Kyoto Protocol had been a wrong approach all along. (Pressetext Austria, 25 November 2000, COP6).
4 i.e. at the expiration date of the Kyoto Protocol.
5 Source: www.ec.gc.ca.
6 "Ad Hoc Working Group on Long-term Cooperative Action under the Convention"
7 Source: http://unfcc.int/meetings.
8 Cancun Climate Conference, November 2010. United Nations Framework Convention on Climate Change.
9 "Working Together. Saving Tomorrow Today". COP17/CMP7.United Nations Climate Conference in Durban South Africa. November/December 2011.
10 See *Negotiating the Rapids. The Dynamics of Regime Formation*, In Spector, Betram I. and Zartman, I. William (2003).

Bibliography

Depledge, J. (2005) *The Organization of Global Negotiations. Constructing the Climate Change Regime.* London and Sterling: Earthscan.

Hohmeyer O. and Rennings K. (eds.) (1999) *Man-Made Climate Change. Economic Aspects and Policy Options.* Heidelberg: Zew Economic Studies, Physica Verlag.

Jasanoff, S. and Long Mertello, M. (eds.) (2004) *Earthly Politics. Local and Global in Environmental Governance.* Cambridge, MA/UK: Massachussetts Institute of Technology.

Oberthuer S. and Ott, H. (1999) *The Kyoto Protocol, International Climate Policy for the 21st century.* Berlin: Springer.

Okonski K. (ed.) (2003) *Adapt or Die. The science, politics and economics of climate change.* London: Profile Books Ltd.

Spector B. and Zartman, I. W. (eds.) (2003) *Getting it Done; Post-Agreement Negotiations and International Regimes.* Washington, DC: United States Institute of Peace.

Werksman, J. (2005) The Negotiation of a Kyoto Compliance System, In Hovi, J., Schram Stokke O. and Ulfstein G. (eds.) *Implementing the Climate Regime.* London and Sterling: Earthscan, 99. 17–37.

Woeldman E. (2002) *Implementing the Kyoto Mechanisms.* University of Groeningen (dissertation), The Netherlands.

Index